ADJUSTMENT COMPUTATIONS

ADJUSTMENT COMPUTATIONS
Spatial Data Analysis

Fourth Edition

CHARLES D. GHILANI, Ph.D.
Professor of Engineering
Surveying Engineering Program
Pennsylvania State University

PAUL R. WOLF, Ph.D.
Professor Emeritus
Department of Civil and Environmental Engineering
University of Wisconsin–Madison

JOHN WILEY & SONS, INC.

This book is printed on acid-free paper. ∞

Copyright © 2006 by John Wiley & Sons, Inc. All rights reserved.

Published by John Wiley & Sons, Inc., Hoboken, New Jersey
Published simultaneously in Canada.

No part of this publication may be reproduced, stored in a retrieval system, or transmitted in any form or by any means, electronic, mechanical, photocopying, recording, scanning, or otherwise, except as permitted under Section 107 or 108 of the 1976 United States Copyright Act, without either the prior written permission of the Publisher, or authorization through payment of the appropriate per-copy fee to the Copyright Clearance Center, Inc., 222 Rosewood Drive, Danvers, MA 01923, (978) 750-8400, fax (978) 750-4470, or on the web at www.copyright.com. Requests to the Publisher for permission should be addressed to the Permissions Department, John Wiley & Sons, Inc., 111 River Street, Hoboken, NJ 07030, (201) 748-6011, fax (201) 748-6008, or online at http://www.wiley.com/go/permission.

Limit of Liability/Disclaimer of Warranty: While the publisher and author have used their best efforts in preparing this book, they make no representations or warranties with respect to the accuracy or completeness of the contents of this book and specifically disclaim any implied warranties of merchantability or fitness for a particular purpose. No warranty may be created or extended by sales representatives or written sales materials. The advice and strategies contained herein may not be suitable for your situation. You should consult with a professional where appropriate. Neither the publisher nor author shall be liable for any loss of profit or any other commercial damages, including but not limited to special, incidental, consequential, or other damages.

For general information on our other products and services or for technical support, please contact our Customer Care Department within the United States at (800) 762-2974, outside the United States at (317) 572-3993 or fax (317) 572-4002.

Wiley also publishes its books in a variety of electronic formats. Some content that appears in print may not be available in electronic books. For more information about Wiley products, visit our web site at www.wiley.com.

Library of Congress Cataloging-in-Publication Data:

Ghilani, Charles D.
 Adjustment computations : spatial data analysis / Charles D. Ghilani, Paul R. Wolf.—4th ed.
   p. cm.
   Prev. ed. entered under Wolf.
   ISBN-13 978-0-471-69728-2 (cloth)
   ISBN-10 0-471-69728-1 (cloth)
  1. Surveying—Mathematics. 2. Spatial analysis (Statistics) I. Wolf, Paul R. II. Title.
   TA556.M38W65 2006
   526.9—dc22
                                                                    2005028948

Printed in the United States of America

10 9 8 7 6 5 4 3 2 1

CONTENTS

PREFACE   xix
ACKNOWLEDGMENTS   xxiii

1 INTRODUCTION   1

    1.1   Introduction / 1
    1.2   Direct and Indirect Measurements / 2
    1.3   Measurement Error Sources / 2
    1.4   Definitions / 3
    1.5   Precision versus Accuracy / 4
    1.6   Redundant Measurements in Surveying and Their Adjustment / 7
    1.7   Advantages of Least Squares Adjustment / 8
    1.8   Overview of the Book / 10
    Problems / 10

2 OBSERVATIONS AND THEIR ANALYSIS   12

    2.1   Introduction / 12
    2.2   Sample versus Population / 12
    2.3   Range and Median / 13
    2.4   Graphical Representation of Data / 14
    2.5   Numerical Methods of Describing Data / 17
    2.6   Measures of Central Tendency / 17
    2.7   Additional Definitions / 18

2.8 Alternative Formula for Determining Variance / 21
2.9 Numerical Examples / 23
2.10 Derivation of the Sample Variance (Bessel's Correction) / 28
2.11 Programming / 29
Problems / 30

3 RANDOM ERROR THEORY — 33

3.1 Introduction / 33
3.2 Theory of Probability / 33
3.3 Properties of the Normal Distribution Curve / 36
3.4 Standard Normal Distribution Function / 38
3.5 Probability of the Standard Error / 41
    3.5.1 50% Probable Error / 42
    3.5.2 95% Probable Error / 42
    3.5.3 Other Percent Probable Errors / 43
3.6 Uses for Percent Errors / 43
3.7 Practical Examples / 44
Problems / 47

4 CONFIDENCE INTERVALS — 50

4.1 Introduction / 50
4.2 Distributions Used in Sampling Theory / 52
    4.2.1 χ^2 Distribution / 52
    4.2.2 t (Student) Distribution / 54
    4.2.3 F Distribution / 55
4.3 Confidence Interval for the Mean: t Statistic / 56
4.4 Testing the Validity of the Confidence Interval / 59
4.5 Selecting a Sample Size / 60
4.6 Confidence Interval for a Population Variance / 61
4.7 Confidence Interval for the Ratio of Two Population Variances / 63
Problems / 65

5 STATISTICAL TESTING — 68

5.1 Hypothesis Testing / 68
5.2 Systematic Development of a Test / 71
5.3 Test of Hypothesis for the Population Mean / 72
5.4 Test of Hypothesis for the Population Variance / 74

5.5 Test of Hypothesis for the Ratio of Two Population Variances / 77

Problems / 81

6 PROPAGATION OF RANDOM ERRORS IN INDIRECTLY MEASURED QUANTITIES   84

6.1 Basic Error Propagation Equation / 84
    6.1.1 Generic Example / 88
6.2 Frequently Encountered Specific Functions / 88
    6.2.1 Standard Deviation of a Sum / 88
    6.2.2 Standard Deviation in a Series / 89
    6.2.3 Standard Deviation of the Mean / 89
6.3 Numerical Examples / 89
6.4 Conclusions / 94

Problems / 95

7 ERROR PROPAGATION IN ANGLE AND DISTANCE OBSERVATIONS   99

7.1 Introduction / 99
7.2 Error Sources in Horizontal Angles / 99
7.3 Reading Errors / 100
    7.3.1 Angles Observed by the Repetition Method / 100
    7.3.2 Angles Observed by the Directional Method / 101
7.4 Pointing Errors / 102
7.5 Estimated Pointing and Reading Errors with Total Stations / 103
7.6 Target Centering Errors / 104
7.7 Instrument Centering Errors / 106
7.8 Effects of Leveling Errors in Angle Observations / 110
7.9 Numerical Example of Combined Error Propagation in a Single Horizontal Angle / 112
7.10 Use of Estimated Errors to Check Angular Misclosure in a Traverse / 114
7.11 Errors in Astronomical Observations for an Azimuth / 116
7.12 Errors in Electronic Distance Observations / 121
7.13 Use of Computational Software / 123

Problems / 123

8 ERROR PROPAGATION IN TRAVERSE SURVEYS — 127

8.1 Introduction / 127
8.2 Derivation of Estimated Error in Latitude and Departure / 128
8.3 Derivation of Estimated Standard Errors in Course Azimuths / 129
8.4 Computing and Analyzing Polygon Traverse Misclosure Errors / 130
8.5 Computing and Analyzing Link Traverse Misclosure Errors / 135
8.6 Conclusions / 140
Problems / 140

9 ERROR PROPAGATION IN ELEVATION DETERMINATION — 144

9.1 Introduction / 144
9.2 Systematic Errors in Differential Leveling / 144
  9.2.1 Collimation Error / 144
  9.2.2 Earth Curvature and Refraction / 146
  9.2.3 Combined Effects of Systematic Errors on Elevation Differences / 147
9.3 Random Errors in Differential Leveling / 148
  9.3.1 Reading Errors / 148
  9.3.2 Instrument Leveling Errors / 148
  9.3.3 Rod Plumbing Error / 148
  9.3.4 Estimated Errors in Differential Leveling / 150
9.4 Error Propagation in Trigonometric Leveling / 152
Problems / 156

10 WEIGHTS OF OBSERVATIONS — 159

10.1 Introduction / 159
10.2 Weighted Mean / 161
10.3 Relation between Weights and Standard Errors / 163
10.4 Statistics of Weighted Observations / 164
  10.4.1 Standard Deviation / 164
  10.4.2 Standard Error of Weight w and Standard Error of the Weighted Mean / 164
10.5 Weights in Angle Observations / 165
10.6 Weights in Differential Leveling / 166

10.7 Practical Examples / 167
Problems / 170

11 PRINCIPLES OF LEAST SQUARES 173

11.1 Introduction / 173
11.2 Fundamental Principle of Least Squares / 174
11.3 Fundamental Principle of Weighted Least Squares / 176
11.4 Stochastic Model / 177
11.5 Functional Model / 177
11.6 Observation Equations / 179
    11.6.1 Elementary Example of Observation Equation Adjustment / 179
11.7 Systematic Formulation of the Normal Equations / 181
    11.7.1 Equal-Weight Case / 181
    11.7.2 Weighted Case / 183
    11.7.3 Advantages of the Systematic Approach / 184
11.8 Tabular Formation of the Normal Equations / 184
11.9 Using Matrices to Form the Normal Equations / 185
    11.9.1 Equal-Weight Case / 185
    11.9.2 Weighted Case / 187
11.10 Least Squares Solution of Nonlinear Systems / 188
11.11 Least Squares Fit of Points to a Line or Curve / 191
    11.11.1 Fitting Data to a Straight Line / 192
    11.11.2 Fitting Data to a Parabola / 194
11.12 Calibration of an EDM Instrument / 195
11.13 Least Squares Adjustment Using Conditional Equations / 196
11.14 Example 11.5 Using Observation Equations / 198
Problems / 200

12 ADJUSTMENT OF LEVEL NETS 205

12.1 Introduction / 205
12.2 Observation Equation / 205
12.3 Unweighted Example / 206
12.4 Weighted Example / 209
12.5 Reference Standard Deviation / 211
    12.5.1 Unweighted Example / 212
    12.5.2 Weighted Example / 213

12.6 Another Weighted Adjustment / 213
Problems / 216

13 PRECISION OF INDIRECTLY DETERMINED QUANTITIES 221

13.1 Introduction / 221
13.2 Development of the Covariance Matrix / 221
13.3 Numerical Examples / 225
13.4 Standard Deviations of Computed Quantities / 226
Problems / 229

14 ADJUSTMENT OF HORIZONTAL SURVEYS: TRILATERATION 233

14.1 Introduction / 233
14.2 Distance Observation Equation / 235
14.3 Trilateration Adjustment Example / 237
14.4 Formulation of a Generalized Coefficient Matrix for a More Complex Network / 243
14.5 Computer Solution of a Trilaterated Quadrilateral / 244
14.6 Iteration Termination / 248
  14.6.1 Method of Maximum Iterations / 249
  14.6.2 Maximum Correction / 249
  14.6.3 Monitoring the Adjustment's Reference Variance / 249
Problems / 250

15 ADJUSTMENT OF HORIZONTAL SURVEYS: TRIANGULATION 255

15.1 Introduction / 255
15.2 Azimuth Observation Equation / 255
  15.2.1 Linearization of the Azimuth Observation Equation / 256
15.3 Angle Observation Equation / 258
15.4 Adjustment of Intersections / 260
15.5 Adjustment of Resections / 265
  15.5.1 Computing Initial Approximations in the Resection Problem / 266
15.6 Adjustment of Triangulated Quadrilaterals / 271
Problems / 275

16 ADJUSTMENT OF HORIZONTAL SURVEYS: TRAVERSES AND NETWORKS   283

16.1 Introduction to Traverse Adjustments / 283
16.2 Observation Equations / 283
16.3 Redundant Equations / 284
16.4 Numerical Example / 285
16.5 Minimum Amount of Control / 291
16.6 Adjustment of Networks / 291
16.7 χ^2 Test: Goodness of Fit / 300
Problems / 301

17 ADJUSTMENT OF GPS NETWORKS   310

17.1 Introduction / 310
17.2 GPS Observations / 311
17.3 GPS Errors and the Need for Adjustment / 314
17.4 Reference Coordinate Systems for GPS Observations / 314
17.5 Converting between the Terrestrial and Geodetic Coordinate Systems / 316
17.6 Application of Least Squares in Processing GPS Data / 321
17.7 Network Preadjustment Data Analysis / 322
  17.7.1 Analysis of Fixed Baseline Measurements / 322
  17.7.2 Analysis of Repeat Baseline Measurements / 324
  17.7.3 Analysis of Loop Closures / 325
  17.7.4 Minimally Constrained Adjustment / 326
17.8 Least Squares Adjustment of GPS Networks / 327
Problems / 332

18 COORDINATE TRANSFORMATIONS   345

18.1 Introduction / 345
18.2 Two-Dimensional Conformal Coordinate Transformation / 345
18.3 Equation Development / 346
18.4 Application of Least Squares / 348
18.5 Two-Dimensional Affine Coordinate Transformation / 350
18.6 Two-Dimensional Projective Coordinate Transformation / 353

18.7 Three-Dimensional Conformal Coordinate Transformation / 356
18.8 Statistically Valid Parameters / 362
Problems / 364

19 ERROR ELLIPSE 369

19.1 Introduction / 369
19.2 Computation of Ellipse Orientation and Semiaxes / 371
19.3 Example Problem of Standard Error Ellipse Calculations / 376
  19.3.1 Error Ellipse for Station Wisconsin / 376
  19.3.2 Error Ellipse for Station Campus / 377
  19.3.3 Drawing the Standard Error Ellipse / 378
19.4 Another Example Problem / 378
19.5 Error Ellipse Confidence Level / 379
19.6 Error Ellipse Advantages / 381
  19.6.1 Survey Network Design / 381
  19.6.2 Example Network / 383
19.7 Other Measures of Station Uncertainty / 385
Problems / 386

20 CONSTRAINT EQUATIONS 388

20.1 Introduction / 388
20.2 Adjustment of Control Station Coordinates / 388
20.3 Holding Control Station Coordinates and Directions of Lines Fixed in a Trilateration Adjustment / 394
  20.3.1 Holding the Direction of a Line Fixed by Elimination of Constraints / 395
20.4 Helmert's Method / 398
20.5 Redundancies in a Constrained Adjustment / 403
20.6 Enforcing Constraints through Weighting / 403
Problems / 406

21 BLUNDER DETECTION IN HORIZONTAL NETWORKS 409

21.1 Introduction / 409
21.2 A Priori Methods for Detecting Blunders in Observations / 410
  21.2.1 Use of the K Matrix / 410
  21.2.2 Traverse Closure Checks / 411

- 21.3 A Posteriori Blunder Detection / 412
- 21.4 Development of the Covariance Matrix for the Residuals / 414
- 21.5 Detection of Outliers in Observations / 416
- 21.6 Techniques Used in Adjusting Control / 418
- 21.7 Data Set with Blunders / 420
- 21.8 Some Further Considerations / 428
  - 21.8.1 Internal Reliability / 429
  - 21.8.2 External Reliability / 429
- 21.9 Survey Design / 430

Problems / 432

22 GENERAL LEAST SQUARES METHOD AND ITS APPLICATION TO CURVE FITTING AND COORDINATE TRANSFORMATIONS 437

- 22.1 Introduction to General Least Squares / 437
- 22.2 General Least Squares Equations for Fitting a Straight Line / 437
- 22.3 General Least Squares Solution / 439
- 22.4 Two-Dimensional Coordinate Transformation by General Least Squares / 443
  - 22.4.1 Two-Dimensional Conformal Coordinate Transformation / 444
  - 22.4.2 Two-Dimensional Affine Coordinate Transformation / 447
  - 22.4.3 Two-Dimensional Projective Transformation / 448
- 22.5 Three-Dimensional Conformal Coordinate Transformation by General Least Squares / 449

Problems / 451

23 THREE-DIMENSIONAL GEODETIC NETWORK ADJUSTMENT 454

- 23.1 Introduction / 454
- 23.2 Linearization of Equations / 456
  - 23.2.1 Slant Distance Observations / 457
  - 23.2.2 Azimuth Observations / 457
  - 23.2.3 Vertical Angle Observations / 459
  - 23.2.4 Horizontal Angle Observations / 459
  - 23.2.5 Differential Leveling Observations / 460
  - 23.2.6 Horizontal Distance Observations / 460

23.3 Minimum Number of Constraints / 462
23.4 Example Adjustment / 462
    23.4.1 Addition of Slant Distances / 464
    23.4.2 Addition of Horizontal Angles / 465
    23.4.3 Addition of Zenith Angles / 466
    23.4.4 Addition of Observed Azimuths / 467
    23.4.5 Addition of Elevation Differences / 467
    23.4.6 Adjustment of Control Stations / 468
    23.4.7 Results of Adjustment / 469
    23.4.8 Updating Geodetic Coordinates / 469
23.5 Building an Adjustment / 471
23.6 Comments on Systematic Errors / 471
Problems / 474

24 COMBINING GPS AND TERRESTRIAL OBSERVATIONS   478

24.1 Introduction / 478
24.2 Helmert Transformation / 480
24.3 Rotations between Coordinate Systems / 484
24.4 Combining GPS Baseline Vectors with Traditional Observations / 484
24.5 Other Considerations / 489
Problems / 489

25 ANALYSIS OF ADJUSTMENTS   492

25.1 Introduction / 492
25.2 Basic Concepts, Residuals, and the Normal Distribution / 492
25.3 Goodness-of-Fit Test / 496
25.4 Comparison of Residual Plots / 499
25.5 Use of Statistical Blunder Detection / 501
Problems / 502

26 COMPUTER OPTIMIZATION   504

26.1 Introduction / 504
26.2 Storage Optimization / 504
26.3 Direct Formation of the Normal Equations / 507
26.4 Cholesky Decomposition / 508
26.5 Forward and Back Solutions / 511

26.6   Using the Cholesky Factor to Find the Inverse of the Normal Matrix / 512

26.7   Spareness and Optimization of the Normal Matrix / 513

Problems / 518

APPENDIX A   INTRODUCTION TO MATRICES   520

A.1   Introduction / 520
A.2   Definition of a Matrix / 520
A.3   Size or Dimensions of a Matrix / 521
A.4   Types of Matrices / 522
A.5   Matrix Equality / 523
A.6   Addition or Subtraction of Matrices / 524
A.7   Scalar Multiplication of a Matrix / 524
A.8   Matrix Multiplication / 525
A.9   Computer Algorithms for Matrix Operations / 528
    A.9.1   Addition or Subtraction of Two Matrices / 528
    A.9.2   Matrix Multiplication / 529
A.10 Use of the MATRIX Software / 531

Problems / 531

APPENDIX B   SOLUTION OF EQUATIONS BY MATRIX METHODS 534

B.1   Introduction / 534
B.2   Inverse Matrix / 534
B.3   Inverse of a 2 × 2 Matrix / 535
B.4   Inverses by Adjoints / 537
B.5   Inverses by Row Transformations / 538
B.6   Example Problem / 542

Problems / 543

APPENDIX C   NONLINEAR EQUATIONS AND TAYLOR'S THEOREM 546

C.1   Introduction / 546
C.2   Taylor Series Linearization of Nonlinear Equations / 546
C.3   Numerical Example / 547
C.4   Using Matrices to Solve Nonlinear Equations / 549

xvi CONTENTS

C.5 Simple Matrix Example / 550
C.6 Practical Example / 551
Problems / 554

APPENDIX D NORMAL ERROR DISTRIBUTION CURVE AND OTHER STATISTICAL TABLES   **556**

D.1 Development of the Normal Distribution Curve Equation / 556
D.2 Other Statistical Tables / 564
    D.2.1 χ^2 Distribution / 564
    D.2.2 t Distribution / 566
    D.2.3 F Distribution / 568

APPENDIX E CONFIDENCE INTERVALS FOR THE MEAN   **576**

APPENDIX F MAP PROJECTION COORDINATE SYSTEMS   **582**

F.1 Introduction / 582
F.2 Mathematics of the Lambert Conformal Conic Map Projection / 583
    F.2.1 Zone Constants / 584
    F.2.2 Direct Problem / 585
    F.2.3 Inverse Problem / 585
F.3 Mathematics of the Transverse Mercator / 586
    F.3.1 Zone Constants / 587
    F.3.2 Direct Problem / 588
    F.3.3 Inverse Problem / 588
F.4 Reduction of Observations / 590
    F.4.1 Reduction of Distances / 590
    F.4.2 Reduction of Geodetic Azimuths / 593

APPENDIX G COMPANION CD   **595**

G.1 Introduction / 595
G.2 File Formats and Memory Matters / 596
G.3 Software / 596
    G.3.1 ADJUST / 596
    G.3.2 STATS / 597
    G.3.3 MATRIX / 598
    G.3.4 Mathcad Worksheets / 598

G.4 Using the Software as an Instructional Aid / 599

BIBLIOGRAPHY   600

INDEX   603

PREFACE

No measurement is ever exact. As a corollary, every measurement contains error. These statements are fundamental and universally accepted. It follows logically, therefore, that surveyors, who are measurement specialists, should have a thorough understanding of errors. They must be familiar with the different types of errors, their sources, and their expected magnitudes. Armed with this knowledge they will be able to (1) adopt procedures for reducing error sizes when making their measurements and (2) account rigorously for the presence of errors as they analyze and adjust their measured data. This book is devoted to creating a better understanding of these topics.

In recent years, the least squares method of adjusting spatial data has been rapidly gaining popularity as the method used for analyzing and adjusting surveying data. This should not be surprising, because the method is the most rigorous adjustment procedure available. It is soundly based on the mathematical theory of probability; it allows for appropriate weighting of all observations in accordance with their expected precisions; and it enables complete statistical analyses to be made following adjustments so that the expected precisions of adjusted quantities can be determined. Procedures for employing the method of least squares and then statistically analyzing the results are major topics covered in this book.

In years past, least squares was seldom used for adjusting surveying data because the time required to set up and solve the necessary equations was too great for hand methods. Now computers have eliminated that disadvantage. Besides advances in computer technology, some other recent developments have also led to increased use of least squares. Prominent among these are the global positioning system (GPS) and geographic information systems and land information systems (GISs and LISs). These systems rely heavily on rigorous adjustment of data and statistical analysis of the results. But perhaps the most compelling of all reasons for the recent increased interest in least squares adjustment is that new accuracy standards for surveys are being developed that are based on quantities obtained from least squares adjustments. Thus, surveyors of the future will not be able to test their measurements for compliance with these standards unless they adjust their data using least squares. Clearly, modern surveyors must be able to apply the method of least squares to adjust their measured data, and they must also be able to perform a statistical evaluation of the results after making the adjustments.

This book originated in 1968 as a set of lecture notes for a course taught to a group of practicing surveyors in the San Francisco Bay area by Professor Paul R. Wolf. The notes were subsequently bound and used as the text for formal courses in adjustment computations taught at both the University of California–Berkeley and the University of Wisconsin–Madison. In 1980, a second edition was produced that incorporated many changes and suggestions from students and others who had used the notes. The second edition, published by Landmark Enterprises, has been distributed widely to practicing surveyors and has also been used as a textbook for adjustment computations courses in several colleges and universities.

For the fourth edition, new chapters on the three-dimensional geodetic network adjustments, combining GPS baseline vectors and terrestrial observations in an adjustment, the Helmert transformation, analysis of adjustments, and state plane coordinate computations are added. These are in keeping with the modern survey firm that collects data in three dimensions and needs to analyze large data sets. Additionally, Chapter 4 of the third edition has been divided into two new chapters on confidence intervals and statistical testing. This edition has greatly expanded and modified the number of problems for each chapter to provide readers with ample practice problems. For instructors who adopt this book in their classes, a *Solutions Manual to Accompany Adjustment Computations* is also available from the publisher.

Two new appendixes have been added, including one on map projection coordinate systems and another on the companion CD. The software included on the CD for this book has also been greatly expanded and updated. A Mathcad electronic book added to the companion CD demonstrates the computations for many of the example problems in the text. To obtain a greater understanding of the material contained in this text, these electronic worksheets allow the reader to explore the intermediate computations in more detail. For readers not having the Mathcad software package, hypertext markup language (html) files are included on the CD for browsing.

The software STATS, ADJUST, and MATRIX are now Windows-based and will run on a PC-compatible computer. The first package, called STATS, performs basic statistical analyses. For any given set of measured data, it will compute the mean, median, mode, and standard deviation, and develop and plot the histogram and normal distribution curve. The second package, called

ADJUST, contains programs for performing specific least-squares adjustments. Level nets, horizontal surveys (trilateration, triangulation, traverses, and horizontal network surveys), GPS networks, and traditional three-dimensional surveys can be adjusted using software in this package. It also contains programs to compute the least-squares solution for a variety of coordinate transformations, and to determine the least squares fit of a line, parabola, or circle to a set of data points. Each of these programs computes residuals and standard deviations following the adjustment. The third program package, called MATRIX, performs a collection of basic matrix operations, such as addition, subtraction, transpose, multiplication, inverse, and more. Using this program, systems of simultaneous linear equations can be solved quickly and conveniently, and the basic algorithm for doing least squares adjustments can be solved in a stepwise fashion. For those who wish to develop their own software, the book provides several helpful computer algorithms in the languages of BASIC, C, FORTRAN, and PASCAL. Additionally, the Mathcad worksheets demonstrate the use of functions in developing modular programs.

This current edition now consists of 26 chapters and several appendixes. The chapters are arranged in the order found most convenient in teaching college courses on adjustment computations. It is believed that this order also best facilitates practicing surveyors who use the book for self-study. In earlier chapters we define terms and introduce students to the fundamentals of errors and methods for analyzing them. The next several chapters are devoted to the subject of error propagation in the various types of traditional surveying measurements. Then chapters follow that describe observation weighting and introduce the least-squares method for adjusting observations. Application of least squares in adjusting basic types of surveys are then presented in separate chapters. Adjustment of level nets, trilateration, triangulation, traverses, and horizontal networks, GPS networks, and traditional three-dimensional surveys are included. The subject of error ellipses is covered in a separate chapter. Procedures for applying least squares in curve fitting and in computing coordinate transformations are also presented. The more advanced topics of blunder detection, the method of general least squares, and computer optimization are covered in the last chapters.

As with previous editions, matrix methods, which are so well adapted to adjustment computations, continue to be used in this edition. For those students who have never studied matrices, or those who wish to review this topic, an introduction to matrix methods is given in Appendixes A and B. Those students who have already studied matrices can conveniently skip this subject.

Least-squares adjustments often require the formation and solution of nonlinear equations. Procedures for linearizing nonlinear equations by Taylor's theorem are therefore important in adjustment computations, and this topic is presented in Appendix C. Appendix D contains several statistical tables including the standard normal error distribution, the χ^2 distribution, Student's t

distribution, and a set of F-distribution tables. These tables are described at appropriate locations in the text, and their use is demonstrated with example problems.

Basic courses in statistics and calculus are necessary prerequisites to understanding some of the theoretical coverage and equation derivations given herein. Nevertheless, those who do not have these courses as background but who wish to learn how to apply least squares in adjusting surveying observations can study Chapters 1 through 3, skip Chapters 4 through 8, and then proceed with the remaining chapters.

Besides being appropriate for use as a textbook in college classes, this book will be of value to practicing surveyors and geospatial information managers. The authors hope that through the publication of this book, least squares adjustment and rigorous statistical analyses of surveying data will become more commonplace, as they should.

ACKNOWLEDGMENTS

Through the years many people have contributed to the development of this book. As noted in the preface, the book has been used in continuing education classes taught to practicing surveyors as well as in classes taken by students at the University of California–Berkeley, the University of Wisconsin–Madison, and the Pennsylvania State University–Wilkes-Barre. The students in these classes have provided data for many of the example problems and have supplied numerous helpful suggestions for improvements throughout the book. The authors gratefully acknowledge their contributions.

Earlier editions of the book benefited specifically from the contributions of Mr. Joseph Dracup of the National Geodetic Survey, Professor Harold Welch of the University of Michigan, Professor Sandor Veress of the University of Washington, Mr. Charles Schwarz of the National Geodetic Survey, Mr. Earl Burkholder of the New Mexico State University, Dr. Herbert Stoughton of Metropolitan State College, Dr. Joshua Greenfeld of New Jersey Institute of Technology, Dr. Steve Johnson of Purdue University, Mr. Brian Naberezny of Pennsylvania State University, and Professor David Mezera of the University of Wisconsin–Madison. The suggestions and contributions of these people were extremely valuable and are very much appreciated.

To improve future editions, the author will gratefully accept any constructive criticisms of this edition and suggestions for its improvement.

ADJUSTMENT COMPUTATIONS

CHAPTER 1

INTRODUCTION

1.1 INTRODUCTION

We currently live in what is often termed the *information age.* Aided by new and emerging technologies, data are being collected at unprecedented rates in all walks of life. For example, in the field of surveying, total station instruments, global positioning system (GPS) equipment, digital metric cameras, and satellite imaging systems are only some of the new instruments that are now available for rapid generation of vast quantities of measured data.

Geographic Information Systems (GISs) have evolved concurrently with the development of these new data acquisition instruments. GISs are now used extensively for management, planning, and design. They are being applied worldwide at all levels of government, in business and industry, by public utilities, and in private engineering and surveying offices. Implementation of a GIS depends upon large quantities of data from a variety of sources, many of them consisting of observations made with the new instruments, such as those noted above.

Before data can be utilized, however, whether for surveying and mapping projects, for engineering design, or for use in a geographic information system, they must be processed. One of the most important aspects of this is to account for the fact that *no measurements are exact.* That is, *they always contain errors.*

The steps involved in accounting for the existence of errors in measurements consist of (1) performing statistical analyses of the observations to assess the magnitudes of their errors and to study their distributions to determine whether or not they are within acceptable tolerances; and if the observations are acceptable, (2) adjusting them so that they conform to exact geometric conditions or other required constraints. Procedures for performing these two steps in processing measured data are principal subjects of this book.

1.2 DIRECT AND INDIRECT MEASUREMENTS

Measurements are defined as observations made to determine unknown quantities. They may be classified as either direct or indirect. *Direct measurements* are made by applying an instrument directly to the unknown quantity and observing its value, usually by reading it directly from graduated scales on the device. Determining the distance between two points by making a direct measurement using a graduated tape, or measuring an angle by making a direct observation from the graduated circle of a theodolite or total station instrument, are examples of direct measurements.

Indirect measurements are obtained when it is not possible or practical to make direct measurements. In such cases the quantity desired is determined from its mathematical relationship to direct measurements. Surveyors may, for example, measure angles and lengths of lines between points directly and use these measurements to compute station coordinates. From these coordinate values, other distances and angles that were not measured directly may be derived indirectly by computation. During this procedure, the errors that were present in the original direct observations are *propagated* (distributed) by the computational process into the indirect values. Thus, the indirect measurements (computed station coordinates, distances, and angles) contain errors that are functions of the original errors. This distribution of errors is known as *error propagation*. The analysis of how errors propagate is also a principal topic of this book.

1.3 MEASUREMENT ERROR SOURCES

It can be stated unconditionally that (1) *no measurement is exact,* (2) *every measurement contains errors,* (3) *the true value of a measurement is never known,* and thus (4) *the exact sizes of the errors present are always unknown.* These facts can be illustrated by the following. If an angle is measured with a scale divided into degrees, its value can be read only to perhaps the nearest tenth of a degree. If a better scale graduated in minutes were available and read under magnification, however, the same angle might be estimated to tenths of a minute. With a scale graduated in seconds, a reading to the nearest tenth of a second might be possible. From the foregoing it should be clear that no matter how well the observation is taken, a better one may be possible. Obviously, in this example, observational accuracy depends on the division size of the scale. But accuracy depends on many other factors, including the overall reliability and refinement of the equipment used, environmental conditions that exist when the observations are taken, and human limitations (e.g., the ability to estimate fractions of a scale division). As better equipment is developed, environmental conditions improve, and observer ability increases, observations will approach their true values more closely, but they can never be exact.

By definition, an *error* is the difference between a measured value for any quantity and its true value, or

$$\varepsilon = y - \mu \tag{1.1}$$

where ε is the error in an observation, y the measured value, and μ its true value.

As discussed above, errors stem from three sources, which are classified as instrumental, natural, and personal:

1. *Instrumental errors.* These errors are caused by imperfections in instrument construction or adjustment. For example, the divisions on a theodolite or total station instrument may not be spaced uniformly. These error sources are present whether the equipment is read manually or digitally.
2. *Natural errors.* These errors are caused by changing conditions in the surrounding environment, including variations in atmospheric pressure, temperature, wind, gravitational fields, and magnetic fields.
3. *Personal errors.* These errors arise due to limitations in human senses, such as the ability to read a micrometer or to center a level bubble. The sizes of these errors are affected by the personal ability to see and by manual dexterity. These factors may be influenced further by temperature, insects, and other physical conditions that cause humans to behave in a less precise manner than they would under ideal conditions.

1.4 DEFINITIONS

From the discussion thus far, it can be stated with absolute certainty that all measured values contain errors, whether due to lack of refinement in readings, instabilities in environmental conditions, instrumental imperfections, or human limitations. Some of these errors result from physical conditions that cause them to occur in a systematic way, whereas others occur with apparent randomness. Accordingly, errors are classified as either systematic or random. But before defining systematic and random errors, it is helpful to define mistakes.

1. *Mistakes.* These are caused by confusion or by an observer's carelessness. They are not classified as errors and must be removed from any

set of observations. Examples of mistakes include (a) forgetting to set the proper parts per million (ppm) correction on an EDM instrument, or failure to read the correct air temperature, (b) mistakes in reading graduated scales, and (c) blunders in recording (i.e., writing down 27.55 for 25.75). Mistakes are also known as *blunders* or *gross errors*.

2. *Systematic errors.* These errors follow some physical law, and thus these errors can be predicted. Some systematic errors are removed by following correct measurement procedures (e.g., balancing backsight and foresight distances in differential leveling to compensate for Earth curvature and refraction). Others are removed by deriving corrections based on the physical conditions that were responsible for their creation (e.g., applying a computed correction for Earth curvature and refraction on a trigonometric leveling observation). Additional examples of systematic errors are (a) temperature not being standard while taping, (b) an index error of the vertical circle of a theodolite or total station instrument, and (c) use of a level rod that is not of standard length. Corrections for systematic errors can be computed and applied to observations to eliminate their effects. Systematic errors are also known as *biases*.

3. *Random errors.* These are the errors that remain after all mistakes and systematic errors have been removed from the measured values. In general, they are the result of human and instrument imperfections. They are generally small and are as likely to be negative as positive. They usually do not follow any physical law and therefore must be dealt with according to the mathematical laws of probability. Examples of random errors are (a) imperfect centering over a point during distance measurement with an EDM instrument, (b) bubble not centered at the instant a level rod is read, and (c) small errors in reading graduated scales. It is impossible to avoid random errors in measurements entirely. Although they are often called accidental errors, their occurrence should not be considered an accident.

1.5 PRECISION VERSUS ACCURACY

Due to errors, repeated observation of the same quantity will often yield different values. A *discrepancy* is defined as the algebraic difference between two observations of the same quantity. When small discrepancies exist between repeated observations, it is generally believed that only small errors exist. Thus, the tendency is to give higher credibility to such data and to call the observations *precise*. However, precise values are not necessarily accurate values. To help understand the difference between precision and accuracy, the following definitions are given:

1. *Precision* is the degree of consistency between observations based on the sizes of the discrepancies in a data set. The degree of precision

attainable is dependent on the stability of the environment during the time of measurement, the quality of the equipment used to make the observations, and the observer's skill with the equipment and observational procedures.

2. *Accuracy* is the measure of the absolute nearness of a measured quantity to its true value. Since the true value of a quantity can never be determined, accuracy is always an unknown.

The difference between precision and accuracy can be demonstrated using distance observations. Assume that the distance between two points is paced, taped, and measured electronically and that each procedure is repeated five times. The resulting observations are:

Observation	Pacing, p	Taping, t	EDM, e
1	571	567.17	567.133
2	563	567.08	567.124
3	566	567.12	567.129
4	588	567.38	567.165
5	557	567.01	567.114

The arithmetic means for these data sets are 569, 567.15, and 567.133, respectively. A line plot illustrating relative values of the electronically measured distances denoted by e, and the taped distances, denoted by t, is shown in Figure 1.1. Notice that although the means of the EDM data set and of the taped observations are relatively close, the EDM set has smaller discrepancies. This indicates that the EDM instrument produced a higher precision. However, this higher precision does not necessarily prove that the mean of the electronically measured data set is implicitly more accurate than the mean of the taped values. In fact, the opposite may be true if the reflector constant was entered incorrectly causing a large systematic error to be present in all the electronically measured distances. Because of the larger discrepancies, it is unlikely that the mean of the paced distances is as accurate as either of the other two values. But its mean could be more accurate if large systematic errors were present in both the taped and electronically measured distances.

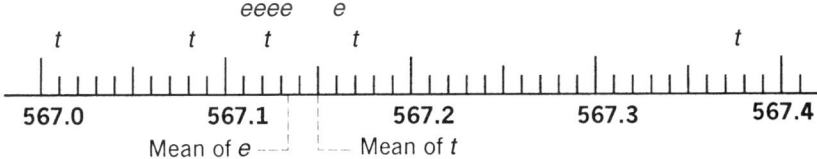

Figure 1.1 Line plot of distance quantities.

6 INTRODUCTION

Another illustration explaining differences between precision and accuracy involves target shooting, depicted in Figure 1.2. As shown, four situations can occur. If accuracy is considered as closeness of shots to the center of a target at which a marksman shoots, and precision as the closeness of the shots to each other, (1) the data may be both precise and accurate, as shown in Figure 1.2(*a*); (2) the data may produce an accurate mean but not be precise, as shown in Figure 1.2(*b*); (3) the data may be precise but not accurate, as shown in Figure 1.2(*c*); or (4) the data may be neither precise nor accurate, as shown in Figure 1.2(*d*).

Figure 1.2(*a*) is the desired result when observing quantities. The other cases can be attributed to the following situations. The results shown in Figure 1.2(*b*) occur when there is little refinement in the observational process. Someone skilled at pacing may achieve these results. Figure 1.2(*c*) generally occurs when systematic errors are present in the observational process. For example, this can occur in taping if corrections are not made for tape length and temperature, or with electronic distance measurements when using the wrong combined instrument–reflector constant. Figure 1.2(*d*) shows results obtained when the observations are not corrected for systematic errors and are taken carelessly by the observer (or the observer is unskilled at the particular measurement procedure).

In general, when making measurements, data such as those shown in Figure 1.2(*b*) and (*d*) are undesirable. Rather, results similar to those shown in Figure 1.2(*a*) are preferred. However, in making measurements the results of Figure 1.2(*c*) can be just as acceptable if proper steps are taken to correct for the presence of systematic errors. (This correction would be equivalent to a

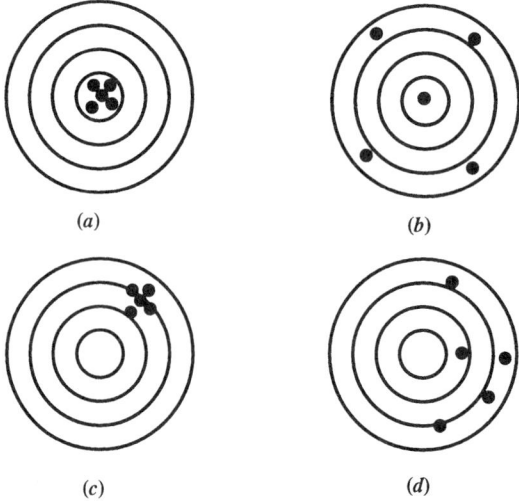

Figure 1.2 Examples of precision versus accuracy.

marksman realigning the sights after taking shots.) To make these corrections, (1) the specific types of systematic errors that have occurred in the observations must be known, and (2) the procedures used in correcting them must be understood.

1.6 REDUNDANT MEASUREMENTS IN SURVEYING AND THEIR ADJUSTMENT

As noted earlier, errors exist in all observations. In surveying, the presence of errors is obvious in many situations where the observations must meet certain conditions. For example, in level loops that begin and close on the same benchmark, the elevation difference for the loop must equal zero. However, in practice, this is hardly ever the case, due to the presence of random errors. (For this discussion it is assumed that all mistakes have been eliminated from the observations and appropriate corrections have been applied to remove all systematic errors.) Other conditions that disclose errors in surveying observations are that (1) the three measured angles in a plane triangle must total 180°, (2) the sum of the angles measured around the horizon at any point must equal 360°, and (3) the algebraic sum of the latitudes (and departures) must equal zero for closed polygon traverses that begin and end on the same station. Many other conditions could be cited; however, in any of them, the observations rarely, if ever, meet the required conditions, due to the presence of random errors.

The examples above not only demonstrate that errors are present in surveying observations but also the importance of *redundant observations;* those measurements made that are in excess of the minimum number that are needed to determine the unknowns. For example, two measurements of the length of a line yield one redundant observation. The first observation would be sufficient to determine the unknown length, and the second is redundant. However, this second observation is very valuable. First, by examining the discrepancy between the two values, an assessment of the size of the error in the observations can be made. If a large discrepancy exists, a blunder or large error is likely to have occurred. In that case, measurements of the line would be repeated until two values having an acceptably small discrepancy were obtained. Second, the redundant observation permits an adjustment to be made to obtain a final value for the unknown line length, and that final adjusted value will be more precise statistically than either of the individual observations. In this case, if the two observations were of equal precision, the adjusted value would be the simple mean.

Each of the specific conditions cited in the first paragraph of this section involves one redundant observation. For example, there is one redundant observation when the three angles of a plane triangle are observed. This is true because with two observed angles, say A and B, the third could be computed as $C = 180° - A - B$, and thus observation of C is unnecessary. However, measuring angle C enables an assessment of the errors in the angles and also makes an adjustment possible to obtain final angles with statistically improved precisions. Assuming that the angles were of equal precision, the adjustment would enforce a 180° sum for the three angles by distributing the total discrepancy in equal parts to each angle.

Although the examples cited here are indeed simple, they help to define redundant measurements and to illustrate their importance. In large surveying networks, the number of redundant observations can become extremely large, and the adjustment process is somewhat more involved than it is for the simple examples given here.

Prudent surveyors always make redundant observations in their work, for the two important reasons indicated above: (1) to make it possible to assess errors and make decisions regarding acceptance or rejection of observations, and (2) to make possible an adjustment whereby final values with higher precisions are determined for the unknowns.

1.7 ADVANTAGES OF LEAST SQUARES ADJUSTMENT

As indicated in Section 1.6, in surveying it is recommended that redundant observations always be made and that adjustments of the observations always be performed. These adjustments account for the presence of errors in the observations and increase the precision of the final values computed for the unknowns. When an adjustment is completed, all observations are corrected so that they are consistent throughout the survey network [i.e., the same values for the unknowns are determined no matter which corrected observation(s) are used to compute them].

Many different methods have been derived for making adjustments in surveying; however, the method of least squares should be used because it has significant advantages over all other rule-of-thumb procedures. The advantages of least squares over other methods can be summarized with the following four general statements: (1) it is the most rigorous of adjustments; (2) it can be applied with greater ease than other adjustments; (3) it enables rigorous postadjustment analyses to be made; and (4) it can be used to perform presurvey planning. These advantages are discussed further below.

Least squares adjustment is rigorously based on the theory of mathematical probability, whereas in general, the other methods do not have this rigorous base. As described later in the book, in a least squares adjustment, the following condition of mathematical probability is enforced: *The sum of the squares of the errors times their respective weights is minimized.* By enforcing this condition in any adjustment, the set of errors that is computed has the highest probability of occurrence. Another aspect of least squares adjustment that adds to its rigor is that it permits all observations, regardless of their number or type, to be entered into the adjustment and used simultaneously in the computations. Thus, an adjustment can combine distances, horizontal angles, azimuths, zenith or vertical angles, height differences, coordinates, and even GPS observations. One important additional asset of least squares adjustment is that it enables "relative weights" to be applied to the observations in accordance with their estimated relative reliabilities. These reliabilities are based on estimated precisions. Thus, if distances were observed in the same survey by pacing, taping, and using an EDM instrument, they could all be combined in an adjustment by assigning appropriate relative weights.

Years ago, because of the comparatively heavy computational effort involved in least squares, nonrigorous or "rule-of-thumb" adjustments were most often used. However, now because computers have eliminated the computing problem, the reverse is true and least squares adjustments are performed more easily than these rule-of-thumb techniques. Least squares adjustments are less complicated because the same fundamental principles are followed regardless of the type of survey or the type of observations. Also, the same basic procedures are used regardless of the geometric figures involved (e.g., triangles, closed polygons, quadrilaterals, or more complicated networks). On the other hand, rules of thumb are not the same for all types of surveys (e.g., level nets use one rule and traverses use another), and they vary for different geometric shapes. Furthermore, the rule of thumb applied for a particular survey by one surveyor may be different from that applied by another surveyor. A favorable characteristic of least squares adjustments is that there is only one rigorous approach to the procedure, and thus no matter who performs the adjustment for any particular survey, the same results will be obtained.

Least squares has the advantage that after an adjustment has been finished, a complete statistical analysis can be made of the results. Based on the sizes and distribution of the errors, various tests can be conducted to determine if a survey meets acceptable tolerances or whether the observations must be repeated. If blunders exist in the data, these can be detected and eliminated. Least squares enables precisions for the adjusted quantities to be determined easily and these precisions can be expressed in terms of error ellipses for clear and lucid depiction. Procedures for accomplishing these tasks are described in subsequent chapters.

Besides its advantages in adjusting survey data, least squares can be used to plan surveys. In this application, prior to conducting a needed survey, simulated surveys can be run in a trial-and-error procedure. For any project, an initial trial geometric figure for the survey is selected. Based on the figure, trial observations are either computed or scaled. Relative weights are assigned to the observations in accordance with the precision that can be estimated using different combinations of equipment and field procedures. A least squares adjustment of this initial network is then performed and the results analyzed. If goals have not been met, the geometry of the figure and the observation precisions are varied and the adjustment performed again. In this process different types of observations can be used, and observations can be added or deleted. These different combinations of geometric figures and observations are varied until one is achieved that produces either optimum or satisfactory results. The survey crew can then proceed to the field, confident that if the project is conducted according to the design, satisfactory results will be obtained. This technique of applying least squares in survey planning is discussed in later chapters.

1.8 OVERVIEW OF THE BOOK

In the remainder of the book the interrelationship between observational errors and their adjustment is explored. In Chapters 2 through 5, methods used to determine the reliability of observations are described. In these chapters, the ways that errors of multiple observations tend to be distributed are illustrated, and techniques used to compare the quality of different sets of measured values are examined. In Chapters 6 through 9 and in Chapter 13, methods used to model error propagation in observed and computed quantities are discussed. In particular, error sources present in traditional surveying techniques are examined, and the ways in which these errors propagate throughout the observational and computational processes are explained. In the remainder of the book, the *principles of least squares* are applied to adjust observations in accordance with random error theory and techniques used to locate mistakes in observations are examined.

PROBLEMS

1.1 Describe an example in which directly measured quantities are used to obtain an indirect measurement.

1.2 Identify the direct and indirect measurements used in computing traverse station coordinates.

1.3 Explain the difference between systematic and random errors.

1.4 List possible systematic and random errors when measuring:
 (a) a distance with a tape.
 (b) a distance with an EDM.
 (c) an angle with a total station.
 (d) the difference in elevation using an automatic level.

1.5 List three examples of mistakes that can be made when measuring an angle with total station instruments.

1.6 Identify each of the following errors as either systematic or random.
   (a) Reading a level rod.
   (b) Not holding a level rod plumb.
   (c) Leveling an automatic leveling instrument.
   (d) Using a level rod that has one foot removed from the bottom of the rod.

1.7 In your own words, define the difference between precision and accuracy.

1.8 Identify each of the following errors according to its source (natural, instrumental, personal):
   (a) Level rod length.
   (b) EDM–reflector constant.
   (c) Air temperature in an EDM observation.
   (d) Reading a graduation on a level rod.
   (e) Earth curvature in leveling observations.
   (f) Horizontal collimation error of an automatic level.

1.9 The calibrated length of a particular line is 400.012 m. A length of 400.015 m is obtained using an EDM. What is the error in the observation?

1.10 In Problem 1.9, if the length observed is 400.007 m, what is the error in the observation?

1.11 Why do surveyors measure angles using both faces of a total station (i.e., direct and reversed)?

1.12 Give an example of compensating systematic errors in a vertical angle observation when the angle is measured using both faces of the instrument.

1.13 What systematic errors exist in taping the length of a line?

1.14 Discuss the importance of making redundant observations in surveying.

1.15 List the advantages of making adjustments by the method of least squares.

CHAPTER 2

OBSERVATIONS AND THEIR ANALYSIS

2.1 INTRODUCTION

Sets of data can be represented and analyzed using either graphical or numerical methods. Simple graphical analyses to depict trends commonly appear in newspapers and on television. A plot of the daily variation of the closing Dow Jones industrial average over the past year is an example. A bar chart showing daily high temperatures over the past month is another. Data can also be presented in numerical form and be subjected to numerical analysis. As a simple example, instead of using a bar chart, the daily high temperatures could be tabulated and the mean computed. In surveying, observational data can also be represented and analyzed either graphically or numerically. In this chapter some rudimentary methods for doing so are explored.

2.2 SAMPLE VERSUS POPULATION

Due to time and financial constraints, generally only a small data sample is collected from a much larger, possibly infinite population. For example, political parties may wish to know the percentage of voters who support their candidate. It would be prohibitively expensive to query the entire voting population to obtain the desired information. Instead, polling agencies select a sample subset of voters from the voting population. This is an example of *population sampling*.

As another example, suppose that an employer wishes to determine the relative measuring capabilities of two prospective new employees. The candidates could theoretically spend days or even weeks demonstrating their abilities. Obviously, this would not be very practical, so instead, the employer could have each person record a sample of readings and from the readings predict the person's abilities. The employer could, for instance, have each candidate read a micrometer 30 times. The 30 readings would represent a *sample* of the entire *population* of possible readings. In fact, in surveying, every time that distances, angles, or elevation differences are measured, samples are being collected from a population of measurements.

From the preceding discussion, the following definitions can be made:

1. *Population.* A population consists of all possible measurements that can be made on a particular item or procedure. Often, a population has an infinite number of data elements.
2. *Sample.* A sample is a subset of data selected from the population.

2.3 RANGE AND MEDIAN

Suppose that a 1-second (1″) micrometer theodolite is used to read a direction 50 times. The seconds portions of the readings are shown in Table 2.1. These readings constitute what is called a *data set*. How can these data be organized to make them more meaningful? How can one answer the question: Are the data representative of readings that should reasonably be expected with this instrument and a competent operator? What statistical tools can be used to represent and analyze this data set?

One quick numerical method used to analyze data is to compute its *range*, also called *dispersion*. Range is the difference between the highest and lowest values. It provides an indication of the precision of the data. From Table 2.1, the lowest value is 20.1 and the highest is 26.1. Thus, the range is 26.1–20.1, or 6.0. The range for this data set can be compared with ranges of other sets, but this comparison has little value when the two sets differ in size. For

TABLE 2.1 Fifty Readings

22.7	25.4	24.0	20.5	22.5
22.3	24.2	24.8	23.5	22.9
25.5	24.7	23.2	22.0	23.8
23.8	24.4	23.7	24.1	22.6
22.9	23.4	25.9	23.1	21.8
22.2	23.3	24.6	24.1	23.2
21.9	24.3	23.8	23.1	25.2
26.1	21.2	23.0	25.9	22.8
22.6	25.3	25.0	22.8	23.6
21.7	23.9	22.3	25.3	20.1

instance, would a set of 100 data points with a range of 8.5 be better than the set in Table 2.1? Clearly, other methods of analyzing data sets statistically would be useful.

To assist in analyzing data, it is often helpful to list the values in order of increasing size. This was done with the data of Table 2.1 to produce the results shown in Table 2.2. By looking at this ordered set, it is possible to determine quickly the data's middle value or *midpoint*. In this example it lies between the values of 23.4 and 23.5. The midpoint value is also known as the *median*. Since there are an even number of values in this example, the median is given by the average of the two values closest to (which straddle) the midpoint. That is, the median is assigned the average of the 25th and 26th entries in the ordered set of 50 values, and thus for the data set of Table 2.2, the median is the average of 23.4 and 23.5, or 23.45.

2.4 GRAPHICAL REPRESENTATION OF DATA

Although an ordered numerical tabulation of data allows for some data distribution analysis, it can be improved with a *frequency histogram,* usually called simply a *histogram*. Histograms are bar graphs that show the frequency distributions in data. To create a histogram, the data are divided into *classes*. These are subregions of data that usually have a uniform range in values, or *class width*. Although there are no universally applicable rules for the selection of class width, generally 5 to 20 classes are used. As a rule of thumb, a data set of 30 values may have only five or six classes, whereas a data set of 100 values may have 10 or more classes. In general, the smaller the data set, the lower the number of classes used.

The histogram *class width* (range of data represented by each histogram bar) is determined by dividing the total range by the number of classes to be used. Consider, for example, the data of Table 2.2. If they were divided into seven classes, the *class width* would be the range divided by the number of classes, or $6.0/7 = 0.857 \approx 0.86$. The first *class interval* is found by adding

TABLE 2.2 Arranged Data

20.1	20.5	21.2	21.7	21.8
21.9	22.0	22.2	22.3	22.3
22.5	22.6	22.6	22.7	22.8
22.8	22.9	22.9	23.0	23.1
23.1	23.2	23.2	23.3	23.4
23.5	23.6	23.7	23.8	23.8
23.8	23.9	24.0	24.1	24.1
24.2	24.3	24.4	24.6	24.7
24.8	25.0	25.2	25.3	25.3
25.4	25.5	25.9	25.9	26.1

the class width to the lowest data value. For the data in Table 2.2, the first class interval is from 20.1 to (20.1 + 0.86), or 20.96. This class interval includes all data from 20.1 up to (but not including) 20.96. The next class interval is from 20.96 up to (20.96 + 0.86), or 21.82. Remaining class intervals are found by adding the class width to the upper boundary value of the preceding class. The class intervals for the data of Table 2.2 are listed in column (1) of Table 2.3.

After creating class intervals, the number of data values in each interval, called the *class frequency,* is tallied. Obviously, having data ordered consecutively as shown in Table 2.2 aids greatly in this counting process. Column (2) of Table 2.3 shows the class frequency for each class interval of the data in Table 2.2.

Often, it is also useful to calculate the *class relative frequency* for each interval. This is found by dividing the class frequency by the total number of observations. For the data in Table 2.2, the class relative frequency for the first class interval is $2/50 = 0.04$. Similarly, the class relative frequency of the fourth interval (from 22.67 to 23.53) is $13/50 = 0.26$. The class relative frequencies for the data of Table 2.2 are given in column (3) of Table 2.3. *Notice that the sum of all class relative frequencies is always 1.* The class relative frequency enables easy determination of percentages. For instance, the class interval from 21.82 to 22.67 contains 16% (0.16 × 100%) of the sample observations.

A *histogram* is a bar graph plotted with either class frequencies or relative class frequencies on the ordinate, versus values of the class interval bounds on the abscissa. Using the data from Table 2.3, the histogram shown in Figure 2.1 was constructed. Notice that in this figure, relative frequencies have been plotted as ordinates.

Histograms drawn with the same ordinate and abscissa scales can be used to compare two data sets. If one data set is more precise than the other, it will have comparatively tall bars in the center of the histogram, with relatively

TABLE 2.3 Frequency Count

(1) Class Interval	(2) Class Frequency	(3) Class Relative Frequency
20.10–20.96	2	2/50 = 0.04
20.96–21.82	3	3/50 = 0.06
21.82–22.67	8	8/50 = 0.16
22.67–23.53	13	13/50 = 0.26
23.53–24.38	11	11/50 = 0.22
24.38–25.24	6	6/50 = 0.12
25.24–26.10	7	7/50 = 0.14
		50/50 = 1

16 OBSERVATIONS AND THEIR ANALYSIS

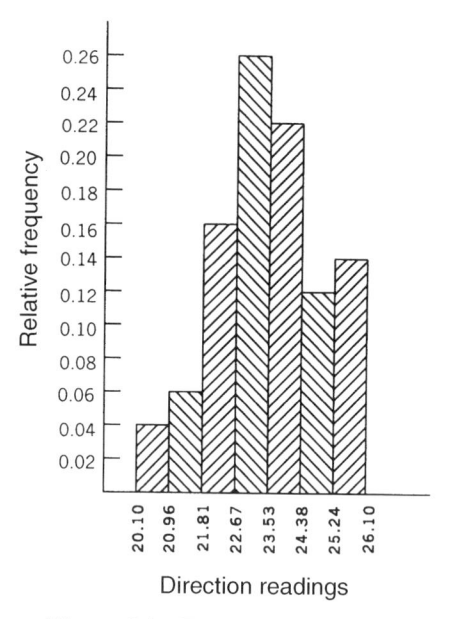

Figure 2.1 Frequency histogram.

short bars near its edges. Conversely, the less precise data set will yield a wider range of abscissa values, with shorter bars at the center.

A summary of items easily seen on a histogram include:

- Whether the data are symmetrical about a central value
- The range or dispersion in the measured values
- The frequency of occurrence of the measured values
- The steepness of the histogram, which is an indication of measurement precision

Figure 2.2 shows several possible histogram shapes. Figure 2.2(*a*) depicts a histogram that is symmetric about its central value with a single peak in the middle. Figure 2.2(*b*) is also symmetric about the center but has a steeper slope than Figure 2.2(*a*), with a higher peak for its central value. Assuming

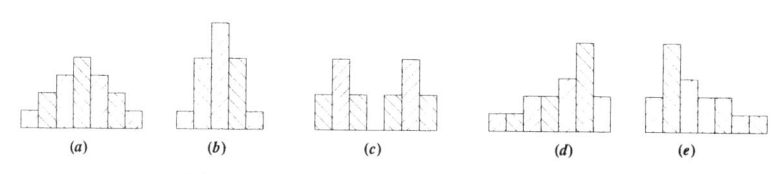

Figure 2.2 Common histogram shapes.

the ordinate and abscissa scales to be equal, the data used to plot Figure 2.2(b) are more precise than those used for Figure 2.2(a). Symmetric histogram shapes are common in surveying practice as well as in many other fields. In fact, they are so common that the shapes are said to be examples of a *normal distribution*. In Chapter 3, reasons why these shapes are so common are discussed.

Figure 2.2(c) has two peaks and is said to be a *bimodal histogram*. In the histogram of Figure 2.2(d), there is a single peak with a long tail to the left. This results from a *skewed* data set, and in particular, these data are said to be *skewed to the right*. The data of histogram Figure 2.2(e) are *skewed to the left*.

In surveying, the varying histogram shapes just described result from variations in personnel, physical conditions, and equipment: for example, repeated observations of a long distance made with an EDM instrument and by taping. An EDM would probably produce data having a very narrow range, and thus the resulting histogram would be narrow and steep with a tall central bar such as that in Figure 2.2(b). The histogram of the same distance measured by tape and plotted at the same scales would probably be wider, with the sides not as steep nor the central value as great, as shown in Figure 2.2(a). Since observations in surveying practice tend to be normally distributed, bimodal or skewed histograms from measured data are not expected. The appearance of such a histogram should lead to an investigation for the cause of this shape. For instance, if a data set from an EDM calibration plots as a bimodal histogram, it could raise questions about whether the instrument or reflector were moved during the measuring process, or if atmospheric conditions changed dramatically during the session. Similarly, a skewed histogram in EDM work may indicate the appearance of a weather front that stabilized over time. The existence of multipath errors in GPS observations could also produce these types of histogram plots.

2.5 NUMERICAL METHODS OF DESCRIBING DATA

Numerical descriptors are values computed from a data set that are used to interpret its precision or quality. Numerical descriptors fall into three categories: (1) *measures of central tendency,* (2) *measures of data variation,* and (3) *measures of relative standing*. These categories are all called *statistics*. Simply described, a statistic is a numerical descriptor computed from sample data.

2.6 MEASURES OF CENTRAL TENDENCY

Measures of central tendency are computed statistical quantities that give an indication of the value within a data set that tends to exist at the center. The arithmetic mean, the median, and the mode are three such measures. They are described as follows:

1. *Arithmetic mean.* For a set of n observations, $y_1, y_2, \ldots, y_n$, this is the average of the observations. Its value, $\bar{y}$, is computed from the equation

$$\bar{y} = \frac{\sum_{i=1}^{n} y_i}{n} \qquad (2.1)$$

   Typically, the symbol $\bar{y}$ is used to represent a sample's arithmetic mean and the symbol μ is used to represent the population mean. Otherwise, the same equation applies. Using Equation (2.1), the mean of the observations in Table 2.2 is 23.5.

2. *Median.* As mentioned previously, this is the midpoint of a sample set when arranged in ascending or descending order. One-half of the data are above the median and one-half are below it. When there are an odd number of quantities, only one such value satisfies this condition. For a data set with an even number of quantities, the average of the two observations that straddle the midpoint is used to represent the median.

3. *Mode.* Within a sample of data, the mode is the most frequently occurring value. It is seldom used in surveying because of the relatively small number of values observed in a typical set of observations. In small sample sets, several different values may occur with the same frequency, and hence the mode can be meaningless as a measure of central tendency. The mode for the data in Table 2.2 is 23.8.

2.7 ADDITIONAL DEFINITIONS

Several other terms, which are pertinent to the study of observations and their analysis, are listed and defined below.

1. *True value,* μ: a quantity's theoretically correct or exact value. As noted in Section 1.3, the true value can never be determined.
2. *Error,* ε: the difference between a measured quantity and its true value. The true value is simply the population's arithmetic mean. Since the true value of a measured quantity is indeterminate, errors are also indeterminate and are therefore only theoretical quantities. As given in Equation (1.1), repeated for convenience here, errors are expressed as

$$\varepsilon_i = y_i - \mu \qquad (2.2)$$

   where y_i is the individual observation associated with ε_i and μ is the true value for that quantity.

2.7 ADDITIONAL DEFINITIONS

3. *Most probable value, $\bar{y}$:* that value for a measured quantity which, based on the observations, has the highest probability of occurrence. It is derived from a sample set of data rather than the population and is simply the mean if the repeated measurements have the same precision.
4. *Residual, v:* The difference between any individual measured quantity and the most probable value for that quantity. Residuals are the values that are used in adjustment computations since most probable values can be determined. The term *error* is frequently used when *residual* is meant, and although they are very similar and behave in the same manner, there is this theoretical distinction. The mathematical expression for a residual is

$$v_i = \bar{y} - y_i \quad (2.3)$$

where v_i is the residual in the *i*th observation, y_i, and $\bar{y}$ is the most probable value for the unknown.

5. *Degrees of freedom:* the number of observations that are in excess of the number necessary to solve for the unknowns. In other words, the number of degrees of freedom equals the number of *redundant* observations (see Section 1.6). As an example, if a distance between two points is measured three times, one observation would determine the unknown distance and the other two would be redundant. These redundant observations reveal the discrepancies and inconsistencies in observed values. This, in turn, makes possible the practice of adjustment computations for obtaining the most probable values based on the measured quantities.
6. *Variance, σ^2:* a value by which the precision for a set of data is given. *Population variance* applies to a data set consisting of an entire *population*. It is the mean of the squares of the *errors* and is given by

$$\sigma^2 = \frac{\sum_{i=1}^{n} \varepsilon_i^2}{n} \quad (2.4)$$

Sample variance applies to a sample set of data. It is an unbiased estimate for the population variance given in Equation (2.4) and is calculated as

$$S^2 = \frac{\sum_{i=1}^{n} v_i^2}{n - 1} \quad (2.5)$$

Note that Equations (2.4) and (2.5) are identical except that ε has been changed to v and n has been changed to $n - 1$ in Equation (2.5). The validity of these modifications is demonstrated in Section 2.10.

It is important to note that the simple algebraic average of all errors in a data set cannot be used as a meaningful precision indicator. This is because random errors are as likely to be positive as negative, and thus the algebraic average will equal zero. This fact is shown for a population of data in the following simple proof. Summing Equation (2.2) for n samples gives

$$\sum_{i=1}^{n} \varepsilon_i^2 = \sum_{i=1}^{n} (y_i - \mu) = \sum_{i=1}^{n} y_i - \sum_{i=1}^{n} y_i - n\mu \qquad (a)$$

Then substituting Equation (2.1) into Equation (a) yields

$$\sum_{i=1}^{n} \varepsilon_i = \sum_{i=1}^{n} y_i - n \frac{\sum_{i=1}^{n} y_i}{n} = \sum_{i=1}^{n} y_i - \sum_{i=1}^{n} y_i = 0 \qquad (b)$$

Similarly, it can be shown that the mean of all residuals of a sample data set equals zero.

7. *Standard error*, σ: the square root of the population variance. From Equation (2.4) and this definition, the following equation is written for the standard error:

$$\sigma = \sqrt{\frac{\sum_{i=1}^{n} \varepsilon_i^2}{n}} \qquad (2.6)$$

where n is the number of observations and $\sum_{i=1}^{n} \varepsilon_i^2$ is the sum of the squares of the errors. Note that both the population variance, σ^2, and the standard error, σ, are indeterminate because true values, and hence errors, are indeterminate.

As explained in Section 3.5, 68.3% of all observations in a population data set lie within $\pm \sigma$ of the true value, μ. Thus, the larger the standard error, the more dispersed are the values in the data set and the less precise is the measurement.

8. *Standard deviation, S:* the square root of the *sample variance*. It is calculated using the expression

$$S = \sqrt{\frac{\sum_{i=1}^{n} v_i^2}{n - 1}} \qquad (2.7)$$

where S is the standard deviation, $n - 1$ the degrees of freedom or number of redundancies, and $\sum_{i=1}^{n} v_i^2$ the sum of the squares of the residuals. Standard deviation is an estimate for the standard error of the population. Since the standard error cannot be determined, the standard deviation is a practical expression for the precision of a *sample* set of data. Residuals are used rather than errors because they can be calculated from most probable values, whereas errors cannot be determined. Again, as discussed in Section 3.5, for a sample data set, 68.3% of the observations will theoretically lie between the most probable value plus and minus the standard deviation, S. The meaning of this statement will be clarified in an example that follows.

9. *Standard deviation of the mean:* the error in the mean computed from a sample set of measured values that results because all measured values contain errors. The standard deviation of the mean is computed from the sample standard deviation according to the equation

$$S_{\bar{y}} = \pm \frac{S}{\sqrt{n}} \tag{2.8}$$

Notice that as $n \to \infty$, then $S_{\bar{y}} \to 0$. This illustrates that as the size of the sample set approaches the total population, the computed mean $\bar{y}$ will approach the true mean μ. This equation is derived in Chapter 4.

2.8 ALTERNATIVE FORMULA FOR DETERMINING VARIANCE

From the definition of residuals, Equation (2.5) is rewritten as

$$S^2 = \frac{\sum_{i=1}^{n} (\bar{y} - y_i)^2}{n - 1} \tag{2.9}$$

Expanding Equation (2.9) yields

$$S^2 = \frac{1}{n-1} [(\bar{y} - y_1)^2 + (\bar{y} - y_2)^2 + \cdots + (\bar{y} - y_n)^2] \tag{c}$$

Substituting Equation (2.1) for $\bar{y}$ into Equation (c) and dropping the bounds for the summation yields

$$S^2 = \frac{1}{n-1} \left[\left(\frac{\sum y_i}{n} - y_1\right)^2 + \left(\frac{\sum y_i}{n} - y_2\right)^2 + \cdots + \left(\frac{\sum y_i}{n} - y_n\right)^2 \right] \tag{d}$$

Expanding Equation (d) gives us

22 OBSERVATIONS AND THEIR ANALYSIS

$$S^2 = \frac{1}{n-1}\left[\left(\frac{\Sigma y_i}{n}\right)^2 - 2y_1\frac{\Sigma y_i}{n} + y_1^2 + \left(\frac{\Sigma y_i}{n}\right)^2 - 2y_2\frac{\Sigma y_i}{n}\right.$$
$$\left. + y_2^2 + \cdots + \left(\frac{\Sigma y_i}{n}\right)^2 - 2y_n\frac{\Sigma y_i}{n} + y_n^2\right] \quad (e)$$

Rearranging Equation (e) and recognizing that $(\Sigma y_i/n)^2$ occurs n times in Equation (e) yields

$$S^2 = \frac{1}{n-1}\left[n\left(\frac{\Sigma y_i}{n}\right)^2 - 2\frac{\Sigma y_i}{n}(y_1 + y_2 + \cdots + y_n) + y_1^2 + y_2^2 + \cdots + y_n^2\right] \quad (f)$$

Adding the summation symbol to Equation (f) yields

$$S^2 = \frac{1}{n-1}\left[n\left(\frac{\Sigma y_i}{n}\right)^2 - \frac{2}{n}\left(\Sigma y_i\right)^2 + \Sigma y_i^2\right] \quad (g)$$

Factoring and regrouping similar summations in Equation (g) produces

$$S^2 = \frac{1}{n-1}\left[\Sigma y_i^2 - \left(\frac{2}{n} - \frac{1}{n}\right)\left(\Sigma y_i\right)^2\right] = \frac{1}{n-1}\left[\Sigma y_i^2 - \frac{1}{n}\left(\Sigma y_i\right)^2\right] \quad (h)$$

Multiplying the last term in Equation (h) by n/n yields

$$S^2 = \frac{1}{n-1}\left[\Sigma y_i^2 - n\left(\frac{\Sigma y_i}{n}\right)^2\right] \quad (i)$$

Finally, by substituting Equation (2.1) in Equation (i), the following expression for the variance results:

$$S^2 = \frac{\Sigma y_i^2 - n\bar{y}^2}{n-1} \quad (2.10)$$

Using Equation (2.10), the variance of a sample data set can be computed by subtracting n times the square of the data's mean from the summation of the squared individual observations. With this equation, the variance and the standard deviation can be computed directly from the data. However, it should be stated that with large numerical values, Equation (2.10) may overwhelm a handheld calculator or a computer working in single precision. If this problem should arise, the data should be *centered* or Equation (2.5) used. Cen-

2.9 NUMERICAL EXAMPLES

tering a data set involves subtracting a constant value (usually, the arithmetic mean) from all values in a data set. By doing this, the values are modified to a smaller, more manageable size.

2.9 NUMERICAL EXAMPLES

Example 2.1 Using the data from Table 2.2, determine the sample set's mean, median, and mode and the standard deviation using both Equations (2.7) and (2.10). Also plot its histogram. (Recall that the data of Table 2.2 result from the seconds' portion of 50 theodolite directions.)

SOLUTION

Mean: From Equation (2.1) and using the $\Sigma\, y_i$ value from Table 2.4, we have

$$\bar{y} = \frac{\sum_{i=1}^{50} y_i}{50} = \frac{1175}{50} = 23.5''$$

Median: Since there is an even number of observations, the data's midpoint lies between the values that are the 25th and 26th numerically from the beginning of the ordered set. These values are 23.4″ and 23.5″, respectively. Averaging these observations yields 23.45″.

Mode: The mode, which is the most frequently occurring value, is 23.8″. It appears three times in the sample.

Range, class width, histogram: These data were developed in Section 2.4, with the histogram plotted in Figure 2.1.

Standard deviation: Table 2.4 lists the residuals [computed using Equation (2.3)] and their squares for each observation.

From Equation (2.7) and using the value of 92.36 from Table 2.4 as the sum of the squared residuals, the standard deviation for the sample set is computed as

$$S = \sqrt{\frac{\Sigma\, v_i^2}{n - 1}} = \sqrt{\frac{92.36}{50 - 1}} = \pm 1.37''$$

Summing the squared *y*-values of Table 2.4 yields

$$\Sigma\, y_i^2 = 27{,}704.86$$

Using Equation (2.10), the standard deviation for the sample set is

TABLE 2.4 Data Arranged for the Solution of Example 2.1

No.	y	v	v^2	No.	y	v	v^2	No.	y	v	v^2	No.	y	v	v^2
1	20.1	3.4	11.56	13	22.6	0.9	0.81	25	23.4	0.1	0.01	38	24.4	−0.9	0.81
2	20.5	3.0	9.00	14	22.7	0.8	0.64	26	23.5	0.0	0.00	39	24.6	−1.1	1.21
3	21.2	2.3	5.29	15	22.8	0.7	0.49	27	23.6	−0.1	0.01	40	24.7	−1.2	1.44
4	21.7	1.8	3.24	16	22.8	0.7	0.49	28	23.7	−0.2	0.04	41	24.8	−1.3	1.69
5	21.8	1.7	2.89	17	22.9	0.6	0.36	29	23.8	−0.3	0.09	42	25.0	−1.5	2.25
6	21.9	1.6	2.56	18	22.9	0.6	0.36	30	23.8	−0.3	0.09	43	25.2	−1.7	2.89
7	22.0	1.5	2.25	19	23.0	0.5	0.25	31	23.8	−0.3	0.09	44	25.3	−1.8	3.24
8	22.2	1.3	1.69	20	23.1	0.4	0.16	32	23.9	−0.4	0.16	45	25.3	−1.8	3.24
9	22.3	1.2	1.44	21	23.1	0.4	0.16	33	24.0	−0.5	0.25	46	25.4	−1.9	3.61
10	22.3	1.2	1.44	22	23.2	0.3	0.09	34	24.1	−0.6	0.36	47	25.5	−2.0	4.00
11	22.5	1.0	1.00	23	23.2	0.3	0.09	35	24.1	−0.6	0.36	48	25.9	−2.4	5.76
12	22.6	0.9	0.81	24	23.3	0.2	0.04	36	24.2	−0.7	0.49	49	25.9	−2.4	5.76
								37	24.3	−0.8	0.64	50	26.1	−2.6	6.76
													1175	0.0	92.36

$$S = \sqrt{\frac{\Sigma y_i^2 - n\bar{y}^2}{n-1}} = \sqrt{\frac{27{,}704.86 - 50(23.5)^2}{50-1}} = \sqrt{\frac{92.36}{49}} = \pm 1.37''$$

By demonstration in Example 2.1, it can be seen that Equations (2.7) and (2.10) will yield the same standard deviation for a sample set. Notice that the number of observations within a single standard deviation of the mean, that is, between (23.5″ − 1.37″) and (23.5″ + 1.37″), or between 22.13″ and 24.87″, is 34. This represents 34/50 × 100%, or 68%, of all observations in the sample and matches the theory noted earlier. Also note that the algebraic sum of residuals is zero, as was demonstrated by Equation (b).

The histogram shown in Figure 2.1 plots class relative frequencies versus class values. Notice how the values tend to be grouped about the central point. This is an example of a precise data set.

Example 2.2 The data set shown below also represents the seconds' portion of 50 theodolite observations of a direction. Compute the mean, median, and mode, and use Equation (2.10) to determine the standard deviation. Also construct a histogram. Compare the data of this example with those of Example 2.1.

34.2	33.6	35.2	30.1	38.4	34.0	30.2	34.1	37.7	36.4
37.9	33.0	33.5	35.9	35.9	32.4	39.3	32.2	32.8	36.3
35.3	32.6	34.1	35.6	33.7	39.2	35.1	33.4	34.9	32.6
36.7	34.8	36.4	33.7	36.1	34.8	36.7	30.0	35.3	34.4
33.7	34.1	37.8	38.7	33.6	32.6	34.7	34.7	36.8	31.8

SOLUTION Table 2.5, which arranges each observation and its square in ascending order, is first prepared.

Mean: $\bar{y} = \Sigma y_i/n = 1737.0/50 = 34.74''$

Median: The median is between the 25th and 26th values, which are both 34.7″. Thus, the median is 34.7″.

Mode: The data have three different values that occur with a frequency of 3. Thus, the modes for the data set are the three values 32.6″, 33.7″, and 34.1″.

Range: The range of the data is 39.3″ − 30.0″ = 9.3″.

Class width: For comparison purposes, the class width of 0.86 is taken because it was used for the data in Table 2.2. Since it is desired that the histogram be centered about the data's mean value, the central interval is determined by adding and subtracting one-half of the class width (0.43) to the mean. Thus, the central interval is from (34.74″ − 0.43″), or 34.31″, to (34.74″ + 0.43″), or 35.17″. To compute the remaining class intervals, the

TABLE 2.5 Data Arranged for the Solution of Example 2.2

No.	y	y^2	No.	y	y^2	No.	y	y^2	No.	y	y^2
1	30.0	900.00	13	33.5	1122.25	25	34.7	1204.09	38	36.3	1317.69
2	30.1	906.01	14	33.6	1128.96	26	34.7	1204.09	39	36.4	1324.96
3	30.2	312.04	15	33.6	1128.96	27	34.8	1211.04	40	36.4	1324.96
4	31.8	1011.24	16	33.7	1135.69	28	34.8	1211.04	41	36.7	1346.89
5	32.2	1036.84	17	33.7	1135.69	29	34.9	1218.01	42	36.7	1346.89
6	32.4	1049.76	18	33.7	1135.36	30	35.1	1232.01	43	36.8	1354.24
7	32.6	1062.76	19	34.0	1156.00	31	35.2	1239.04	44	37.7	1421.29
8	32.6	1062.76	20	34.1	1162.81	32	35.3	1246.09	45	37.8	1428.84
9	32.6	1062.76	21	34.1	1162.81	33	35.3	1246.09	46	37.9	1436.41
10	32.8	1075.84	22	34.1	1162.81	34	35.6	1267.36	47	38.4	1474.56
11	33.0	1089.00	23	34.2	1169.64	35	35.9	1288.81	48	38.7	1497.69
12	33.4	1115.56	24	34.4	1183.36	36	35.9	1288.81	49	39.2	1536.64
						37	36.1	1303.21	50	39.3	1544.49
										1737.0	60,584.48

class width is subtracted, or added, to the bounds of the computed intervals as necessary until all the data are contained within the bounds of the intervals. Thus, the interval immediately preceding the central interval will be from (34.31″ − 0.86″), or 33.45″, to 34.31″, and the interval immediately following the central interval will be from 35.17″ to (35.17″ + 0.86″), or 36.03″. In a similar fashion, the remaining class intervals were determined and a class frequency chart was constructed as shown in Table 2.6. Using this table, the histogram of Figure 2.3 was constructed.

Variance: By Equation (2.10), using the sum of observations squared in Table 2.5, the sample variance is

$$S^2 = \frac{\Sigma y_i^2 - n\bar{y}^2}{n-1} = \frac{60{,}584.48 - 50(34.74)^2}{50 - 1} = 4.92$$

and the sample *standard deviation* is

$$S = \sqrt{4.92} = \pm 2.22''$$

The number of observations that actually fall within the bounds of the mean $\pm S$ (i.e., between 34.74″ ± 2.22″) is 30. This is 60% of all the observations, and closely approximates the theoretical value of 68.3%. These bounds and the mean value are shown as dashed lines in Figure 2.3.

Comparison: The data set of Example 2.2 has a larger standard deviation (±2.22″) than that of Example 2.1 (±1.37″). The range for the data of Example 2.2 (9.3″) is also larger than that of Example 2.1 (6.0″). Thus, the data set of Example 2.2 is less precise than that of Example 2.1. A comparison of

TABLE 2.6 Frequency Table for Example 2.2

Class	Class Frequency	Relative Class Frequency
29.15–30.01	1	0.02
30.01–30.87	2	0.04
30.87–31.73	0	0.00
31.73–32.59	3	0.06
32.59–33.45	6	0.12
33.45–34.31	11	0.22
34.31–35.17	7	0.14
35.17–36.03	6	0.12
36.03–36.89	7	0.14
36.89–37.75	1	0.02
37.75–38.61	3	0.06
38.61–39.47	3	0.06
	50	1.00

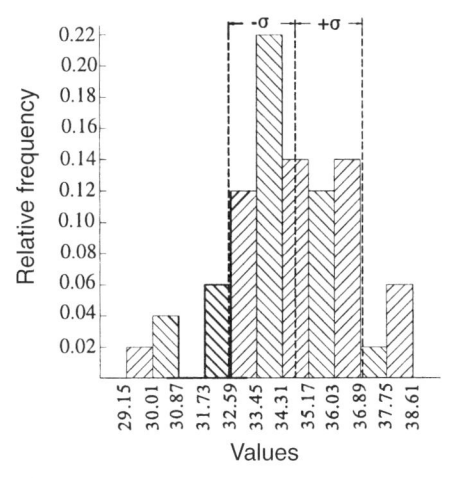

Figure 2.3 Histogram for Example 2.2.

the two histograms shows this precision difference graphically. Note, for example, that the histogram in Figure 2.1 is narrower in width and taller at the center than the histogram in Figure 2.3.

2.10 DERIVATION OF THE SAMPLE VARIANCE (BESSEL'S CORRECTION)

Recall from Section 2.7 that the denominator of the equation for sample variance was $n - 1$, whereas the denominator of the population variance was n. A simple explanation for this difference is that one observation is necessary to compute the mean ($\bar{y}$), and thus only $n - 1$ observations remain for the computation of the variance. A derivation of Equation (2.5) will clarify.

Consider a sample size of n drawn from a population with a mean, μ, and standard error of σ. Let y_i be an observation from the sample. Then

$$y_i - \mu = (y_i - \bar{y}) + (\bar{y} - \mu)$$
$$= (y_i - \bar{y}) + \varepsilon \quad (j)$$

where $\varepsilon = \bar{y} - \mu$ is the *error* or deviation of the sample mean. Squaring and expanding Equation (j) yields

$$(y_i - \mu)^2 = (y_i - \bar{y})^2 + \varepsilon^2 + 2\varepsilon(y_i - \bar{y})$$

Summing all the observations in the sample from i equaling 1 to n yields

$$\sum_{i=1}^{n} (y_i - \mu)^2 = \sum_{i=1}^{n} (y_i - \bar{y})^2 + n\varepsilon^2 + 2\varepsilon \sum_{i=1}^{n} (y_i - \bar{y}) \tag{k}$$

Since by definition of the sample mean $\bar{y}$

$$\sum_{i=1}^{n} (y_i - \bar{y}) = \sum_{i=1}^{n} y_i - n\bar{y} = \sum_{i=1}^{n} y_i - \sum_{i=1}^{n} y_i = 0 \tag{l}$$

Equation (k) becomes

$$\sum_{i=1}^{n} (y_i - \mu)^2 = \sum_{i=1}^{n} (y_i - \bar{y})^2 + n\varepsilon^2 \tag{m}$$

Repeating this calculation for many samples, the mean value of the left-hand side of Equation (m) will (by definition of σ^2) tend to $n\sigma^2$. Similarly, by Equation (2.8), the mean value of $n\varepsilon^2 = n(\mu - \bar{y})^2$ will tend to n times the variance of $\bar{y}$ since ε represents the deviation of the sample mean from the population mean. Thus, $n\varepsilon^2 \rightarrow n(\sigma^2/n)$, where σ^2/n is the variance in $\bar{y}$ as $n \rightarrow \infty$. The discussion above, Equation (m) results in

$$n\sigma^2 \rightarrow \sum_{i=1}^{n} (y_i - \bar{y})^2 + \sigma^2 \tag{n}$$

Rearranging Equation (n) produces

$$\sum_{i=1}^{n} (y_i - \bar{y})^2 \rightarrow (n-1)\sigma^2 \tag{o}$$

Thus, from Equation (o) and recognizing the left side of the equation as $(n-1)S^2$ for a sample set of data, it follows that

$$S^2 = \frac{\sum_{i=1}^{n} (y_i - \bar{y})^2}{n-1} \rightarrow \sigma^2 \tag{p}$$

In other words for a large number of random samples, the value of $\sum_{i=1}^{n} (y_i - \bar{y})^2/(n-1)$ tends to σ^2. That is, S^2 is an unbiased estimate of the population's variance.

2.11 PROGRAMMING

STATS, a Windows-based statistical software package, is included on the CD accompanying this book. It can be used to quickly perform statistical analysis of data sets as presented in this chapter. Directions regarding its use are provided on its help screen.

30 OBSERVATIONS AND THEIR ANALYSIS

   Additionally, an electronic book is provided on the CD accompanying this book. To view the electronic book interactively, Mathcad software is required. However, for those of you who do not have a copy of Mathcad, html files of the electronic book are included on the CD. The electronic book demonstrates most of the numerical examples given in the book.

   Many chapters include programming problems following the problem sets at the end of the chapters. The electronic book demonstrates the rudiments of programming these problems. Other programs on the CD include MATRIX and ADJUST. MATRIX can be used to solve problems in the book that involve matrices. ADJUST has examples of working least squares adjustment programs. ADJUST can be used to check solutions to many of the examples in the book.

   When you select the desired installation options, the installation program provided on the CD will load the files to your computer. The installation package will install each option as the option is selected. This software does not remove (uninstall) the packages. This can be done using the "Add/Remove programs" options in your computer's control panel.

PROBLEMS

2.1 The optical micrometer of a precise differential level is set and read 10 times as 8.801, 8.803, 8.798, 8.801, 8.799, 8.802, 8.802, 8.804, 8.800, and 8.802. What value would you assign to the operator's ability to set the micrometer on this instrument?

2.2 The seconds' part of 50 pointings and readings for a particular direction made using a 1″ total station with a 0.1″ display are

26.7, 26.4, 24.8, 27.4, 25.8, 27.0, 26.3, 27.8, 26.7, 26.0, 25.9, 25.4,

28.0, 27.2, 25.3, 27.2, 27.0, 27.7, 27.3, 24.8, 26.7, 25.3, 26.9, 25.5,

27.4, 25.4, 25.8, 25.5, 27.4, 27.2, 27.1, 27.4, 26.6, 26.2, 26.3, 25.3,

25.1, 27.3, 27.3, 28.1, 27.4, 27.2, 27.2, 26.4, 28.2, 25.5, 26.5, 25.9,

26.1, 26.3

   (a) What is the mean of the data set?
   (b) Construct a frequency histogram of the data using seven uniform-width class intervals.
   (c) What are the variance and standard deviation of the data?
   (d) What is the standard deviation of the mean?

2.3 An EDM instrument and reflector are set at the ends of a baseline that is 400.781 m long. Its length is measured 24 times, with the following results:

400.787 400.796 400.792 400.787 400.787 400.786 400.792 400.794

400.790 400.788 400.797 400.794 400.789 400.785 400.791 400.791

400.793 400.791 400.792 400.787 400.788 400.790 400.798 400.789

    (a) What are the mean, median, and standard deviation of the data?
    (b) Construct a histogram of the data with five intervals and describe its properties. On the histogram, lay off the sample standard deviation from both sides of the mean.
    (c) How many observations are between $\bar{y} \pm S$, and what percentage of observations does this represent?

2.4 Answer Problem 2.3 with the following additional observations: 400.784, 400.786, 400.789, 400.794, 400.792, and 400.789.

2.5 Answer Problem 2.4 with the following additional observations: 400.785, 400.793, 400.791, and 400.789.

2.6 A distance was measured in two parts with a 100-ft steel tape and then in its entirety with a 200-ft steel tape. Five repetitions were made by each method. What are the mean, variance, and standard deviation for each method of measurement?

Distances measured with a 100-ft tape:
Section 1: 100.006, 100.004, 100.001, 100.006, 100.005
Section 2: 86.777, 86.779, 86.785, 86.778, 86.774

Distances measured with a 200-ft tape:
186.778, 186.776, 186.781, 186.766, 186.789

2.7 Repeat Problem 2.6 using the following additional data for the 200-ft taped distance: 186.781, 186.794, 186.779, 186.778, and 186.776.

2.8 During a triangulation project, an observer made 16 readings for each direction. The seconds' portion of the directions to Station Orion are listed as 43.0, 41.2, 45.0, 43.4, 42.4, 52.5, 53.6, 50.9, 52.0, 50.8, 51.9, 49.5, 51.6, 51.2, 51.8, and 50.2.
    (a) Using a 1" class interval, plot the histogram using relative frequencies for the ordinates.
    (b) Analyze the data and note any abnormalities.
    (c) As a supervisor, would you recommend that the station be reobserved?

2.9 Use the program STATS to compute the mean, median, mode, and standard deviation of the data in Table 2.2 and plot a centered histogram of the data using nine intervals.

2.10 The particular line in a survey is measured three times on four separate occasions. The resulting 12 observations in units of meters are 536.191,

536.189, 536.187, 536.202, 536.200, 536.203, 536.202, 536.201, 536.199, 536.196, 536.205, and 536.202.

(a) Compute the mean, median, and mode of the data.
(b) Compute the variance and standard deviation of the data.
(c) Using a class width of 0.004 m, plot a histogram of the data and note any abnormalities that may be present.

Use the program STATS to do:

2.11 Problem 2.2.

2.12 Problem 2.3.

2.13 Problem 2.4.

2.14 Problem 2.5.

2.15 Problem 2.10. Use a class width of 0.003 m in part (c).

Practical Exercises

2.16 Using a total station, point and read a horizontal circle to a well-defined target. With the tangent screw or jog-shuttle mechanism, move the instrument of the point and repoint on the same target. Record this reading. Repeat this process 50 times. Perform the calculations of Problem 2.2 using this data set.

2.17 Determine your EDM–reflector constant, K, by observing the distances between the following three points:

■————————————————■————————————————————————————————■
A   B   C

Distance AC should be roughly 1 mile long, with B situated at some location between A and C. From measured values AC, AB, and BC, the constant K can be determined as follows: Since

$$AC + K = (AB + K) + (BC + K)$$

thus

$$K = AC - (AB + BC)$$

When establishing the line, be sure that $AB \neq BC$ and that all three points are precisely on a straight line. Use three tripods and tribrachs to minimize setup errors and be sure that all are in adjustment. Measure each line 20 times with the instrument in the metric mode. Be sure to adjust the distances for the appropriate temperature and pressure and for differences in elevation. Determine the 20 values of K and analyze the sample set. What is the mean value for K and its standard deviation?

CHAPTER 3

RANDOM ERROR THEORY

3.1 INTRODUCTION

As noted earlier, the adjustment of measured quantities containing random errors is a major concern to people involved in the geospatial sciences. In the remaining chapters it is assumed that all systematic errors have been removed from the measured values and that only random errors and blunders (which have escaped detection) remain. In this chapter the general theory of random errors is developed, and some simple methods that can be used to isolate remaining blunders in sets of data are discussed.

3.2 THEORY OF PROBABILITY

Probability is the ratio of the number of times that an event should occur to the total number of possibilities. For example, the probability of tossing a two with a fair die is 1/6 since there are six total possibilities (faces on a die) and only one of these is a two. When an event can occur in m ways and fail to occur in n ways, the probability of its occurrence is $m/(m + n)$, and the probability of its failure is $n/(m + n)$.

Probability is always a fraction ranging between zero and one. Zero denotes impossibility, and one indicates certainty. Since an event must either occur or fail to occur, the sum of all probabilities for any event is 1, and thus if 1/6 is the probability of throwing a two with one throw of a die, then $1 - 1/6$, or 5/6, is the probability that a two will not appear.

In probability terminology, a *compound event* is the simultaneous occurrence of two or more independent events. This is the situation encountered most frequently in surveying. For example, random errors from angles and distances (compound events) cause traverse misclosures. The probability of the simultaneous occurrence of two independent events is the product of their individual probabilities.

To illustrate this condition, consider the simple example of having two boxes containing combinations of red and white balls. Box A contains four balls, one red and three white. Box B contains five balls, two red and three white. What is the probability that two red balls would be drawn if one ball is drawn randomly from each box? The total number of possible pairs is 4 × 5 (20) since by drawing one ball from box A, any of the five balls in box B would complete the pair. There are only two ways to draw two red balls; that is, box A's red ball can be matched with either red ball from box B. Therefore, the probability of obtaining two red balls simultaneously is 2/20. Thus, the probability of this compound event is the product of the individual probabilities of drawing a red ball from each box, or

$$P = 1/4 \times 2/5 = 2/20$$

Similarly, the probability of drawing two white balls simultaneously is 3/4 × 3/5, or 9/20, and the probability of getting one red ball and one white ball is 1 − (2/20 + 9/20), or 9/20.

From the foregoing it is seen that the probability of the simultaneous occurrence of two independent events is the product of the individual probabilities of those two events. This principle is extended to include any number of events:

$$P = P_1 \times P_2 \times \cdots \times P_n \tag{3.1}$$

where P is the probability of the simultaneous occurrence of events having individual probabilities $P_1, P_2, \ldots, P_n$.

To develop the principle of how random errors occur, consider a very simple example where a single tape measurement is taken between points A and B. Assume that this measurement contains a single random error of size 1. Since the error is random, there are two possibilities for its value +1 or −1. Let t be the number of ways that each error can occur, and let T be the total number of possibilities, which is two. The probability of obtaining +1, which can occur only one way (i.e., $t = 1$), is t/T or 1/2. This is also the probability of obtaining −1. Suppose now that in measuring a distance AE, the tape must be placed end to end so that the result depends on the combination of two of these tape measurements. Then the possible error combinations in the result are −1 *and* −1, −1 *and* +1, +1 *and* −1, and +1 *and* +1, with $T = 4$. The final errors are −2, 0, and +2, and their t values are 1, 2, and 1, respectively. This produces probabilities of 1/4, 1/2, and 1/4, respectively. In general, as n, the number of single combined measurements, is increased, T increases according to the function $T = 2^n$, and thus for three combined measurements, $T = 2^3 = 8$, and for four measurements, $T = 2^4 = 16$.

The analysis of the preceding paragraph can be continued to obtain the results shown in Table 3.1. Figure 3.1(*a*) through (*e*) are histogram plots of the results in Table 3.1, where the values of the errors are plotted as the abscissas and the probabilities are plotted as ordinates of equal-width bars.

If the number of combining measurements, n, is increased progressively to larger values, the plot of error sizes versus probabilities would approach a smooth curve of the characteristic bell shape shown in Figure 3.2. This curve is known as the *normal error distribution curve*. It is also called the *probability density function of a normal random variable*. Notice that when n is 4, as illustrated in Figure 3.1(*d*), and when $n = 5$, as shown in Figure 3.1(*e*), the dashed lines are already beginning to take on this form.

It is important to notice that the total area of the vertical bars for each plot equals 1. This is true no matter the value of n, and thus *the area under the smooth normal error distribution curve is equal to 1*. If an event has a probability of 1, it is certain to occur, and therefore *the area under the curve represents the sum of all the probabilities of the occurrence of errors.*

TABLE 3.1 Occurrence of Random Errors

(1) Number of Combining Measurements	(2) Value of Resulting Error	(3) Frequency, t	(4) Total Number of Possibilities, T	(5) Probability
1	+1	1	2	1/2
	−1	1		1/2
2	+2	1	4	1/4
	0	2		1/2
	−2	1		1/4
3	+3	1	8	1/8
	+1	3		3/8
	−1	3		3/8
	−3	1		1/8
4	+4	1	16	1/16
	+2	4		1/4
	0	6		3/8
	−2	4		1/4
	−4	1		1/16
5	+5	1	32	1/32
	+3	5		5/32
	+1	10		5/16
	−1	10		5/16
	−3	5		5/32
	−5	1		1/32

36 RANDOM ERROR THEORY

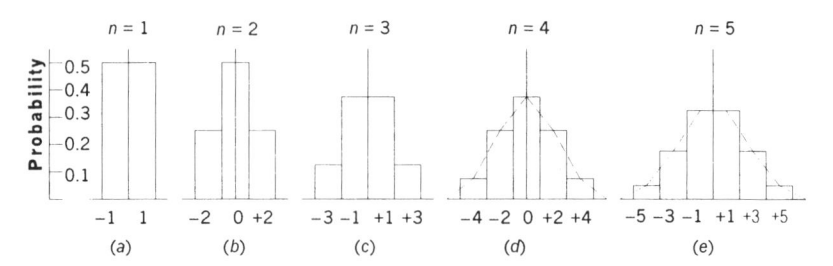

Figure 3.1 Plots of probability versus size of errors.

As derived in Section D.1, the equation of the normal distribution curve, also called the *normal probability density function,* is

$$f(x) = \frac{1}{\sigma\sqrt{2\pi}} e^{-x^2/2\sigma^2} \tag{3.2}$$

where $f(x)$ is the probability density function, e the base of natural logarithms, x the error, and σ the standard error as defined in Chapter 2.

3.3 PROPERTIES OF THE NORMAL DISTRIBUTION CURVE

In Equation (3.2), $f(x)$ is the probability of occurrence of an error of size between x and $x + dx$, where dx is an infinitesimally small value. The error's probability is equivalent to the area under the curve between the limits of x and $x + dx$, which is shown crosshatched in Figure 3.3. As stated previously, the total area under the probability curve represents the total probability, which is 1. This is represented in equation form as

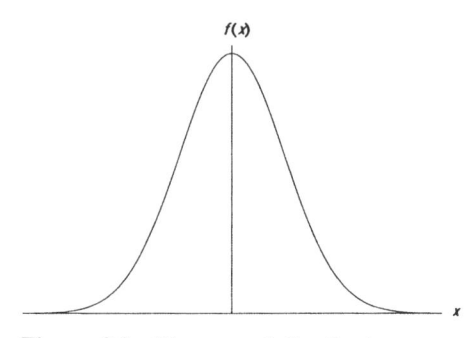

Figure 3.2 The normal distribution curve.

3.3 PROPERTIES OF THE NORMAL DISTRIBUTION CURVE

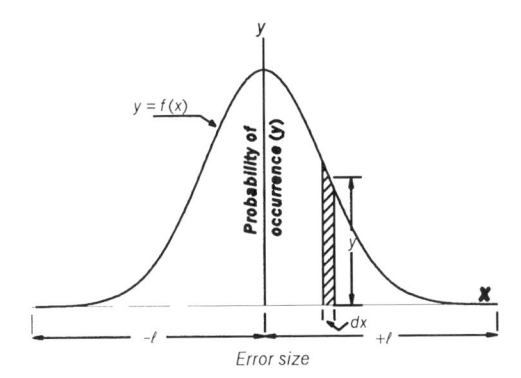

Figure 3.3 Normal density function.

$$\text{area} = \int_{-\infty}^{\infty} \frac{1}{\sigma\sqrt{2\pi}} e^{-x^2/2\sigma^2} \, dx = 1 \tag{3.3}$$

Let y represent $f(x)$ in Equation (3.2) and differentiate:

$$\frac{dy}{dx} = -\frac{x}{\sigma^2}\left(\frac{1}{\sigma\sqrt{2\pi}} e^{x^2/2\sigma^2}\right) \tag{3.4}$$

Recognizing the term in parentheses in Equation (3.4) as y gives

$$\frac{dy}{dx} = -\frac{x}{\sigma^2} y \tag{3.5}$$

Taking the second derivative of Equation (3.2), we obtain

$$\frac{d^2y}{dx^2} = -\frac{x}{\sigma^2}\frac{dy}{dx} - \frac{y}{\sigma^2} \tag{3.6}$$

Substituting Equation (3.5) into Equation (3.6) yields

$$\frac{d^2y}{dx^2} = \frac{x^2}{\sigma^4} y - \frac{y}{\sigma^2} \tag{3.7}$$

Equation (3.7) can be simplified to

$$\frac{d^2y}{dx^2} = \frac{y}{\sigma^2}\left(\frac{x^2}{\sigma^2} - 1\right) \tag{3.8}$$

From calculus, the first derivative of a function yields the slope of the function when evaluated at a point. In Equation (3.5), $dy/dx = 0$ when the values of x or y equal zero. This implies that the curve is parallel to the x axis at the center of the curve when x is zero and is asymptotic to the x axis as y approaches zero.

Also from calculus, a function's second derivative provides the rate of change in a slope when evaluated at a point. The curve's inflection points (points where the algebraic sign of the slope changes) can be located by finding where the function's second derivative equals zero. In Equation (3.8), $d^2y/dx^2 = 0$ when $x^2/\sigma^2 - 1 = 0$, and thus the curve's inflection point occurs when x equals $\pm\sigma$.

Since $e^0 = 1$, if x is set equal to zero, Equation (3.2) gives us

$$y = \frac{1}{\sigma\sqrt{2\pi}} \tag{3.9}$$

This is the curve's central ordinate, and as can be seen, it is inversely proportional to σ. According to Equation (3.9), a group of measurements having small σ must have a large central ordinate. Thus, the area under the curve will be concentrated near the central ordinate, and the errors will be correspondingly small. This indicates that the set of measurements is precise. Since σ bears this inverse relationship to the precision, it is a numerical measure for the precision of a measurement set. In Section 2.7 we defined σ as the *standard error* and gave equations for computing its value.

3.4 STANDARD NORMAL DISTRIBUTION FUNCTION

In Section 3.2 we defined the probability density function of a normal random variable as $f(x) = 1/(\sigma\sqrt{2\pi})e^{-x^2/2\sigma^2}$. From this we develop the *normal distribution function*

$$F_x(t) = \int_{-\infty}^{t} \frac{1}{\sigma\sqrt{2\pi}} e^{-x^2/2\sigma^2} \, dx \tag{3.10}$$

where t is the upper bound of integration, as shown in Figure 3.4.

As stated in Section 3.3, the area under the normal density curve represents the probability of occurrence. Furthermore, integration of this function yields the area under the curve. Unfortunately, the integration called for in Equation (3.10) cannot be carried out in closed form, and thus numerical integration techniques must be used to tabulate values for this function. This has been done for the function when the mean is zero ($\mu = 0$) and the variance is 1 ($\sigma^2 = 1$). The results of this integration are shown in the *standard normal distribution table,* Table D.1. In this table the leftmost column with a heading

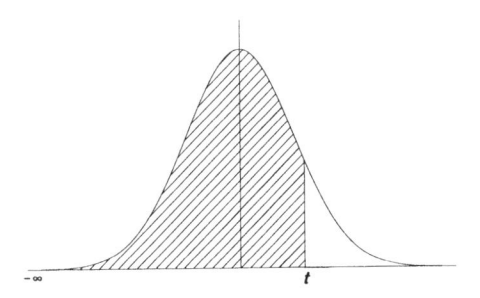

Figure 3.4 Area under the distribution curve determined by Equation (3.10).

of t is the value shown in Figure 3.4 in units of σ. The top row (with headings 0 through 9) represents the hundredths decimal places for the t values. The values in the body of Table D.1 represent areas under the standard normal distribution curve from $-\infty$ to t. For example, to determine the area under the curve from $-\infty$ to 1.68, first find the row with 1.6 in the t column. Then scan along the row to the column with a heading of 8. At the intersection of row 1.6 and column 8 (1.68), the value 0.95352 occurs. This is the area under the standard normal distribution curve from $-\infty$ to a t value of 1.68. Similarly, other areas under the standard normal distribution curve can be found for various values for t. Since the area under the curve represents probability and its maximum area is 1, this means that there is a 95.352% (0.95352 × 100%) probability that t is less than or equal to 1.68. Alternatively, it can be stated that there is a 4.648% [(1 × 0.95352) × 100%] probability that t is greater than 1.68.

Once available, Table D.1 can be used to evaluate the distribution function for any mean, μ, and variance, σ^2. For example, if y is a normal random variable with a mean of μ and a variance of σ^2, an equivalent normal random variable $z = (y - \mu)/\sigma$ can be defined that has a mean of zero and a variance of 1. Substituting the definition for z with $\mu = 0$ and $\sigma^2 = 1$ into Equation (3.2), its density function is

$$N_z(z) = \frac{1}{\sqrt{2\pi}} e^{-z^2/2} \tag{3.11}$$

and its distribution function, known as the *standard normal distribution function*, becomes

$$N_z(z) = \int_{-\infty}^{t} \frac{1}{\sqrt{2\pi}} e^{-z^2/2} \, dz \tag{3.12}$$

For any group of normally distributed measurements, the probability of the normal random variable can be computed by analyzing the integration of the

40   RANDOM ERROR THEORY

distribution function. Again, as stated previously, the area under the curve in Figure 3.4 represents probability. Let z be a normal random variable, then the probability that z is less than some value of t is given by

$$P(z < t) = N_z(t) \tag{3.13}$$

To determine the area (probability) between t values of a and b (the cross-hatched area in Figure 3.5), the difference in the areas between a and b, respectively, can be computed. By Equation (3.13), the area from $-\infty$ to b is $P(z < b) = N_z(b)$. By the same equation, the area from $-\infty$ to a is $P(z < a) = N_z(a)$. Thus, the area between a and b is the difference in these values and is expressed as

$$P(a < z < b) = N_z(b) - N_z(a) \tag{3.14}$$

If the bounds are equal in magnitude but opposite in sign (i.e., $-a = b = t$), the probability is

$$P(|z| < t) = N_z(t) - N_z(-t) \tag{3.15}$$

From the symmetry of the normal distribution in Figure 3.6 it is seen that

$$P(z > t) = P(z < -t) \tag{3.16}$$

for any $t > 0$. This symmetry can also be shown with Table D.1. The tabular value (area) for a t value of -1.00 is 0.15866. Furthermore, the tabular value for a t value of $+1.00$ is 0.84134. Since the maximum probability (area) is 1, the area above $+1.00$ is $1 - 0.84134$, or 0.15866, which is the same as the area below -1.00. Thus, since the total probability is always 1, we can define the following relationship:

$$1 - N_z(t) = N_z(-t) \tag{3.17}$$

Now substituting Equation (3.17) into Equation (3.15), we have

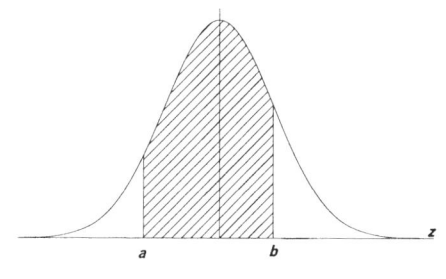

Figure 3.5 Area representing the probability in Equation (3.14).

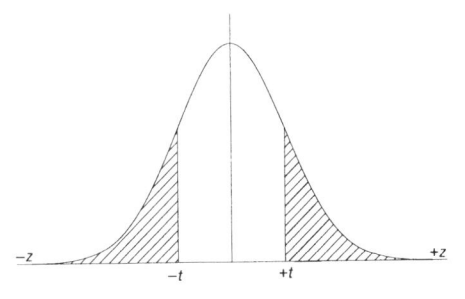

Figure 3.6 Area representing the probability in Equation (3.16).

$$P(|z| < t) = 2N_z(t) - 1 \tag{3.18}$$

3.5 PROBABILITY OF THE STANDARD ERROR

The equations above can be used to determine the probability of the standard error, which from previous discussion is the area under the normal distribution curve between the limits of $\pm\sigma$. For the standard normal distribution when σ^2 is 1, it is necessary to locate the values of $t = -1$ ($\sigma = -1$) and $t = +1$ ($\sigma = 1$) in Table D.1. As seen previously, the appropriate value from the table for $t = -1.00$ is 0.15866. Also, the tabular value for $t = 1.00$ is 0.84134, and thus, according to Equation (3.15), the area between $-\sigma$ and $+\sigma$ is

$$P(-\sigma < z < +\sigma) = N_z(+\sigma) - N_z(-\sigma)$$
$$= 0.84134 - 0.15866 = 0.68268$$

From this it has been determined that about 68.3% of all measurements from any data set are expected to lie between $-\sigma$ and $+\sigma$. It also means that for any group of measurements there is approximately a 68.3% chance that any single observation has an error between plus σ and minus σ. The cross-hatched area of Figure 3.7 illustrates that approximately 68.3% of the area

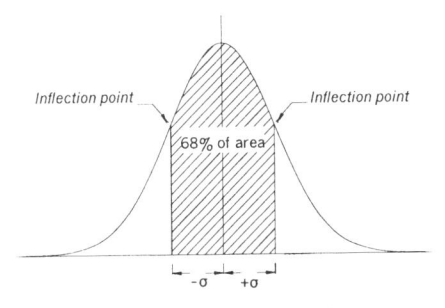

Figure 3.7 Normal distribution curve.

42 RANDOM ERROR THEORY

exists between plus σ and minus σ. This is true for any set of measurements having normally distributed errors. Note that as discussed in Section 3.3, the inflection points of the normal distribution curve occur at ±σ. This is illustrated in Figure 3.7.

3.5.1 50% Probable Error

For any group of observations, the 50% probable error establishes the limits within which 50% of the errors should fall. In other words, any measurement has the same chance of coming within these limits as it has of falling outside them. Its value can be obtained by multiplying the standard deviation of the observations by the appropriate t *value*. Since the 50% probable error has a probability of 1/2, Equation (3.18) is set equal to 0.50 and the t value corresponding to this area is determined as

$$P(|z| < t) = 0.5 = 2N_z(t) - 1$$

$$1.5 = 2N_z(t)$$

$$0.75 = N_z(t)$$

From Table D.1 it is apparent that 0.75 is between a t value of 0.67 and 0.68; that is,

$$N_z(0.67) = 0.7486 \quad \text{and} \quad N_z(0.68) = 0.7517$$

The t value can be found by linear interpolation, as follows:

$$\frac{\Delta t}{0.68 - 0.67} = \frac{0.75 - 0.7486}{0.7517 - 0.7486} = \frac{0.0014}{0.0031} = 0.4516$$

$$\Delta t = 0.01 \times 0.4516$$

and $t = 0.67 + 0.0045 = 0.6745$.

For any set of observations, therefore, the 50% probable error can be obtained by computing the standard error and then multiplying it by 0.6745:

$$E_{50} = 0.6745\sigma \tag{3.19}$$

3.5.2 95% Probable Error

The 95% probable error, E_{95}, is the bound within which, theoretically, 95% of the observation group's errors should fall. This error category is popular with surveyors for expressing precision and checking for outliers in data.

Using the same reasoning as in developing the equation for the 50% probable error, substituting into Equation (3.18) gives

$$0.95 = P(|z| < t) = 2N_z(t) - 1$$

$$1.95 = 2N_z(t)$$

$$0.975 = N_z(t)$$

Again from the Table D.1, it is determined that 0.975 occurs with a t value of 1.960. Thus, to find the 95% probable error for any group of measurements, the following equation is used:

$$E_{95} = 1.960\sigma \qquad (3.20)$$

3.5.3 Other Percent Probable Errors

Using the same computational techniques as in Sections 3.5.1 and 3.5.2, other percent probable errors can be calculated. One other percent error worthy of particular note is $E_{99.7}$. It is obtained by multiplying the standard error by 2.968:

$$E_{99.7} = 2.968\sigma \qquad (3.21)$$

This value is often used for detecting blunders, as discussed in Section 3.6. A summary of probable errors with varying percentages, together with their multipliers, is given in Table 3.2.

3.6 USES FOR PERCENT ERRORS

Standard errors and errors of other percent probabilities are commonly used to evaluate measurements for acceptance. Project specifications and contracts

TABLE 3.2 Multipliers for Various Percent Probable Errors

Symbol	Multiplier	Percent Probable Errors
E_{50}	0.6745σ	50
E_{90}	1.645σ	90
E_{95}	1.960σ	95
E_{99}	2.576σ	99
$E_{99.7}$	2.968σ	99.7
$E_{99.9}$	3.29σ	99.9

44 RANDOM ERROR THEORY

often require that acceptable errors be within specified limits, such as the 90% and 95% errors. The 95% error, sometimes called the *two-sigma* (2σ) *error* because it is computed as approximately 2σ, is most often specified. Standard error is also frequently used. The probable error, E_{50}, is seldom employed.

Higher percent errors are used to help isolate outliers (very large errors) and blunders in data sets. Since outliers seldom occur in a data set, measurements outside a selected high percentage range can be rejected as possible blunders. Generally, any data that differ from the mean by more than 3σ can be considered as blunders and removed from a data set. As seen in Table 3.2, rejecting observations greater that 3σ means that about 99.7% of all measurements should be retained. In other words, only about 0.3% of the measurements in a set of normally distributed random errors (or 3 observations in 1000) should lie outside the range $\pm 3\sigma$.

Note that as explained in Chapter 2, standard error and standard deviation are often used interchangeably, when in practice it is actually the standard deviation that is computed, not the standard error. Thus, for practical applications, σ in the equations of the preceding sections is replaced by S to distinguish between these two related values.

3.7 PRACTICAL EXAMPLES

Example 3.1 Suppose that the following values (in feet) were obtained in 15 independent distance observations, D_i: 212.22, 212.25, 212.23, 212.15, 212.23, 212.11, 212.29, 212.34, 212.22, 212.24, 212.19, 212.25, 212.27, 212.20, and 212.25. Calculate the mean, S, E_{50}, E_{95}, and check for any observations outside the 99.7% probability level.

SOLUTION   From Equation (2.1), the mean is

$$\overline{D} = \frac{\Sigma D_i}{n} = \frac{3183.34}{15} = 212.22 \text{ ft}$$

From Equation (2.10), S is

$$S = \sqrt{\frac{675{,}576.955 - 15(212.223^2)}{15 - 1}} = \sqrt{\frac{0.051298}{14}} = \pm 0.055 \text{ ft}$$

where $\Sigma D_i = 675{,}576.955$. By scanning the data, it is seen that 10 observations are between 212.22 ± 0.06 ft or within the range (212.16, 212.28).[1]

---

[1] The expression (x, y) represents a range between x and y. That is, the population mean lies between 212.16 and 212.28 in this example.

3.7 PRACTICAL EXAMPLES   45

This corresponds to 10/15 × 100, or 66.7% of the observations. For the set, this is what is expected if it conforms to normal error distribution theory.

From Equation (3.19), E_{50} is

$$E_{50} = 0.6745S = \pm 0.6745 (0.055) = \pm 0.04 \text{ ft}$$

Again by scanning the data, nine observations lie in the range 212.22 ± 0.04 ft. That is, they are within the range (212.18, 212.26) ft. This corresponds to 9/15 × 100%, or 60% of the observations. Although this should be 50% and thus is a little high for a normal distribution, it must be remembered that this is only a sample of the population and should not be considered a reason to reject the entire data set. (In Chapter 4, statistical intervals involving sample sets are discussed.)

From Equation (3.20), E_{95} is

$$E_{95} = 1.960S = \pm 1.960(0.055) = \pm 0.11 \text{ ft}$$

Note that 14 of the observations lie in the range 212.22 ± 0.11 (212.11, 212.33) ft, or 93% of the data is in the range.

At the 99.7% level of confidence, the range $\pm 2.968S$ corresponds to an interval of ±0.16 ft. With this criterion for rejection of outliers, all values in the data lie in this range. Thus, there is no reason to believe that any observation is a blunder or outlier.

Example 3.2 The seconds portion of 50 micrometer readings from a 1" theodolite are listed below. Find the mean, standard deviation, and E_{95}. Check the observations at a 99% level of certainty for blunders.

41.9 46.3 44.6 46.1 42.5 45.9 45.0 42.0 47.5 43.2 43.0 45.7 47.6
49.5 45.5 43.3 42.6 44.3 46.1 45.6 52.0 45.5 43.4 42.2 44.3 44.1
42.6 47.2 47.4 44.7 44.2 46.3 49.5 46.0 44.3 42.8 47.1 44.7 45.6
45.5 43.4 45.5 43.1 46.1 43.6 41.8 44.7 46.2 43.2 46.8

SOLUTION The sum of the 50 observations is 2252, and thus the mean is 2252/50 = 45.04". Using Equation (2.10), the standard deviation is

$$S = \sqrt{\frac{101{,}649.94 - 50(45.04)^2}{50 - 1}} = \pm 2.12''$$

where $\Sigma y^2 = 101{,}649.94$. There are 35 observations in the range $45.04'' \pm 2.12''$, or from $42.92''$ to $47.16''$. This corresponds to $35/50 \times 100 = 70\%$ of the observations and correlates well with the anticipated level of 68.3%.

From Equation (3.20), $E_{95} = \pm 1.960(2.12'') = \pm 4.16''$. The data actually contain three values that deviate from the mean by more than $4.16''$ (i.e., that are outside the range $40.88''$ to $49.20''$). They are $49.5''$ (two values) and $52.0''$. No values are less than $40.88''$, and therefore $47/50 \times 100\%$, or 94% of the observations lie in the E_{95} range.

From Equation (3.21), $E_{99} = \pm 2.576(2.12'') = \pm 5.46''$, and thus 99% of the data should fall in the range $45.04'' \pm 5.46''$, or ($39.58''$, $50.50''$). Actually, one value is greater than $50.50''$, and thus 98% of all the observations fall in this range.

By the analysis above it is seen that the data set is skewed to the left. That is, values higher than the range always fell on the right side of the data. The histogram shown in Figure 3.8 depicts this skewness. This suggests that it may be wise to reject the value of $52.0''$ as a mistake. The recomputed values for the data set (minus $52.0''$) are

$$\text{mean} = \frac{2252'' - 52''}{49} = 44.90''$$

$$\Sigma y^2 = 101{,}649.94 - 52.0^2 = 98{,}945.94$$

$$S = \sqrt{\frac{98{,}945.94 - 49(44.89795918)^2}{49 - 1}} = \pm 1.88''$$

Now after recomputing errors, 32 observations lie between plus S or minus S, which represents 65.3% of the observations, 47 observations lie in the E_{95}

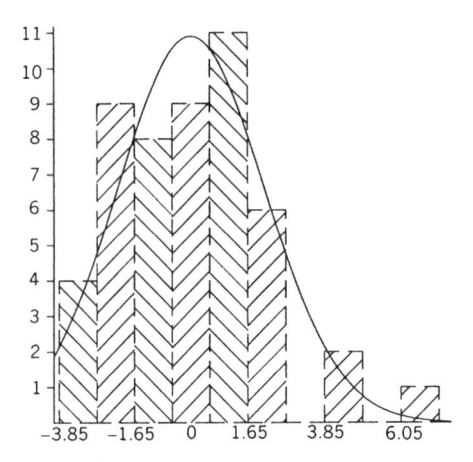

Figure 3.8 Skewed data set.

range, which represents 95.9% of the data, and no values are outside the E_{99} range. Thus, there is no reason to reject any additional data at a 99% level of confidence.

PROBLEMS

3.1 Determine the t value for E_{80}.

3.2 Determine the t value for E_{75}.

3.3 Use STATS to determine t for E_{90}.

3.4 Use STATS to determine t for $E_{99.9}$.

3.5 Assuming a normal distribution, explain the statement: "As the standard deviation of the group of observations decreases, the precision of the group increases."

3.6 If the mean of a population is 2.456 and its variance is 2.042, what is the peak value for the normal distribution curve and the points of inflection?

3.7 If the mean of a population is 13.4 and its variance is 5.8, what is the peak value for the normal distribution curve and the points of inflection?

3.8 Plot the curve in Problem 3.6 using Equation (3.2) to determine ordinate and abscissa values.

3.9 Plot the curve in Problem 3.7 using Equation (3.2) to determine ordinate and abscissa values.

3.10 The following data represent 60 planimeter observations of the area within a plotted traverse.

1.677 1.676 1.657 1.667 1.673 1.671 1.673 1.670 1.675 1.664 1.664 1.668

1.664 1.651 1.663 1.665 1.670 1.671 1.651 1.665 1.667 1.662 1.660 1.667

1.660 1.667 1.667 1.652 1.664 1.690 1.649 1.671 1.675 1.653 1.654 1.665

1.668 1.658 1.657 1.690 1.666 1.671 1.664 1.685 1.667 1.655 1.679 1.682

1.662 1.672 1.667 1.667 1.663 1.670 1.667 1.669 1.671 1.660 1.683 1.663

(a) Calculate the mean and standard deviation.
(b) Plot the relative frequency histogram (of residuals) for the data above using a class interval of one-half of the standard deviation.
(c) Calculate the E_{50} and E_{90} intervals.

(d) Can any observations be rejected at a 99% level of certainty?
(e) What is the peak value for the normal distribution curve, and where are the points of inflection on the curve?

3.11 Discuss the normality of each set of data below and whether any observations may be removed at the 99% level of certainty as blunders or outliers. Determine which set is more precise after apparent blunders and outliers are removed. Plot the relative frequency histogram to defend your decisions.

Set 1:

468.09 468.13 468.11 468.13 468.10 468.13 468.12 468.09 468.14
468.10 468.10 468.12 468.14 468.16 468.12 468.10 468.10 468.11
468.13 468.12 468.18

Set 2:

750.82 750.86 750.83 750.88 750.88 750.86 750.86 750.85 750.86
750.86 750.88 750.84 750.84 750.88 750.86 750.87 750.86 750.83
750.90 750.84 750.86

3.12 Using the following data set, answer the questions that follow.

17.5	15.0	13.4	23.9	25.2	19.5	25.8	30.0	22.5	35.3
39.5	23.5	26.5	21.3	22.3	21.6	27.2	21.1	24.0	23.5
32.5	32.2	24.2	35.7	28.0	24.0	16.8	21.1	19.0	30.7
30.2	33.7	19.7	19.7	25.1	27.9	28.5	22.7	31.0	28.4
31.2	24.6	30.2	16.8	26.9	23.3	21.5	18.8	21.4	20.7

(a) What are the mean and standard deviation of the data set?
(b) Construct a centered relative frequency histogram of the data using seven intervals and discuss whether it appears to be a normal data set.
(c) What is the E_{95} interval for this data set?
(d) Would there be any reason to question the validity of any observation at the 95% level?

3.13 Repeat Problem 3.12 using the following data:

2.898 2.918 2.907 2.889 2.901 2.901 2.899 2.899 2.911 2.909
2.904 2.905 2.895 2.920 2.899 2.896 2.907 2.897 2.900 2.897

Use STATS to do each problem.

3.14 Problem 3.10
3.15 Problem 3.11
3.16 Problem 3.12
3.17 Problem 3.13

Programming Problems

3.18 Create a computational package that solves Problem 3.11.
3.19 Create a computational package that solves Problem 3.12.

CHAPTER 4

CONFIDENCE INTERVALS

4.1 INTRODUCTION

Table 4.1 contains a discrete population of 100 values. The mean (μ) and variance (σ^2) of that population are 26.1 and 17.5, respectively. By randomly selecting 10 values from Table 4.1, an estimate of the mean and variance of the population can be determined. However, it should not be expected that these estimates ($\bar{y}$ and S^2) will exactly match the mean and variance of the population. Sample sets of 10 values each could continue to be selected from the population to determine additional estimates for the mean and variance of the population. However, it is just as unlikely that these additional values would match those obtained from either the population or the first sample set.

As the sample size is increased, the mean and variance of the sample should approach the values of the population. In fact, as the sample size becomes very large, the mean and variance of the samples should be close to those of the population. This procedure was carried out for various sample sizes starting at 10 values and increasing the sample by 10 values, with the results shown in Table 4.2. Note that the value computed for the mean of the sample approaches the value of the population as the sample size is increased. Similarly, the value computed for the variance of the sample tends to approach the value of the population as the sample size is increased.

Since the mean of a sample set $\bar{y}$ and its variance S^2 are computed from random variables, they are also random variables. This means that even if the size of the sample is kept constant, varying values for the mean and variance can be expected from the samples, with greater confidence given to larger samples. Also, it can be concluded that the values computed from a sample also contain errors. To illustrate this, an experiment was run for four randomly

TABLE 4.1 Population of 100 Values

18.2	26.4	20.1	29.9	29.8	26.6	26.2
25.7	25.2	26.3	26.7	30.6	22.6	22.3
30.0	26.5	28.1	25.6	20.3	35.5	22.9
30.7	32.2	22.2	29.2	26.1	26.8	25.3
24.3	24.4	29.0	25.0	29.9	25.2	20.8
29.0	21.9	25.4	27.3	23.4	38.2	22.6
28.0	24.0	19.4	27.0	32.0	27.3	15.3
26.5	31.5	28.0	22.4	23.4	21.2	27.7
27.1	27.0	25.2	24.0	24.5	23.8	28.2
26.8	27.7	39.8	19.8	29.3	28.5	24.7
22.0	18.4	26.4	24.2	29.9	21.8	36.0
21.3	28.8	22.8	28.5	30.9	19.1	28.1
30.3	26.5	26.9	26.6	28.2	24.2	25.5
30.2	18.9	28.9	27.6	19.6	27.9	24.9
21.3	26.7					

selected sets of 10 values from Table 4.1. Table 4.3 lists the samples, their means, and variances. Notice the variation in the values computed for the four sets. As discussed above, this variation is expected.

Fluctuations in the means and variances computed from varying sample sets raises questions about the ability of these values to estimate the population values reliably. For example, a higher confidence is likely to be placed on a sample set with a small variance than on one with a large variance. Thus, in Table 4.3, because of its small variance, one is more likely to believe that the mean of the second sample set is a more reliable estimate than the others for the mean of the population. In reality, this is not the case, since the means of the other three sets are actually closer to the population mean of 26.1.

As noted earlier, the size of the sample should also be considered when determining the reliability of a computed mean or variance. If the mean were

TABLE 4.2 Increasing Sample Sizes

No.	$\bar{y}$	S^2
10	26.9	28.1
20	25.9	21.9
30	25.9	20.0
40	26.5	18.6
50	26.6	20.0
60	26.4	17.6
70	26.3	17.1
80	26.3	18.4
90	26.3	17.8
100	26.1	17.5

TABLE 4.3 Random Sample Sets from a Population

Set 1: 29.9, 18.2, 30.7, 24.4, 36.0, 25.6, 26.5, 29.9, 19.6, 27.9	$\bar{y} = 26.9$, $S^2 = 28.1$
Set 2: 26.9, 28.1, 29.2, 26.2, 30.0, 27.1, 26.5, 30.6, 28.5, 25.5	$\bar{y} = 27.9$, $S^2 = 2.9$
Set 3: 32.2, 22.2, 23.4, 27.9, 27.0, 28.9, 22.6, 27.7, 30.6, 26.9	$\bar{y} = 26.9$, $S^2 = 10.9$
Set 4: 24.2, 36.0, 18.2, 24.3, 24.0, 28.9, 28.8, 30.2, 28.1, 29.0	$\bar{y} = 27.2$, $S^2 = 23.0$

computed from a sample of five values, and another computed from a sample of 30 values, more confidence is likely to be placed on the values derived from the larger sample set than on those from the smaller one, even if both sample sets have the same mean and standard deviation.

In statistics, this relationship between the sample sets, the number of samples, and the values computed for the means and variances is part of *sampling distribution theory*. This theory recognizes that estimates for the mean and variance do vary from sample to sample. *Estimators* are the functions used to compute these estimates. Examples of estimator functions are Equations (2.1) and (2.5), which are used to compute estimates of the mean and variance for a population, respectively. As demonstrated and discussed, these estimates vary from sample to sample and thus have their own population distributions. In Section 4.2, three distributions are defined that are used for describing or quantifying the reliability of mean and variance estimates. By applying these distributions, statements can be written for the reliability at any given level of confidence of the estimates computed. In other words, a range called the *confidence interval* can be determined within which the population mean and population variance can be expected to fall for varying levels of probability.

4.2 DISTRIBUTIONS USED IN SAMPLING THEORY

4.2.1 χ^2 Distribution

The *chi-square distribution*, symbolized as χ^2, compares the relationship between the population variance and the variance of a sample set based on the number of redundancies, v, in the sample. If a random sample of n observations, $y_1, y_2, \ldots, y_n$, is selected from a population that has a normal distribution with mean μ and variance σ^2, then, by definition, the χ^2 sampling distribution is

$$\chi^2 = \frac{vS^2}{\sigma^2} \qquad (4.1)$$

where v is the number of degrees of freedom in the sample and the other terms are as defined previously.

A plot of the distribution is shown in Figure 4.1. The number of redundancies (degrees of freedom) in sample set statistics such as those for the mean or variance are $v = n - 1$; in later chapters on least squares it will be shown that the number of redundancies is based on the number of independent observations and unknown parameters. In the case of the mean, one observation is necessary for its determination, thus leaving $n - 1$ values as redundant observations. Table D.2 is a tabulation of χ^2 distribution curves that have from 1 to 120 degrees of freedom. To find the area under the upper tail of the curve (right side, shown hatched in Figure 4.1), we start at some specific χ^2 value and, going to infinity (∞), intersect the row corresponding to the appropriate degrees of freedom, v, with the column corresponding to the desired area under the curve. For example, to find the specific χ^2 value relating to 1% ($\alpha = 0.010$) of the area under a curve having 10 degrees of freedom, we intersect the row headed by 10 with the column headed by 0.010 and find a χ^2 value of 23.21. This means that 1% of the area under this curve is between the values of 23.21 and ∞.

Due to the asymmetric nature of the distribution shown in Figure 4.1, the percentage points[1] (α) of the lower tail (left side of the curve) must be computed from those tabulated for the upper tail. A specific area under the left side of the curve starting at zero and going to a specific χ^2 value is found by subtracting the tabulated α (right-side area) from 1. This can be done since the table lists α (areas) starting at the χ^2 value and going to ∞, and the total area under the curve is 1. For example, if there are 10 degrees of freedom and the χ^2 value relating to 1% of the area under the left side of the curve is needed, the row corresponding to v equal to 10 is intersected with the column headed by $\alpha = 0.990$ $(1 - 0.010)$, and a value of 2.56 is obtained. This means that 1% of the area under the curve occurs from 0 to 2.56.

The χ^2 distribution is used in sampling statistics to determine the range in which the variance of the population can be expected to occur based on (1) some specified percentage probability, (2) the variance of a sample set, and (3) the number of degrees of freedom in the sample. In an example given in

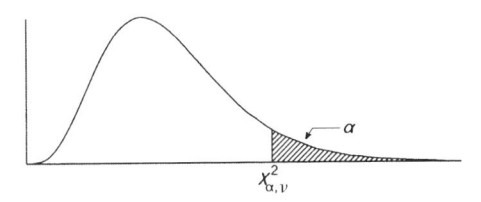

Figure 4.1 χ^2 distribution.

---
[1] Percentage points are decimal equivalents of percent probability; that is, a percent probability of 95% is equivalent to 0.95 percentage points.

Section 4.6, this distribution is used to construct probability statements about the variance of the population being in a range centered about the variance S^2 of a sample having v degrees of freedom. In Section 5.4 a statistical test is presented using the χ^2 distribution to check if the variance of a sample is a valid estimate for the population variance.

4.2.2 *t* (Student) Distribution

The *t* distribution is used to compare a population mean with the mean of a sample set based on the number of redundancies (v) in the sample set. It is similar to the normal distribution (discussed in Chapter 3) except that the normal distribution applies to an entire population, whereas the *t* distribution applies to a sampling of the population. The *t* distribution is preferred over the normal distribution when the sample contains fewer than 30 values. Thus, it is an important distribution in analyzing surveying data.

If z is a standard normal random variable as defined in Section 3.4, χ^2 is a chi-square random variable with v degrees of freedom, and z and χ^2 are both independent variables, then by definition

$$t = \frac{z}{\sqrt{\chi^2/v}} \tag{4.2}$$

The *t* values for selected upper-tail percentage points (hatched area in Figure 4.2) versus the *t distributions* with various degrees of freedom v are listed in Table D.3. For specific degrees of freedom (v) and percentage points (α), the table lists specific *t* values that correspond to the areas α under the curve between the tabulated *t* values and ∞. Similar to the normal distribution, the *t* distribution is symmetric. Generally in statistics, only percentage points in the range 0.0005 to 0.4 are necessary. These *t* values are tabulated in Table D.3. To find the *t* value relating to $\alpha = 0.01$ for a curve developed with 10 degrees of freedom ($v = 10$), intersect the row corresponding to $v = 10$ with the row corresponding to $\alpha = 0.01$. At this intersection a *t* value of 2.764 is obtained. This means that 1% ($\alpha = 0.01$) of the area exists under the *t* distribution curve having 10 degrees of freedom in the interval between 2.764 and ∞. Due to the symmetry of this curve, it can also be stated that 1% (α

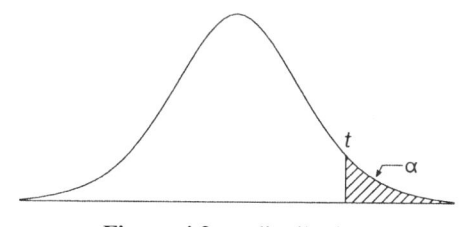

Figure 4.2 *t* distribution.

= 0.01) of the area under the curve developed for 10 degrees of freedom also lies between $-\infty$ and -2.764.

As described in Section 4.3, this distribution is used to construct confidence intervals for the population mean (μ) based on the mean ($\bar{y}$) and variance (S^2) of a sample set v degrees of freedom. An example in that section illustrates the procedure. Furthermore, in Section 5.3 it is shown that this distribution can be used to develop statistical tests about the population mean.

4.2.3 *F* Distribution

The *F* distribution is used when comparing the variances computed from two sample sets. If χ_1^2 and χ_2^2 are two chi-square random variables with v_1 and v_2 degrees of freedom, respectively, and both variables are independent, then by definition

$$F = \frac{\chi_1^2/v_1}{\chi_2^2/v_2} \tag{4.3}$$

Various percentage points (areas under the upper tail of the curve shown hatched in Figure 4.3) of the *F* distribution are tabulated in Table D.4. Notice

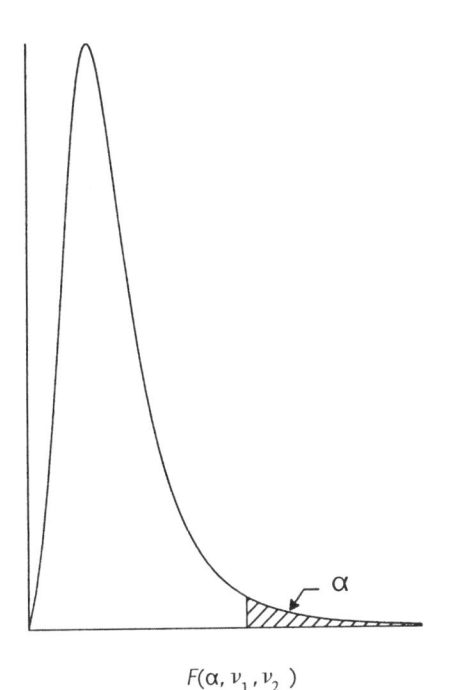

Figure 4.3 *F* distribution.

that this distribution has v_1 numerator degrees of freedom and v_2 denominator degrees of freedom, which correspond to the two sample sets. Thus, unlike the χ^2 and t distributions, each desired α percentage point must be represented in a separate table. In Appendix D, tables for the more commonly used values of α (0.20, 0.10, 0.05, 0.025, 0.01, 0.005, and 0.001) are listed.

To illustrate the use of the tables, suppose that the F value for 1% of the area under the upper tail of the curve is needed. Also assume that 5 is the numerator degrees of freedom relating to S_1 and 10 is the denominator degrees of freedom relating to S_2. In this example, α equals 0.01 and thus the F table in Table D.4 that is written for $\alpha = 0.01$ must be used. In that table, intersect the row headed by v_2 equal to 10 with the column headed by v_1 equal to 5, and find the F value of 5.64. This means that 1% of the area under the curve constructed using these degrees of freedom lies in the region from 5.64 to ∞.

To determine the area in the lower tail of this distribution, use the following functional relationship:

$$F_{\alpha, v_1, v_2} = \frac{1}{F_{1-\alpha, v_2, v_1}} \tag{4.4}$$

The critical F value for the data in the preceding paragraph [v_1 equal to 5 and v_2 equal to 10 with α equal to 0.99 (0.01 in the lower tail)] is determined by going to the intersection of the row headed by 5 with the column headed by 10 in the section $\alpha = 0.01$. The intersection is at F equal to 2.19. According to Equation (4.4), the critical $F_{0.99, 5, 10}$ is $1/F_{0.01, 10, 5} = 1/2.19 = 0.457$. Thus, 1% of the area is under the F-distribution curve from $-\infty$ to 0.457.

The F distribution is used to answer the question of whether two sample sets come from the same population. For example, suppose that two samples have variances of S_1^2 and S_2^2. If these two sample variances represent the same population variance, the ratio of their population variances (σ_1^2/σ_2^2) should equal 1 (i.e., $\sigma_1^2 = \sigma_2^2$). As discussed in Section 4.7, this distribution enables confidence intervals to be established for the ratio of the population variances. Also, as discussed in Section 5.5, the distribution can be used to test whether the ratio of the two variances is statistically equal to 1.

4.3 CONFIDENCE INTERVAL FOR THE MEAN: t STATISTIC

In Chapter 3 the standard normal distribution was used to predict the range in which the mean of a population can exist. This was based on the mean and standard deviation for a sample set. However, as noted previously, the normal distribution is based on an entire population, and as was demonstrated, variations from the normal distribution are expected from sample sets having a small number of values. From this expectation, the t *distribution* was developed. As demonstrated later in this section by an example, the t distribution (in Table D.3) for samples having an infinite number of values uses the same t values as those listed in Table 3.2 for the normal distribution. It is generally accepted that when the number of observations is greater than about 30, the values in Table 3.2 are valid for constructing intervals about the population mean. However, when the sample set has fewer than 30 values, a t value from the t distribution should be used to construct the confidence interval for the population mean.

To derive an expression for a confidence interval of the population mean, a sample mean ($\bar{y}$) is computed from a sample set of a normally distributed population having a mean of μ and a variance in the mean of σ^2/n. Let $z = (\bar{y} - \mu)/(\sigma/\sqrt{n})$ be a normal random variable. Substituting it and Equation (4.1) into Equation (4.2) yields

$$t = \frac{z}{\sqrt{\chi^2/\nu}} = \frac{(\bar{y} - \mu)/(\sigma/\sqrt{n})}{\sqrt{(\nu S^2/\sigma^2)/\nu}} = \frac{(\bar{y} - \mu)/(\sigma/\sqrt{n})}{S/\sigma} = \frac{\bar{y} - \mu}{S/\sqrt{n}} \quad (4.5)$$

To compute a confidence interval for the population mean (μ) given a sample set mean and variance, it is necessary to determine the area of a $1 - \alpha$ region. For example, in a 95% confidence interval (nonhatched area in Figure 4.4), center the percentage point of 0.95 on the t distribution. This leaves 0.025 in each of the upper- and lower-tail areas (hatched areas in Figure 4.4). The t value that locates an $\alpha/2$ area in both the upper and lower tails of the distribution is given in Table D.3 as $t_{\alpha/2,\nu}$. For sample sets having a mean of $\bar{y}$ and a variance of S^2, the correct probability statement to locate this area is

$$P(|z| < t) = 1 - \alpha \quad (a)$$

Substituting Equation (4.5) into Equation (a) yields

$$P\left(\left|\frac{\bar{y} - \mu}{S/\sqrt{n}}\right| < t\right) = 1 - \alpha$$

which after rearranging yields

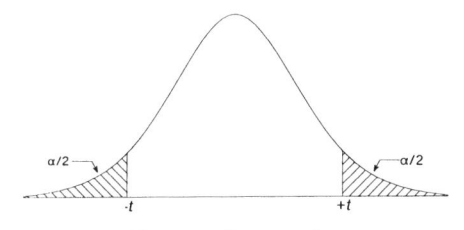

Figure 4.4 $t_{\alpha/2}$ plot.

58 CONFIDENCE INTERVALS

$$P\left(\bar{y} - t_{\alpha/2}\frac{S}{\sqrt{n}} < \mu < \bar{y} + t_{\alpha/2}\frac{S}{\sqrt{n}}\right) = 1 - \alpha \quad (4.6)$$

Thus, given $\bar{y}$, $t_{\alpha/2,\nu}$, n, and S, it is seen from Equation (4.6) that a $1 - \alpha$ probable error interval for the population mean μ is computed as

$$\bar{y} - t_{\alpha/2}\frac{S}{\sqrt{n}} < \mu < \bar{y} + t_{\alpha/2}\frac{S}{\sqrt{n}} \quad (4.7)$$

where $t_{\alpha/2}$ is the t value from the t distribution based on ν degrees of freedom and $\alpha/2$ percentage points.

The following example illustrates the use of Equation (4.7) and Table D.3 for determining the 95% confidence interval for the population mean based on a sample set having a small number of values (n) with a mean of $\bar{y}$ and a variance of S.

Example 4.1 In carrying out a control survey, 16 directional readings were measured for a single line. The mean (seconds' portion only) of the readings was 25.4", with a standard deviation of ± 1.3". Determine the 95% confidence interval for the population mean. Compare this with the interval determined by using a t value determined from the standard normal distribution tables (Table 3.2).

SOLUTION In this example the confidence level $1 - \alpha$ is 0.95, and thus α is 0.05. Since the interval is to be centered about the population mean μ, a value of $\alpha/2$ in Table D.3 is used. This yields equal areas in both the lower and upper tails of the distribution, as shown in Figure 4.4. Thus, for this example, $\alpha/2$ is 0.025. The appropriate t value for this percentage point with ν equal to 15 (16 − 1) degrees of freedom is found in Table D.3 as follows:

Step 1: In the leftmost column of Table D.3, find the row with the correct number of degrees of freedom (ν) for the sample. In this case it is 16 − 1, or 15.
Step 2: Find the column headed by 0.025 for $\alpha/2$.
Step 3: Locate the value at the intersection of this row and column, which is 2.131.
Step 4: Then by Equation (4.7), the appropriate 95% confidence interval is

$$24.7 = 25.4 - 2.131\left(\frac{1.3}{\sqrt{16}}\right) = \bar{y} - t_{0.025}\frac{S}{\sqrt{n}}$$

$$< \mu < \bar{y} + t_{0.025}\frac{S}{\sqrt{n}} = 25.4 + 2.131\left(\frac{1.3}{\sqrt{16}}\right) = 26.1$$

This computation can be written more compactly as

$$\bar{y} \pm t_{0.025}\frac{S}{\sqrt{n}} \quad \text{or} \quad 25.4 \pm 2.131\left(\frac{1.3}{\sqrt{16}}\right) = 25.4 \pm 0.7$$

After making the calculation above, it can be stated that for this sample, with 95% confidence, the population mean (μ) lies in the range (24.7, 26.1). If this were a large sample, the t value from Table 3.2 could be used for 95%. That t value for 95% is 1.960, and the standard error of the mean then would be $\pm 1.3/\sqrt{16} = \pm 0.325$. Thus, the population's mean would be in the range $25.4 \pm 1.960 \times 0.325''$, or (24.8, 26.0). Notice that due to the small sample size, the t distribution gives a larger range for the population mean than does the standard normal distribution. Notice also that in the t distribution of Table D.3, for a sample of infinite size (i.e., $\nu = \infty$), the t value tabulated for α equal to 0.025 is 1.960, which matches Table 3.2.

The t distribution is often used to isolate outliers or blunders in observations. To do this, a percent confidence interval is developed about the mean for a single observation as

$$\bar{y} - t_{\alpha/2}S \le y_i \le \bar{y} + t_{\alpha/2}S \tag{4.8}$$

Using the data from Example 4.1 and Equation (4.8), the 95% range for the 16 directional readings is

$$25.4'' - 2.131(1.3'') = 22.63'' \le y_i \le 28.17'' = 25.4'' + 2.131(1.3'')$$

Thus, 95% of the data should be in the range (22.6″, 28.2″). Any data values outside this range can be considered as outliers and rejected with a 95% level of confidence. It is important to note that if the normal distribution value of 1.960 was used to compute this interval, the range would be smaller, (22.85″, 27.95″). Using the normal distribution could result in discarding more observations than is justified when using sample estimates of the mean and variance. It is important to note that this will become more significant as the number of observations in the sample becomes smaller. For example, if only four directional readings are obtained, the t-distribution multiplier would become 3.183. The resulting 95% confidence interval for a single observation would be 1.6 times larger than that derived using a normal distribution t value.

4.4 TESTING THE VALIDITY OF THE CONFIDENCE INTERVAL

A test that demonstrates the validity of the theory of the confidence interval is illustrated as follows. Using a computer and normal random number generating software, 1000 sample data sets of 16 values each were collected randomly from a population with mean $\mu = 25.4$ and standard error $\sigma = \pm 1.3$. Using a 95% confidence interval ($\alpha = 0.05$) and Equation (4.7), the interval for the population mean derived for each sample set was computed and compared with the actual population mean. If the theory is valid, the interval constructed would be expected to contain the population's mean 95% of the time based on the confidence level of 0.05. Appendix E shows the 95% intervals computed for the 1000 samples. The intervals not containing the population mean of 25.4 are marked with an asterisk. From the data tabulated it is seen that 50 of 1000 sample sets failed to contain the population mean. This corresponds to exactly 5% of the samples. In other words, the proportion of samples that do enclose the mean is exactly 95%. This demonstrates that the bounds calculated by Equation (4.7) do, in fact, enclose the population mean μ at the confidence level selected.

4.5 SELECTING A SAMPLE SIZE

A common problem encountered in surveying practice is to determine the number of repeated observations necessary to meet a specific precision. In practice, the size of S cannot be controlled absolutely. Rather, as seen in Equation (4.7), the confidence interval can be controlled only by varying the number of repeated observations. In general, the larger the sample size, the smaller the confidence interval. From Equation (4.7), the range in which the population mean (μ) resides at a selected level of confidence (α) is

$$\bar{y} \pm t_{\alpha/2} \frac{S}{\sqrt{n}} \qquad (b)$$

Now let I represent one-half of the interval in which the population mean lies. Then from Equation (b), I is

$$I = t_{\alpha/2} \frac{S}{\sqrt{n}} \qquad (4.9)$$

Rearranging Equation (4.9) yields

$$n = \left(\frac{t_{\alpha/2} S}{I}\right)^2 \qquad (4.10)$$

In Equation (4.10), n is the number of repeated measurements, I the desired confidence interval, $t_{\alpha/2}$ the t value based on the number of degrees of freedom (ν), and S the sample set standard deviation. In the practical application of Equation (4.10), $t_{\alpha/2}$ and S are unknown since the data set has yet to be collected. Also, the number of measurements, and thus the number of redundancies, is unknown, since they are the computational objectives in this problem. Therefore, Equation (4.10) must be modified to use the standard normal random variable, z, and its value for t, which is not dependent on v or n; that is,

$$n = \left(\frac{t_{\alpha/2}\sigma}{I}\right)^2 \tag{4.11}$$

where n is the number of repetitions, $t_{\alpha/2}$ the t value determined from the standard normal distribution table (Table D.1), σ an estimated value for the standard error of the measurement, and I the desired confidence interval.

Example 4.2 From the preanalysis of a horizontal control network, it is known that all angles must be measured to within $\pm 2''$ at the 95% confidence level. How many repetitions will be needed if the standard deviation for a single angle measurement has been determined to be $\pm 2.6''$?

SOLUTION In this problem, a final 95% confidence interval of $\pm 2''$ is desired. From previous experience or analysis,[2] the standard error for a single angle observation is estimated as $\pm 2.6''$. From Table 3.2, the multiplier (or t value) for a 95% confidence level is found to be 1.960. Substituting this into Equation (4.11) yields

$$n = \left(\frac{1.960 \times 2.6}{2}\right)^2 = 6.49$$

Thus, eight repetitions are selected, since this is the closest even number above 6.49. [Note that to eliminate instrumental systematic errors it is necessary to select an even number of repetitions, because an equal number of face-left (direct) and face-right (reverse) readings must be taken.]

4.6 CONFIDENCE INTERVAL FOR A POPULATION VARIANCE

From Equation (4.1), $\chi^2 = vS^2/\sigma^2$, and thus confidence intervals for the variance of the population, σ^2, are based on the χ^2 statistic. Percentage points (areas) for the upper and lower tails of the χ^2 distribution are tabulated in Table D.2. This table lists values (denoted by χ_α^2) that determine the upper boundary for areas from χ_α^2 to $+\infty$ of the distribution, such that

---

[2] See Chapter 6 for a methodology to estimate the variance in an angle observation.

62 CONFIDENCE INTERVALS

$$P(\chi^2 > \chi_\alpha^2) = \alpha$$

for a given number of redundancies, ν. Unlike the normal distribution and the t distribution, the χ^2 distribution is not symmetric about zero. To locate an area in the lower tail of the distribution, the appropriate value of $\chi^2_{1-\alpha}$ must be found, where $P(\chi^2 > \chi^2_{1-\alpha}) = 1 - \alpha$. These facts are used to construct a probability statement for χ^2 as

$$P(\chi^2_{1-\alpha/2} < \chi^2 < \chi^2_{\alpha/2}) = 1 - \alpha \qquad (4.12)$$

where $\chi^2_{1-\alpha/2}$ and $\chi^2_{\alpha/2}$ are tabulated in Table D.2 by the number of redundant observations. Substituting Equation (4.1) into Equation (4.12) yields

$$P\left(\chi^2_{1-\alpha/2} < \frac{\nu S^2}{\sigma^2} < \chi^2_{\alpha/2}\right) = P\left(\frac{\chi^2_{1-\alpha/2}}{\nu S^2} < \frac{1}{\sigma^2} < \frac{\chi^2_{\alpha/2}}{\nu S^2}\right) \qquad (4.13)$$

Recalling a property of mathematical inequalities—that in taking the reciprocal of a function, the inequality is reversed—it follows that

$$P\left(\frac{\nu S^2}{\chi^2_{\alpha/2}} < \sigma^2 < \frac{\nu S^2}{\chi^2_{1-\alpha/2}}\right) = 1 - \alpha \qquad (4.14)$$

Thus, the $1 - \alpha$ confidence interval for the population variance (σ^2) is

$$\frac{\nu S^2}{\chi^2_{\alpha/2}} < \sigma^2 < \frac{\nu S^2}{\chi^2_{1-\alpha/2}} \qquad (4.15)$$

Example 4.3 An observer's pointing and reading error with a 1″ theodolite is estimated by collecting 20 readings while pointing at a well-defined distant target. The sample standard deviation is determined to be $\pm 1.8″$. What is the 95% confidence interval for σ^2?

SOLUTION For this example the desired area enclosed by the confidence interval $1 - \alpha$ is 0.95. Thus, α is 0.05 and $\alpha/2$ is 0.025. The values of $\chi^2_{0.025}$ and $\chi^2_{0.975}$ with ν equal to 19 degrees of freedom are needed. They are found in the χ^2 table (Table D.2) as follows:

Step 1: Find the row with 19 degrees of freedom and intersect it with the column headed by 0.975. The value at the intersection is 8.91.
Step 2: Follow this procedure for 19 degrees of freedom and 0.025. The value is 32.85. Using Equation (4.15), the 95% confidence interval for σ^2 is

$$\frac{(20-1)1.8^2}{32.85} < \sigma^2 < \frac{(20-1)1.8^2}{8.91}$$

$$1.87 < \sigma^2 < 6.91$$

Thus, 95% of the time, the population's variance should lie between 1.87 and 6.91.

4.7 CONFIDENCE INTERVAL FOR THE RATIO OF TWO POPULATION VARIANCES

Another common statistical procedure is used to compare the ratio of two population variances. The sampling distribution of the ratio σ_1^2/σ_2^2 is well known when samples are collected randomly from a normal population. The confidence interval for σ_1^2/σ_2^2 is based on the F distribution using Equation (4.3) as

$$F = \frac{\chi_1^2/\nu_1}{\chi_2^2/\nu_2}$$

Substituting Equation (4.1) and reducing yields

$$F = \frac{(\nu_1 S_1^2/\sigma_1^2)/\nu_1}{(\nu_2 S_2^2/\sigma_2^2)/\nu_2} = \frac{S_1^2/\sigma_1^2}{S_2^2/\sigma_2^2} = \frac{S_1^2}{S_2^2}\frac{\sigma_2^2}{\sigma_1^2} \quad (4.16)$$

To establish a confidence interval for the ratio, the lower and upper values corresponding to the tails of the distribution must be found. A probability statement to find the confidence interval for the ratio is constructed as follows:

$$P(F_{1-\alpha/2,\nu_1,\nu_2} < F < F_{\alpha/2,\nu_1,\nu_2}) = 1 - \alpha$$

Substituting in Equation (4.16) and rearranging yields

$$P(F_l < F < F_u) = P\left(F_l < \frac{S_1^2}{S_2^2} \times \frac{\sigma_2^2}{\sigma_1^2} < F_u\right)$$

$$= P\left(F_l \frac{S_2^2}{S_1^2} < \frac{\sigma_2^2}{\sigma_1^2} < \frac{S_2^2}{S_1^2} F_u\right)$$

$$= P\left(\frac{1}{F_u}\frac{S_1^2}{S_2^2} < \frac{\sigma_1^2}{\sigma_2^2} < \frac{S_1^2}{S_2^2}\frac{1}{F_l}\right) = 1 - \alpha \quad (4.17)$$

Substituting Equation (4.4) into (4.17) yields

$$P\left(\frac{1}{F_{\alpha/2,v_1,v_2}}\frac{S_1^2}{S_2^2} < \frac{\sigma_1^2}{\sigma_2^2} < \frac{S_1^2}{S_2^2}\frac{1}{F_{1-\alpha/2,v_1v_2}}\right)$$

$$= P\left(\frac{1}{F_{\alpha/2,v_1,v_2}}\frac{S_1^2}{S_2^2} < \frac{\sigma_1^2}{\sigma_2^2} < \frac{S_1^2}{S_2^2}F_{\alpha/2,v_2,v_1}\right) = 1 - \alpha \qquad (4.18)$$

Thus, from Equation (4.18), the $1 - \alpha$ confidence interval for the σ_1^2/σ_2^2 ratio is

$$\frac{1}{F_{\alpha/2,v_1v_2}}\frac{S_1^2}{S_2^2} < \frac{\sigma_1^2}{\sigma_2^2} < \frac{S_1^2}{S_2^2}F_{\alpha/2,v_2,v_1} \qquad (4.19)$$

Notice that the degrees of freedom for the upper and lower limits in Equation (4.19) are opposite each other, and thus v_2 is the numerator degrees of freedom and v_1 is the denominator degrees of freedom in the upper limit.

An important situation where Equation (4.19) can be applied occurs in the analysis and adjustment of horizontal control surveys. During least squares adjustments of these types of surveys, control stations fix the data in space both positionally and rotationally. When observations tie into more than a minimal number of control stations, the control coordinates must be mutually consistent. If they are not, any attempt to adjust the observations to the control will warp the data to fit the discrepancies in the control. A method for isolating control stations that are not consistent is first to do a least squares adjustment using only enough control to fix the data in space both positionally and rotationally. This is known as a *minimally constrained adjustment*. In horizontal surveys, this means that one station must have fixed coordinates and one line must be fixed in direction. This adjustment is then followed with an adjustment using all available control. If the control information is consistent, the reference variance (S_1^2) from the minimally constrained adjustment should be statistically equivalent to the reference variance (S_2^2) obtained when using all the control information. That is, the ratio of S_1^2/S_2^2 should be 1.

Example 4.4 Assume that a minimally constrained trilateration network adjustment with 24 degrees of freedom has a reference variance of 0.49 and that the fully constrained network adjustment with 30 degrees of freedom has a reference variance of 2.25. What is the 95% $1 - \alpha$ confidence interval for the ratio of the variances and does this interval contain the numerical value 1? Stated in another way, is there reason to be concerned about the control having values that are not consistent?

SOLUTION In this example the objective is to determine whether the two reference variances are statistically equal. To solve the problem, let the variance in the numerator be 2.25 and that in the denominator be 0.49. Thus, the numerator has 30 degrees of freedom ($v_1 = 30$) and corresponds to an adjustment using all the control. The denominator has 24 degrees of freedom ($v_2 = 24$) and corresponds to the minimally constrained adjustment.[3] With α equal to 0.05 and using Equation (4.19), the 95% confidence interval for this ratio is

$$2.08 = \frac{2.25}{0.49}\left(\frac{1}{2.21}\right) < \frac{\sigma_1^2}{\sigma_2^2} < \frac{2.25}{0.49}(2.14) = 9.83$$

Note from the calculations above that 95% of the time, the ratio of the population variances is in the range (2.08, 9.83). Since this interval does not contain 1, it can be said that $\sigma_1^2/\sigma_2^2 \neq 1$ and $\sigma_1^2 \neq \sigma_2^2$ at a 95% level of confidence. Recalling from Equation (2.4) that the size of the variance depends on the size of the errors, it can be stated that the fully constrained adjustment revealed discrepancies between the observations and the control. This could be caused by inconsistencies in the coordinates of the control stations or by the presence of uncorrected systematic errors in the observations. An example of an uncorrected systematic error is the failure to reduce distance observations to a mapping grid before the adjustment. (See Appendix F.4 for a discussion of the reduction of distance observations to the mapping grid.)

PROBLEMS

4.1 Use the χ^2-distribution table (Table D.2) to determine the values of $\chi^2_{\alpha/2}$ that would be used to construct confidence intervals for a population variance for the following combinations:
(a) $\alpha = 0.10$, $v = 25$
(b) $\alpha = 0.05$, $v = 15$
(c) $\alpha = 0.05$, $v = 10$
(d) $\alpha = 0.01$, $v = 30$

4.2 Use the t-distribution table (Table D.3) to determine the values of $t_{\alpha/2}$ that would be used to construct confidence intervals for a population mean for each of the following combinations:
(a) $\alpha = 0.10$, $v = 25$
(b) $\alpha = 0.05$, $v = 15$
(c) $\alpha = 0.01$, $v = 10$
(d) $\alpha = 0.01$, $v = 40$

---

[3] For confidence intervals, it is not important which variance is selected as the numerator. In this case, the larger variance was selected arbitrarily as the numerator, to match statistical testing methods discussed in Chapter 5.

4.3
Use the F-distribution table (Table D.4) to determine the values of F_{α, v_1, v_2} that would be used to construct confidence intervals for a population mean for each of the following combinations:
(a) $\alpha = 0.20$, $v_1 = 24$, $v_2 = 2$
(b) $\alpha = 0.01$, $v_1 = 15$, $v_2 = 8$
(c) $\alpha = 0.05$, $v_1 = 60$, $v_2 = 20$
(d) $\alpha = 0.80$, $v_1 = 2$, $v_2 = 24$

Use STATS to do Problems 4.4 through 4.6.

4.4 Problem 4.1

4.5 Problem 4.2

4.6 Problem 4.3

4.7 A least squares adjustment is computed twice on a data set. When the data are minimally constrained with 10 degrees of freedom, a variance of 1.07 is obtained. In the second run, the fully constrained network has 12 degrees of freedom with a standard deviation of 1.53. The a priori estimates for the reference variances in both adjustments are 1; that is, $\sigma_1^2 = \sigma_2^2 = 1$.
(a) What is the 95% confidence interval for the ratio of the two variances? Is there reason to be concerned about the consistency of the control? Justify your response statistically.
(b) What is the 95% confidence interval for the reference variance in the minimally constrained adjustment? The population variance is 1. Does this interval contain 1?
(c) What is the 95% confidence interval for the reference variance in the fully constrained adjustment? The population variance is 1. Does this interval contain 1?

4.8 The calibrated length of a baseline is 402.167 m. An average distance of 402.151 m with a standard deviation of ±0.0055 m is computed after the line is observed five times with an EDM.
(a) What is the 95% confidence interval for the measurement?
(b) At a 95% level of confidence, can you state that the EDM is working properly? Justify your response statistically.
(c) At a 90% level of confidence can you state that the EDM is working properly? Justify your response statistically.

4.9 An observer's pointing and reading standard deviation is determined to be ±1.8″ after pointing and reading the circles of a particular instrument six times ($n = 6$). What is the 99% confidence interval for the population variance?

4.10 Using sample statistics and the data in Example 3.1, construct a 90% confidence interval:
   (a) for a single observation, and identify any observations that may be identified as possible outliers.
   (b) for the population variance.

4.11 Using sample statistics and the data in Example 3.2, construct a 90% confidence interval:
   (a) for a single observation, and identify any observations that may be identified as possible outliers.
   (b) for the population variance.

4.12 Using sample statistics and the data from Problem 3.10 construct a 95% confidence interval:
   (a) for a single observation, and identify any observations that may be identified as possible outliers.
   (b) for the population variance.

4.13 For the data in Problem 3.11, construct a 95% confidence interval for the ratio of the two variances for sets 1 and 2. Are the variances equal statistically at this level of confidence? Justify your response statistically.

4.14 Using sample statistics and the data from Problem 3.12, construct a 95% confidence interval:
   (a) for a single observation, and identify any observations that may be identified as possible outliers in the data.
   (b) for the mean.

CHAPTER 5

STATISTICAL TESTING

5.1 HYPOTHESIS TESTING

In Example 4.4 we were not concerned about the actual bounds of the interval constructed, but rather, whether the interval contained the expected ratio of the variances. This is often the case in statistics. That is, the actual values of the interval are not as important as is answering the question: Is the sample statistic consistent with what is expected from the population? The procedures used to test the validity of a statistic are known as *hypothesis testing*. The basic elements of hypothesis testing are

1. The *null hypothesis*, H_0, is a statement that compares a population statistic with a sample statistic. This implies that the sample statistic is what is "expected" from the population. In Example 4.4, this would be that the ratio of the variances is statistically 1.
2. The *alternative hypothesis*, H_a, is what is accepted when a decision is made to reject the null hypothesis, and thus represents an alternative population of data from which the sample statistic was derived. In Example 4.4 the alternative hypothesis would be that the ratio of the variances is not equal to 1 and thus the variance did not come from the same population of data.
3. The *test statistic* is computed from the sample data and is the value used to determine whether the null hypothesis should be rejected. When the null hypothesis is rejected, it can be said that the sample statistic computed is not consistent with what is expected from the population. In Example 4.4 a rejection of the null hypothesis would occur when the ratio of the variances is not statistically equivalent to 1.

4. The *rejection region* is the value for the test statistic where the null hypothesis is rejected. In reference to confidence intervals, this number takes the place of the confidence interval bounds. That is, when the test statistic computed is greater than the value defining the rejection region, it is equivalent to the sample statistic of the null hypothesis being outside the bounds of the confidence interval. That is, when the rejection criterion is true, the null hypothesis is rejected.

Whenever a decision is made concerning the null hypothesis, there is a possibility of making a wrong decision since we can never be 100% certain about a statistic or a test. Returning to Example 4.4, a confidence interval of 95% was constructed. With this interval, there is a 5% chance that the decision was wrong. That is, it is possible that the larger-than-expected ratio of the variances is consistent with the population of observations. This reasoning suggests that further analysis of statistical testing is needed.

Two basic errors can occur when a decision is made about a statistic. A valid statistic could be rejected, or an invalid statistic could be accepted. These two errors can be stated in terms of statistical testing elements as Type I and Type II errors. If the null hypothesis is rejected when in fact it is true, a *Type I error* is committed. If the null hypothesis is not rejected when in fact it is false, a *Type II error* occurs. Since these errors are not from the same population, the probability of committing each error is not directly related. A decision must be made as to the type of error that is more serious for the situation, and the decision should be based on the consequences of committing each error. For instance, if a contract calls for positional accuracies on 95% of the stations to be within ± 0.3 ft, the surveyor is more inclined to commit a Type I error to ensure that the contract specifications are met. However, the same surveyor, needing only 1-ft accuracy on control to support a small-scale mapping project, may be more inclined to commit a Type II error. In either case, it is important to compute the probabilities of committing both Type I and Type II errors to assess the reliability of the inferences derived from a hypothesis test. For emphasis, the two basic hypothesis-testing errors are repeated.

- *Type I error:* rejecting the null hypothesis when it is, in fact, true (symbolized by α)
- *Type II error:* not rejecting the null hypothesis when it is, in fact, false (symbolized by β)

Table 5.1 shows the relationship between the decision, the probabilities of α and β, and the acceptance or rejection of the null hypothesis, H_0. In Figure 5.1 the left distribution represents the data from which the null hypothesis is derived. That is, this distribution represents a true null hypothesis. Similarly, the distribution on the right represents the distribution of data for the true alternative. These two distributions could be attributed to measurements that

TABLE 5.1 Relationships in Statistical Testing

	Decision	
Situation	Accept H_0	Reject H_0
H_0 true	Correct decision: $P = 1 - \alpha$ (confidence level)	Type I error: $P = \alpha$ (significance level)
H_0 false (H_a true)	Type II error: $P = \beta$	Correct decision: $P = 1 - \beta$ (power of test)

contain only random errors (left distribution) versus measurements containing blunders (right distribution). In the figure it is seen that valid measurements in the α region of the left distribution are being rejected at a significance level of α. Thus, α represents the probability of committing a Type I error. This is known as the *significance level of the test*. Furthermore, data from the right distribution are being accepted at a β level of significance. The *power of the test* is $1 - \beta$ and corresponds to a true alternative hypothesis. Methods of computing β or $1 - \beta$ are not clear, or are often difficult, since nothing is generally known about the distribution of the alternative. Consequently, in statistical testing, the objective is to prove the alternative hypothesis true by showing that the data do not support the statistic coming from the null hypothesis distribution. In doing this, only a Type I error can be made, for which a known probability of making an incorrect decision is α.

Example 5.1 Assume that for a population of 10,000 people, a flu virus test has a 95% confidence level and thus a significance level, α, of 0.05. Suppose that 9200 people test negative for the flu virus and 800 test positive. Of the 800 people who tested positive, 5%, or 40 people, will test incorrectly (false positive). That is, they will test positive for the flu but do not have it. This is an example of committing a Type I error at an α level of significance. Similarly assume that 460 people test negative for the flu when, in fact, they do have it (a false-negative case). This is an example of a Type II error at a

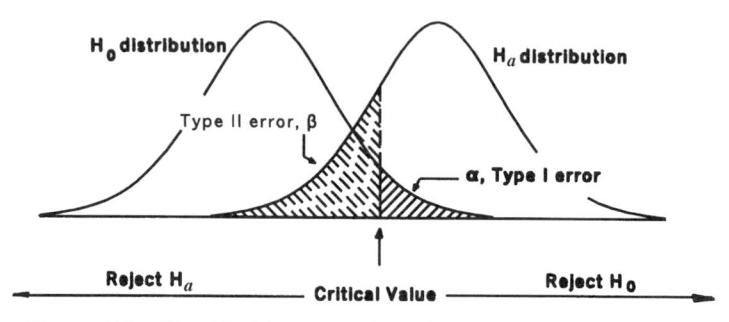

Figure 5.1 Graphical interpretation of Type I and Type II errors.

probability of β, that is equal to 0.046 (460/10,000). The power of the test is 1 − β or 0.954.

From the foregoing it is seen that it is possible to set the probability of committing a Type I error for a given H_0. However, for a fixed level of α and sample size n, the probability of a Type II error, β, is generally unknown. If the null hypothesis, H_0, and α are fixed, the power of the test can only be increased by increasing the sample size, n. Since the power of the test may be low or unknown, statisticians always say that the test *failed to reject* the null hypothesis, rather than making any statements about its acceptance. This is an important statistical concept. That is, it should never be stated that the null hypothesis is accepted since the power of the test is unknown. It should only be said that "there is no statistical evidence to reject the null hypothesis." Because of this small but important distinction, it is important to construct a test that rejects the null hypothesis whenever possible.

A similar situation exists with surveying measurements. If a distance measurement contains a large systematic error, it is possible to detect this with a fully constrained adjustment and thus reject the null hypothesis. However, if a distance contains a very small systematic error, the ability to detect the systematic error may be low. Thus, although some confidence can be placed in the rejection of the null hypothesis, it can never be said that the null hypothesis should be accepted since the probability of undetected small systematic errors or blunders cannot be determined. What we strive to do is minimize the size of these errors so that they have little effect on the computed results.

5.2 SYSTEMATIC DEVELOPMENT OF A TEST

When developing a statistical test, the statistician must determine the test variables and the type of test to perform. This book will look at statistical tests for the mean, variance, and ratio of two sample variances. The t test is used when comparing a sample mean versus a population mean. This test compares the mean of a set of observations against a known calibrated value. The χ^2 test is used when comparing a sample variance against a population variance. As discussed in Section 16.7, this test is used in a least squares adjustment when comparing the reference variance from an adjustment against its population value. Finally, when comparing variances from two different sample sets, the F test is used. As discussed in Section 21.6, this test is used in least squares adjustments when comparing the reference variances from a minimally constrained and fully constrained adjustment. Table 5.2 lists the test variables of these three statistical tests.

TABLE 5.2 Test Variables and Statistical Tests

Variable 1, Test Statistic	Variable 2, Sample Statistic	Null Hypothesis	Test Statistic
Population mean, μ	Sample mean, $\bar{y}$	H_0: $\mu = \bar{y}$	t
Population variance, σ^2	Sample variance, S^2	H_0: $\sigma^2 = S^2$	χ^2
Ratio of sample variances equals 1	S_1^2/S_2^2	H_0: $S_1^2/S_2^2 = 1$	F

A test can take two forms based on the distributions. A one-tailed test uses the critical value from either the left or right side of a distribution, whereas the two-tailed test is much like a confidence interval, with the critical value divided equally on both sides of the distribution.

In the *one-tailed test,* the concern is whether the sample statistic is either greater or less than the statistic being tested. In the *two-tailed test,* the concern is whether the sample statistic is different from the statistic being tested. For example, when checking the angle-reading capabilities of a total station against the manufacturer's specifications, a surveyor would not be concerned if the instrument were working at a level better than the manufacturer's stated accuracy. However, the surveyor would probably send the instrument in for repairs if it was performing at a level below the manufacturer's stated accuracy. In this case, it would be appropriate to perform a one-tailed test. On the other hand, when checking the mean distance observed using an EDM against a known calibration baseline length, the surveyor wants to know if the mean length is statistically different from the calibrated length. In this case it is appropriate to perform a two-tailed test. In the following sections it is important (1) identify the appropriate test statistic and (2) the type of test to perform.

In all forms of statistical testing, a test statistic is developed from the data. The test statistic is then compared against a critical value from the distribution. If the rejection region statement is true, the null hypothesis is rejected at the level of significance selected. As stated earlier, this is the goal of a well-developed test since only Type I error occurs at the selected level of significance. If the rejection region statement is false, the test fails to reject null hypothesis. Because of the possibility of Type II error and due to the lack of knowledge about the alternative distribution, no statement about the validity of the null hypothesis can be made; at best it can be stated that there is no reason to reject the null hypothesis.

5.3 TEST OF HYPOTHESIS FOR THE POPULATION MEAN

At times it may be desirable to test a sample mean against a known value. The t distribution is used to build this test. The null hypothesis for this test can take two forms: one- or two-tailed tests. In the *one-tailed test,* the concern

5.3 TEST OF HYPOTHESIS FOR THE POPULATION MEAN

is whether the sample mean is either statistically greater or less than the population mean. In the *two-tailed test*, the concern is whether the sample mean is statistically different from the population mean. These two tests are shown below.

	One-Tailed Test	Two-Tailed Test
Null hypothesis:	$H_a: \mu = \bar{y}$	$H_0: \mu = \bar{y}$
Alternative hypothesis:	$H_a: \mu > \bar{y} (\mu < \bar{y})$	$H_a: \mu \neq \bar{y}$

The test statistic is

$$t = \frac{\bar{y} - \mu}{S/\sqrt{n}} \tag{5.1}$$

The region where the null hypothesis is rejected is

$$t > t_\alpha \text{ (or } t < t_\alpha) \qquad |t| > t_{\alpha/2}$$

It should be stated that for large samples ($n > 30$), the t value can be replaced by the standard normal variate, z.

Example 5.2 A baseline of calibrated length 400.008 m is observed repeatedly with an EDM instrument. After 20 observations, the average of the observed distances is 400.012 m with a standard deviation of ±0.002 m. Is the distance observed significantly different from the distance calibrated at a 0.05 level of significance?

SOLUTION Assuming that proper field and office procedures were followed, the fundamental question is whether the EDM is working within its specifications and thus providing distance observations in a population of calibrated values. To answer this question, a two-tailed test is used to determine whether the distance is the same or is different from the distance calibrated at a 0.05 level of significance. That is, the mean of the distances observed will be rejected if it is statistically either too short or too long to be considered the same as the calibration value. The rationale behind using a two-tailed test is similar to that used when constructing a confidence interval, as in Example 4.1. That is, 2.5% of the area from the lower and upper tails of the t distribution is to be excluded from the interval constructed, or in this case, the test.

The null hypothesis is

$$H_0: \mu = 400.012$$

and the alternative hypothesis is

$$H_a: \mu \neq 400.012$$

By Equation (5.1), the test statistic is

$$t = \frac{\bar{y} - \mu}{S/\sqrt{n}} = \frac{400.012 - 400.008}{0.002/\sqrt{20}} = 8.944$$

and the rejection region is

$$t = 8.944 > t_{\alpha/2}$$

Since a two-tailed test is being done, the $\alpha/2$ (0.025) column in the t-distribution table is intersected with the $\nu = n - 1$, or 19 degrees of freedom, row. From the t distribution (Table D.3), $t_{0.025, 19}$ is found to be 2.093, and thus the rejection region is satisfied. In other words, the value computed for t is greater than the tabulated value, and thus the null hypothesis can be rejected at a 95% level of confidence. That is,

$$t = 8.944 > t_{\alpha/2} = 2.093$$

Based on the foregoing, there is reason to believe that the average length observed is significantly different from its calibrated value at a 5% significance level. This implies that at least 5% of the time, the decision will be wrong. As stated earlier, a 95% confidence interval for the population mean could also have been constructed to derive the same results. Using Equation (4.7), that interval would yield

$$400.011 = 400.012 - 2.093\left(\frac{0.002}{\sqrt{20}}\right)$$
$$\leq \mu \leq 400.012 + 2.093\left(\frac{0.002}{\sqrt{20}}\right) = 400.013$$

Note that the 95% confidence interval fails to contain the baseline value of 400.008, and similarly, there is reason to be concerned about the calibration status of the instrument. That is, it may not be working properly and should be repaired.

5.4 TEST OF HYPOTHESIS FOR THE POPULATION VARIANCE

In Example 5.2, the procedure for checking whether an observed length compares favorably with a calibrated value was discussed. The surveyor may also

5.4 TEST OF HYPOTHESIS FOR THE POPULATION VARIANCE

want to check if the instrument is measuring at its published precision. The χ^2 distribution is used when comparing the variance of a sample set against that of a population. This test involves checking the variance computed from a sample set of observations against the published value (the expected variance of the population).

As shown in Table 5.2, the χ^2 distribution checks the sample variance against a population variance. By using Equation (4.1), the following statistical test is written.

One-Tailed Test *Two-Tailed Test*

Null hypothesis:

$H_0: S^2 = \sigma^2$ $H_0: S^2 = \sigma^2$

Alternative hypothesis:

$H_a: S^2 > \sigma^2$ (or $H_a: S^2 < \sigma^2$) $H_a: S^2 \neq \sigma^2$

The test statistic is

$$\chi^2 = \frac{\nu S^2}{\sigma^2} \tag{5.2}$$

from which the null hypothesis is rejected when the following statement is satisfied:

$\chi^2 > \chi_\alpha^2$ (or $\chi^2 < \chi_{1-\alpha}^2$) $\chi^2 < \chi_{1-\alpha/2}^2$ (or $\chi^2 > \chi_{\alpha/2}^2$)

The rejection region is determined from Equation (4.13). Graphically, the null hypothesis is rejected in the one-tailed test when the χ^2 value computed is greater than the value tabulated. This rejection region is the shaded region shown in Figure 5.2(a). In the two-tailed test, the null hypothesis is rejected when the value computed is either less than $\chi_{1-\alpha/2}^2$ or greater than $\chi_{\alpha/2}^2$. This is similar to the computed variance being outside the constructed confidence interval for the population variance. Again in the two-tailed test, the proba-

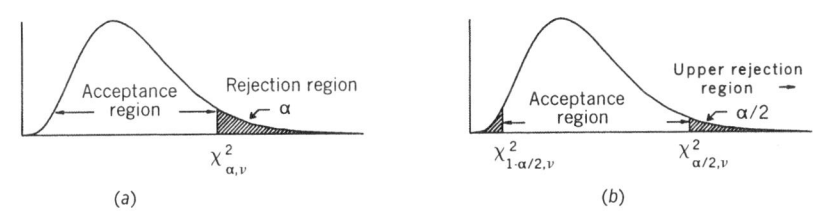

Figure 5.2 Graphical interpretation of (*a*) one- and (*b*) two-tailed tests.

bility selected is evenly divided between the upper and lower tails of the distribution such that the acceptance region is centered on the distribution. These rejection regions are shown graphically in Figure 5.2(b).

Example 5.3 The owner of a surveying firm wants all surveying technicians to be able to read a particular instrument to within $\pm 1.5''$. To test this value, the owner asks the senior field crew chief to perform a reading test with the instrument. The crew chief reads the circle 30 times and obtains $\sigma_r = \pm 0.9''$. Does this support the 1.5″ limit at a 5% level of significance?

SOLUTION In this case the owner wishes to test the hypothesis that the computed sample variance is the same as the population variance rather than being greater than the population variance. That is, all standard deviations that are equal to or less than 1.5″ will be accepted. Thus, a one-tailed test is constructed as follows (note that $v = 30 - 1$, or 29): The null hypothesis is

$$H_0: S^2 = \sigma^2$$

and the alternative hypothesis is

$$H_a: S^2 > \sigma^2$$

The test statistic is

$$\chi^2 = \frac{(30-1)0.9^2}{1.5^2} = 10.44$$

The null hypothesis is rejected when the computed test statistic exceeds the tabulated value, or when the following statement is true:

$$\chi^2 = 10.44 > \chi^2_{\alpha,v} = \chi^2_{0.05,29} = 42.56$$

where 42.56 is from Table D.2 for $\chi^2_{0.05,29}$. Since the computed χ^2 value (10.44) is less than the tabulated value (42.56), the null hypothesis cannot be rejected.

However, simply failing to reject the null hypothesis does not mean that the value of $\pm 1.5''$ is valid. This example demonstrates a common problem in statistical testing when results are interpreted incorrectly. A valid sample set from the population of all surveying employees cannot be obtained by selecting only one employee. Furthermore, the test is flawed since every instrument reads differently and thus new employees may initially have problems reading an instrument, due to their lack of experience with the instrument. To account properly for this lack of experience, the employer could test a random sample of prospective employees during the interview process, and again after several months of employment. The owner could then check for a correlation between the company's satisfaction with the employee, and the employee's initial ability to read the instrument. However, it is unlikely that any correlation would be found. This is an example of misusing statistics.

Example 5.3 illustrates an important point to be made when using statistics. The interpretation of statistical testing requires judgment by the person performing the test. It should always be remembered that with a test, the objective is to reject and not accept the null hypothesis. Furthermore, a statistical test should be used only where appropriate.

5.5 TEST OF HYPOTHESIS FOR THE RATIO OF TWO POPULATION VARIANCES

When adjusting data, surveyors have generally considered control to be absolute and without error. However, like any other quantities derived from observations, it is a known fact that control may contain error. As discussed in Example 4.4, one method of detecting both errors in control and possible systematic errors in horizontal network measurements is to do both a minimally constrained and a fully constrained least squares adjustment with the data. After doing both adjustments, the post-adjustment reference variances can be compared. If the control is without error and no systematic errors are present in the data, the ratio of the two reference variances should be close to 1. Using Equation (4.18), a hypothesis test can be constructed to compare the ratio of variances for two sample sets as follows:

One-Tailed Test *Two-Tailed Test*

Null hypothesis:

$$H_0: \frac{S_1^2}{S_2^2} = 1 \quad (\text{i.e., } S_1^2 = S_2^2) \quad H_0: \frac{S_1^2}{S_2^2} = 1 \quad (\text{i.e., } S_1^2 = S_2^2)$$

Alternative hypothesis:

$$H_a: \frac{S_1^2}{S_2^2} > 1 \quad (\text{i.e., } S_1^2 > S_2^2) \quad H_a: \frac{S_1^2}{S_2^2} \neq 1 \quad (\text{i.e., } S_1^2 \neq S_2^2)$$

or

$$H_a: \frac{S_1^2}{S_2^2} < 1 \quad (\text{i.e., } S_1^2 < S_2^2)$$

The test statistic that will be used to determine rejection of the null hypothesis is

$$F = \frac{S_1^2}{S_2^2} \quad \text{or} \quad F = \frac{S_2^2}{S_1^2} \qquad F = \frac{\text{larger sample variance}}{\text{smaller sample variance}}$$

The null hypothesis should be rejected when the following statement is satisfied:

$$F > F_\alpha \qquad\qquad F > F_{\alpha/2}$$

F_α and $F_{\alpha/2}$ are values that locate the α and $\alpha/2$ areas, respectively, in the upper tail of the F distribution with v_1 numerator degrees of freedom and v_2 denominator degrees of freedom. Notice that in the two-tailed test, the degrees of freedom of the numerator are taken from the numerically larger sample variance, and the degrees of freedom of the denominator are from the smaller variance.

Example 5.4 Using the same data as presented in Example 4.4, would the null hypothesis be rejected?

SOLUTION In this example, a two-tailed test is appropriate since the only concern is whether the two reference variances are equal statistically. In this problem, the interval is centered on the F distribution with an $\alpha/2$ area in the lower and upper tails. In the analysis, 30 degrees of freedom in the numerator correspond to the larger sample variance, and 24 degrees of freedom in the denominator to the smaller variance, so that the following test is constructed: The null hypothesis is

$$H_0: \frac{S_1^2}{S_2^2} = 1$$

and the alternative hypothesis is

$$H_a: \frac{S_1^2}{S_2^2} \neq 1$$

The test statistic for checking rejection of the null hypothesis is

$$F = \frac{2.25}{0.49} = 4.59$$

Rejection of the null hypothesis occurs when the following statement is true:

$$F = 4.59 > F_{\alpha/2, v1, v2} = F_{0.025, 30, 24} = 2.21$$

5.5 TEST OF HYPOTHESIS FOR THE RATIO OF TWO POPULATION VARIANCES

Here it is seen that the F value computed (4.59) is greater than its value from Table D.4 (2.21). Thus, the null hypothesis can be rejected. In other words, the fully constrained adjustment does not have the same variance as its minimally constrained counterpart at the 0.05 level of significance. Notice that the same result was obtained here as was obtained in Example 4.4 with the 95% confidence interval. Again, the network should be inspected for the presence of systematic errors, followed by an analysis of possible errors in the control stations. This post-adjustment analysis is revisited in greater detail in Chapter 20.

Example 5.5 Ron and Kathi continually debate who measures angles more precisely with a particular total station. After listening to enough of this debate, their supervisor describes a test where each is to measure a particular direction by pointing and reading the instrument 51 times. They must then compute the variance for their data. At the end of the 51 readings, Kathi determines her variance to be 0.81 and Ron finds his to be 1.21. Is Kathi a better instrument operator at a 0.01 level of significance?

SOLUTION   In this situation, even though Kathi's variance implies that her observations are more precise than Ron's, a determination must be made to see if the reference variances are statistically equal versus Kathi's being better than the Ron's. This test requires a one-tailed F test with a significance level of $\alpha = 0.01$. The null hypothesis is

$$H_0: \frac{S_R^2}{S_K^2} = 1 \qquad (S_R^2 = S_K^2)$$

and the alternative hypothesis is

$$H_a: \frac{S_R^2}{S_K^2} > 1 \qquad (S_R^2 > S_K^2)$$

The test statistic is

$$F = \frac{1.21}{0.81} = 1.49$$

The null hypothesis is rejected when the computed value for F (1.49) is greater than the tabulated value for $F_{0.01,50,50}$ (1.95). Here it is seen that the value computed for F is less than its tabulated value, and thus the test statistic does not satisfy the rejection region. That is,

80 STATISTICAL TESTING

$$F = 1.49 > F_{\alpha, 50, 50} = 1.95 \text{ is false}$$

Therefore, there is no statistical reason to believe that Kathi is better than Ron at a 0.01 level of significance.

Example 5.6 A baseline is observed repeatedly over a period of time using an EDM instrument. Each day, 10 observations are taken and averaged. The variances for the observations are listed below. At a significance level of 0.05, are the results of day 2 significantly different from those of day 5?

Day	1	2	3	4	5
Variance, S^2 (mm²)	50.0	61.0	51.0	53.0	54.0

SOLUTION This problem involves checking whether the variances of days 2 and 5 are statistically equal or are different. This is the same as constructing a confidence interval involving the ratio of the variances. Because the concern is about equality or inequality, this will require a two-tailed test. Since 10 observations were collected each day, both variances are based on 9 degrees of freedom (v_1 and v_2). Assume that the variance for day 2 is S_2^2 and the variance for day 5 is S_5^2. The test is constructed as follows: The null hypothesis is

$$H_0: \frac{S_2^2}{S_5^2} = 1$$

and the alternative hypothesis is

$$H_a: \frac{S_2^2}{S_5^2} \neq 1$$

The test statistic is

$$F = \frac{61}{54} = 1.13$$

The null hypothesis is rejected when the computed F value (1.13) is greater than the value in Table D.4 (4.03). In this case the rejection region is $F = 1.13 > F_{0.025, 9, 9} = 4.03$ and is not satisfied. Consequently, the test fails to reject the null hypothesis, and there is no statistical reason to believe that the data of day 2 are statistically different from those of day 5.

PROBLEMS

5.1 In your own words, state why the null hypothesis can never be accepted.

5.2 Explain why medical tests on patients are performed several times in a laboratory before the results of the test are returned to the doctor.

5.3 In your own words, discuss when it is appropriate to use:
   (a) a one-tailed test.
   (b) a two-tailed test.

5.4 Match the following comparison with the appropriate test.
   (a) A calibration baseline length against a value measured using an EDM.
   (b) Two sample variances.
   (c) The reference variance of a fully constrained adjustment against a minimally constrained adjustment.
   (d) The reference variance of a least squares adjustment against its a priori value of 1.

5.5 Compare the variances of days 2 and 5 in Example 5.6 at a level of significance of 0.20 ($\alpha = 0.20$). Would testing the variances of days 1 and 2 result in a different finding?

5.6 Compare the variances of days 1 and 4 in Example 5.6 at a level significance of 0.05 ($\alpha = 0.05$). Would testing the variances of days 1 and 3 result in a different finding?

5.7 Using the data given in Example 5.5, determine if Kathi is statistically better with the equipment than Ron at a significance level of:
   (a) 0.05.
   (b) 0.10.

5.8 The population value for the reference variance from a properly weighted least squares adjustment is 1. After running a minimally constrained adjustment having 15 degrees of freedom, the reference variance is computed 1.52. Is this variance statistically equal to 1 at:
   (a) a 0.01 level of significance?
   (b) a 0.05 level of significance?
   (c) a 0.10 level of significance?

5.9 When all the control is added to the adjustment in Problem 5.8, the reference variance for the fully constrained adjustment with 15 degrees of freedom is found to be 1.89. Are the reference variances from the minimally constrained and fully constrained adjustments statistically equal at:

(a) a 0.01 level of significance?
(b) a 0.05 level of significance?
(c) a 0.10 level of significance?

5.10 The calibrated length of a baseline is 402.267 m. A mean observation for the distance is 402.251 m with a standard deviation of ±0.0052 m after six readings with an EDM.
(a) Is the distance observed statistically different from the length calibrated at a 5% level of significance?
(b) Is the distance observed statistically different from the length calibrated at a 10% level of significance?

5.11 A mean length of 1023.573 m with a standard deviation of ±0.0056 m is obtained for a distance after five observations. Using the technical specifications, it is found that the standard deviation for this observation should be ±0.0043 m.
(a) Perform a statistical test to check the repeatability of the instrument at a level of significance of 0.05.
(b) Perform a statistical test to check the repeatability of the instrument at a level of significance of 0.01.

5.12 A least squares adjustment is computed twice on a data set. When the data are minimally constrained with 24 degrees of freedom, a reference variance of 0.89 is obtained. In the second run, the fully constrained network, which also has 24 degrees of freedom, has a reference variance of 1.15. The a priori estimate for the reference variance in both adjustments is 1; that is, $\sigma_1^2 = \sigma_2^2 = 1$.
(a) Are the two variances statistically equal at a 0.05 level of significance?
(b) Is the minimally constrained adjustment reference variance statistically equal to 1 at a 0.05 level of significance?
(c) Is the fully constrained adjustment reference variance statistically equal to 1 at a 0.05 level of significance?
(d) Is there a statistical reason to be concerned about the presence of errors in either the control or the observations?

5.13 A total station with a manufacturer's specified angular accuracy of ±5″ was used to collect the data in Problem 5.12. Do the data warrant this accuracy at a 0.05 level of significance? Develop a statistical test to validate your response.

5.14 A surveying company decides to base a portion of their employees' salary raises on improvement in their use of equipment. To determine the improvement, the employees measure their ability to point on a target and read the circles of a theodolite every six months. One employee is tested six weeks after starting employment and obtains a standard deviation of ±1.5″ with 25 measurements. Six months later the employee obtains a standard deviation of ±1.2″ with 30 measurements. Did the employee improve statistically over six months at a 5% level of significance? Is this test an acceptable method of determining improvements in quality? What suggestion, if any, would you give to modify the test?

5.15 An EDM is placed on a calibration baseline and the distance between two monuments is determined to be 1200.012 m ± 0.047 m after 10 observations. The length between the monuments is calibrated as 1200.005 m. Is the instrument measuring the length properly at:

(a) a 0.01 level of significance?
(b) a 0.05 level of significance?
(c) a 0.10 level of significance?

CHAPTER 6

PROPAGATION OF RANDOM ERRORS IN INDIRECTLY MEASURED QUANTITIES

6.1 BASIC ERROR PROPAGATION EQUATION

As discussed in Section 1.2, unknown values are often determined indirectly by making direct measurements of other quantities which are functionally related to the desired unknowns. Examples in surveying include computing station coordinates from distance and angle observations, obtaining station elevations from rod readings in differential leveling, and determining the azimuth of a line from astronomical observations. As noted in Section 1.2, since all quantities that are measured directly contain errors, any values computed from them will also contain errors. This intrusion, or *propagation,* of errors that occurs in quantities computed from direct measurements is called *error propagation.* This topic is one of the most important discussed in this book.

In this chapter it is assumed that all systematic errors have been eliminated, so that only random errors remain in the direct observations. To derive the basic error propagation equation, consider the simple function, $z = a_1 x_1 + a_2 x_2$, where x_1 and x_2 are two independently observed quantities with standard errors σ_1 and σ_2, and a_1 and a_2 are constants. By analyzing how errors propagate in this function, a general expression can be developed for the propagation of random errors through any function.

Since x_1 and x_2 are two independently observed quantities, they each have different probability density functions. Let the errors in n determinations of x_1 be $\varepsilon_1^i, \varepsilon_1^{ii}, \ldots, \varepsilon_1^n$ and the errors in n determinations of x_2 be $\varepsilon_2^i, \varepsilon_2^{ii}, \ldots, \varepsilon_2^n$; then z_T, the true value of z for each independent measurement, is

6.1 BASIC ERROR PROPAGATION EQUATION

$$z_T = \begin{cases} a_1(x_1^i - \varepsilon_1^i) + a_2(x_2^i - \varepsilon_2^i) = a_1 x_1^i + a_2 x_2^i - (a_1 \varepsilon_1^i + a_2 \varepsilon_2^i) \\ a_1(x_1^{ii} - \varepsilon_1^{ii}) + a_2(x_2^{ii} - \varepsilon_2^{ii}) = a_1 x_1^{ii} + a_2 x_2^{ii} - (a_1 \varepsilon_1^{ii} + a_2 \varepsilon_2^{ii}) \\ a_1(x_1^{iii} - \varepsilon_1^{iii}) + a_2(x_2^{iii} - \varepsilon_2^{iii}) = a_1 x_1^{iii} + a_2 x_2^{iii} - (a_1 \varepsilon_1^{iii} + a_2 \varepsilon_2^{iii}) \\ \vdots \end{cases} \quad (6.1)$$

The values for z computed from the observations are

$$z^i = a_1 x_1^i + a_2 x_2^i$$
$$z^{ii} = a_1 x_1^{ii} + a_2 x_2^{ii} \quad (6.2)$$
$$z^{iii} = a_1 x_1^{iii} + a_2 x_2^{iii}$$
$$\vdots$$

Substituting Equations (6.2) into Equations (6.1) and regrouping Equations (6.1) to isolate the errors for each computed value yields

$$z^i - z_T = a_1 \varepsilon_1^i + a_2 \varepsilon_2^i$$
$$z^{ii} - z_T = a_1 \varepsilon_1^{ii} + a_2 \varepsilon_2^{ii} \quad (6.3)$$
$$z^{iii} - z_T = a_1 \varepsilon_1^{iii} + a_2 \varepsilon_2^{iii}$$
$$\vdots$$

From Equation (2.4) for the variance in a population, $n\sigma^2 = \sum_{i=1}^{n} \varepsilon^2$, and thus for the case under consideration, the sum of the squared errors for the value computed is

$$\sum_{i=1}^{n} \varepsilon_i^2 = (a_1 \varepsilon_1^i + a_2 \varepsilon_2^i)^2 + (a_1 \varepsilon_1^{ii} + a_2 \varepsilon_2^{ii})^2 + (a_1 \varepsilon_1^{iii} + a_2 \varepsilon_2^{iii})^2 + \cdots = n\sigma_z^2$$

(6.4)

Expanding the terms in Equation (6.4) yields

$$n\sigma_z^2 = (a_1 \varepsilon_1^i)^2 + 2 a_1 a_2 \varepsilon_1^i \varepsilon_2^i + (a_2 \varepsilon_2^i)^2 + (a_1 \varepsilon_1^{ii})^2 + 2 a_1 a_2 \varepsilon_1^{ii} \varepsilon_2^{ii} + (a_2 \varepsilon_2^{ii})^2 + \cdots$$

(6.5)

Factoring terms in Equation (6.5) gives

$$n\sigma_z^2 = a_1^2(\varepsilon_1^{i^2} + \varepsilon_1^{ii^2} + \varepsilon_1^{iii^2} + \cdots) + a_2^2(\varepsilon_2^{i^2} + \varepsilon_2^{ii^2} + \varepsilon_2^{iii^2} + \cdots)$$
$$+ 2 a_1 a_2 (\varepsilon_1^i \varepsilon_2^i + \varepsilon_1^{ii} \varepsilon_2^{ii} + \varepsilon_1^{iii} \varepsilon_2^{iii} + \cdots) \quad (6.6)$$

Inserting summation symbols for the error terms in Equation (6.6) results in

86 PROPAGATION OF RANDOM ERRORS IN INDIRECTLY MEASURED QUANTITIES

$$\sigma_z^2 = a_1^2 \left(\frac{\sum_{i=1}^n \varepsilon_1^2}{n} \right) + 2a_1 a_2 \left(\frac{\sum_{i=1}^n \varepsilon_1 \varepsilon_2}{n} \right) + a_2^2 \left(\frac{\sum_{i=1}^n \varepsilon_2^2}{n} \right) \tag{6.7}$$

Recognizing that the terms in parentheses in Equation (6.7) are by definition: $\sigma_{x_1}^2$, $\sigma_{x_1 x_2}$, and $\sigma_{x_2}^2$, respectively, Equation (6.7) can be rewritten as

$$\sigma_z^2 = a_1^2 \sigma_{x_1}^2 + 2a_1 a_2 \sigma_{x_1 x_2} + a_2^2 \sigma_{x_2}^2 \tag{6.8}$$

In Equation (6.8), the middle term, $\sigma_{x_1 x_2}$, is known as the *covariance*. This term shows the interdependence between the two unknown variables, x_1 and x_2. As the covariance term decreases, the interdependence of the variables also decreases. When these terms are zero, the variables are said to be *mathematical independent*. Its importance is discussed in more detail in later chapters.

Equations (6.7) and (6.8) can be written in matrix form as

$$\Sigma_{zz} [a_1 \quad a_2] \begin{bmatrix} \sigma_{x_1}^2 & \sigma_{x_1 x_2} \\ \sigma_{x_1 x_2} & \sigma_{x_2}^2 \end{bmatrix} \begin{bmatrix} a_1 \\ a_2 \end{bmatrix} \tag{6.9}$$

where Σ_{zz} is the variance–covariance matrix for the function z. It follows logically from this derivation that, in general, if z is a function of n independently measured quantities, $x_1, x_2, \ldots, x_n$, then Σ_{zz} is

$$\Sigma_{zz} = [a_1 \quad a_2 \quad \cdots \quad a_n] \begin{bmatrix} \sigma_{x_1}^2 & \sigma_{x_1 x_2} & \cdots & \sigma_{x_1 x_n} \\ \sigma_{x_2 x_1} & \sigma_{x_2}^2 & & \sigma_{x_2 x_n} \\ \vdots & & \ddots & \\ \sigma_{x_n x_1} & \sigma_{x_n x_2} & & \sigma_{x_n}^2 \end{bmatrix} \begin{bmatrix} a_1 \\ a_2 \\ \vdots \\ a_n \end{bmatrix} \tag{6.10}$$

Further, for a set of m functions with n independently measured quantities, $x_1, x_2, \ldots, x_n$, Equation (6.10) expands to

$$\Sigma_{zz} = \begin{bmatrix} a_{11} & a_{12} & \cdots & a_{1n} \\ a_{21} & a_{22} & \cdots & a_{2n} \\ \vdots & \vdots & \cdots & \vdots \\ a_{m1} & a_{m2} & \cdots & a_{mn} \end{bmatrix} \begin{bmatrix} \sigma_{x_1}^2 & \sigma_{x_1 x_2} & \cdots & \sigma_{x_1 x_n} \\ \sigma_{x_1 x_2} & \sigma_{x_2}^2 & \cdots & \sigma_{x_2 x_n} \\ \vdots & \vdots & \ddots & \vdots \\ \sigma_{x_n x_1} & \sigma_{x_2 x_n} & \cdots & \sigma_{x_n}^2 \end{bmatrix} \begin{bmatrix} a_{11} & a_{21} & \cdots & a_{m1} \\ a_{12} & a_{22} & \cdots & a_{m2} \\ \vdots & \vdots & \cdots & \vdots \\ a_{1n} & a_{2n} & \cdots & a_{mn} \end{bmatrix}$$
(6.11)

Similarly, if the functions are nonlinear, a first-order Taylor series expansion can be used to linearize them.[1] Thus, a_{11}, a_{12}, ... are replaced by the partial derivatives of Z_1, Z_2, ... with respect to the unknown parameters, x_1,

---

[1] Readers who are unfamiliar with solving nonlinear equations should refer to Appendix C.

$x_2, \ldots$. Therefore, after linearizing a set of nonlinear equations, the matrix for the function of Z can be written in linear form as

$$\Sigma_{zz} = \begin{bmatrix} \frac{\partial Z_1}{\partial x_1} & \frac{\partial Z_1}{\partial x_2} & \cdots & \frac{\partial Z_1}{\partial x_n} \\ \frac{\partial Z_2}{\partial x_1} & \frac{\partial Z_2}{\partial x_2} & \cdots & \frac{\partial Z_2}{\partial x_n} \\ \vdots & \vdots & \cdots & \vdots \\ \frac{\partial Z_m}{\partial x_1} & \frac{\partial Z_m}{\partial x_2} & \cdots & \frac{\partial Z_m}{\partial x_n} \end{bmatrix} \begin{bmatrix} \sigma_{x_1}^2 & \sigma_{x_1 x_2} & \cdots & \sigma_{x_1 x_n} \\ \sigma_{x_1 x_2} & \sigma_{x_2}^2 & \cdots & \sigma_{x_2 x_n} \\ \vdots & \vdots & \ddots & \vdots \\ \sigma_{x_n x_1} & \sigma_{x_2 x_n} & \cdots & \sigma_{x_n}^2 \end{bmatrix} \begin{bmatrix} \frac{\partial Z_1}{\partial x_1} & \frac{\partial Z_2}{\partial x_1} & \cdots & \frac{\partial Z_m}{\partial x_1} \\ \frac{\partial Z_1}{\partial x_2} & \frac{\partial Z_2}{\partial x_2} & \cdots & \frac{\partial Z_m}{\partial x_2} \\ \vdots & \vdots & \cdots & \vdots \\ \frac{\partial Z_1}{\partial x_n} & \frac{\partial Z_2}{\partial x_n} & \cdots & \frac{\partial Z_m}{\partial x_n} \end{bmatrix}$$
(6.12)

Equations (6.11) and (6.12) are known as the *general law of propagation of variances* (GLOPOV) for linear and nonlinear equations, respectively. Equations (6.11) and (6.12) can be written symbolically in matrix notation as

$$\Sigma_{zz} = A\Sigma A^T \qquad (6.13)$$

where Σ_{zz} is the covariance matrix for the function Z. For a nonlinear set of equations that is linearized using Taylor's theorem, the coefficient matrix (A) is called a *Jacobian matrix*, a matrix of partial derivatives with respect to each unknown, as shown in Equation (6.12).

If the measurements $x_1, x_2, \ldots, x_n$ are unrelated (i.e., are statistically independent), the covariance terms $\sigma_{x_1 x_2}, \sigma_{x_1 x_3}, \ldots$ equal to zero, and thus the right-hand sides of Equations (6.10) and (6.11) can be rewritten, respectively, as

$$\Sigma_{zz} = \begin{bmatrix} a_{11} & a_{12} & \cdots & a_{1n} \\ a_{21} & a_{22} & \cdots & a_{2n} \\ \vdots & \vdots & \cdots & \vdots \\ a_{m1} & a_{m2} & \cdots & a_{mn} \end{bmatrix} \begin{bmatrix} \sigma_{x_1}^2 & 0 & \cdots & 0 \\ 0 & \sigma_{x_2}^2 & \cdots & 0 \\ \vdots & \vdots & \ddots & \vdots \\ 0 & 0 & \cdots & \sigma_{x_n}^2 \end{bmatrix} \begin{bmatrix} a_{11} & a_{21} & \cdots & a_{m1} \\ a_{12} & a_{22} & \cdots & a_{m2} \\ \vdots & \vdots & \cdots & \vdots \\ a_{1n} & a_{2n} & \cdots & a_{mn} \end{bmatrix}$$
(6.14)

$$\Sigma_{zz} = \begin{bmatrix} \frac{\partial Z_1}{\partial x_1} & \frac{\partial Z_1}{\partial x_2} & \cdots & \frac{\partial Z_1}{\partial x_n} \\ \frac{\partial Z_2}{\partial x_1} & \frac{\partial Z_2}{\partial x_2} & \cdots & \frac{\partial Z_2}{\partial x_n} \\ \vdots & \vdots & \cdots & \vdots \\ \frac{\partial Z_m}{\partial x_1} & \frac{\partial Z_m}{\partial x_2} & \cdots & \frac{\partial Z_m}{\partial x_n} \end{bmatrix} \begin{bmatrix} \sigma_{x_1}^2 & 0 & \cdots & 0 \\ 0 & \sigma_{x_2}^2 & \cdots & 0 \\ \vdots & \vdots & \ddots & \vdots \\ 0 & 0 & \cdots & \sigma_{x_n}^2 \end{bmatrix} \begin{bmatrix} \frac{\partial Z_1}{\partial x_1} & \frac{\partial Z_2}{\partial x_1} & \cdots & \frac{\partial Z_m}{\partial x_1} \\ \frac{\partial Z_1}{\partial x_2} & \frac{\partial Z_2}{\partial x_2} & \cdots & \frac{\partial Z_m}{\partial x_2} \\ \vdots & \vdots & \cdots & \vdots \\ \frac{\partial Z_1}{\partial x_n} & \frac{\partial Z_2}{\partial x_n} & \cdots & \frac{\partial Z_m}{\partial x_n} \end{bmatrix}$$
(6.15)

If there is only one function Z, involving n unrelated quantities, x_1, x_2, ..., x_n, Equation (6.15) can be rewritten in algebraic form as

$$\sigma_Z = \sqrt{\left(\frac{\partial Z}{\partial x_1}\sigma_{x_1}\right)^2 + \left(\frac{\partial Z}{\partial x_2}\sigma_{x_2}\right)^2 + \cdots + \left(\frac{\partial Z}{\partial x_n}\sigma_{x_n}\right)^2} \qquad (6.16)$$

Equations (6.14), (6.15), and (6.16) express the *special law of propagation of variances* (SLOPOV). These equations govern the manner in which errors from statistically independent measurements (i.e., $\sigma_{x_i x_j} = 0$) propagate in a function. In these equations, individual terms $(\partial Z/\partial x_i)\sigma_{x_i}$ represent the individual contributions to the total error that occur as the result of observational errors in each independent variable. When the size of a function's estimated error is too large, inspection of these individual terms will indicate the largest contributors to the error. The most efficient way to reduce the overall error in the function is to closely examine ways to reduce the largest error terms in Equation (6.16).

6.1.1 Generic Example

Let $A = B + C$, and assume that B and C are independently observed quantities. Note that $\partial A/\partial B = 1$ and $\partial Z/\partial C = 1$. Substituting these into Equation (6.16) yields

$$\sigma_A = \sqrt{(1\sigma_B)^2 + (1\sigma_C)^2} \qquad (6.17)$$

Using Equation (6.15) yields

$$\Sigma_{AA} = \begin{bmatrix} 1 & 1 \end{bmatrix} \begin{bmatrix} \sigma_B^2 & 0 \\ 0 & \sigma_C^2 \end{bmatrix} \begin{bmatrix} 1 \\ 1 \end{bmatrix} = [\sigma_B^2 + \sigma_C^2]$$

Equation (6.17) yields the same results as Equation (6.16) after the square root of the single element is determined. In the equations above, standard error (σ) and standard deviation (S) can be used interchangeably.

6.2 FREQUENTLY ENCOUNTERED SPECIFIC FUNCTIONS

6.2.1 Standard Deviation of a Sum

Let $A = B_1 + B_2 + \cdots + B_n$, where the B's are n independently observed quantities having standard deviations of $S_{B_1}, S_{B_2}, \ldots, S_{B_n}$. Then by Equation (6.16),

$$S_A = \sqrt{S_{B_1}^2 + S_{B_2}^2 + \cdots + S_{B_n}^2} \tag{6.18}$$

6.2.2 Standard Deviation in a Series

Assume that the error for each observed value in Equation (6.18) is equal; that is, $S_{B_1}, S_{B_2}, \ldots, S_{B_n} = S_B$ then Equation (6.18) simplifies to

$$S_A = S_B\sqrt{n} \tag{6.19}$$

6.2.3 Standard Deviation of the Mean

Let $\bar{y}$ be the mean obtained from n independently observed quantities $y_1, y_2, \ldots, y_n$, each of which has the same standard deviation S. As given in Equation (2.1), the mean is expressed as

$$\bar{y} = \frac{y_1 + y_2 + \cdots + y_n}{n}$$

An equation for $S_{\bar{y}}$, the standard deviation of $\bar{y}$, is obtained by substituting the expression above into Equation (6.16). Since the partial derivatives of $\bar{y}$ with respect to the observed quantities, $y_1, y_2, \ldots, y_n$, is $\partial \bar{y}/\partial y_1 = \partial \bar{y}/\partial y_2 = \cdots = \partial \bar{y}/\partial y_n = 1/n$, the resulting error in $\bar{y}$ is

$$S_{\bar{y}} = \sqrt{\left(\frac{1}{n}S_{y_1}\right)^2 + \left(\frac{1}{n}S_{y_2}\right)^2 + \cdots + \left(\frac{1}{n}S_{y_n}\right)^2} = \sqrt{\frac{nS^2}{n^2}} = \frac{S}{\sqrt{n}} \tag{6.20}$$

Note that Equation (6.20) is the same as Equation (2.8).

6.3 NUMERICAL EXAMPLES

Example 6.1 The dimensions of the rectangular tank shown in Figure 6.1 are measured as

$$L = 40.00 \text{ ft} \quad S_L = \pm 0.05 \text{ ft}$$
$$W = 20.00 \text{ ft} \quad S_W = \pm 0.03 \text{ ft}$$
$$H = 15.00 \text{ ft} \quad S_H = \pm 0.02 \text{ ft}$$

Find the tank's volume and the standard deviation in the volume using the measurements above.

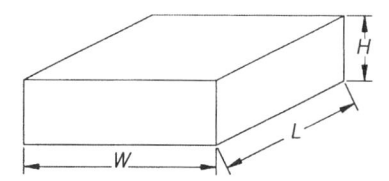

Figure 6.1 Rectangular tank.

SOLUTION The volume of the tank is found using the formula

$$V = LWH = 40.00(20.00)(15.00) = 12,000 \text{ ft}^3$$

Given that $\partial V/\partial L = WH$, $\partial V/\partial W = LH$, and $\partial V/\partial H = LW$, the standard deviation in the volume is determined by using Equation (6.16), which yields

$$\begin{aligned} S_V &= \sqrt{\left(\frac{\partial V}{\partial L}S_L\right)^2 + \left(\frac{\partial V}{\partial W}S_W\right)^2 + \left(\frac{\partial V}{\partial H}S_H\right)^2} \\ &= \sqrt{(WH)^2(0.05)^2 + (LH)^2(0.03)^2 + (LW)^2(0.02)^2} \\ &= \sqrt{(300 \times 0.05)^2 + (600 \times 0.03)^2 + (800 \times 0.02)^2} \\ &= \sqrt{225 + 324 + 256} = \sqrt{805} = \pm 28 \text{ ft}^3 \end{aligned} \qquad (a)$$

In Equation (a), the second term is the largest contributor to the total error, and thus to reduce the overall error in the computed volume, it would be prudent first to try to make S_W smaller. This would yield the greatest effect in the error of the function.

Example 6.2 As shown in Figure 6.2, the vertical angle α to point B is observed at point A as 3°00′, with S_α being $\pm 1'$. The slope distance D from A to B is observed as 1000.00 ft, with S_D being ± 0.05 ft. Compute the horizontal distance and its standard deviation.

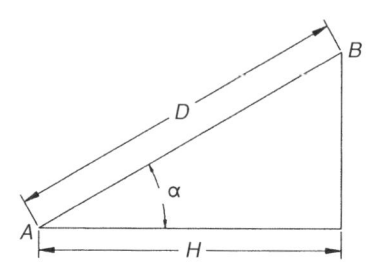

Figure 6.2 Horizontal distance from slope observations.

SOLUTION The horizontal distance is determined using the equation

$$H = D \cos \alpha = 1000.00 \cos(3°00') = 998.63 \text{ ft}$$

Given that $\partial H/\partial D = \cos \alpha$ and $\partial H/\partial \alpha = -D \sin \alpha$, the error in the function is determined by using Equation (6.16) as

$$S_H = \sqrt{\left(\frac{\partial H}{\partial D} S_D\right)^2 + \left(\frac{\partial H}{\partial \alpha} S_\alpha\right)^2} \qquad (6.21)$$

In Equation (6.21), S_α must be converted to its equivalent radian value to achieve agreement in the units. Thus,

$$S_H = \sqrt{(\cos \alpha \times 0.05)^2 + \left(-\sin \alpha \times D \times \frac{60''}{206{,}264.8''/\text{rad}}\right)^2}$$

$$= \sqrt{(0.9986 \times 0.05)^2 + \left(\frac{-0.0523 \times 1000 \times 60''}{206{,}264.8''}\right)^2}$$

$$= \sqrt{0.04993^2 + 0.0152^2} = \pm 0.052 \text{ ft}$$

Notice in this example that the major contributing error source (largest number under the radical) is 0.04993^2. This is the error associated with the distance measurement, and thus if the resulting error of ± 0.052 ft is too large, the logical way to improve the results (reduce the overall error) is to adopt a more precise method of measuring the distance.

Example 6.3 The elevation of point C on the chimney shown in Figure 6.3 is desired. Field angles and distances are observed. Station A has an elevation of 1298.65 ± 0.006 ft, and station B has an elevation of 1301.53 ± 0.004

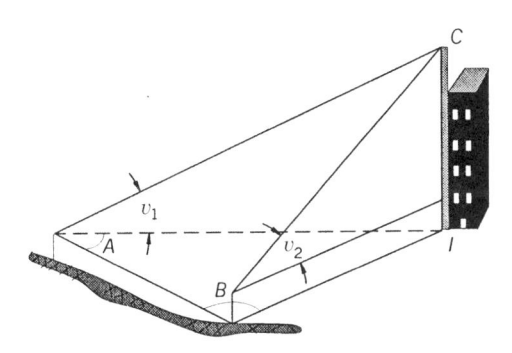

Figure 6.3 Elevation of a chimney determined using intersecting angles.

92 PROPAGATION OF RANDOM ERRORS IN INDIRECTLY MEASURED QUANTITIES

ft. The instrument height, hi_A, at station A is 5.25 ± 0.005 ft, and the instrument height, hi_B, at station B is 5.18 ± 0.005 ft. The other observations and their errors are

$$AB = 136.45 \pm 0.018$$

$$A = 44°12'34'' \pm 8.6'' \quad B = 39°26'56'' \pm 11.3''$$

$$v_1 = 8°12'47'' \pm 4.1'' \quad v_2 = 5°50'10'' \pm 5.1''$$

What are the elevation of the chimney and the error in this computed value?

SOLUTION Normally, this problem is worked in several steps. The steps include computing distances AI and BI and then solving for the average elevation of C using observations obtained from stations A and B. However, caution must be exercised when doing error analysis in a stepwise fashion since the computed values could be correlated and the stepwise method might lead to an incorrect analysis of the errors. To avoid this, either GLOPOV can be used or a single function can be derived that includes all quantities observed when the elevation is calculated. The second method is demonstrated as follows.

From the sine law, the solution of AI and BI can be derived as

$$AI = \frac{AB \sin B}{\sin[180° - (A + B)]} = \frac{AB \sin B}{\sin(A + B)} \tag{6.22}$$

$$BI = \frac{AB \sin A}{\sin(A + B)} \tag{6.23}$$

Using Equations (6.22) and (6.23), the elevations for C from stations A and B are

$$\text{Elev}_{C_A} = AI \tan v_1 + \text{Elev}_A + hi_A \tag{6.24}$$

$$\text{Elev}_{C_B} = BI \tan v_2 + \text{Elev}_B + hi_B \tag{6.25}$$

Thus, the chimney's elevation is computed as the average of Equations (6.24) and (6.25), or

$$\text{Elev}_C = \tfrac{1}{2}(\text{Elev}_{C_A} + \text{Elev}_{C_B}) \tag{6.26}$$

Substituting Equations (6.22) through (6.25) into (6.26), a single expression for the chimney elevation can be written as

$$\text{Elev}_C$$
$$= \frac{1}{2}\left[\text{Elev}_A + hi_A + \frac{AB \sin B \tan v_1}{\sin(A+B)} + \text{Elev}_B + hi_B + \frac{AB \sin A \tan v_2}{\sin(A+B)}\right]$$
(6.27)

From Equation (6.27), the elevation of C is 1316.49 ft. To perform the error analysis, Equation (6.16) is used. In this complex problem, it is often easier to break the problem into smaller parts. This can be done by numerically solving each partial derivative necessary for Equation (6.16) before squaring and summing the results. From Equation (6.26),

$$\frac{\partial \text{Elev}_C}{\partial \text{Elev}_A} = \frac{\partial \text{Elev}_C}{\partial \text{Elev}_B} = \frac{1}{2}$$

$$\frac{\partial \text{Elev}_C}{\partial hi_A} = \frac{\partial \text{Elev}_C}{\partial hi_B} = \frac{1}{2}$$

From Equation (6.27),

$$\frac{\partial \text{Elev}_C}{\partial AB} = \frac{1}{2}\left(\frac{\sin B \tan v_1 + \sin A \tan v_2}{\sin(A+B)}\right) = 0.08199$$

$$\frac{\partial \text{Elev}_C}{\partial A} = \frac{AB}{2}\left[\frac{-\cos(A+B)(\sin B \tan v_1 + \sin A \tan v_2)}{\sin^2(A+B)} + \frac{\cos A + \tan v_2}{\sin(A+B)}\right]$$
$$= 3.78596$$

$$\frac{\partial \text{Elev}_C}{\partial B} = \frac{AB}{2}\left[\frac{-\cos(A+B)(\sin B \tan v_1 + \sin A \tan v_2)}{\sin^2(A+B)} + \frac{\cos B + \tan v_1}{\sin(A+B)}\right]$$
$$= 6.40739$$

$$\frac{\partial \text{Elev}_C}{\partial v_1} = \frac{AB \sin B}{2 \sin(A+B)\cos^2 v_1} = 44.52499$$

$$\frac{\partial \text{Elev}_C}{\partial v_2} = \frac{AB \sin A}{2 \sin(A+B)\cos^2 v_2} = 48.36511$$

Again for compatibility of the units in this problem, all angular errors are converted to their radian equivalents by dividing each by 206,264.8″/rad. Finally, using Equation (6.16), the error in the elevation computed is

$$S^2_{\text{Elev}_C} = \left(\frac{\partial \text{Elev}_C}{\partial \text{Elev}_A} S_{\text{Elev}_A}\right)^2 + \left(\frac{\partial \text{Elev}_C}{\partial \text{Elev}_B} S_{\text{Elev}_B}\right)^2 + \left(\frac{\partial \text{Elev}_C}{\partial hi_A} S_{hi_A}\right)^2$$
$$+ \left(\frac{\partial \text{Elev}_C}{\partial hi_B} S_{hi_B}\right)^2 + \left(\frac{\partial \text{Elev}_C}{\partial AB} S_{AB}\right)^2 + \left(\frac{\partial \text{Elev}_C}{\partial A} S_A\right)^2$$
$$+ \left(\frac{\partial \text{Elev}_C}{\partial B} S_B\right)^2 + \left(\frac{\partial \text{Elev}_C}{\partial v_1} S_{v_1}\right)^2 + \left(\frac{\partial \text{Elev}_C}{\partial v_2} S_{v_2}\right)^2$$

$$S_{\text{Elev}_C}$$
$$= \left[\begin{array}{l}\left(\frac{0.006}{2}\right)^2 + \left(\frac{0.004}{2}\right)^2 + 2\left(\frac{1}{2}0.005\right)^2 + (0.08199 \times 0.018)^2 \\ + (3.78596 \times 4.1693 \times 10^{-5})^2 + (6.40739 \times 5.4783 \times 10^{-5})^2 \\ + (44.52499 \times 1.9877 \times 10^{-5})^2 + (48.36511 \times 2.4725 \times 10^{-5})^2\end{array}\right]^{1/2}$$
$$= \pm 0.0055 \text{ ft} \approx \pm 0.01 \text{ ft}$$

Thus, the elevation of point C is 1316.49 ± 0.01 ft.

6.4 CONCLUSIONS

Errors associated with any indirect measurement problem can be analyzed as described above. Besides being able to compute the estimated error in a function, the sizes of the individual errors contributing to the functional error can also be analyzed. This identifies those observations whose errors are most critical in reducing the functional error. An alternative use of the error propagation equation involves computing the error in a function of observed values prior to fieldwork. The calculation can be based on the geometry of the problem and the observations that are included in the function. The estimated errors in each value can be varied to correspond with those expected using different combinations of available equipment and field procedures. The particular combination that produces the desired accuracy in the final computed function can then be adopted in the field. This analysis falls under the heading of *survey planning and design*. This topic is discussed further in Chapters 7 and 19.

The computations in this chapter can be time consuming and tedious, often leading to computational errors in the results. It is often more efficient to program these equations in a computational package. The programming of the examples in this chapter is demonstrated in the electronic book on the CD that accompanies this book.

PROBLEMS

6.1 In running a line of levels, 18 instrument setups are required, with a backsight and foresight taken from each. For each rod reading, the error estimated is ±0.005 ft. What is the error in the measured elevation difference between the origin and the terminus?

6.2 In Problem 2.6, compute the estimated error in the overall distance as measured by both the 100- and 200-ft tapes. Which tape produced the smallest error?

6.3 Determine the estimated error in the length of AE, which was measured in sections as follows:

Section	Measured Length (ft)	Standard Deviation (ft)
AB	416.24	±0.02
BC	1044.16	±0.05
CD	590.03	±0.03
DE	714.28	±0.04

6.4 A slope distance is observed as 1506.843 ± 0.009 m. The zenith angle is observed as 92°37′29″ ± 8.8″. What is the horizontal distance and its uncertainty?

6.5 A rectangular parcel has dimensions of 538.056 ± 0.005 m by 368.459 ± 0.004 m. What is the area of the parcel and the uncertainty in this area?

6.6 The volume of a cone is given by $V = \frac{1}{2}\pi D^2 h$. The cone's measured height is 8.5 in., with $S_h = \pm 0.15$ in. Its measured diameter is 5.98 in., with $S_D = \pm 0.05$ in. What are the cone's volume and standard deviation?

6.7 An EDM instrument manufacturer publishes the instrument's accuracy as ±(3 mm + 3 ppm). [*Note:* 3 ppm means 3 parts per million. This is a scaling error and is computed as (distance × 3/1,000,000).]
  (a) What formula should be used to determine the error in a distance observed with this instrument?
  (b) What is the error in a 1864.98-ft distance measured with this EDM?

6.8 As shown in Figure P6.8, a racetrack is measured in three simple components: a rectangle and two semicircles. Using an EDM with a manufacturer's specified accuracy of ±(5 mm + 5 ppm), the rectangle's dimensions measured at the inside of the track are 5279.95 ft by 840.24 ft. Assuming only errors in the distance observations, what is:

(a) the area enclosed by the track?
(b) the length of the track?
(c) the standard deviation in each track dimension?
(d) the standard deviation in the perimeter of the track?
(e) the standard deviation in the area enclosed by the track?

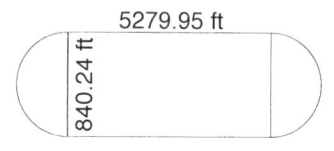

5279.95 ft

840.24 ft

Figure P6.8

6.9 Using an EDM instrument, the rectangular dimensions of a large building 1435.67 ± 0.025 ft by 453.67 ± 0.01 ft are laid out. Assuming only errors in distance observations, what is:
(a) the area enclosed by the building and its standard deviation?
(b) the perimeter of the building and its standard deviation?

6.10 A particular total station's reading error is determined to be ±2.5″. After pointing repeatedly on a distant target with the same instrument, an observer determines an error due to both pointing and reading the circles of ±3.6″. What is the observer's pointing error?

6.11 For each tape correction formula noted below, express the error propagation formula in the form of Equation (6.16) using the variables listed.
(a) $H = L \cos \alpha$, where L is the slope length and α is the slope angle. Determine the error with respect to L and α.
(b) $C_T = k(T_f - T)L$, where k is the coefficient of thermal expansion, T_f the tape's field temperature, T the calibrated temperature of the tape, and L the measured length. Determine the error with respect to T_f.
(c) $C_P = (P_f - P)L/AE$, where P_f is the field tension, P the tension calibrated for the tape, A its cross-sectional area, E the modulus of elasticity, and L the measured length. Determine the error with respect to P_f.
(d) $C_S = -w^2 l_s^3/24P_f^2$, where w is the weight per unit length of the tape, l_s the length between supports, and P_f the field tension. Determine the error with respect to P_f.

6.12 Compute the corrected distance and its expected error if the measured distance is 145.67 ft. Assume that $T_f = 45 ± 5°F$, $P_f = 16 ± 1$ lb, there was a reading error of ±0.01 ft, and that the distance was mea-

sured as two end-support distances of 100.00 ft and 86.87 ft.
(*Reminder:* Do not forget the correction for length: $C_L = [(l - l')/l']L$, where l is the actual tape length, l' its nominal length, and L the measured line length.) The tape calibration data are given as follows.

$$A = 0.004 \text{ in}^2 \qquad l = 100.012 \text{ ft}$$

$$w = 0.015 \text{ lb} \qquad l' = 100 \text{ ft}$$

$$k = 0.00000645°F \qquad E = 29{,}000{,}000 \text{ lb/in}^2$$

$$P = 10 \text{ lb} \qquad T = 68°F$$

6.13 Show that Equation (6.12) is equivalent to Equation (6.11) for linear equations.

6.14 Derive an expression similar to Equation (6.9) for the function $z = a_1x_1 + a_2x_2 + a_3x_3$.

6.15 The elevation of point C on the chimney shown in Figure 6.3 is desired. Field angles and distances are observed. Station A has an elevation of 345.618 ± 0.008 m and station B has an elevation of 347.758 ± 0.008 m. The instrument height, hi_A, at station A is 1.249 ± 0.003 m, and the instrument height, hi_B, at station B is 1.155 ± 0.003 m. Zenith angles are read in the field. The other observations and their errors are

$$AB = 93.505 \pm 0.006 \text{ m}$$

$$A = 44°12'34'' \pm 7.9'' \qquad B = 39°26'56'' \pm 9.8''$$

$$z_1 = 81°41'06'' \pm 12.3'' \qquad z_2 = 84°10'25'' \pm 11.6''$$

What are the elevation of the chimney and the standard deviation in this elevation?

Practical Exercises

6.16 With an engineer's scale, measure the radius of the circle in the Figure P6.16 ten times using different starting locations on the scale. Use a magnifying glass and interpolate the readings on the scale to a tenth of the smallest graduated reading on the scale.
   (a) What are the mean radius of the circle and its standard deviation?
   (b) Compute the area of the circle and its standard deviation.

(c) Calibrate a planimeter by measuring a 2-in. square. Calculate the mean constant for the planimeter (k = units/4 in^2), and based on 10 measurements, determine the standard deviation in the constant.

(d) Using the same planimeter, measure the area of the circle and determine its standard deviation.

Figure P6.16

6.17 Develop a computational worksheet that solves Problem 6.11.

CHAPTER 7

ERROR PROPAGATION IN ANGLE AND DISTANCE OBSERVATIONS

7.1 INTRODUCTION

All surveying observations are subject to errors from varying sources. For example, when observing an angle, the major error sources include instrument placement and leveling, target placement, circle reading, and target pointing. Although great care may be taken in observing the angle, these error sources will render inexact results. To appreciate fully the need for adjustments, surveyors must be able to identify the major observational error sources, know their effects on the measurements, and understand how they can be modeled. In this chapter, emphasis is placed on analyzing the errors in observed horizontal angles and distances. In Chapter 8 the manner in which these errors propagate to produce traverse misclosures is studied. In Chapter 9 the propagation of angular errors in elevation determination is covered.

7.2 ERROR SOURCES IN HORIZONTAL ANGLES

Whether a transit, theodolite, or total station instrument is used, errors are present in every horizontal angle observation. Whenever an instrument's circles are read, a small error is introduced into the final angle. Also, in pointing to a target, a small amount of error always occurs. Other major error sources in angle observations include instrument and target setup errors and the instrument leveling error. Each of these sources produces random errors. They may be small or large, depending on the instrument, the operator, and the conditions at the time of the angle observation. The effects of reading, pointing, and leveling errors can be reduced by increasing the number of angle repetitions. However, the effects of instrument and target setup errors can be reduced only by increasing sight distances.

7.3 READING ERRORS

Errors in reading conventional transits and theodolites depend on the quality of the instrument's optics, the size of the smallest division of the circle, and the operator's abilities: for example, the ability to set and read a transit vernier, or to set and read the micrometer of a theodolite. Typical reading errors for a 1" micrometer theodolite can range from tenths of a second to several seconds. Reading errors also occur with digital instruments, their size being dependent on the sensitivity of the particular electronic angular resolution system. Manufacturers quote the estimated combined pointing and reading precision for an individual direction measured face I (direct) and face II (reversed) with their instruments in terms of standard deviations. Typical values range from ±1" for the more precise instruments to ±10" for the less expensive ones. These errors are random, and their effects on an angle depend on the observation method and the number of repeated observations.

7.3.1 Angles Observed by the Repetition Method

When observing a horizontal angle by repetition using a repeating instrument, the circle is first zeroed so that angles can be accumulated on the horizontal circle. The angle is turned a number of times, and finally, the cumulative angle is read and divided by the number of repetitions, to determine the average angular value. In this method, a reading error exists in just two positions, regardless of the number of repetitions. The first reading error occurs when the circle is zeroed and the second when reading the final cumulative angle. For this procedure, the average angle is computed as

$$\alpha = \frac{\alpha_1 + \alpha_2 + \cdots + \alpha_n}{n} \qquad (a)$$

where α is the average angle, and $\alpha_1, \alpha_2, \ldots, \alpha_n$ are the n repetitions of the angle. Recognizing that readings occur only when zeroing the plates and reading the final direction α_n and applying Equation (6.16) to Equation (a), the standard error in reading the angle using the repetition method is

$$\sigma_{\alpha_r} = \frac{\sqrt{\sigma_0^2 + \sigma_r^2}}{n} \qquad (7.1)$$

where σ_{α_r} is the error in the average angle due to reading, σ_0 the estimated error in setting zero on the circle, σ_r the estimated error in the final reading, and n the number of repetitions of the angle. Note that the number of repetitions should always be an even number, with half being turned face I (direct) and half face II (reversed). This procedure compensates for systematic instrumental errors.

Assuming that the observer's ability to set zero and to read the circle are equal, Equation (7.1) is simplified to

$$\sigma_{\alpha_r} = \frac{\sigma_r \sqrt{2}}{n} \tag{7.2}$$

Example 7.1 Suppose that an angle is turned six times using the repetition method. For an observer having a personal reading error of $\pm 1.5''$, what is the error in the final angle due to circle reading?

SOLUTION From Equation (7.2),

$$\sigma_{\alpha_r} = \pm \frac{1.5'' \sqrt{2}}{6} = \pm 0.4''$$

7.3.2 Angles Observed by the Directional Method

When a horizontal angle is observed by the directional method, the horizontal circle is read in both the backsight and foresight directions. The angle is then the difference between the two readings. Multiple observations of the angle are made, with the circle being advanced prior to each reading to compensate for the systematic errors. The final angle is taken as the average of all values observed. Again, an even number of repetitions are made, with half taken in the face I and half in the face II position. Since each repetition of the angle requires two readings, the error in the average angle due to the reading error is computed using Equation (6.16), which yields

$$\sigma_{\alpha_r} = \frac{\sqrt{(\sigma_{r_{1b}}^2 + \sigma_{r_{1f}}^2) + (\sigma_{r_{2b}}^2 + \sigma_{r_{2f}}^2) + \cdots + (\sigma_{r_{nb}}^2 + \sigma_{r_{nf}}^2)}}{n} \tag{7.3}$$

where $\sigma_{r_{ib}}$ and $\sigma_{r_{if}}$ are the estimated errors in reading the circle for the backsight and foresight directions, respectively, and n is the number of repetitions. Assuming that one's ability to read the circle is independent of the particular direction, so that $\sigma_{r_{ib}} = \sigma_{r_{if}} = \sigma_r$, Equation (7.3) simplifies to

$$\sigma_{\alpha_r} = \frac{\sigma_r \sqrt{2}}{\sqrt{n}} \tag{7.4}$$

Example 7.2 Using the same parameters of six repetitions and an estimated observer reading error of $\pm 1.5''$ as given in Example 7.1, find the error estimated in the average angle due to reading when the directional method is used.

SOLUTION

$$\sigma_{\alpha_p} = \pm \frac{1.5'' \sqrt{2}}{\sqrt{6}} = \pm 0.9''$$

Note that the additional readings required in the directional method produce a larger error in the angle than that obtained using the repetition method.

7.4 POINTING ERRORS

Accuracy in pointing to a target depends on several factors. These include the optical qualities of the instrument, target size, the observer's personal ability to place the crosswires on a target, and the weather conditions at the time of observation. Pointing errors are random, and they will occur in every angle observation no matter the method used. Since each repetition of an angle consists of two pointings, the pointing error for an angle that is the mean of n repetitions can be estimated using Equation (6.16) as

$$\sigma_{\alpha_p} = \frac{\sqrt{2\sigma_{p1}^2 + 2\sigma_{p2}^2 + \cdots + 2\sigma_{pn}^2}}{n} \tag{7.5}$$

where σ_{α_p} is the error due to pointing and $\sigma_{p1}, \sigma_{p2}, \ldots, \sigma_{pn}$ are the estimated errors in pointings for the first repetition, second repetition, and so on. Again for a given instrument and observer, the pointing error can be assumed the same for each repetition (i.e., $\sigma_{p1} = \sigma_{p2} = \cdots = \sigma_{pn} = \sigma_p$), and Equation (7.5) simplifies to

$$\sigma_{\alpha_p} = \frac{\sigma_p \sqrt{2}}{\sqrt{n}} \tag{7.6}$$

Example 7.3 An angle is observed six times by an observer whose ability to point on a well-defined target is estimated to be $\pm 1.8''$. What is the estimated error in the average angle due to the pointing error?

SOLUTION From Equation (7.6),

$$\sigma_{\alpha_p} = \pm \frac{1.8''\sqrt{2}}{\sqrt{6}} = \pm 1.0''$$

7.5 ESTIMATED POINTING AND READING ERRORS WITH TOTAL STATIONS

With the introduction of electronic theodolites and subsequently, total station instruments, new standards were developed for estimating errors in angle observations. The new standards, called DIN 18723, provide values for estimated errors in the mean of two direction observations, one each in the face I and face II positions. Thus, in terms of a single pointing and reading error, σ_{pr}, the DIN value, σ_{DIN}, can be expressed as

$$\sigma_{DIN} = \frac{\sigma_{pr}\sqrt{2}}{2} = \frac{\sigma_{pr}}{\sqrt{2}}$$

Using this equation, the expression for the estimated error in the observation of a single direction due to pointing and reading with an electronic theodolite is

$$\sigma_{pr} = \sigma_{DIN}\sqrt{2} \qquad (b)$$

Using a procedure similar to that given in Equation (7.6), the estimated error in an angle measured n times and averaged due to pointing and reading is

$$\sigma_{\alpha_{pr}} = \frac{\sigma_{pr}\sqrt{2}}{\sqrt{n}} \qquad (c)$$

Substituting Equation (b) into Equation (c) yields

$$\sigma_{\alpha_{pr}} = \frac{2\sigma_{DIN}}{\sqrt{n}} \qquad (7.7)$$

Example 7.4 An angle is observed six times by an operator with a total station instrument having a published DIN 18723 value for the pointing and reading error of $\pm 5''$. What is the estimated error in the angle due to the pointing and reading error?

104 ERROR PROPAGATION IN ANGLE AND DISTANCE OBSERVATIONS

SOLUTION From Equation (7.7),

$$\sigma_{\alpha_{pr}} = \frac{2 \times 5''}{\sqrt{6}} = \pm 4.1''$$

7.6 TARGET CENTERING ERRORS

Whenever a target is set over a station, there will be some error due to faulty centering. This can be attributed to environmental conditions, optical plummet errors, quality of the optics, plumb bob centering error, personal abilities, and so on. When care is taken, the instrument is usually within 0.001 to 0.01 ft of the true station location. Although these sources produce a constant centering error for any particular angle, it will appear as random in the adjustment of a network involving many stations since targets and instruments will center differently over a point. This error will also be noticed in resurveys of the same points.

An estimate of the effect of this error in an angle observation can be made by analyzing its contribution to a single direction. As shown in Figure 7.1, the angular error due to the centering error depends on the position of the target. If the target is on line but off center, as shown in Figure 7.1(a), the target centering error does not contribute to the angular error. However, as the target moves to either side of the sight line, the error size increases. As shown in Figure 7.1(d), the largest error occurs when the target is offset perpendicular to the line of sight. Letting σ_d represent the distance the target

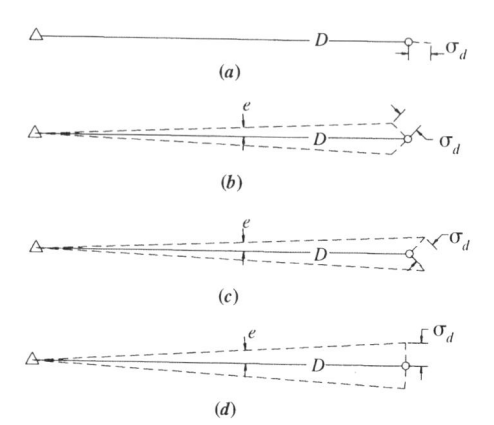

Figure 7.1 Possible target locations.

is from the *true* station location, from Figure 7.1(*d*), the maximum error in an individual direction due to the target centering error is

$$e = \pm \frac{\sigma_d}{D} \quad \text{rad} \tag{7.8}$$

where e is the uncertainty in the direction due to the target centering error, σ_d the amount of a centering error at the time of pointing, and as shown in Figure 7.2, D is the distance from the instrument center to the target.

Since two directions are required for each angle observation, the contribution of the target centering error to the total angular error is

$$\sigma_{\alpha_t} = \sqrt{\left(\frac{\sigma_{d_1}}{D_1}\right)^2 + \left(\frac{\sigma_{d_2}}{D_2}\right)^2} \tag{7.9}$$

where σ_{α_t} is the angular error due to the target centering error, σ_{d_1} and σ_{d_2} are the target centering errors at stations 1 and 2, respectively, and D_1 and D_2 are the distances from the target to the instrument at stations 1 and 2, respectively. Assuming the ability to center the target over a point is independent of the particular direction, it can be stated that $\sigma_{d_1} = \sigma_{d_2} = \sigma_t$. Finally, the results of Equation (7.9) are unitless. To convert the result to arc seconds, it must be multiplied by the constant ρ (206,264.8″/rad), which yields

$$\sigma''_{\alpha_t} = \pm \frac{\sqrt{D_1^2 + D_2^2}}{D_1 D_2} \sigma_t \rho \tag{7.10}$$

Notice that the same target centering error occurs on each pointing. Thus, it cannot be reduced in size by taking multiple pointings, and therefore Equation (7.10) is not divided by the number of angle repetitions. This makes the target centering error one of the more significant errors in angle observations.

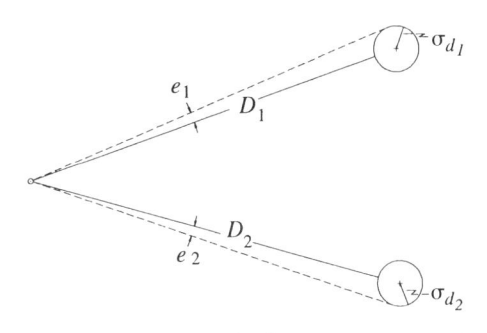

Figure 7.2 Error in an angle due to target centering error.

It also shows that the only method to decrease the size of this error is to increase the sight distances.

Example 7.5 An observer's estimated ability at centering targets over a station is ±0.003 ft. For a particular angle observation, the backsight and foresight distances from the instrument station to the targets are approximately 250 ft and 450 ft, respectively. What is the angular error due to the error in target centering?

SOLUTION   From Equation (7.10), the estimated error is

$$\sigma''_{\alpha_t} = \pm \frac{\sqrt{250^2 + 450^2}}{250 \times 450} 0.003 \times 206{,}264.8''/\text{rad} = \pm 2.8''$$

If handheld range poles were used in this example with an estimated centering error of ±0.01 ft, the estimated angular error due to the target centering would be

$$\sigma''_{\alpha_t} = \pm \frac{\sqrt{250^2 + 450^2}}{250 \times 450} 0.01 \times 206{,}264.8''/\text{rad} = \pm 9.4''$$

Obviously, this is a significant error source if care is not taken in target centering.

7.7   INSTRUMENT CENTERING ERRORS

Every time an instrument is centered over a point, there is some error in its position with respect to the true station location. This error is dependent on the quality of the instrument and the state of adjustment of its optical plummet, the quality of the tripod, and the skill of surveyor. The error can be compensating, as shown in Figure 7.3(a), or it can be maximum when the instrument is on the angle bisector, as shown in Figure 7.3(b) and (c). For any individual setup, this error is a constant; however, since the instrument's location is random with respect to the true station location, it will appear to be random in the adjustment of a network involving many stations. Like the target centering error, it will appear also during a resurvey of the points. From Figure 7.3, the true angle α is

$$\alpha = (P_2 + \varepsilon_2) - (P_1 + \varepsilon_1) = (P_2 - P_1) + (\varepsilon_2 - \varepsilon_1)$$

where P_1 and P_2 are the true directions and ε_1 and ε_2 are errors in those directions due to faulty instrument centering. The error size for any setup is

7.7 INSTRUMENT CENTERING ERRORS   107

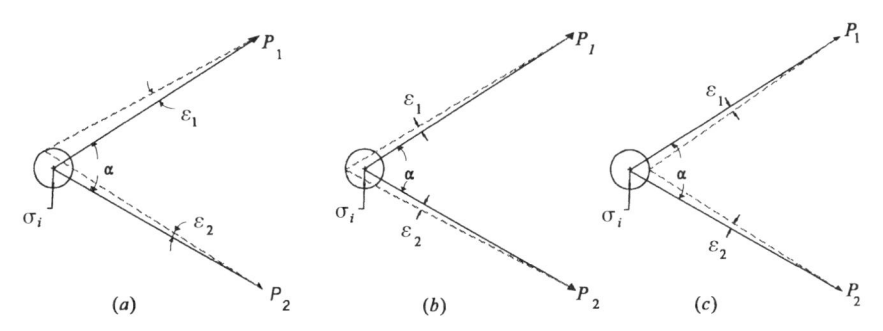

Figure 7.3 Error in angle due to error in instrument centering.

$$\varepsilon = \varepsilon_2 - \varepsilon_1 \tag{7.11}$$

The error in the observed angle due to instrument centering errors is analyzed by propagating errors in a formula based on (x,y) coordinates. In Figure 7.4 a coordinate system has been constructed with the x axis going from the true station to the foresight station. The y axis passes through the instrument's vertical axis and is perpendicular to the x axis. From the figure the following equations can be derived:

$$ih = ip - qr \tag{7.12}$$
$$ih = iq \cos \alpha - sq \sin \alpha$$

Letting $sq = x$ and $iq = y$, Equation (7.12) can be rewritten as

$$ih = y \cos \alpha - x \sin \alpha \tag{7.13}$$

Furthermore, in Figure 7.4,

$$\varepsilon_1 = \frac{ih}{D_1} = \frac{y \cos \alpha - x \sin \alpha}{D_1} \tag{7.14}$$

$$\varepsilon_2 = \frac{y}{D_2} \tag{7.15}$$

By substituting Equations (7.14) and (7.15) into Equation (7.11), the error in an observed angle due to the instrument centering error is

$$\varepsilon = \frac{y}{D_2} - \frac{y \cos \alpha - x \sin \alpha}{D_1} \tag{7.16}$$

Reorganizing Equation (7.16) yields

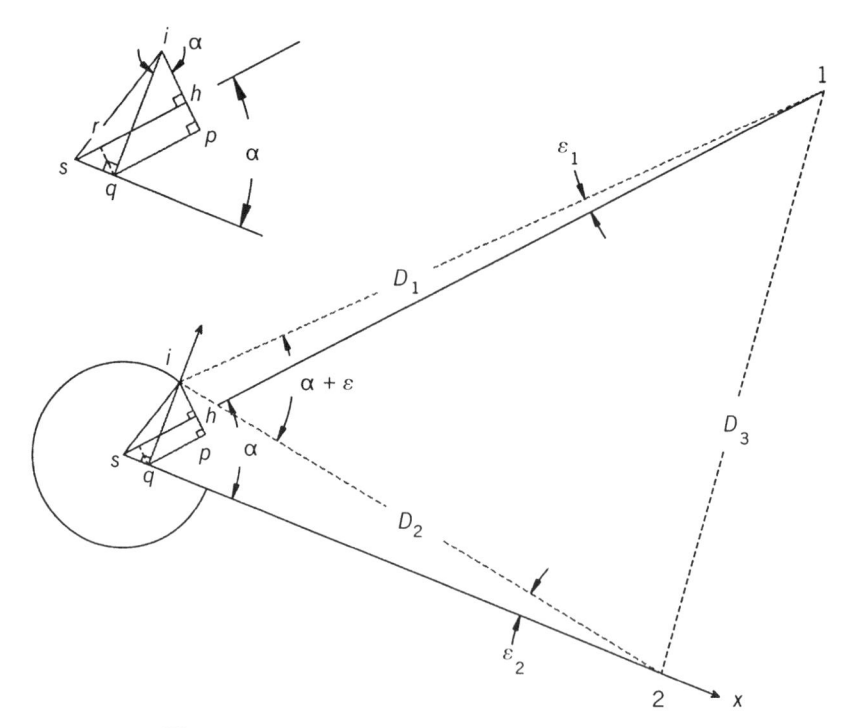

108 ERROR PROPAGATION IN ANGLE AND DISTANCE OBSERVATIONS

Figure 7.4 Analysis of instrument centering error.

$$\varepsilon = \frac{D_1 y + D_2 x \sin \alpha - D_2 y \cos \alpha}{D_1 D_2} \tag{7.17}$$

Now because the instrument's position is truly random, Equation (6.16) can be used to find the angular uncertainty due to the instrument centering error. Taking the partial derivative of Equation (7.17) with respect to both x and y gives

$$\frac{\partial \varepsilon}{\partial x} = \frac{D_2 \sin \alpha}{D_1 D_2}$$
$$\frac{\partial \varepsilon}{\partial y} = \frac{D_1 - D_2 \cos \alpha}{D_1 D_2} \tag{7.18}$$

Now substituting the partial derivatives in Equation (7.18) into Equation (6.16) gives

$$\sigma_\varepsilon^2 = \frac{D_2 \sin \alpha}{D_1 D_2} \sigma_x^2 + \frac{D_1 - D_2 \cos \alpha}{D_1 D_2} \sigma_y^2 \tag{7.19}$$

7.7 INSTRUMENT CENTERING ERRORS

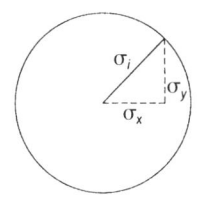

Figure 7.5 Instrument centering errors at a station.

Because this error is a constant for a setup, the mean angle has the same error as a single angle, and thus it is not reduced by taking several repetitions. The estimated error in the position of a station is derived from a *bivariate distribution*,[1] where the coordinate components are independent and have equal magnitudes. Assuming that estimated errors in the x and y axes are σ_x and σ_y, from Figure 7.5 it is seen that

$$\sigma_x = \sigma_y = \frac{\sigma_i}{\sqrt{2}}$$

Letting $\sigma_\varepsilon = \sigma_{\alpha_i}$, expanding the squares of Equation (7.19), and rearranging yields

$$\sigma_{\alpha_i} = \frac{D_1^2 + D_2^2 (\cos^2\alpha + \sin^2\alpha) - 2D_1 D_2 \cos\alpha}{D_1^2 D_2^2} \frac{\sigma_i^2}{2} \tag{7.20}$$

Making the trigonometric substitutions of $\cos^2\alpha + \sin^2\alpha = 1$ and $D_1^2 + D_2^2 - 2D_1 D_2 \cos\alpha = D_3^2$ in Equation (7.20), taking the square root of both sides, and multiplying by ρ (206,264.8″/rad) to convert the results to arc seconds yields

$$\sigma''_{\alpha_i} = \pm \frac{D_3}{D_1 D_2} \frac{\sigma_i}{\sqrt{2}} \rho \tag{7.21}$$

Example 7.6 An observer centers the instrument to within ±0.005 ft of a station for an angle with backsight and foresight distances of 250 ft and 450 ft, respectively. The angle observed is 50°. What is the error in the angle due to the instrument centering error?

---
[1] The bivariate distribution is discussed in Chapter 19.

SOLUTION Using the cosine law, $D_3^2 = D_1^2 + D_2^2 - 2D_1D_2 \cos \angle$, and substituting in the appropriate values, we find D_3 to be

$$D_3 = \sqrt{250^2 + 450^2 - 2 \times 250 \times 450 \times \cos 50°} = 346.95 \text{ ft}$$

Substituting this value into Equation (7.21), the estimated contribution of the instrument centering error to the overall angular error is

$$\sigma''_{\alpha_i} = \pm \frac{346.95}{250 \times 450} \frac{0.005}{\sqrt{2}} 206{,}264.8''/\text{rad} = \pm 2.2''$$

7.8 EFFECTS OF LEVELING ERRORS IN ANGLE OBSERVATIONS

If an instrument is imperfectly leveled, its vertical axis is not truly vertical and its horizontal circle and horizontal axis are both inclined. If while an instrument is imperfectly leveled it is used to measure horizontal angles, the angles will be observed in a plane other than horizontal. Errors that result from this error source are most severe when the backsights and foresights are steeply inclined: for example, in making astronomical observations or traversing over mountains. If the bubble of a theodolite were to remain off center by the same amount during the entire angle-observation process, the resulting error would be systematic. However, because an operator normally monitors the bubble carefully and attempts to keep it centered while turning angles, the amount and direction by which the instrument is out of level becomes random, and hence the resulting errors tend to be random. Even if the operator does not monitor the instrument's level, this error will appear to be random in a resurvey.

In Figure 7.6, ε represents the angular error that occurs in either the backsight or foresight of a horizontal angle observation made with an instrument out of level and located at station I. The line of sight IS is elevated by the vertical angle v. In the figure, IS is shown perpendicular to the instrument's horizontal axis. The amount by which the instrument is out of level is $f_d\mu$, where f_d is the number of fractional divisions the bubble is off center and μ is the sensitivity of the bubble. From the figure,

$$SP = D \tan v \qquad (d)$$

and

$$PP' = D\varepsilon \qquad (e)$$

7.8 EFFECTS OF LEVELING ERRORS IN ANGLE OBSERVATIONS

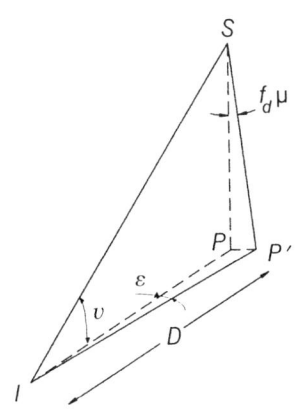

Figure 7.6 Effects of instrument leveling error.

where D is the horizontal component of the sighting distance and the angular error ε is in radians. Because the amount of leveling error is small, PP' can be approximated as a circular arc, and thus

$$PP' = f_d\mu(SP) \qquad (f)$$

Substituting Equation (d) into Equation (f) yields

$$PP' = f_d\mu D \tan v \qquad (7.22)$$

Now substituting Equation (7.22) into Equation (e) and reducing, the error in an individual pointing due to the instrument leveling error is

$$\varepsilon = f_d\mu \tan v \qquad (7.23)$$

As noted above, Figure 7.6 shows the line of sight oriented perpendicular to the instrument's horizontal axis. Also, the direction in which a bubble runs is random. Thus, Equation (6.18) can be used to compute the combined angular error that results from n repetitions of an angle made with an imperfectly leveled instrument (note that each angle measurement involves both backsight and foresight pointings):

$$\sigma_{\alpha l} = \pm \frac{\sqrt{(f_d\mu \tan v_b)^2 + (f_d\mu \tan v_f)^2}}{\sqrt{n}} \qquad (7.24)$$

where v_b and v_f are the vertical angles to the backsight and foresight targets, respectively, and n is the number of repetitions of the angle.

112 ERROR PROPAGATION IN ANGLE AND DISTANCE OBSERVATIONS

Example 7.7 A horizontal angle is observed on a mountainside where the backsight is to the peak and the foresight is in the valley. The average zenith angles to the backsight and foresight are 80° and 95°, respectively. The instrument has a level bubble with a sensitivity of 30″/div and is leveled to within 0.3 div. For the average angle obtained from six repetitions, what is the contribution of the leveling error to the overall angular error?

SOLUTION The zenith angles converted to vertical angles are +10° and −5°, respectively. Substituting the appropriate values into Equation (7.24) yields

$$\sigma_{\alpha l} = \pm \frac{\sqrt{[0.3 \text{ div}(30''/\text{div}) \tan 10°]^2 + [0.3 \text{ div}(30''/\text{div}) \tan(-5°)]^2}}{\sqrt{6}}$$

$$= \pm 0.7''$$

This error is generally small for traditional surveying work when normal care is taken in leveling the instrument. Thus, it can generally be ignored for all but the most precise work. However, as noted earlier, for astronomical observations this error can become quite large, due to the steeply inclined sights to celestial objects. Thus, for astronomical observations it is extremely important to keep the instrument leveled precisely for each observation.

7.9 NUMERICAL EXAMPLE OF COMBINED ERROR PROPAGATION IN A SINGLE HORIZONTAL ANGLE

Example 7.8 Assume that an angle is observed four times with a directional-type instrument. The observer has an estimated reading error of ±1″ and a pointing error of ±1.5″. The targets are well defined and placed on an optical plummet tribrach with an estimated centering error of ±0.003 ft. The instrument is in adjustment and centered over the station to within ±0.003 ft. The horizontal distances from the instrument to the backsight and foresight targets are approximately 251 ft and 347 ft, respectively. The average angle is 65°37′12″. What is the estimated error in the angle observation?

SOLUTION The best way to solve this type of problem is to computed estimated errors for each item in Sections 7.3 to 7.8 individually, and then apply Equation (6.18).
  Error due to reading: Substituting the appropriate values into Equation (7.4) yields

7.9 NUMERICAL EXAMPLE OF COMBINED ERROR PROPAGATION

$$\sigma_{\alpha_r} = \pm \frac{1''\sqrt{2}}{\sqrt{4}} = \pm 0.71''$$

Error due to pointing: Substituting the appropriate values into Equation (7.6) yields

$$\sigma_{\alpha_p} = \pm \frac{1.5''\sqrt{2}}{\sqrt{4}} = \pm 1.06''$$

Error due to target centering: Substituting the appropriate values into Equation (7.10) yields

$$\sigma''_{\alpha_t} = \frac{\sqrt{251^2 + 347^2}}{251 \times 347} (0.003) 206{,}264.8''/\text{rad} = \pm 3.04''$$

Error due to instrument centering: From the cosine law we have

$$D_3^2 = 251^2 + 347^2 - 2(251)(347) \cos(65°37'12'')$$

$$D_3 = 334 \text{ ft}$$

Substituting the appropriate values into Equation (7.21) yields

$$\sigma''_{\alpha_i} = \pm \frac{334}{251 \times 347} \frac{0.003}{\sqrt{2}} 206{,}264.8''/\text{rad} = \pm 1.68''$$

Combined error: From Equation (6.18), the estimated angular error is

$$\sigma_\alpha = \sqrt{0.71^2 + 1.06^2 + 3.04^2 + 1.68^2} = \pm 3.7''$$

In Example 7.8, the largest error sources are due to the target and instrument centering errors, respectively. This is true even when the estimated error in centering the target and instrument are only ±0.003 ft. Unfortunately, many surveyors place more confidence in their observations than is warranted. Since these two error sources do not decrease with increased repetitions, there is a limit to what can be expected from any survey. For instance, assume that the targets were handheld reflector poles with an estimated centering error of ±0.01 ft. Then the error due to the target centering error becomes ±10.1''. This results in an estimated angular error of ±10.3''. If a 99% probable error were computed, a value as large as ±60'' would be possible!

7.10 USE OF ESTIMATED ERRORS TO CHECK ANGULAR MISCLOSURE IN A TRAVERSE

When a traverse is closed geometrically, the angles are generally checked for misclosure. By computing the errors for each angle in the traverse as described in Section 7.9 and summing the results with Equation (6.18), an estimate for the size of the angular misclosure is obtained. The procedure is best demonstrated with an example.

Example 7.9 Assume that each of the angles in Figure 7.7 was observed using four repetitions (twice direct and twice reverse) and their estimated errors were computed as shown in Table 7.1. Does this traverse meet acceptable angular closure at a 95% level of confidence?

SOLUTION The *actual* angular misclosure of the traverse is 30″. The *estimated* angular misclosure of the traverse is found by applying Equation (6.18) with the errors computed for each angle. That is, the estimated angular misclosure is

$$\sigma_{\Sigma\angle} = \sqrt{8.9^2 + 12.1^2 + 13.7^2 + 10.0^2 + 9.9^2} = \pm 24.7''$$

Thus, the actual angular misclosure of 30″ is greater than the value estimated (24.7″) at a 68.3% probable error level. However, since each angle was turned only four times, a 95% probable error must be computed by using the appropriate t value from Table D.3.

This problem begs the question of what the appropriate number of degrees of freedom is for the summation of the angles. Only four of the angles are required in the summation since the fifth angle can be computed from the other four and thus is redundant. Since each angle is turned four times, it can be argued that there are 16 redundant observations: that is, 12 angles at the first four stations and four at the fifth station. However, this assumes that instrumental systematic errors were not present in the observational process since only the average of a face I and a face II reading can eliminate system-

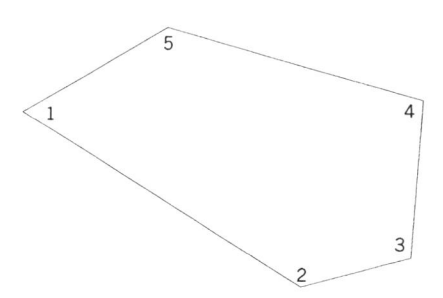

Figure 7.7 Close polygon traverse.

7.10 USE OF ESTIMATED ERRORS TO CHECK ANGULAR MISCLOSURE IN A TRAVERSE

TABLE 7.1 Data for Example 7.9

Angle	Observed Value	Computed Angular Error
1	60°50'48"	±8.9"
2	134°09'24"	±12.1"
3	109°00'12"	±13.7"
4	100°59'54"	±10.0"
5	135°00'12"	±9.9"

atic errors in the instrument. If there are n angles and each angle is turned r times, the total number of redundant observations would be $n(r - 1) + 1$. In this case it would be $5(4 - 1) + 1 = 16$.

A second approach is to account for instrumental systematic errors when counting redundant observations. This method requires that an angle exists only if it is observed with both faces of the instrument. In this case there is one redundant angle at each of the first four stations, with the fifth angle having two redundant observations, for a total of six redundant observations. Using this argument, the number of redundant angles in the traverse would be $n(r/2 - 1) + 1$. In this example it would be $5(4/2 - 1) + 1 = 6$.

A third approach would be to consider each mean angle observed at each station to be a single observation, since only mean observations are being used in the computations. In this case there would be only one redundant angle for the traverse. However, had horizon closures been observed at each station, the additional angles would add n redundant observations.

A fourth approach would be to determine the 95% probable error at each station and then use Equation (6.18) to sum these 95% error values. In this example, each station has three redundant observations. In general, there would be $r - 1$ redundant angle observations, where r represents the number of times that the angle was repeated during the observation process.

The last two methods are the most conservative since they allow the most error in the sum of the angles. The fourth method is used in this book. However, a surveyor must decide which method is most appropriate for his or her practice. As stated in Chapter 5, the statistician must make decisions when performing any test. Using the fourth method, there are three redundant observation at each station. To finish the problem, we construct a 95% confidence interval, or perform a two-tailed test to determine the range of error that is statistically equal to zero. In this case, $t_{0.025,3} = 3.183$, the 95% probable error for the angular sum is

$$\sigma_{95\%} = 3.183 \times 24.7" = \pm 78.6"$$

Thus, the traverse angles are well within the range of allowable error. We cannot reject the null hypothesis that the error in the angles in not statistically equal to zero. Thus, the survey meets the minimum level of angular closure at a 95% probable error. However, it must be remembered that because of the possibility of Type II errors, we can only state that there is no statistical reason to believe that there is a blunder in the angle observations.

Example 7.9 presents another question for the statistician or surveyor. That is, should a surveyor allow a field crew to have this large an angular misclosure in the traverse? Statistically, the answer would seem to be yes, but it must remembered that the target and instrument centering errors affect angle observations only if the instrument and targets are reset after each observation. Since this is never done in practice, these two errors should not be included in the summation of the angles. Instead, the allowable angular misclosure should be based solely on pointing and reading errors. For example, if the angles were observed with a total station having a DIN 18723 standard of $\pm 1''$, by Equation (7.7) the pointing and reading error for each angle would be

$$\sigma_{\alpha_{pr}} = \pm \frac{2 \times 1''}{\sqrt{2}} = \pm 1.4''$$

By Equation (6.19), the error in the summation of the five angles would be $\pm 1.4'' \sqrt{5} = \pm 3.2''$. Using the same critical t value of 3.183, the allowable error in the angular misclosure should only be $3.183 \times 3.2'' = \pm 10''$. If this instrument had be used in Example 7.9, the field-observed angular closure of $30''$ would be unacceptable and would warrant reobservation of some or all of the angles.

As stated in Sections 7.6 and 7.7, the angular misclosure of $78.6''$ computed in Example 7.9 will be noticed only when the target and instrument are reset on a survey. This will happen during the resurvey, when the centering errors of the target and instrument from the original survey will be present in the *record* directions. Thus, record azimuths and/or bearings could disagree from those determined in the resurvey by this amount, assuming that the equipment used in the resurvey is comparable or of higher quality than that used in the original survey.

7.11 ERRORS IN ASTRONOMICAL OBSERVATIONS FOR AN AZIMUTH

The total error in an azimuth determined from astronomical observations depends on errors from several sources, including those in timing, the observer's latitude and longitude, the celestial object's position at observation time, timing accuracy, observer response time, instrument optics, atmospheric condi-

7.11 ERRORS IN ASTRONOMICAL OBSERVATIONS FOR AN AZIMUTH

tions, and others, as identified in Section 7.2. The error in an astronomical observation can be estimated by analyzing the hour–angle formula, which is

$$z = \tan^{-1} \frac{\sin t}{\cos \phi \tan \delta - \sin \phi \cos t} \qquad (7.25)$$

In Equation (7.25), z is the azimuth of the celestial object at the time of the observation, t the t angle of the *PZS* triangle at the time of observation, ϕ the observer's latitude, and δ the object's declination at the time of the observation.

The t angle is a function of the *local hour angle* (LHA) of the sun or a star at the time of observation. That is, when the LHA $< 180°$, $t =$ LHA; otherwise, $t = 360° -$ LHA. Furthermore, LHA is a function of the *Greenwich hour angle* (GHA) of the celestial body and the observer's longitude; that is,

$$\text{LHA} = \text{GHA} + \lambda \qquad (7.26)$$

where λ is the observer's longitude, considered positive for eastern longitude and negative for western longitude. The GHA increases approximately $15°$ per hour of time, and thus an estimate of the error in the GHA is approximately

$$\sigma_t = 15° \times \sigma_T$$

where σ_T is the estimated error in time (in hours). Similarly, by using the declination at 0^h and 24^h, the amount of change in declination per second can be derived and thus the estimated error in declination determined.

Using Equation (6.16), the error in a star's azimuth is estimated by taking the partial derivative of Equation (7.25) with respect to t, δ, ϕ, and λ. To do this, simplify Equation (7.25) by letting

$$F = \cos \phi \tan \delta - \sin \phi \cos t \qquad (7.27a)$$

and

$$u = \sin t \times F^{-1} \qquad (7.27b)$$

Substituting in Equations (7.27), Equation (7.25) is rewritten as

$$z = \tan^{-1} \frac{\sin t}{F} = \tan^{-1} u \qquad (7.28)$$

From calculus it is known that

118 ERROR PROPAGATION IN ANGLE AND DISTANCE OBSERVATIONS

$$\frac{d \tan^{-1} u}{dx} = \frac{1}{1 + u^2} \frac{du}{dx}$$

Applying this fundamental relation to Equation (7.28) and letting G represent GHA yields

$$\frac{\partial z}{\partial G} = \frac{1}{1 + [\sin(G - \lambda)/F]^2} \frac{du}{dG} = \frac{F^2}{F^2 + \sin^2(G - \lambda)} \frac{du}{dG} \quad (7.29)$$

Now du/dG is

$$\frac{du}{dG} = \frac{\cos(G - \lambda)}{F} - \frac{\sin(G - \lambda)}{F^2} \sin \phi \sin(G - \lambda)$$

$$= \frac{\cos(G - \lambda)}{F} - \frac{\sin^2(G - \lambda) \sin \phi}{F^2}$$

and thus,

$$\frac{du}{dG} = \frac{F \cos(G - \lambda) - \sin^2(G - \lambda) \sin \phi}{F^2} \quad (7.30)$$

Substituting Equation (7.30) into Equation (7.29) and substituting in t for $G - \lambda$ yields

$$\frac{\partial z}{\partial G} = \frac{F \cos t - \sin^2 t \sin \phi}{F^2 + \sin^2 t} \quad (7.31)$$

In a similar fashion, the following partial derivatives are developed from Equation (7.25):

$$\frac{dz}{d\delta} = -\frac{\sin t \cos \phi}{\cos^2 \delta (F^2 + \sin^2 t)} \quad (7.32)$$

$$\frac{\partial z}{\partial \phi} = \frac{\sin t \cos t \cos \phi + \sin t \sin \phi \tan \delta}{F^2 + \sin^2 t} \quad (7.33)$$

$$\frac{\partial z}{\partial \lambda} = \frac{\sin^2 t \sin \phi - F \cos t}{F^2 + \sin^2 t} \quad (7.34)$$

where t is the t angle of the *PZS* triangle, z the celestial object's azimuth, δ the celestial object's declination, ϕ the observer's latitude, λ the observer's longitude, and $F = \cos \phi \tan \delta - \sin \phi \cos t$.

7.11 ERRORS IN ASTRONOMICAL OBSERVATIONS FOR AN AZIMUTH

If the horizontal angle, H, is the angle to the right observed from the line to the celestial body, the equation for a line's azimuth is

$$Az = z + 360° - H$$

Therefore, the error contributions from the horizontal angle observation must be included in computing the overall error in the azimuth. Since the distance to the star is considered infinite, the estimated contribution to the angular error due to the instrument centering error can be determined with a formula similar to that for the target centering error with one pointing. That is,

$$\sigma_{\alpha_i} = \frac{\sigma_i}{D} \qquad (7.35)$$

where σ_i is the centering error in the instrument and D is the length of the azimuth line in the same units. Note that the results of Equation (7.35) are in radian units and must be multiplied by ρ to yield a value in arc seconds.

Example 7.10 Using Equation (7.25), the azimuth to Polaris was found to be $0°01'31.9''$. The observation time was 1:00:00 UTC with an estimated error of $\sigma_T = \pm 0.5^s$. The Greenwich hour angles to the star at 0^h and 24^h UTC were $243°27'05.0''$ and $244°25'50.0''$, respectively. The LHA at the time of the observation was $181°27'40.4''$. The declinations at 0^h and 24^h were $89°13'38.18''$ and $89°13'38.16''$, respectively. At the time of observation, the declination was $89°13'38.18''$. The clockwise horizontal angle measured from the backsight to a target 450.00 ft was $221°25'55.9''$. The observer's latitude and longitude were scaled from a map as $40°13'54''$N and $77°01'51.5''$W, respectively, with estimated errors of $\pm 1''$. The vertical angle to the star was $39°27'33.1''$. The observer's estimated errors in reading and pointing are $\pm 1''$ and $\pm 1.5''$, respectively, and the instrument was leveled to within 0.3 of a division with a bubble sensitivity of $25''$/div. The estimated error in instrument and target centering is ± 0.003 ft. What are the azimuth of the line and its estimated error? What is the error at the 95% level of confidence?

SOLUTION The azimuth of the line is $Az = 0°01'31.9'' + 360 - 221°25'55.9'' = 138°35'36''$. Using the Greenwich hour angles at 0^h and 24^h, an error of 0.5^s time will result in an estimated error in the GHA of

$$\pm \frac{360° + (244°25'50.0'' - 243°27'05.0'')}{24^h \times 3600^{s/h}} \, 0.05^s = \pm 7.52''$$

Since $t = 360° - LHA = 178°32'19.6''$, F in Equations (7.29) through (7.34) is

$$F = \cos(40°13'54'') \tan(89°13'38.18'') - \sin(40°13'54'') \cos(178°32'19.6'')$$
$$= 57.249$$

The error in the observed azimuth can be estimated by computing the individual error terms as follows:

(a) From Equation (7.31), the error with respect to the GHA, G, is

$$\frac{\partial z}{\partial G} \sigma_G =$$

$$\frac{57.249 \cos(178°32'19.6'') - \sin^2(178°32'19.6'') \sin(40°13'54'')}{57.249^2 + \sin^2(178°32'19.6'')} (7.52'')$$

$$= \pm 0.13''$$

(b) By observing the change in declination, it is obvious that for this observation, the error in a time of 0.5^s is insignificant. In fact, for the entire day, the declination changes only 0.02''. This situation is common for stars. However, the sun's declination may change from only a few seconds daily to more than 23 minutes per day, and thus for solar observations, this error term should not be ignored.

(c) From Equation (7.33) the error with respect to latitude, ϕ, is

$$\frac{\partial z}{\partial \phi} \sigma_\phi = \pm \frac{\sin t \cos t \cos \phi + \sin t \sin \phi \tan \delta}{F^2 + \sin^2 t} \sigma_\phi = \pm 0.0004''$$

(d) From Equation (7.34) the error with respect to longitude, λ, is

$$\frac{\partial z}{\partial \lambda} =$$

$$\pm \frac{\sin^2(178°32'19.6'') \sin(40°13'54'') - 57.249 \cos(178°32'19.6'')}{57.249^2 + \sin^2(178°32'19.6'')} (1'')$$

$$= \pm 0.02''$$

(e) The circles are read both when pointing on the star and on the azimuth mark. Thus, from Equation (7.2), the reading contribution to the estimated error in the azimuth is

$$\sigma_{\alpha_r} = \pm \sigma_r \sqrt{2} = \pm 1'' \sqrt{2} = \pm 1.41''$$

(f) Using Equation (7.6), the estimated error in the azimuth due to pointing is

$$\sigma_{\alpha_p} = \pm \sigma_p \sqrt{2} = \pm 1.5'' \sqrt{2} = \pm 2.12''$$

(g) From Equation (7.8), the estimated error in the azimuth due to target centering is

$$\sigma''_{\alpha_t} = \pm \frac{d}{D} = \pm \left(\frac{0.003}{450}\right) 206{,}264.8''/\text{rad} = \pm 1.37''$$

(h) Using Equation (7.35), the estimated error in the azimuth due to instrument centering is

$$\sigma''_{\alpha_i} = \pm \frac{d}{D} = \pm \left(\frac{0.003}{450}\right) 206{,}264.8''/\text{rad} = \pm 1.37''$$

(i) From Equation (7.23), the estimated error in the azimuth due to the leveling error is

$$\sigma_{\alpha_b} = \pm f_d \mu \tan v = \pm 0.3 \times 25'' \tan(39°27'33.1'') = \pm 6.17''$$

Parts (a) through (i) are the errors for each individual error source. Using Equation (6.18), the estimated error in the azimuth observation is

$$\sigma_{AZ} = \sqrt{\begin{array}{c}(0.13'')^2 + (1.32'')^2 + (0.02'')^2 + (0.0004'')^2 \\ + (2.12'')^2 + 2(1.37'')^2 + (6.17'')^2\end{array}}$$

$$= \pm 7.0''$$

Using the appropriate t value of $t_{0.025,1}$ from Table D.3, the 95% error is

$$\sigma_{Az} = \pm 12.705 \times 7.0'' = \pm 88.9''$$

Notice that in this problem, the largest error source in the azimuth error is caused by the instrument leveling error.

7.12 ERRORS IN ELECTRONIC DISTANCE OBSERVATIONS

All EDM observations are subject to instrumental errors that manufacturers list as constant, a, and scalar, b, error. A typical specified accuracy is $\pm(a + b \text{ ppm})$. In this expression, a is generally in the range 1 to 10 mm, and b is a scalar error that typically has the range 1 to 10 ppm. Other errors involved in electronic distance observations stem from the target and instrument centering errors. Since in any survey involving several stations these errors tend to be random, they should be combined using Equation (6.18). Thus, the estimated error in an EDM observed distance is

$$\sigma_D = \sqrt{\sigma_i^2 + \sigma_t^2 + a^2 + (D \times b \text{ ppm})^2} \qquad (7.36)$$

where σ_D is the error in the observed distance D, σ_i the instrument centering error, σ_t the reflector centering error, and a and b the instrument's specified accuracy parameters.

Example 7.11 A distance of 453.87 ft is observed using an EDM with a manufacturer's specified accuracy of $\pm(5 \text{ mm} + 10 \text{ ppm})$. The instrument is centered over the station with an estimated error of ± 0.003 ft, and the reflector, which is mounted on a handheld prism pole, is centered with an estimated error of ± 0.01 ft. What is the error in the distance? What is the E_{95} value?

SOLUTION Converting millimeters to feet using the survey foot[2] definition gives us

$$0.005 \text{ m} \times 39.37 \text{ in.}/12 \text{ in.} = 0.0164 \text{ ft}$$

The scalar portion of the manufacturer's estimated standard error is computed as

$$\frac{\text{distance} \times b}{1,000,000}$$

In this example, the error is $453.87 \times 10/1,000,000 = 0.0045$ ft. Thus, according to Equation (7.36), the distance error is

$$\sigma = \sqrt{(0.003)^2 + (0.01)^2 + (0.164)^2 + (0.0045)^2} = \pm 0.02 \text{ ft}$$

Using the appropriate t value from Table 3.2, the 95% probable error is

$$E_{95} = 1.6449 \sigma = \pm 0.03 \text{ ft}$$

Notice in this example that the instrument's constant error is the largest single contributor to the overall error in the observed distance, and it is followed closely by the target centering error. Furthermore, since both errors are constants, their contribution to the total error is unchanged regardless of the distance. Thus, for this particular EDM instrument, distances under 200 ft could probably be observed more accurately with a calibrated steel tape. However, this statement depends on the terrain and the skill of the surveyors in using a steel tape.

---

[2] The survey foot definition is 1 meter = 39.37 inches, exactly.

7.13 USE OF COMPUTATIONAL SOFTWARE

The computations demonstrated in this chapter are rather tedious and time consuming when done by hand, and this often leads to mistakes. This problem, and many others in surveying that involve repeated computations of a few equations with different values, can be done conveniently with a spreadsheet, worksheet, or program. On the CD that accompanies this book, the electronic book prepared with Mathcad demonstrates the programming of the computational examples in this chapter. When practicing the following problems, the reader should consider writing software to perform the aforementioned computations.

PROBLEMS

7.1 Plot a graph of vertical angles from 0° to 50° versus the error in horizontal angle measurement due to an instrument leveling error of 5″.

7.2 For a direction with sight distances to the target of 100, 200, 300, 400, 600, 1000, and 1500 ft, construct:
 (a) a table of estimated standard deviations due to target centering when $\sigma_d = \pm 0.005$ ft.
 (b) a plot of distance versus the standard deviations computed in part (a).

7.3 For an angle of size 125° with equal sight distances to the target of 100, 200, 300, 400, 600, 1000, and 1500 ft, construct:
 (a) a table of standard deviations due to instrument centering when $\sigma_i = \pm 0.005$ ft.
 (b) a plot of distance versus the standard deviations computed in part (a).

7.4 Assuming setup errors of $\sigma_i = \pm 0.002$ m and $\sigma_t = \pm 0.005$ m, what is the estimated error in a distance of length 684.326 m using an EDM with stated accuracies of 3 mm + 3 ppm?

7.5 Repeat Problem 7.4 for a distance of length 1304.597 m.

7.6 Assuming setup errors of $\sigma_i = \pm 0.005$ ft and $\sigma_t = \pm 0.005$ ft, what is the estimated error in a distance of length 1234.08 ft using an EDM with stated accuracies of 3 mm + 3 ppm?

7.7 Repeat Problem 7.6 for a target centering error of 0.02 ft and a distance of length 423.15 ft.

7.8 A 67°13′46″ angle having a backsight length of 312.654 m and a foresight length of 205.061 m is observed, with the total station twice having a stated DIN 18723 accuracy of ±3". Assuming instrument and target centering errors of $\sigma_i = \pm 0.002$ m and $\sigma_t = \pm 0.005$ m, what is the estimated error in the angle?

7.9 Repeat Problem 7.8 for an angle of 107°07'39" observed four times, a backsight length of 306.85 ft, a foresight length of 258.03 ft, instrument and target centering errors of $\sigma_i = \pm 0.005$ ft and $\sigma_t = \pm 0.01$ ft, respectively, and an instrument with a stated DIN 18723 accuracy of ±2".

7.10 For the following traverse data, compute the estimated error for each angle if $\sigma_{DIN} = \pm 2"$, $\sigma_i = \pm 0.005$ ft, $\sigma_t = \pm 0.01$ ft, and the angles were each measured four times (twice direct and twice reverse). Does the traverse meet acceptable angular closures at a 95% level of confidence?

Station	Angle	Distance (ft)
A	62°33'11"	221.85
B	124°56'19"	346.55
C	60°44'08"	260.66
D	111°46'07"	349.17
A		

7.11 A total station with a DIN 18723 value of ±3" was used to turn the angles in Problem 7.10. Do the problem assuming the same estimated errors in instrument and target centering.

7.12 A total station with a DIN 18723 value of ±5" was used to turn the angles in Problem 7.10. Do the problem assuming the same estimated errors in instrument and target centering.

7.13 For the following traverse data, compute the estimated error in each angle if $\sigma_r = \pm 3"$, $\sigma_p = \pm 2"$, $\sigma_i = \sigma_t = \pm 0.005$ ft, and the angles were observed four times (twice direct and twice reverse) using the repetition method. Does the traverse meet acceptable angular closures at a 95% level of confidence?

Station	Angle	Distance (ft)
A	38°58'24"	321.31
B	148°53'30"	276.57
C	84°28'06"	100.30
D	114°40'24"	306.83
E	152°59'18"	255.48
A		

7.14 A total station with a DIN 18723 value of ±2″ was used to turn the angles in Problem 7.13. Repeat the problem for this instrument.

7.15 An EDM was used to measure the distances in Problem 7.10. The manufacturer's specified errors for the instrument are ±(3 mm + 3 ppm). Using $\sigma_i = \pm 0.005$ ft and $\sigma_r = \pm 0.01$ ft, calculate the error in each distance.

7.16 Repeat Problem 7.15 for the distances in Problem 7.13. The manufacturer's specified error for the instrument was ±(5 mm + 5 ppm). Use σ_i and σ_r from Problem 7.13.

7.17 The following observations and calculations were made on a sun observation to determine the azimuth of a line:

Observation No.	UTC	Horizontal Angle	Vertical Angle	δ	LHA	z
1	16:30:00	41°02′33″	39°53′08″	−3°28′00.58″	339°54′05.5″	153°26′51.8″
2	16:35:00	42°35′28″	40°16′49″	−3°28′05.43″	341°09′06.5″	154°59′42.4″
3	16:40:00	44°09′23″	40°39′11″	−3°28′10.27″	342°24′07.5″	156°33′39.0″
4	16:45:00	45°44′25″	41°00′05″	−3°28′15.11″	343°39′08.5″	158°08′39.2″
5	16:50:00	47°20′21″	41°19′42″	−3°28′19.96″	344°54′09.5″	159°44′40.2″
6	16:55:00	48°57′24″	41°37′47″	−3°28′24.80″	346°09′10.5″	161°21′38.9′

The Greenwich hour angles for the day were 182°34′06.00″ at 0^h UT, and 182°38′53.30″ at 24^h UT. The declinations were −3°12′00.80″ at 0^h and −3°35′16.30″ at 24^h. The observer's latitude and longitude were scaled from a map as 43°15′22″ and 90°13′18″, respectively, with an estimated standard error of ±1″ for both values. Stopwatch times were assumed to be correct to within a error of ±0.5^s. A Roelof's prism was used to take pointings on the center of the sun. The target was 535 ft from the observer's station. The observer's estimated reading and pointing errors were ±1.2″ and ±1.8″, respectively. The instrument was leveled to within 0.3 div on a level bubble with a sensitivity of 20″/div. The target was centered to within an estimated error of ±0.003 ft of the station. What is:

(a) the average azimuth of the line and its standard deviation?

(b) the estimated error of the line at 95% level of confidence?

(c) the largest error contributor in the observation?

7.18 The following observations were made on the sun.

Pointing	UT Time	Horizontal Angle	Zenith Angle
1	13:01:27	179°16'35"	56°00'01"
2	13:03:45	179°40'25"	55°34'11"
3	13:08:58	180°35'19"	54°35'36"
4	13:11:03	180°57'28"	54°12'12"
5	13:16:53	182°00'03"	53°06'47"
6	13:18:23	182°16'23"	52°50'05"

The Greenwich hour angles for the day were 178°22'55.20" at 0^h and 178°22'58.70" at 24^h. The declinations were 19°25'44.40° at 0^h and 19°12'18.80" at 24^h. The observer's latitude and longitude were scaled from a map as 41°18'06" and 75°00'01", respectively, with an estimated error of ±1" for both. Stopwatch times were assumed to be correct to within an estimated error of ±0.2^s. A Roelof's prism was used to take pointings on the center of the sun. The target was 335 ft from the observer's station. The observer's estimated reading error was ±1.1" and the estimated pointing error was ±1.6". The instrument was leveled to within 0.3 div on a level bubble with a sensitivity of 30"/div. The target was centered to within an estimated error of ±0.003 ft of the station. What is:

(a) the average azimuth of the line and its standard deviation?

(b) the estimated error of the line at 95% level of confidence?

(c) the largest error contributor in the observation?

Programming Problems

7.19 Create a computational package that will compute the errors in angle observations. Use the package to compute the estimated errors for the angles in Problem 7.11.

7.20 Create a computational package that will compute the errors in EDM observed distances. Use the package to solve Problem 7.16.

7.21 Create a computational package that will compute the reduced azimuths and their estimated errors from astronomical observations. Use the package to solve Problem 7.19.

CHAPTER 8

ERROR PROPAGATION IN TRAVERSE SURVEYS

8.1 INTRODUCTION

Even though the specifications for a project may allow lower accuracies, the presence of blunders in observations is never acceptable. Thus, an important question for every surveyor is: How can I tell when blunders are present in the data? In this chapter we begin to address that question, and in particular, stress traverse analysis. The topic is discussed further in Chapter 20.

In Chapter 6 it was shown that the estimated error in a function of observations depends on the individual errors in the observations. Generally, observations in horizontal surveys (e.g., traverses) are independent. That is, the measurement of a distance observation is independent of the azimuth observation. But the latitude and departure of a line, which are computed from the distance and azimuth observations, are not independent. Figure 8.1 shows the effects of errors in distance and azimuth observations on the computed latitude and departure. In the figure it can be seen that there is correlation between the latitude and departure; that is, if either distance or azimuth observation changes, it causes changes in both latitude and departure.

Because the observations from which latitudes and departures are computed are assumed to be independent with no correlation, the SLOPOV approach [Equation (6.16)] can be used to determine the estimated error in these computed values. However, for proper computation of estimated errors in functions that use these computed values (i.e., latitudes and departures), the effects of correlation must be considered, and thus the GLOPOV approach [Equation (6.13)] will be used.

128 ERROR PROPAGATION IN TRAVERSE SURVEYS

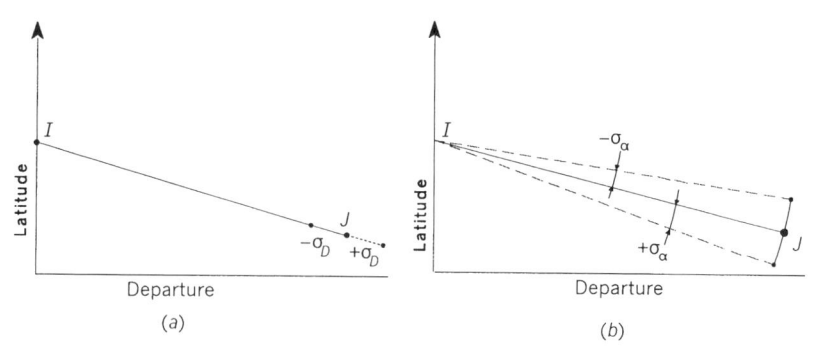

Figure 8.1 Latitude and departure uncertainties due to (a) the distance error (σ_D) and (b) the azimuth error σ_{Az}. Note that if either the distance or azimuth changes, both the latitude and departure of the course are affected.

8.2 DERIVATION OF ESTIMATED ERROR IN LATITUDE AND DEPARTURE

When computing the latitude and departure of a line, the following well-known equations are used:

$$\text{Lat} = D \cos Az \qquad (8.1)$$
$$\text{Dep} = D \sin Az$$

where Lat is the latitude, Dep the departure, Az the azimuth, and D the horizontal length of the line. To derive the estimated error in the line's latitude or departure, the following partial derivatives from Equation (8.1) are required in using Equation (6.16):

$$\frac{\partial \text{Lat}}{\partial D} = \cos Az \qquad \frac{\partial \text{Lat}}{\partial Az} = -D \sin Az$$
$$\frac{\partial \text{Dep}}{\partial D} = \sin Az \qquad \frac{\partial \text{Dep}}{\partial Az} = D \cos Az \qquad (8.2)$$

Example 8.1 A traverse course has a length of 456.87 ± 0.02 ft and an azimuth of 23°35'26" ± 9". What are the latitude and departure and their estimated errors?

SOLUTION Using Equation (8.1), the latitude and departure of the course are

$$\text{Lat} = 456.87 \cos(23°35'26") = 418.69 \text{ ft}$$

$$\text{Dep} = 456.87 \sin(23°35'26") = 182.84 \text{ ft}$$

The estimated errors in these values are solved using matrix Equation (6.16) as

8.3 DERIVATION OF ESTIMATED STANDARD ERRORS IN COURSE AZIMUTHS

$$\Sigma_{\text{Lat,Dep}} = \begin{bmatrix} \dfrac{\partial \text{Lat}}{\partial D} & \dfrac{\partial \text{Lat}}{\partial \text{Az}} \\ \dfrac{\partial \text{Dep}}{\partial D} & \dfrac{\partial \text{Dep}}{\partial \text{Az}} \end{bmatrix} \begin{bmatrix} \sigma_D^2 & 0 \\ 0 & \sigma_{\text{Az}}^2 \end{bmatrix} \begin{bmatrix} \dfrac{\partial \text{Lat}}{\partial D} & \dfrac{\partial \text{Dep}}{\partial D} \\ \dfrac{\partial \text{Lat}}{\partial \text{Az}} & \dfrac{\partial \text{Dep}}{\partial \text{Az}} \end{bmatrix} = \begin{bmatrix} \sigma_{\text{Lat}}^2 & \sigma_{\text{Lat,Dep}} \\ \sigma_{\text{Lat,Dep}} & \sigma_{\text{Dep}}^2 \end{bmatrix}$$

Substituting partial derivatives into the above yields

$$\Sigma_{\text{Lat,Dep}} = \begin{bmatrix} \cos \text{Az} & -D \sin \text{Az} \\ \sin \text{Az} & D \cos \text{Az} \end{bmatrix} \begin{bmatrix} 0.02^2 & 0 \\ 0 & (9''/\rho)^2 \end{bmatrix} \begin{bmatrix} \cos \text{Az} & \sin \text{Az} \\ -D \sin \text{Az} & D \cos \text{Az} \end{bmatrix} \quad (8.3)$$

Entering in the appropriate numerical values into Equation (8.3), the *covariance matrix* is

$$\Sigma_{\text{Lat,Dep}} = \begin{bmatrix} 0.9167 & -456.87(0.4002) \\ 0.4002 & 456.87(0.9164) \end{bmatrix} \begin{bmatrix} 0.0004 & 0 \\ 0 & (9''/\rho)^2 \end{bmatrix}$$

$$\begin{bmatrix} 0.9167 & 0.4002 \\ -456.87(0.4002) & 456.87(0.9164) \end{bmatrix}$$

from which

$$\Sigma_{\text{Lat,Dep}} = \begin{bmatrix} 0.00039958 & 0.00000096 \\ 0.00000096 & 0.00039781 \end{bmatrix} \quad (8.4)$$

In Equation (8.4), σ_{11}^2 is the variance of the latitude, σ_{22}^2 the variance of the departure, and σ_{12} and σ_{21} their covariances. Thus, the standard errors are

$$\sigma_{\text{Lat}} = \sqrt{\sigma_{11}^2} = \sqrt{0.00039958} = \pm 0.020 \text{ ft} \quad \text{and}$$

$$\sigma_{\text{Dep}} = \sqrt{\sigma_{22}^2} = \sqrt{0.00039781} = \pm 0.020 \text{ ft}$$

Note that the off-diagonal of $\Sigma_{\text{Lat,Dep}}$ is not equal to zero, and thus the computed values are correlated as illustrated in Figure 8.1.

8.3 DERIVATION OF ESTIMATED STANDARD ERRORS IN COURSE AZIMUTHS

Equation (8.1) is based on the azimuth of a course. In practice, however, traverse azimuths are normally computed from observed angles rather than being measured directly. Thus, another level of error propagation exists in calculating the azimuths from angular values. In the following analysis, consider that *angles to the right* are observed and that azimuths are computed in a *counterclockwise* direction successively around the traverse using the formula

$$Az_C = Az_P + 180° + \theta_i \tag{8.5}$$

where Az_C is the azimuth for the current course, Az_P the previous course azimuth, and θ_i the appropriate interior angle to use in computing the current course azimuth. By applying Equation (6.18), the error in the current azimuth, Az_C, is

$$\sigma_{Az_C} = \sqrt{\sigma_{Az_P}^2 + \sigma_{\theta_i}^2} \tag{8.6}$$

In Equation (8.6) σ_θ is the error in the appropriate interior angle used in computation of the current azimuth, and the other terms are as defined previously. This equation is also valid for azimuth computations going clockwise around the traverse. The proof of this is left as an exercise.

8.4 COMPUTING AND ANALYZING POLYGON TRAVERSE MISCLOSURE ERRORS

From elementary surveying it is known that the following geometric constraints exist for any closed polygon-type traverse:

$$\Sigma \text{ interior } \angle\text{'s} = (n - 2) \times 180° \tag{8.7}$$

$$\Sigma \text{ Lat} = \Sigma \text{ Dep} = 0 \tag{8.8}$$

Deviations from these conditions, normally called *misclosures*, can be calculated from the observations of any traverse. Statistical analyses can then be performed to determine the acceptability of the misclosures and check for the presence of blunders in the observations. If blunders appear to be present, the measurements must be rejected and the observations repeated. The following example illustrates methods of making these computations for any closed polygon traverse.

Example 8.2 Compute the angular and linear misclosures for the traverse illustrated in Figure 8.2. The observations for the traverse are given in Table 8.1. Determine the estimated misclosure errors at the 95% confidence level, and comment on whether or not the observations contain blunders.

SOLUTION
Angular check: First the angular misclosure is checked to see if it is within the tolerances specified. From Equation (6.18), and using the standard devi-

8.4 COMPUTING AND ANALYZING POLYGON TRAVERSE MISCLOSURE ERRORS

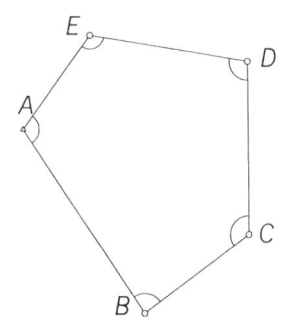

Figure 8.2 Closed polygon traverse.

ations given in Table 8.1, the angular sum should have an error within $\pm\sqrt{\sigma_{\angle 1}^2 + \sigma_{\angle 2}^2 + \cdots + \sigma_{\angle n}^2}$ 68.3% of the time. Since the angles were measured four times, each computed mean has three degrees of freedom, and the appropriate t value from Table D.3 (the t distribution) is $t_{0.025,3}$, which equals 3.183. This is a two-tailed test since we are looking for the range that is statistically equal to zero at the level of confidence selected. If this range contains the actual misclosure, there is no statistical reason to believe that the observations contain a blunder. In this case, the angular misclosure at a 95% confidence level is estimated as

$$\sigma_{\Sigma\text{Angles}} = 3.183\sqrt{3.5^2 + 3.1^2 + 3.6^2 + 3.1^2 + 3.9^2} = \pm 24.6''$$

Using the summation of the angles in Table 8.1, the actual angular misclosure in this problem is

$$540°00'19'' - (5 - 2)180° = 19''$$

Thus, the actual angular misclosure for the traverse (19″) is within its estimated range of error and there is no reason to believe that a blunder exists in the angles.

TABLE 8.1 Distance and Angle Observations for Figure 8.2

Station	Sighted	Distance (ft)	S (ft)	BS	Occupied	FS	Angle[a]	S
A	B	1435.67	0.020	E	A	B	110°24′40″	3.5″
B	C	856.94	0.020	A	B	C	87°36′14″	3.1″
C	D	1125.66	0.020	B	C	D	125°47′27″	3.6″
D	E	1054.54	0.020	C	D	E	99°57′02″	3.1″
E	A	756.35	0.020	D	E	A	116°14′56″	3.9″
							540°00′19″	

[a] Each angle was measured with four repetitions.

132 ERROR PROPAGATION IN TRAVERSE SURVEYS

Azimuth computation: In this problem, no azimuth is given for the first course. To solve the problem, however, the azimuth of the first course can be assumed as 0°00′00″ and to be free of error. This can be done even when the initial course azimuth is observed, since only geometric closure on the traverse is being checked, not the orientation of the traverse. For the data of Table 8.1, and using Equations (8.5) and (8.6), the values for the course azimuths and their estimated errors are computed and listed in Table 8.2.

Computation of estimated linear misclosure: Equation (6.13) properly accounts for correlation in the latitude and departure when computing the linear misclosure of the traverse. Applying the partial derivatives of Equation (8.2) to the latitudes and departures, the Jacobian matrix, A, has the form

$$A = \begin{bmatrix} \cos Az_{AB} & -AB \sin Az_{AB} & 0 & 0 & \cdots & 0 & 0 \\ \sin Az_{AB} & AB \cos Az_{AB} & 0 & 0 & \cdots & 0 & 0 \\ 0 & 0 & \cos Az_{BC} & -BC \sin Az_{BC} & & 0 & 0 \\ 0 & 0 & \sin Az_{BC} & BC \cos Az_{BC} & & 0 & 0 \\ \vdots & \vdots & & & \ddots & & \\ 0 & 0 & 0 & 0 & & \cos Az_{EA} & -EA \sin Az_{EA} \\ 0 & 0 & 0 & 0 & & \sin Az_{EA} & EA \cos Az_{EA} \end{bmatrix}$$

(8.9)

Because the lengths and angles were measured independently, they are uncorrelated. Thus, the appropriate covariance matrix, Σ, for solving this problem using Equation (6.16) is

$$\Sigma = \begin{bmatrix} \sigma_{AB}^2 & 0 & 0 & 0 & 0 & 0 & 0 & 0 & 0 & 0 \\ 0 & \left(\dfrac{\sigma_{Az_{AB}}}{\rho}\right)^2 & 0 & 0 & 0 & 0 & 0 & 0 & 0 & 0 \\ 0 & 0 & \sigma_{BC}^2 & 0 & 0 & 0 & 0 & 0 & 0 & 0 \\ 0 & 0 & 0 & \left(\dfrac{\sigma_{Az_{BC}}}{\rho}\right)^2 & 0 & 0 & 0 & 0 & 0 & 0 \\ 0 & 0 & 0 & 0 & \sigma_{CD}^2 & 0 & 0 & 0 & 0 & 0 \\ 0 & 0 & 0 & 0 & 0 & \left(\dfrac{\sigma_{Az_{CD}}}{\rho}\right)^2 & 0 & 0 & 0 & 0 \\ 0 & 0 & 0 & 0 & 0 & 0 & \sigma_{DE}^2 & 0 & 0 & 0 \\ 0 & 0 & 0 & 0 & 0 & 0 & 0 & \left(\dfrac{\sigma_{Az_{DE}}}{\rho}\right)^2 & 0 & 0 \\ 0 & 0 & 0 & 0 & 0 & 0 & 0 & 0 & \sigma_{EA}^2 & 0 \\ 0 & 0 & 0 & 0 & 0 & 0 & 0 & 0 & 0 & \left(\dfrac{\sigma_{Az_{EA}}}{\rho}\right)^2 \end{bmatrix}$$

(8.10)

8.4 COMPUTING AND ANALYZING POLYGON TRAVERSE MISCLOSURE ERRORS 133

TABLE 8.2 Estimated Errors in the Computed Azimuths of Figure 8.2

From	To	Azimuth	Estimated Error
A	B	0°00'00"	0"
B	C	267°36'14"	±3.1"
C	D	213°23'41"	$\sqrt{3.1^2 + 3.6^2} = \pm 4.8''$
D	E	133°20'43"	$\sqrt{4.8^2 + 3.2^2} = \pm 5.7''$
E	A	69°35'39"	$\sqrt{5.7^2 + 3.9^2} = \pm 6.9''$

Substituting numerical values for this problem into Equations (8.9) and (8.10), the covariance matrix, $\Sigma_{\text{Lat,Dep}}$, is computed for the latitudes and departures $A\Sigma A^T$, or

$$\Sigma_{\text{Lat,Dep}} = \begin{bmatrix} 0.00040 & 0 & 0 & 0 & 0 & 0 & 0 & 0 & 0 & 0 \\ 0 & 0 & 0 & 0 & 0 & 0 & 0 & 0 & 0 & 0 \\ 0 & 0 & 0.00017 & 0.00002 & 0 & 0 & 0 & 0 & 0 & 0 \\ 0 & 0 & 0.00002 & 0.00040 & 0 & 0 & 0 & 0 & 0 & 0 \\ 0 & 0 & 0 & 0 & 0.00049 & 0.00050 & 0 & 0 & 0 & 0 \\ 0 & 0 & 0 & 0 & 0.00050 & 0.00060 & 0 & 0 & 0 & 0 \\ 0 & 0 & 0 & 0 & 0 & 0 & 0.00064 & -0.00062 & 0 & 0 \\ 0 & 0 & 0 & 0 & 0 & 0 & -0.00062 & 0.00061 & 0 & 0 \\ 0 & 0 & 0 & 0 & 0 & 0 & 0 & 0 & 0.00061 & 0.00034 \\ 0 & 0 & 0 & 0 & 0 & 0 & 0 & 0 & 0.00034 & 0.00040 \end{bmatrix}$$

(8.11)

By taking the square roots of the diagonal elements in the $\Sigma_{\text{Lat,Dep}}$ matrix [Equation (8.11)], the errors for the latitude and departure of each course are found. That is, the estimated error in the latitude for course BC is the square root of the (3,3) element in Equation (8.11), and the estimated error in the departure of BC is the square root of the (4,4) element. In a similar fashion, the estimated errors in latitude and departure can be computed for any other course.

The formula for determining the linear misclosure of a closed polygon traverse is

$$LC = \sqrt{(\text{Lat}_{AB} + \text{Lat}_{BC} + \cdots + \text{Lat}_{EA})^2 + (\text{Dep}_{AB} + \text{Dep}_{BC} + \cdots + \text{Dep}_{EA})^2}$$

(8.12)

where LC is the linear misclosure. To determine the estimated error in the linear misclosure, Equation (6.16) is applied to the linear misclosure formula (8.12). The necessary partial derivatives from Equation (8.12) for substitution into Equation (6.16) must first be determined. The partial derivatives with respect to the latitude and departure of course AB are

134 ERROR PROPAGATION IN TRAVERSE SURVEYS

$$\frac{\partial LC}{\partial \text{Lat}_{AB}} = \frac{\Sigma \text{ Lats}}{LC} \qquad \frac{\partial LC}{\partial \text{Dep}_{AB}} = \frac{\Sigma \text{ Deps}}{LC} \qquad (8.13)$$

Notice that these partial derivatives are independent of the course. Also, the other courses have the same partial derivatives as given by Equation (8.13), and thus the Jacobian matrix for Equation (6.16) has the form

$$A = \begin{bmatrix} \dfrac{\Sigma \text{ Lats}}{LC} & \dfrac{\Sigma \text{ Deps}}{LC} & \dfrac{\Sigma \text{ Lats}}{LC} & \dfrac{\Sigma \text{ Deps}}{LC} & \cdots & \dfrac{\Sigma \text{ Lats}}{LC} & \dfrac{\Sigma \text{ Deps}}{LC} \end{bmatrix}$$

(8.14)

As shown in Table 8.3, the sum of the latitudes is -0.083, the sum of the departures is 0.022, and $LC = 0.086$ ft. Substituting these values into Equation (8.14), which in turn is substituted into Equation (6.16), yields

$$\Sigma_{LC} = [-0.9674 \quad 0.2531 \quad -0.9674 \quad 0.2531 \quad \cdots \quad -0.9674 \quad 0.2531]$$

$$\times \Sigma_{\text{Lat,Dep}} \begin{bmatrix} -0.9674 \\ 0.2531 \\ -0.9674 \\ 0.2531 \\ \vdots \\ -0.9674 \\ 0.2531 \end{bmatrix} = [0.00226]$$

(8.15)

In Equation (8.15), Σ_{LC} is a single-element covariance matrix that is the variance of the linear closure and can be called σ^2_{LC}. Also, $\Sigma_{\text{Lat,Dep}}$ is the matrix given by Equation (8.11). To compute the E_{95} confidence interval, a t value from Table D.3 (the t distribution) must be used with $\alpha = 0.025$ and 3 degrees

TABLE 8.3 Latitudes and Departures for Example 8.2

Course	Latitude	Departure
AB	1435.670	0
BC	−35.827	−856.191
CD	−939.812	−619.567
DE	−723.829	766.894
EA	263.715	708.886
	−0.083	0.022

$LC = \sqrt{(-0.083)^2 + (0.022)^2} = 0.086$ ft

of freedom.[1] The misclosure estimated for a traverse at a $1 - \alpha$ level of confidence is $t_{\alpha/2,3}\sigma_{LC}$. Again, we are checking to see if the traverse misclosure falls within a range of errors that are statistically equal to zero. This requires placing $\alpha/2$ into the upper and lower tails of the distribution. Thus, the error estimated in the traverse closure at a 95% level of confidence is

$$\sigma_{LC} = t_{0.025,3}\sqrt{\sigma_1^2} = 3.183\sqrt{0.00226} = \pm 0.15 \text{ ft}$$

This value is well above the actual traverse linear misclosure of 0.086 ft, and thus there is no reason to believe that the traverse contains any blunders.

In Example 8.2 we failed to reject the null hypothesis; that is, there was no statistical reason to believe that there were errors in the data. However, it is important to remember that this does necessarily imply that the observations are error-free. There is always the possibility of a Type II error. For example, if the computations were supposed to be performed on a map projection grid[2] but the observations were not reduced, the traverse would still close within acceptable tolerances. However, the results computed would be incorrect since all the distances would be either too long or too short. Another example of an undetectable systematic error is an incorrectly entered EDM–reflector constant (see Problem 2.17). Again all the distances observed would be either too long or too short, but the traverse misclosure would still be within acceptable tolerances.

Surveyors must always be aware of instrumental systematic errors and follow proper field and office procedures to remove these errors. As discussed, simply passing a statistical test does not imply directly that the observations are error- or mistake-free. However, when the test fails, only a Type I error can occur at an α level of confidence. Depending on the value of α, a failed test can be a strong indicator of problems within the data.

8.5 COMPUTING AND ANALYZING LINK TRAVERSE MISCLOSURE ERRORS

As illustrated in Figure 8.3, a link traverse begins at one station and ends on a different one. Normally, they are used to establish the positions of inter-

---

[1] A closed polygon traverse has $2(n - 1)$ unknown coordinates with $2n + 1$ observations, where n is the number of traverse sides. Thus, the number of degrees of freedom in a simple closed traverse is always $2n + 1 - 2(n - 1) = 3$. For a five-sided traverse there are five angle and five distance observations plus one azimuth. Also, there are four stations each having two unknown coordinates, thus $11 - 8 = 3$ degrees of freedom.

[2] Readers who wish to familiarize themselves with map projection computations should refer to Appendix F and the CD that accompanies this book.

136 ERROR PROPAGATION IN TRAVERSE SURVEYS

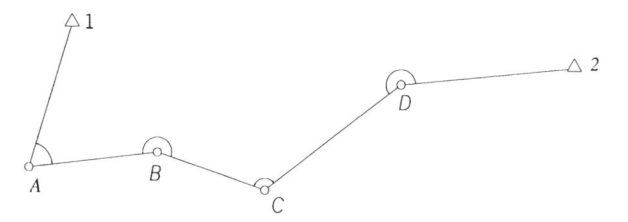

Figure 8.3 Closed link traverse.

mediate stations, as in A through D of the figure. The coordinates at the endpoints, stations 1 and 2 of the figure, are known. Angular and linear misclosures are also computed for these types of traverse, and the resulting values are used as the basis for accepting or rejecting the observations. Example 8.3 illustrates the computational methods.

Example 8.3 Compute the angular and linear misclosures for the traverse illustrated in Figure 8.3. The data observed for the traverse are given in Table 8.4. Determine the estimated misclosures at the 95% confidence level, and comment on whether or not the observations contain blunders.

SOLUTION

Angular misclosure: In a link traverse, angular misclosure is found by computing initial azimuths for each course and then subtracting the final com-

TABLE 8.4 Data for Link Traverse in Example 8.3

Distance observations

From	To	Distance (ft)	S (ft)
1	A	1069.16	±0.021
A	B	933.26	±0.020
B	C	819.98	±0.020
C	D	1223.33	±0.021
D	2	1273.22	±0.021

Angle observations

BS	Occ	FS	Angle	S (")
1	A	B	66°16'35"	±4.9
A	B	C	205°16'46"	±5.5
B	C	D	123°40'19"	±5.1
C	D	2	212°00'55"	±4.6

Control stations

Station	X (ft)	Y (ft)
1	1248.00	3979.00
2	4873.00	3677.00

Azimuth observations

From	To	Azimuth	S (")
1	A	197°04'47"	±4.3
D	2	264°19'13"	±4.1

8.5 COMPUTING AND ANALYZING LINK TRAVERSE MISCLOSURE ERRORS

TABLE 8.5 Computed Azimuths and Their Uncertainties

Course	Azimuth	σ (″)
1A	197°04′47″	±4.3
AB	83°21′22″	±6.5
BC	108°38′08″	±8.5
CD	52°18′27″	±9.9
D2	84°19′22″	±11.0

puted azimuth from its given counterpart. The initial azimuths and their estimated errors are computed using Equations (8.5) and (8.6) and are shown in Table 8.5.

The difference between the azimuth computed for course D2 (84°19′22″) and its actual value (264°19′13″ − 180°) is +9″. Using Equation (6.18), the estimated error in the difference is $\sqrt{11.0^2 + 4.1^2} = \pm 11.7″$, and thus there is no reason to assume that the angles contain blunders.

Linear misclosure: First the actual traverse misclosure is computed using Equation (8.1). From Table 8.6, the total change in latitude for the traverse is −302.128 ft and the total change in departure is 3624.968 ft. From the control coordinates, the cumulative change in X and Y coordinate values is

$$\Delta X = X_2 - X_1 = 4873.00 - 1248.00 = 3625.00 \text{ ft}$$

$$\Delta Y = Y_2 - Y_1 = 3677.00 - 3979.00 = -302.00 \text{ ft}$$

The actual misclosures in departure and latitude are computed as

$$\Delta \text{Dep} = \Sigma \text{ Dep} - (X_2 - X_1) = 3624.968 - 3625.00 = -0.032 \quad (8.16)$$
$$\Delta \text{Lat} = \Sigma \text{ Lat} - (Y_2 - Y_1) = -302.128 - (-302.00) = -0.128$$

TABLE 8.6 Computed Latitudes and Departures

Course	Latitude (ft)	Departure (ft)
1A	−1022.007	−314.014
AB	107.976	926.993
BC	−262.022	776.989
CD	747.973	968.025
D2	125.952	1266.975
	−302.128	3624.968

138 ERROR PROPAGATION IN TRAVERSE SURVEYS

In Equation (8.16), ΔDep represents the misclosure in departure and ΔLat represents the misclosure in latitude. Thus, the linear misclosure for the traverse is

$$LC = \sqrt{\Delta \text{Lat}^2 + \Delta \text{Dep}^2} = \sqrt{(-0.128)^2 + (-0.032)^2} = 0.132 \text{ ft} \quad (8.17)$$

Estimated misclosure for the traverse: Following procedures similar to those described earlier for polygon traverses, the estimated misclosure in this link traverse is computed. The Jacobian matrix of the partial derivative for the latitude and departure with respect to distance and angle observations is

$$A = \begin{bmatrix} \cos Az_{1A} & -1A \sin Az_A & 0 & 0 & \cdots & 0 & 0 \\ \sin Az_{1A} & 1A \cos Az_{1A} & 0 & 0 & \cdots & 0 & 0 \\ 0 & 0 & \cos Az_{AB} & -AB \sin Az_{AB} & & 0 & 0 \\ 0 & 0 & \sin Az_{AB} & AB \cos Az_{AB} & & 0 & 0 \\ \vdots & \vdots & & & \ddots & & \\ 0 & 0 & 0 & 0 & & \cos Az_{D2} & -D2 \sin Az_{D2} \\ 0 & 0 & 0 & 0 & & \sin Az_{D2} & D2 \cos Az_{D2} \end{bmatrix}$$

(8.18)

Similarly, the corresponding covariance matrix Σ in Equation (6.16) has the form

$$\Sigma = \begin{bmatrix} \sigma_{1A}^2 & 0 & 0 & 0 & \cdots & 0 & 0 \\ 0 & \left(\dfrac{\sigma_{Az_{1A}}}{\rho}\right)^2 & 0 & 0 & \cdots & 0 & 0 \\ 0 & 0 & \sigma_{AB}^2 & 0 & \cdots & 0 & 0 \\ 0 & 0 & 0 & \left(\dfrac{\sigma_{Az_{AB}}}{\rho}\right)^2 & & 0 & 0 \\ 0 & 0 & 0 & 0 & \ddots & 0 & 0 \\ 0 & 0 & 0 & 0 & & \sigma_{D2}^2 & 0 \\ 0 & 0 & 0 & 0 & \cdots & 0 & \left(\dfrac{\sigma_{Az_{D2}}}{\rho}\right)^2 \end{bmatrix} \quad (8.19)$$

Substituting the appropriate numerical values into Equations (8.18) and (8.19) and applying Equation (6.16), the covariance matrix is

8.5 COMPUTING AND ANALYZING LINK TRAVERSE MISCLOSURE ERRORS

$\Sigma_{Lat,Dep} =$

$$\begin{bmatrix} 0.00045 & -0.00016 & 0 & 0 & 0 & 0 & 0 & 0 & 0 & 0 \\ -0.00016 & 0.00049 & 0 & 0 & 0 & 0 & 0 & 0 & 0 & 0 \\ 0 & 0 & 0.00086 & -0.00054 & 0 & 0 & 0 & 0 & 0 & 0 \\ 0 & 0 & -0.00054 & 0.00041 & 0 & 0 & 0 & 0 & 0 & 0 \\ 0 & 0 & 0 & 0 & 0.00107 & 0.00023 & 0 & 0 & 0 & 0 \\ 0 & 0 & 0 & 0 & 0.00023 & 0.00048 & 0 & 0 & 0 & 0 \\ 0 & 0 & 0 & 0 & 0 & 0 & 0.00234 & -0.00147 & 0 & 0 \\ 0 & 0 & 0 & 0 & 0 & 0 & -0.00147 & 0.00157 & 0 & 0 \\ 0 & 0 & 0 & 0 & 0 & 0 & 0 & 0 & 0.00453 & -0.00041 \\ 0 & 0 & 0 & 0 & 0 & 0 & 0 & 0 & -0.00041 & 0.00048 \end{bmatrix}$$

To estimate the error in the traverse misclosure, Equation (6.16) must be applied to Equation (8.18). As was the case for closed polygon traverse, the terms of the Jacobian matrix are independent of the course for which they are determined, and thus the Jacobian matrix has the form

$$A = \begin{bmatrix} \dfrac{\Delta \text{Lat}}{LC} & \dfrac{\Delta \text{Dep}}{LC} & \dfrac{\Delta \text{Lat}}{LC} & \dfrac{\Delta \text{Dep}}{LC} & \cdots & \dfrac{\Delta \text{Lat}}{LC} & \dfrac{\Delta \text{Dep}}{LC} \end{bmatrix} \quad (8.20)$$

Following procedures similar to those used in Example 8.2, the estimated standard error in the misclosure of the link traverse is

$$\Sigma_{LC} = A\Sigma_{Lat,Dep}A^T = [0.00411]$$

From these results and using a t value from Table D.3 for 3 degrees of freedom, the estimated linear misclosure error for a 95% level of confidence is

$$\sigma_{95\%} = 3.183\sqrt{0.00411} = \pm 0.20 \text{ ft}$$

Because the actual misclosure of 0.13 ft is within the range of values that are statistically equal to zero at the 95% level (± 0.20 ft), there is no reason to believe that the traverse observations contain any blunders. Again, this test does not remove the possibility of a Type II error occurring.

This example leads to an interesting discussion. When using traditional methods of adjusting link traverse data, such as the compass rule, the control is assumed to be perfect. However, since control coordinates are themselves derived from observations, they contain errors that are not accounted for in these computations. This fact is apparent in Equation (8.20), where the coordinate values are assumed to have no error and thus are not represented.

140   ERROR PROPAGATION IN TRAVERSE SURVEYS

These equations can easily be modified to consider the control errors, but this is left as an exercise for the student.

One of the principal advantages of the least squares adjustment method is that it allows application of varying weights to the observations, and control can be included in the adjustment with appropriate weights. A full discussion of this subject is presented in Section 21.6.

8.6 CONCLUSIONS

In this chapter, propagation of observational errors through traverse computations has been discussed. Error propagation is a powerful tool for the surveyor, enabling an answer to be obtained for the question: What is an acceptable traverse misclosure? This is an example of surveying *engineering*. Surveyors are constantly designing measurement systems and checking their results against personal or legal standards. The subjects of error propagation and detection of measurement blunders are discussed further in later chapters.

PROBLEMS

8.1 Show that Equation (8.6) is valid for clockwise computations about a traverse.

8.2 Explain the significance of the standard error in the azimuth of the first course of a polygon traverse.

8.3 Given a course with an azimuth of $105°27'44''$ with an estimated error of $\pm 5''$ and a distance of 638.37 ft with an estimated error of ± 0.02 ft, what are the latitude and departure and their estimated errors?

8.4 Given a course with an azimuth of $272°14'08''$ with an estimated error of $\pm 9.2''$ and a distance of 215.69 ft with an estimated error of ± 0.016 ft, what are the latitude and departure and their estimated errors?

8.5 Given a course with an azimuth of $328°49'06''$ with an estimated error of $\pm 4.4''$ and a distance of 365.977 m with an estimated error of ± 6.5 mm, what are the latitude and departure and their estimated errors?

8.6 Given a course with an azimuth of $44°56'22''$ with an estimated error of $\pm 6.7''$ and a distance of 138.042 m with an estimated error of ± 5.2 mm, what are the latitude and departure and their estimated errors?

8.7 A traverse meets statistical closures at the 95% level of confidence. In your own words, explain why this does not necessarily imply that the traverse is without error.

8.8 A polygon traverse has the following angle measurements and related standard deviations. Each angle was observed twice (one direct and one reverse). Do the angles meet acceptable closure limits at a 95% level of confidence?

Backsight	Occupied	Foresight	Angle	S
A	B	C	107°53′31″	±2.2″
B	C	D	81°56′44″	±2.4″
C	D	A	92°34′28″	±3.2″
D	A	B	77°35′39″	±2.8″

8.9 Given an initial azimuth for course AB of 36°34′25″, what are the azimuths and their estimated standard errors for the remaining three courses of Problem 8.8?

8.10 Using the distances listed in the following table and the data from Problems 8.8 and 8.9, compute:
(a) the misclosure of the traverse.
(b) the estimated misclosure error.
(c) the 95% misclosure error.

From	To	Distance (ft)	S (ft)
A	B	211.73	±0.016
B	C	302.49	±0.017
C	D	254.48	±0.016
D	A	258.58	±0.016

8.11 Given the traverse misclosures in Problem 8.10, does the traverse meet acceptable closure limits at a 95% level of confidence? Justify your answer statistically.

8.12 Using the data for the link traverse listed below, compute:
(a) the angular misclosure and its estimated error.
(b) the misclosure of the traverse.
(c) the estimated misclosure error.
(d) the 95% error in the traverse misclosure.

Distance observations

From	To	Distance (m)	σ (m)
W	X	185.608	±0.0032
X	Y	106.821	±0.0035
Y	Z	250.981	±0.0028

Angle observations

Back	Occ	For	Angle	σ (″)
W	X	Y	86°27′45″	±2.5
X	Y	Z	199°29′46″	±3.2

142 ERROR PROPAGATION IN TRAVERSE SURVEYS

Control azimuths

From	To	Azimuth	σ (")
W	X	132°26'15"	±9"
Y	Z	58°23'56"	±8"

Control stations

Station	Easting (m)	Northing (m)
W	10,000.000	5000.000
Z	10,417.798	5089.427

8.13 Does the link traverse of Problem 8.12 have acceptable traverse misclosure at a 95% level of confidence? Justify your answer statistically.

8.14 Following are the length and azimuth data for a city lot survey.

Course	Distance (ft)	σ_D (ft)	Azimuth	σ_{Az}
AB	134.58	0.02	83°59'54"	±0"
BC	156.14	0.02	353°59'44"	±20"
CD	134.54	0.02	263°59'54"	±28"
DA	156.10	0.02	174°00'04"	±35"

Compute:
(a) the misclosure of the traverse.
(b) the estimated misclosure error.
(c) the 95% misclosure error.
(d) Does the traverse meet acceptable 95% closure limits? Justify your response with statistically.

8.15 Repeat Problem 8.14 using the data from Problems 7.11 and 7.15.

8.16 Repeat Problem 8.14 using the data from Problems 7.12 and 7.15.

8.17 A survey produces the following set of data. The angles were obtained from the average of four measurements (two direct and two reverse) made with a total station. The estimated uncertainties in the observations are

$$\sigma_{DIN} = \pm 3" \qquad \sigma_t = \pm 0.010 \text{ ft} \qquad \sigma_i = \pm 0.003 \text{ ft}$$

The EDM instrument has a specified accuracy of ±(3 mm + 3 ppm).

Distance observations

From	To	Distance (ft)
1	2	999.99
2	3	801.55
3	4	1680.03
4	5	1264.92
5	1	1878.82

Angle observations

Back	Occ.	For	Angle
5	1	2	191°40'12"
1	2	3	56°42'22"
2	3	4	122°57'10"
3	4	5	125°02'11"
4	5	1	43°38'10"

Control azimuths

From	To	Azimuth	σ (")
1	2	216°52'11"	±3

Control stations

Station	Easting (ft)	Northing (ft)
1	1000.00	1000.00

Compute:
(a) the estimated errors in angles and distances.
(b) the angular misclosure and its 95% probable error.
(c) the misclosure of the traverse.
(d) the estimated misclosure error and its 95% value.
(e) Did the traverse meet acceptable closures? Justify your response statistically.

8.18 Develop new matrices for the link traverse of Example 8.3 that considers the errors in the control. Assume control coordinate standard errors of $\sigma_x = \sigma_y = \pm 0.05$ ft for both stations 1 and 2, and use these new matrices to compute:
(a) the estimated misclosure error.
(b) the 95% misclosure error.
(c) Compare these results with those in the examples.

Programming Problems

8.19 Develop a computational package that will compute the course azimuths and their estimated errors given an initial azimuth and measured angles. Use this package to answer Problem 8.9.

8.20 Develop a computational package that will compute estimated traverse misclosure error given course azimuths and distances and their estimated errors. Use this package to answer Problem 8.10.

8.21 Develop a computational package that will compute estimated traverse misclosure errors given the data of Problem 8.17.

CHAPTER 9

ERROR PROPAGATION IN ELEVATION DETERMINATION

9.1 INTRODUCTION

Differential and trigonometric leveling are the two most commonly employed procedures for finding elevation differences between stations. Both of these methods are subject to systematic and random errors. The primary systematic errors include Earth curvature, atmospheric refraction, and instrument maladjustment. The effects of these systematic errors can be minimized by following proper field procedures. They can also be modeled and corrected for computationally. The random errors in differential and trigonometric leveling occur in instrument leveling, distance observations, and reading graduated scales. These must be treated according to the theory of random errors.

9.2 SYSTEMATIC ERRORS IN DIFFERENTIAL LEVELING

During the differential leveling process, sight distances are held short, and equal, to minimize the effects of systematic errors. Still, it should always be assumed that these errors are present in differential leveling observations, and thus corrective field procedures should be followed to minimize their effects. These procedures are the subjects of discussions that follow.

9.2.1 Collimation Error

Collimation error occurs when the line of sight of an instrument is not truly horizontal and is minimized by keeping sight distances short and balanced.

9.2 SYSTEMATIC ERRORS IN DIFFERENTIAL LEVELING

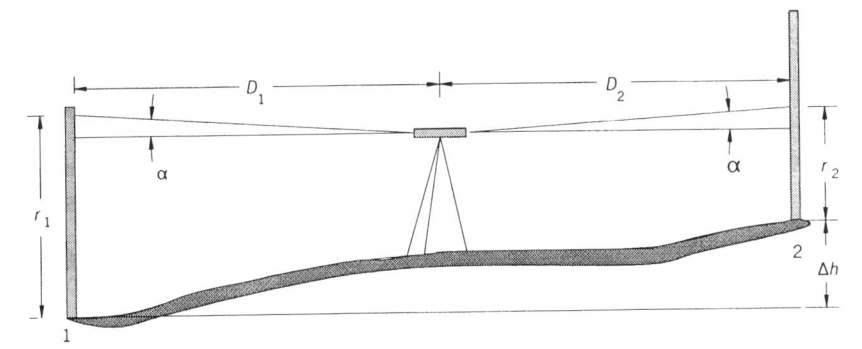

Figure 9.1 Collimation error in differential leveling.

Figure 9.1 shows the effects of a collimation error. For an individual setup, the resulting error in an elevation difference due to collimation is

$$e_C = D_1\alpha - D_2\alpha \tag{9.1}$$

where e_C is the error in elevation due to the presence of a collimation error, D_1 and D_2 the distances to the backsight and foresight rods, respectively, and α the amount of collimation error present at the time of the observation expressed in radian units. Applying Equation (9.1), the collimation error for a line of levels can be expressed as

$$e_C = \alpha[(D_1 - D_2) + (D_3 - D_4) + \cdots + (D_{n-1} - D_n)] \tag{9.2}$$

where $D_1, D_3, \ldots, D_{n-1}$ are the backsight distances and $D_2, D_4, \ldots, D_n$ are the foresight distances. If the backsight and foresight distances are grouped in Equation (9.2),

$$e_C = \alpha \left(\sum D_{BS} - \sum D_{FS} \right) \tag{9.3}$$

The collimation error determined from Equation (9.3) is treated as a correction and thus subtracted from the observed elevation difference to obtain the corrected value.

Example 9.1 A level having a collimation error of 0.04 mm/m is used on a level line where the backsight distances sum to 863 m and the foresight distances sum to 932 m. If the elevation difference observed for the line is 22.865 m, what is the corrected elevation difference?

146 ERROR PROPAGATION IN ELEVATION DETERMINATION

SOLUTION Using Equation (9.3), the error due to collimation is

$$e_C = 0.00004(863 - 932) \text{ m} = -0.0028 \text{ m}$$

Thus, the corrected elevation difference is

$$22.865 - (-0.00276) = 22.868 \text{ m}$$

9.2.2 Earth Curvature and Refraction

As the line of sight extends from an instrument, the level surface curves down and away from it. This condition always causes rod readings to be too high. Similarly, as the line of sight extends from the instrument, refraction bends it toward the Earth surface, causing readings to be too low. The combined effect of Earth curvature and refraction on an individual sight always causes a rod reading to be too high by an amount

$$h_{CR} = CR\left(\frac{D}{1000}\right)^2 \tag{9.4}$$

where h_{CR} is the error in the rod reading (in feet or meters), CR is 0.0675 when D is in units of meters or 0.0206 when D is in units of feet, and D is the individual sight distance.

The effect of this error on a single elevation difference is minimized by keeping backsight and foresight distances short and equal. For unequal sight distances, the resulting error is expressed as

$$e_{CR} = CR\left(\frac{D_1}{1000}\right)^2 - CR\left(\frac{D_2}{1000}\right)^2 \tag{9.5}$$

where e_{CR} is the error due to Earth curvature and refraction on a single elevation difference. Factoring common terms in Equation (9.5) yields

$$e_{CR} = \frac{CR}{1000^2}(D_1^2 - D_2^2) \tag{9.6}$$

To get the corrected value, the curvature and refraction error computed from Equation (9.6) is treated as a correction and thus subtracted from the elevation difference observed.

9.2 SYSTEMATIC ERRORS IN DIFFERENTIAL LEVELING

Example 9.2 An elevation difference between two stations on a hillside is determined to be 1.256 m. What would be the error in the elevation difference and the corrected elevation difference if the backsight distance were 100 m and the foresight distance only 20 m?

SOLUTION Substituting the distances into Equation (9.6) and using CR = 0.0675 gives us

$$e_{CR} = \frac{0.0675}{1000^2}(100^2 - 20^2) = 0.0006 \text{ m}$$

From this, the corrected elevation difference is

$$\Delta h = 1.256 - 0.0006 = 1.255 \text{ m}$$

For a line of differential leveling, the combined effect of this error is

$$e_{CR} = \frac{CR}{1000^2}(D_1^2 - D_2^2 + D_3^2 - D_4^2 + \cdots) \qquad (9.7)$$

Regrouping backsight and foresight distances, Equation (9.7) becomes

$$e_{CR} = \frac{CR}{1000^2}\left(\sum D_{BS}^2 - \sum D_{FS}^2\right) \qquad (9.8)$$

The error due to refraction caused by the vertical gradient of temperature can be large when sight lines are allowed to pass through the lower layers of the atmosphere. Since observing the temperature gradient along the sight line would be cost prohibitive, a field procedure is generally adopted that requires all sight lines to be at least 0.5 m above the ground. This requirement eliminates the lower layers of the atmosphere, where refraction is difficult to model.

9.2.3 Combined Effects of Systematic Errors on Elevation Differences

With reference to Figure 9.1, and by combining Equations (9.1) and (9.5), a corrected elevation difference, Δh, for one instrument setup is

148 ERROR PROPAGATION IN ELEVATION DETERMINATION

$$\Delta h = (r_1 - r_2) - (D_1\alpha - D_2\alpha) - \frac{CR}{1000^2}(D_1^2 - D_2^2) \quad (9.9)$$

where r_1 is the backsight rod reading, r_2 the foresight rod reading, and the other terms are as defined previously.

9.3 RANDOM ERRORS IN DIFFERENTIAL LEVELING

Differential leveling is subject to several sources of random errors. Included are errors in leveling the instrument and in reading the rod. The sizes of these errors are affected by atmospheric conditions, the quality of the optics of the telescope, the sensitivity of the level bubble or compensator, and the graduation scale on the rods. These errors are discussed below.

9.3.1 Reading Errors

The estimated error in rod readings is usually expressed as a ratio of the estimated standard error in the rod reading per unit sight distance length. For example, if an observer's ability to read a rod is within ± 0.005 ft per 100 ft, then $\sigma_{r/D}$ is $\pm 0.005/100 = \pm 0.00005$ ft/ft. Using this, rod reading errors for any individual sight distance D can be estimated as

$$\sigma_r = D\sigma_{r/D} \quad (9.10)$$

where $\sigma_{r/D}$ is the estimated error in the rod reading per unit length of sight distance and D is the length of the sight distance.

9.3.2 Instrument Leveling Errors

The estimated error in leveling for an automatic compensator or level vial is generally given in the technical data for each instrument. For precise levels, this information is usually listed in arc seconds or as an estimated elevation error for a given distance. As an example, the estimated error may be listed as ± 1.5 mm/km, which corresponds to $\pm 1.5/1{,}000{,}000 \times \rho = \pm 0.3''$. A precise level will usually have a compensator accuracy or setting accuracy between $\pm 0.1''$ and $\pm 0.2''$, whereas for a less precise level, the value may be as high as $\pm 10''$.

9.3.3 Rod Plumbing Error

Although a level rod that is held nonvertical always causes the reading to be too high, this error will appear random in a leveling network, due to its presence in all backsight and foresight distances of the network. Thus, the rod plumbing error should be modeled when computing the standard error in

9.3 RANDOM ERRORS IN DIFFERENTIAL LEVELING

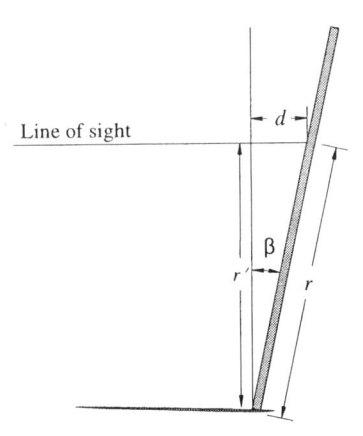

Figure 9.2 Novertical level rod.

an elevation difference. With reference to Figure 9.2 for any rod reading, the rod plumbing error is approximated as

$$e_{LS} = r - r' = \frac{d^2}{2r} \tag{9.11}$$

where d is the linear amount that the rod is out of plumb at the location of the rod reading, r. The size of d is dependent on the rod level bubble centering error and the reading location. If the rod bubble is out of level by β, then d is

$$d = r \sin \beta \tag{9.12}$$

Substituting Equation (9.12) into (9.11) gives

$$e_{LS} = \frac{r}{2} \sin^2 \beta \tag{9.13}$$

Example 9.3 Assume that a rod level bubble is within $\pm 5'$ of level and the rod reading is at 4 m. What is the estimated error in the rod reading?

SOLUTION

$$e_{LS} = \frac{4}{2} \sin^2(5') = 0.004 \text{ mm}$$

150 ERROR PROPAGATION IN ELEVATION DETERMINATION

Since the rod plumbing error occurs on every sighting, backsight errors will tend to cancel foresight errors. Thus, with precise leveling techniques, the combined effect of this error can be written as

$$e = \left(\frac{r_1 \sin^2\beta}{2} - \frac{r_2 \sin^2\beta}{2}\right) + \left(\frac{r_3 \sin^2\beta}{2} - \frac{r_4 \sin^2\beta}{2}\right) + \cdots \quad (9.14)$$

Grouping like terms in Equation (9.13) yields

$$e = \tfrac{1}{2} \sin^2\beta \, (r_1 - r_2 + r_3 - r_4 + \cdots) \quad (9.15)$$

Recognizing that the quantity in parentheses in Equation (9.15) is the elevation difference for the leveling line yields

$$e_{LS} = \frac{\Delta \text{Elev}}{2} \sin^2\beta \quad (9.16)$$

Example 9.4 If a level rod is maintained to within ±5′ of level and the elevation difference is 22.865 m, the estimated error in the final elevation is

$$e_{LS} \frac{22.865}{2} \sin^2(5') = 0.02 \text{ mm}$$

The rod plumbing error can be practically eliminated by carefully centering the bubble of a well-adjusted rod level. It is generally small, as the example illustrates, and thus will be ignored in subsequent computations.

9.3.4 Estimated Errors in Differential Leveling

From the preceding discussion, the major random error sources in differential leveling are caused by random errors in rod readings and instrument leveling. Furthermore, in Equation (9.9), the collimation error is considered to be systematic and is effectively negated by balancing the backsight and foresight distances. However, no matter what method is used to observe the lengths of the sight distances, some random error in these lengths will be present. This causes random errors in the elevation differences, due to the effects of Earth curvature, refraction, and instrumental collimation errors. Equation (6.16) can be applied to Equation (9.9) to model the effects of the random errors in rod readings, leveling, and sighting lengths. The following partial derivatives are needed:

$$\frac{\partial \Delta h}{\partial r_1} = \frac{\partial \Delta h}{\partial r_2} = 1 \qquad \frac{\partial \Delta h}{\partial \alpha_1} = -D_1 \qquad \frac{\partial \Delta h}{\partial \alpha_2} = -D_2 \qquad (9.17)$$

$$\frac{\partial \Delta h}{\partial D_1} = -\left(\alpha + \frac{CR(D_1)}{500{,}000}\right) \qquad \frac{\partial \Delta h}{\partial D_2} = -\left(\alpha + \frac{CR(D_2)}{500{,}000}\right)$$

By substituting Equations (9.17) into (6.16) with their corresponding estimated standard errors, the standard error in a single elevation difference can be estimated as

$$\sigma_{\Delta h} = \sqrt{\begin{array}{l}(D_1\sigma_{r/D})^2 + (D_2\sigma_{r/D})^2 + (-D_1\sigma_{\alpha_1})^2 + (-D_2\sigma_{\alpha_2})^2 \\ + \left[-\left(\alpha + \dfrac{CR(D_1)}{500{,}000}\right)\sigma_{D_1}\right]^2 + \left[\left(\alpha + \dfrac{CR(D_2)}{500{,}000}\right)\sigma_{D_2}\right]^2\end{array}} \qquad (9.18)$$

where $\sigma_{r/D}$ is the estimated error in a rod reading, σ_{α_1} and σ_{α_2} the estimated collimation errors in the backsight and foresight, respectively, and σ_{D_1} and σ_{D_2} the errors estimated in the sight lengths D_1 and D_2, respectively.

In normal differential leveling procedures, $D_1 = D_2 = D$. Also, the estimated standard errors in the sight distances are equal: $\sigma_{D_1} = \sigma_{D_2} = \sigma_D$. Furthermore, the estimated collimation error for the backsight and foresight can be assumed equal: $\sigma_{\alpha_1} = \sigma_{\alpha_2} = \sigma_\alpha$. Thus, Equation (9.18) simplifies to

$$\sigma_{\Delta h} = \sqrt{2D^2 (\sigma_{r/D}^2 + \sigma_\alpha^2) + 2\sigma_D^2 \left(\alpha + \frac{CR(D)}{500{,}000}\right)^2} \qquad (9.19)$$

Equation (9.19) is appropriate for a single elevation difference when the sight distances are approximately equal. In general, if sight distances are kept equal for N instrument setups, the total estimated error in an elevation difference is

$$\sigma_{\Delta h} = \sqrt{2ND^2 (\sigma_{r/D}^2 + \sigma_\alpha^2) + 2N\sigma_D^2 \left(\alpha + \frac{CR(D)}{500{,}000}\right)^2} \qquad (9.20)$$

Since the error estimated in the elevation difference due to Earth curvature and refraction, and the actual collimation error, α, are small, the last term is ignored. Thus, the final equation for the standard error estimated in differential leveling is

$$\sigma_{\Delta h} = D\sqrt{2N(\sigma_{r/D}^2 + \sigma_\alpha^2)} \qquad (9.21)$$

Example 9.5 A level line is run from benchmark A to benchmark B. The standard error estimated in rod readings is ±0.01 mm/m. The instrument is maintained to within ±2.0″ of level. A collimation test shows that the instrument is within 4 mm per 100 m. Fifty-meter sight distances are maintained within an uncertainty of ±2 m. The total line length from A to B is 1000 m. What is the estimated error in the elevation difference between A and B? If A had an elevation of 212.345 ± 0.005 m, what is the estimated error in the computed elevation of B?

SOLUTION The total number of setups in this problem is $1000/(2 \times 50) = 10$ setups. Substituting the appropriate values into Equation (8.20) yields

$$\sigma_{\Delta h} = \sqrt{2(10)50^2\left[\left(\frac{0.01}{1000}\right)^2 + \left(\frac{2.0''}{\rho}\right)^2\right] + 2(10)2^2\left(\frac{0.004}{100} + \frac{0.0675(50)}{500,000}\right)^2}$$

$$= \sqrt{0.0031^2 + 0.0004^2} = 0.0031 \text{ m} = \pm 3.1 \text{ mm}$$

From an analysis of the individual error components in the equation above, it is seen that the error caused by the errors in the sight distances is negligible for all but the most precise leveling. Thus, like the error due to rod bubble centering, this error can be ignored in all but the most precise work. Thus, the simpler Equation (9.21) can be used to solve the problem:

$$\sigma_{\Delta h} = 50\sqrt{2 \times 10\left[\left(\frac{0.01}{1000}\right)^2 + \left(\frac{2.0''}{\rho}\right)^2\right]}$$

$$= \sqrt{0.0031^2} = \pm 0.0031 \text{ m} = \pm 3.1 \text{ mm}$$

The estimated error in the elevation of B is found by applying Equation (6.18) as

$$\sigma_{\text{Elev}_B} = \sqrt{\sigma^2_{\text{Elev}_A} + \sigma^2_{\Delta\text{Elev}}} = \sqrt{5^2 + 3.1^2} = \pm 5.9 \text{ mm}$$

9.4 ERROR PROPAGATION IN TRIGONOMETRIC LEVELING

With the introduction of total station instruments, it is becoming increasingly convenient to observe elevation differences using trigonometric methods. However, in this procedure, because sight distances cannot be balanced, it is important that the systematic effects of Earth curvature and refraction, and inclination in the instrument's line of sight (collimation error), be removed.

9.4 ERROR PROPAGATION IN TRIGONOMETRIC LEVELING

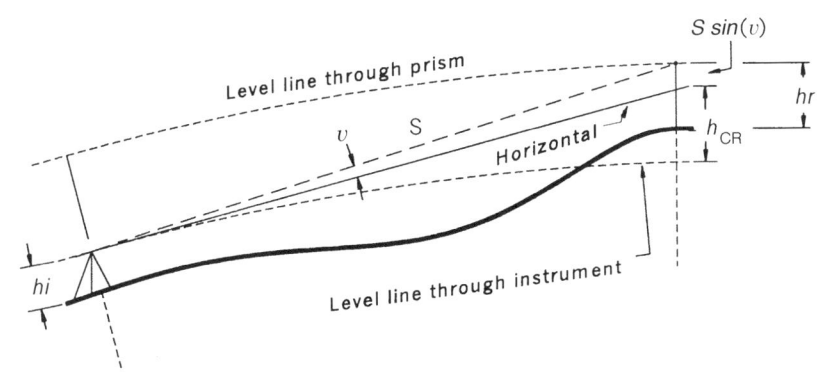

Figure 9.3 Determination of elevation difference by trigonometric leveling.

From Figure 9.3, the corrected elevation difference, Δh, between two points is

$$\Delta h = hi + S \sin v + h_{CR} - hr \qquad (9.22)$$

Equation (9.22) for zenith-angle reading instruments is

$$\Delta h = hi + S \cos v + h_{CR} - hr \qquad (9.23)$$

where hi is the instrument height above ground, S is the slope distance between the two points, v the vertical angle between the instrument and the prism, z the zenith angle between the instrument and the prism, h_{CR} the Earth curvature and refraction correction given in Equation (9.4), and hr the rod reading. Substituting the curvature and refraction formula into Equation (9.23) yields

$$\Delta h = hi + \cos z + CR\left(\frac{S \sin z}{1000}\right)^2 - hr \qquad (9.24)$$

In developing an error propagation formula for Equation (9.24), not only must errors relating to the height of instrument and prism be considered, but also, errors in leveling, pointing, reading, and slope distances as discussed in Chapter 6. Applying Equation (6.16) to Equation (9.24), the following partial derivatives apply:

$$\frac{\partial \Delta h}{\partial hi} = 1$$

$$\frac{\partial \Delta h}{\partial hr} = -1$$

$$\frac{\partial \Delta h}{\partial S} = \cos z + \frac{CR(S) \sin^2 z}{500{,}000}$$

$$\frac{\partial \Delta h}{\partial z} = \frac{CR(S^2) \sin z \cos z}{500{,}000} - S \sin z$$

Entering the partial derivatives and the standard errors of the observations into Equation (6.16), the total error in trigonometric leveling is

$$\sigma_{\Delta h} = \sqrt{\sigma_{hi}^2 + \sigma_{hr}^2 + \left[\left(\cos z + \frac{CR(S) \sin^2 z}{500{,}000}\right)\sigma_S\right]^2 + \left[\left(\frac{CR(S^2) \sin z \cos z}{500{,}000} - S \sin z\right)\frac{\sigma_z}{\rho}\right]^2} \quad (9.25)$$

where z is the zenith angle, CR is 0.0675 if units of meters are used or 0.0206 if units of feet are used, S is the slope distance, and ρ is the radians-to-seconds conversion of 206,264.8″/rad.

In Equation (9.25), errors from several sources make up the estimated error in the zenith angle. These include the operator's ability to point and read the instrument, the accuracy of the vertical compensator or the operator's ability to center the vertical circle bubble, and the sensitivity of the compensator or vertical circle bubble. For best results, zenith angles should be observed using faces of the instrument and an average taken. Using Equations (7.4) and (7.6), the estimated error in a zenith angle that is observed in both positions (face I and face II) with a theodolite is

$$\sigma_z = \sqrt{\frac{2\sigma_r^2 + 2\sigma_p^2 + 2\sigma_B^2}{N}} \quad (9.26a)$$

where σ_r is the error in reading the circle, σ_p the error in pointing, σ_B the error in the vertical compensator or in leveling the vertical circle bubble, and N the number of face-left and face-right observations of the zenith angle. For digital theodolites or total stations, the appropriate formula is

9.4 ERROR PROPAGATION IN TRIGONOMETRIC LEVELING

$$\sigma_z = \sqrt{\frac{2\sigma_{DIN}^2 + 2\sigma_B^2}{N}} \quad (9.26b)$$

where σ_{DIN} is the DIN 18723 value for the instrument and all other values are as above.

Notice that if only a single zenith-angle observation is made (i.e., it is observed only in face I), its estimated error is simply

$$\sigma_z = \sqrt{\sigma_r^2 + \sigma_p^2 + \sigma_B^2} \quad (9.27a)$$

For the digital theodolite and total station, the estimated error for a single observation is

$$\sigma_z = \sqrt{2\sigma_{DIN}^2 + \sigma_B^2} \quad (9.27b)$$

Similarly, the estimated error in the slope distance, S, is computed using Equation (6.36).

Example 9.6 A total station instrument has a vertical compensator accurate to within ±0.3″, a digital reading accuracy of ±5″, and a distance accuracy of ±(5 mm + 5 ppm). The slope distance observed is 1256.78 ft. It is estimated that the instrument is set to within ± 0.005 ft of the station, and the target is set to within ±0.01 ft. The height of the instrument is 5.12 ± 0.01 ft, and the prism height is 6.72 ± 0.01 ft. The zenith angle is observed in only one position and recorded as 88°13′15″. What are the corrected elevation difference and its estimated error?

SOLUTION Using Equation (9.24), the corrected elevation difference is

$$\Delta h = 5.12 + 1256.78 \cos(88°13'15') + 0.0206$$
$$\left(\frac{1256.78 \sin(88°13'15')}{1000}\right)^2 - 6.72$$
$$= 37.39 \text{ ft}$$

With Equation (9.27b), the zenith angle error is estimated as $\sigma_z = \sqrt{2(5)^2 + 0.3^2} = \pm 7.1''$. From Equation (7.24) and converting 5 mm to 0.0164 ft, the estimated error in the distance is

$$\sigma_S = \sqrt{0.005^2 + 0.01^2 + 0.0164^2 + \left(\frac{5}{1{,}000{,}000} 1256.78\right)^2} = \pm 0.021 \text{ ft}$$

Substituting the values into Equation (9.25), the estimated error in the elevation difference is

$$\sigma_{\Delta h} = \sqrt{0.01^2 + 0.01^2 + (0.031 \times 0.021)^2 + \left(-1256.172 \frac{7.1''}{\rho}\right)^2}$$
$$= \sqrt{0.01^2 + 0.01^2 + 0.00065^2 + 0.043^2} = \pm 0.045 \text{ ft}$$

Note in this example that the estimated error in the elevation difference caused by the distance error is negligible (± 0.00065 ft), whereas the error in the zenith angle is the largest (± 0.043 ft). Furthermore, since the vertical angle is not observed with both faces of the instrument, it is possible that uncompensated systematic errors are present in the final value computed. For example, assume that a 10″ indexing error existed on the vertical circle. If the observations are taken using both faces of the instrument, the effects of this error are removed. However, by making only a face I observation, the systematic error due to the vertical indexing error is

$$1256.78 \sin(10'') = 0.061 \text{ ft}$$

The uncompensated systematic error in the final value is considerably larger than the error estimated for the observation. One should always account for systematic errors by using proper field procedures. Failure to do so can only lead to poor results. In trigonometric leveling, a minimum of a face I and face II reading should always be taken.

PROBLEMS

9.1 A collimation error of 0.00005 ft/ft is used in leveling a line that has a sum of 1425 ft for the backsight distances and only 632 ft for the foresight distances. If the elevation difference observed is -15.84 ft, what is the elevation difference corrected for the collimation error?

9.2 Repeat Problem 9.1 for a collimation error of 0.005 mm/m, a sum of the backsight distances of 823 m, a sum of the foresight distance of 1206 m, and an observed elevation difference of 23.475 m.

9.3 To expedite going down a hill, the backsight distances were consistently held to 50 ft while the foresight distances were held to 200 ft.

There were 33 setups. If the elevation difference observed was −306.87 ft, what is the elevation difference corrected for Earth curvature and refraction?

9.4 Repeat Problem 9.3 for backsight distances of 20 m, foresight distances of 60 m, 46 setups, and an elevation difference of −119.603 m.

9.5 How far from vertical must an 8-ft level rod be to create an error of 0.01 ft with a reading at 5.00 ft?

9.6 Repeat Problem 9.5 for a reading at 15.00 ft on a 25-ft rod.

9.7 Repeat Problem 9.5 for a 1-mm error with a reading at 1.330 m on a 2-m rod.

9.8 A line of three-wire differential levels goes from benchmark Gloria to benchmark Carey. The length of the line was determined to be 2097 m. The instrument had a stated compensator accuracy of ±1.4 mm/km. The instrument–rod combination had an estimated reading error of ±0.4 mm per 40 m. Sight distances were kept to approximately 50 ± 5 m. If the observed difference in elevation is 15.601 m, what is the estimated error in the final elevation difference?

9.9 If in Problem 9.8, benchmark Gloria had a fixed elevation of 231.071 m and Carey had a fixed elevation of 246.660 m, did the job meet acceptable closure limits at a 95% level of confidence? Justify your answer statistically.

9.10 Repeat Problem 9.8 for a compensator accuracy of ±1.2 mm/km and an estimated error in reading the rod of 0.4 mm per 100 m.

9.11 Using the data in Problem 9.9, did the leveling circuit meet acceptable closures at a 95% level of confidence? Justify your response statistically.

9.12 An elevation must be established on a benchmark on an island that is 2536.98 ft from the nearest benchmark on the island's shore. The surveyor decides to use a total station that has a stated distance measuring accuracy of ±(3 mm + 3 ppm) and a vertical compensator accurate to within ±0.4″. The height of instrument was 5.37 ft with an estimated error of ±0.05 ft. The prism height was 6.00 ft with an estimated error of ±0.02 ft. The single zenith angle is read as 87°05′32″. The estimated errors in instrument and target centering are ±0.003 ft. If the elevation of the occupied benchmark is 632.27 ft, what is the corrected benchmark elevation on the island? (Assume that the instrument does not correct for Earth curvature and refraction.)

9.13 In Problem 9.12, what is the estimated error in the computed benchmark elevation if the instrument has a DIN 18723 stated accuracy of ±3″.

158 ERROR PROPAGATION IN ELEVATION DETERMINATION

9.14 After completing the job in Problem 9.12, the surveyor discovered that the instrument had a vertical indexing error that caused the sight line to be inclined by 15″.
  (a) How much error would be created in the elevation of the island benchmark if the indexing error were ignored?
  (b) What is the corrected elevation?

9.15 A tilting level is used to run a set of precise levels to a construction project from benchmark DAM, whose elevation is 101.865 m. The line is run along a road that goes up a steep incline. To expedite the job, backsight distances are kept to 50 ± 1 m and foresight distances are 20 ± 1 m. The total length of differential levels is 4410 m. The elevation difference observed is 13.634 m. What is the corrected project elevation if the instrument has a sight line that declines at the rate of 0.3 mm per 50 m?

9.16 In Problem 9.15, the instrument's level is centered to within ±0.4″ for each sight and the rod is read to ±1.2 mm per 50 m.
  (a) What is the estimated error in the elevation difference?
  (b) What is the 95% error in the final established benchmark?

9.17 Which method of leveling presented in this chapter offers the most precision? Defend your answer statistically.

Programming Problems

9.18 Create a computational package that will compute a corrected elevation difference and its estimated error using the method of differential leveling. Use this package to solve Problem 9.15.

9.19 Create a computational package that will compute a corrected elevation difference and its estimated error using the method of trigonometric leveling. Use this package to solve Problem 9.12.

CHAPTER 10

WEIGHTS OF OBSERVATIONS

10.1 INTRODUCTION

When surveying data are collected, they must usually conform to a given set of geometric conditions, and when they do not, the measurements are adjusted to force that geometric closure. For a set of uncorrelated observations, a measurement with high precision, as indicated by a small variance, implies a good observation, and in the adjustment it should receive a relatively small portion of the overall correction. Conversely, a measurement with lower precision, as indicated by a larger variance, implies an observation with a larger error, and should receive a larger portion of the correction.

The weight of an observation is a measure of its relative worth compared to other measurements. *Weights* are used to control the sizes of corrections applied to measurements in an adjustment. The more precise an observation, the higher its weight; in other words, the smaller the variance, the higher the weight. From this analysis it can be stated intuitively that *weights are inversely proportional to variances*. Thus, it also follows that *correction sizes should be inversely proportional to weights*.

In situations where measurements are correlated, weights are related to the inverse of the covariance matrix, Σ. As discussed in Chapter 6, the elements of this matrix are variances and covariances. Since weights are relative, variances and covariances are often replaced by cofactors. A *cofactor* is related to its covariance by the equation

$$q_{ij} = \frac{\sigma_{ij}}{\sigma_0^2} \qquad (10.1)$$

160 WEIGHTS OF OBSERVATIONS

where q_{ij} is the cofactor of the *ij*th measurement, σ_{ij} the covariance of the *ij*th measurement, and σ_0^2 the *reference variance*, a value that can be used for scaling. Equation (10.1) can be expressed in matrix notation as

$$Q = \frac{1}{\sigma_0^2} \Sigma \qquad (10.2)$$

where Q is defined as the *cofactor matrix*. The structure and individual elements of the Σ matrix are

$$\Sigma = \begin{bmatrix} \sigma_{x_1}^2 & \sigma_{x_1 x_2} & \cdots & \sigma_{x_1 x_n} \\ \sigma_{x_2 x_1} & \sigma_{x_2}^2 & & \sigma_{x_2 x_n} \\ \vdots & \vdots & \vdots & \\ \sigma_{x_n x_1} & \sigma_{x_n x_2} & & \sigma_{x_n}^2 \end{bmatrix}$$

From the discussion above, the weight matrix W is

$$W = Q^{-1} = \sigma_0^2 \Sigma^{-1} \qquad (10.3)$$

For uncorrelated measurements, the covariances are equal to zero (i.e., all $\sigma_{x_i x_j} = 0$) and the matrix Σ is diagonal. Thus, Q is also a diagonal matrix with elements equal to $\sigma_{x_i}^2 / \sigma_0^2$. The inverse of a diagonal matrix is also a diagonal matrix, with its elements being the reciprocals of the original diagonals, and therefore Equation (10.3) becomes

$$W = \begin{bmatrix} \dfrac{\sigma_0^2}{\sigma_{x_1}^2} & 0 & \cdots & 0 \\ 0 & \dfrac{\sigma_0^2}{\sigma_{x_2}^2} & & 0 \\ 0 & & \ddots & \\ 0 & 0 & & \dfrac{\sigma_0^2}{\sigma_{x_n}^2} \end{bmatrix} = \sigma_0^2 \Sigma^{-1} \qquad (10.4)$$

From Equation (10.4), any independent measurement with variance equal to σ_i^2 has a weight of

$$w_i = \frac{\sigma_0^2}{\sigma_i^2} \qquad (10.5)$$

If the *i*th observation has a weight $w_i = 1$, then $\sigma_0^2 = \sigma_i^2$, or $\sigma_0^2 = 1$. Thus, σ_0^2 is often called the *variance of an observation of unit weight,* shortened to *variance of unit weight* or simply *unit variance*. Its square root is called the *standard deviation of unit weight*. If σ_0^2 is set equal to 1 in Equation (10.5), then

$$w_i = \frac{1}{\sigma_i^2} \qquad (10.6)$$

Note in Equation (10.6) that as stated earlier, the weight of an observation is inversely proportional to its variance.

With correlated observations, it is possible to have a covariance matrix, Σ, and a cofactor matrix, Q, but not a weight matrix. This occurs when the cofactor matrix is singular, and thus an inverse for Q does not exist. Most situations in surveying involve uncorrelated observations. For the remainder of this chapter, only the uncorrelated case with variance of unit weight is considered.

10.2 WEIGHTED MEAN

If two measurements are taken of a quantity and the first is twice as good as the second, their relative worth can be expressed by giving the first measurement a weight of 2 and the second a weight of 1. A simple adjustment involving these two measurements would be to compute the mean value. In this calculation, the observation of weight 2 could be added twice, and the observation of weight 1 added once. As an illustration, suppose that a distance is measured with a tape to be 151.9 ft, and the same distance is measured with an EDM instrument as 152.5 ft. Assume that experience indicates that the electronically measured distance is twice as good as the taped distance, and accordingly, the taped distance is given a weight of 1 and the electronically measured distance is given a weight of 2. Then one method of computing the mean from these observations is

$$\overline{M} = \frac{151.9 + 152.5 + 152.5}{3} = 152.3$$

As an alternative, the calculation above can be rewritten as

$$\overline{M} = \frac{1(151.9) + 2(152.5)}{1 + 2} + 152.3$$

Note that the weights of 1 and 2 were entered directly into the second computation and that the result of this calculation is the same as the first. Note also that the computed mean tends to be closer to the measured value having the higher weight (i.e., 152.3 is closer to 152.5 than it is to 151.9). A mean value computed from weighted observations is called the *weighted mean*.

To develop a general expression for computing the weighted mean, suppose that we have m independent uncorrelated observations $(z_1, z_2, \ldots, z_m)$ for a quantity z and that each observation has standard deviation σ. Then the mean of the observations is

$$\bar{z} = \frac{\sum_{i=1}^{m} z_i}{m} \tag{10.7}$$

If these m observations were now separated into two sets, one of size m_a and the other m_b such that $m_a + m_b = m$, the means for these two sets would be

$$\bar{z}_a = \frac{\sum_{i=1}^{m_a} z_i}{m_a} \tag{10.8}$$

$$\bar{z}_b = \frac{\sum_{i=m_a+1}^{m} z_i}{m_b} \tag{10.9}$$

The mean $\bar{z}$ is found by combining the means of these two sets as

$$\bar{z} = \frac{\sum_{i=1}^{m_a} z_i + \sum_{i=m_a+1}^{m} z_i}{m} = \frac{\sum_{i=1}^{m_a} z_i + \sum_{i=m_a+1}^{m} z_i}{m_a + m_b} \tag{10.10}$$

But from Equations (10.8) and (10.9),

$$\bar{z}_a m_a = \sum_{i=1}^{m_a} z_i \quad \text{and} \quad \bar{z}_b m_b = \sum_{i=m_a+1}^{m} z_i \tag{10.11}$$

Thus,

$$\bar{z} = \frac{\bar{z}_a m_a + \bar{z}_b m_b}{m_a + m_b} \tag{10.12}$$

Note the correspondence between Equation (10.12) and the second equation used to compute the weighted mean in the simple illustration given earlier. By intuitive comparison it should be clear that m_a and m_b correspond to weights that could be symbolized as w_a and w_b, respectively. Thus, Equation (10.12) can be written as

$$\bar{z} = \frac{w_a \bar{z}_a + w_b \bar{z}_b}{w_a + w_b} = \frac{\sum wz}{\sum w} \tag{10.13}$$

Equation (10.13) is used in calculating the weighted mean for a group of uncorrelated observations having unequal weights. In Chapter 11 it will be shown that the weighted mean is the most probable value for a set of weighted observations.

Example 10.1 As a simple example of computing a weighted mean using Equation (10.13), suppose that a distance d is measured three times, with the following results: 92.61 with weight 3, 92.60 with weight 2, and 92.62 with weight 1. Calculate the weighted mean.

$$\bar{d} = \frac{3(92.61) + 2(92.60) + 1(92.62)}{3 + 2 + 1} = 92.608$$

Note that if weight had been neglected, the simple mean would have been 92.61.

10.3 RELATION BETWEEN WEIGHTS AND STANDARD ERRORS

By applying the special law of propagation of variances [Equation (6.16)] to Equation (10.8), the variance $\bar{z}_a$ in Equation (10.8) is

$$\sigma_{\bar{z}_a}^2 = \left(\frac{\partial \bar{z}_a}{\partial z_1}\right)^2 \sigma^2 + \left(\frac{\partial \bar{z}_a}{\partial z_2}\right)^2 \sigma^2 + \cdots + \left(\frac{\partial \bar{z}_a}{\partial z_{m_a}}\right)^2 \sigma^2 \qquad (10.14)$$

Substituting partial derivatives with respect to the measurements into Equation (10.14) yields

$$\sigma_{\bar{z}_a}^2 = \left(\frac{1}{m_a}\right)^2 \sigma^2 + \left(\frac{1}{m_a}\right)^2 \sigma^2 + \cdots + \left(\frac{1}{m_a}\right)^2 \sigma^2$$

Thus,

$$\sigma_{\bar{z}_a}^2 = m_a \left(\frac{1}{m_a}\right)^2 \sigma^2 = \frac{1}{m_a} \sigma^2 \qquad (10.15)$$

Using a similar procedure, the variance of $\bar{z}_b$ is

$$\sigma_{\bar{z}_b}^2 = \frac{1}{m_b} \sigma^2 \qquad (10.16)$$

In Equations (10.15) and (10.16), σ is a constant, and the weights of $\bar{z}_a$ and $\bar{z}_b$ were established as m_a and m_b, respectively, from Equation (10.13). Since the weights are relative, from Equations (10.15) and (10.16),

$$w_a = \frac{1}{\sigma_{z_a}^2} \quad \text{and} \quad w_b = \frac{1}{\sigma_{z_b}^2} \tag{10.17}$$

Conclusion: With uncorrelated observations, the weights of the observations are inversely proportional to their variances.

10.4 STATISTICS OF WEIGHTED OBSERVATIONS

10.4.1 Standard Deviation

By definition, an observation is said to have a weight w when its precision is equal to that of the mean of w observations of unit weight. Let σ_0 be the standard error of an observation of weight 1, or *unit weight*. If $y_1, y_2, \ldots, y_n$ are observations having standard errors $\sigma_1, \sigma_2, \ldots, \sigma_n$ and weights $w_1, w_2, \ldots, w_n$, then, by Equation (10.5),

$$\sigma_1 = \frac{\sigma_0}{\sqrt{w_1}}, \quad \sigma_2 = \frac{\sigma_0}{\sqrt{w_2}}, \quad \ldots, \quad \sigma_n = \frac{\sigma_0}{\sqrt{w_n}} \tag{10.18}$$

In Section 2.7, the standard error for a group of observations of equal weight was defined as

$$\sigma = \sqrt{\frac{\sum_{i=1}^{n} \varepsilon_i^2}{n}}$$

Now, in the case where the observations are not equal in weight, the equation above becomes

$$\sigma = \sqrt{\frac{w_1\varepsilon_1^2 + w_2\varepsilon_2^2 + \cdots w_n\varepsilon_n^2}{n}} = \sqrt{\frac{\sum_{i=1}^{n} w_i\varepsilon_i^2}{n}} \tag{10.19}$$

When modified for the standard deviation in Equation (2.7), it is

$$S = \sqrt{\frac{w_1v_1^2 + w_2v_2^2 + \cdots + w_nv_n^2}{n-1}} = \sqrt{\frac{\sum_{i=1}^{n} w_iv_i^2}{n-1}} \tag{10.20}$$

10.4.2 Standard Error of Weight *w* and Standard Error of the Weighted Mean

The relationship between standard error and standard error of weight w was given in Equation (10.18). Combining this with Equation (10.19) and dropping the summation limits, equations for standard errors of weight w are obtained in terms of σ_0 as follows:

$$\sigma_1 = \frac{\sigma_0}{\sqrt{w_1}} = \sqrt{\frac{\Sigma w\varepsilon^2}{n}} \frac{1}{\sqrt{w_1}} = \sqrt{\frac{\Sigma w\varepsilon^2}{nw_1}}$$

$$\sigma_2 = \frac{\sigma_0}{\sqrt{w_2}} = \sqrt{\frac{\Sigma w\varepsilon^2}{n}} \frac{1}{\sqrt{w_2}} = \sqrt{\frac{\Sigma w\varepsilon^2}{nw_2}} \quad (10.21)$$

$$\vdots$$

$$\sigma_n = \frac{\sigma_0}{\sqrt{w_n}} = \sqrt{\frac{\Sigma w\varepsilon^2}{n}} \frac{1}{\sqrt{w_n}} = \sqrt{\frac{\Sigma w\varepsilon^2}{nw_n}}$$

Similarly, standard deviations of weight w can be expressed as

$$S_1 = \sqrt{\frac{\Sigma wv^2}{w_1(n-1)}}, \quad S_2 = \sqrt{\frac{\Sigma wv^2}{w_2(n-1)}}, \quad \ldots, \quad S_n = \sqrt{\frac{\Sigma wv^2}{w_n(n-1)}} \quad (10.22)$$

Finally, the reference standard error of the weighted mean is calculated as

$$\sigma_{\overline{M}} = \sqrt{\frac{\Sigma w\varepsilon^2}{n\Sigma w}} \quad (10.23)$$

and the standard deviation of the weighted mean is

$$\sigma_{\overline{M}} = \sqrt{\frac{\Sigma wv^2}{(n-1)\Sigma w}} \quad (10.24)$$

10.5 WEIGHTS IN ANGLE OBSERVATIONS

Suppose that the three angles α_1, α_2, and α_3 in a plane triangle are observed n_1, n_2, and n_3 times, respectively, with the same instrument under the same conditions. What are the relative weights of the angles?

To analyze the relationship between weights and the number of times an angle is turned, let S be the standard deviation of a single angle observation. The means of the three angles are

166 WEIGHTS OF OBSERVATIONS

$$\alpha_1 = \frac{\Sigma\, \alpha_1}{n_1} \qquad \alpha_2 = \frac{\Sigma\, \alpha_2}{n_2} \qquad \alpha_3 = \frac{\Sigma\, \alpha_3}{n_3}$$

The variances of the means, as obtained by applying Equation (6.16), are

$$S_{\alpha_1}^2 = \frac{1}{n_1} S^2 \qquad S_{\alpha_2}^2 = \frac{1}{n_2} S^2 \qquad S_{\alpha_3}^2 = \frac{1}{n_3} S^2$$

Again, since the weights of the observations are inversely proportional to the variances and relative, the weights of the three angles are

$$w_1 = \frac{1}{S_{\alpha_1}^2} = \frac{n_1}{S^2} \qquad w_2 = \frac{1}{S_{\alpha_2}^2} = \frac{n_2}{S^2} \qquad w_3 = \frac{1}{S_{\alpha_3}^2} = \frac{n_3}{S^2}$$

In the expressions above, S is a constant term in each of the weights, and because the weights are relative, it can be dropped. Thus, the weights of the angles are $w_1 = n_1$, $w_2 = n_2$, and $w_3 = n_3$.

In summary, it has been shown that when all conditions in angle observation are equal except for the number of turnings, *angle weights are proportional to the number of times the angles are turned.*

10.6 WEIGHTS IN DIFFERENTIAL LEVELING

Suppose that for the leveling network shown in Figure 10.1, the lengths of lines 1, 2, and 3 are 2, 3, and 4 miles, respectively. For these varying lengths of lines, it can be expected that the errors in their elevation differences will vary, and thus the weights assigned to the elevation differences should also be varied. What relative weights should be used for these lines?

To analyze the relationship of weights and level line lengths, recall from Equation (9.20) that the variance in Δh is

$$\sigma_{\Delta h}^2 = D^2[2N(\sigma_{r/D}^2 + \sigma_\alpha^2)] \qquad (a)$$

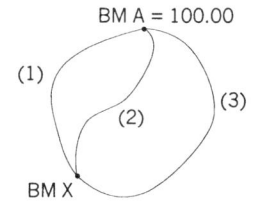

Figure 10.1 Differential leveling network.

where D is the length of the individual sights, N the number of setups, $\sigma_{r/D}$ the estimated error in a rod reading, and σ_α the estimated collimation error for each sight. Let l_i be the length of the ith course between benchmarks; then

$$N = \frac{l_i}{2D} \tag{b}$$

Substituting Equation (b) into Equation (a) yields

$$\sigma_{\Delta h}^2 = l_i D(\sigma_{r/D}^2 + \sigma_\alpha^2) \tag{c}$$

However, D, $\sigma_{r/D}$, and σ_α are constants, and thus by letting $k = D(\sigma_{r/D}^2 + \sigma_\alpha^2)$, Equation (c) becomes

$$\sigma_{\Delta h}^2 = l_i k \tag{d}$$

For this example, it can be said that the weights are

$$w_1 = \frac{1}{l_1 k} \quad w_2 = \frac{1}{l_2 k} \quad w_3 = \frac{1}{l_3 k} \tag{e}$$

Now since k is a constant and weights are relative, Equation (e) can be simplified to

$$w_1 = \frac{1}{l_1} \quad w_2 = \frac{1}{l_2} \quad w_3 = \frac{1}{l_3}$$

In summary, it has been shown that *weights of differential leveling lines are inversely proportional to their lengths,* and since any course length is proportional to its number of setups, *weights are also inversely proportional to the number of setups.*

10.7 PRACTICAL EXAMPLES

Example 10.2 Suppose that the angles in an equilateral triangle ABC were observed by the same operator using the same instrument, but the number of repetitions for each angle varied. The results were $A = 45°15'25''$, $n = 4$; $B = 83°37'22''$, $n = 8$; and $51°07'39''$, $n = 6$. Adjust the angles.

SOLUTION Weights proportional to the number of repetitions are assigned and corrections are made in inverse proportion to those weights. The sum of the three angles is $180°00'26''$, and thus the misclosure that must be adjusted is $26''$. The correction process is demonstrated in Table 10.1.

168 WEIGHTS OF OBSERVATIONS

TABLE 10.1 Adjustment of Example 10.2

Angle	n (Weight)	Correction Factor	Correction	Corrected Angle
A	4	(1/4) × 24 = 6	(6/13) × 26 = 12"	45°15'13"
B	8	(1/8) × 24 = 3	(3/13) × 26 = 06"	83°37'16"
C	6	(1/6) × 24 = 4	(4/13) × 26 = 08"	51°07'31"
		13	26"	180°00'00"

Note that a multiplier of 24 was used for convenience to avoid fractions in computing correction factors. Because weights are relative, this did not alter the adjustment. Note also that two computational checks are secured in the solution above; the sum of the individual corrections totaled 26", and the sum of the corrected angles totaled 180°00'00".

Example 10.3   In the leveling network of Figure 10.1, recall that the lengths of lines 1, 2, and 3 were 2, 3, and 4 miles, respectively. If the observed elevation differences in lines 1, 2, and 3 were +21.20 ft, +21.23 ft, and +21.29 ft, respectively, find the weighted mean for the elevation difference and the adjusted elevation of *BMX*. (*Note:* All level lines were run from *BMA* to *BMX*.)

SOLUTION   The weights of lines 1, 2, and 3 are 1/2, 1/3, and 1/4, respectively. Again since weights are relative, these weights can arbitrarily be multiplied by 12 to obtain weights of 6, 4, and 3, respectively. Applying Equation (10.13), the weighted mean is

$$\text{mean } \Delta\text{Elev} = \frac{6(21.20) + 4(21.23) + 3(21.29)}{6 + 4 + 3} = +21.23$$

Thus, the elevation of *BMX* = 100.00 + 21.23 = 123.23 ft. Note that if the weights had been neglected, the simple average would have given a mean elevation difference of +21.24.

Example 10.4   A distance is measured as 625.79 ft using a cloth tape and a given weight of 1; it is measured again as 625.71 ft using a steel tape and assigned a weight of 2; and finally, it is measured a third time as 625.69 ft with an EDM instrument and given a weight of 4. Calculate the most probable value of the length (*weighted mean*), and find the standard deviation of the weighted mean.

10.7 PRACTICAL EXAMPLES

SOLUTION By Equation (10.13), the weighted mean is

$$\overline{M} = \frac{1(625.79) + 2(623.71) + 4(625.69)}{1 + 2 + 4} = 625.71 \text{ ft}$$

By Equation (10.24), the standard deviation of the weighted mean is

$$S_{\overline{M}} = \sqrt{\frac{\Sigma w v^2}{(n-1)\Sigma w}} = \sqrt{\frac{0.0080}{(2)7}} = \pm 0.024 \text{ ft}$$

where

$v_1 = 625.71 - 625.79 = -0.08$	$w_1 v_1^2 = 1(-0.08)^2 =$	0.0064
$v_2 = 625.71 - 625.71 = 0.00$	$w_2 v_2^2 = 2(0.00)^2 =$	0.0000
$v_3 = 625.71 - 625.69 = +0.02$	$w_3 v_3^2 = 4(+0.02)^2 =$	$\underline{0.0016}$
		0.0080

Example 10.5 In leveling from benchmark A to benchmark B, four different routes of varying length are taken. The data of Table 10.2 are obtained. (Note that the weights were computed as $18/l$ for computational convenience only.)

Calculate the most probable elevation difference (weighted mean), the standard deviation of unit weight, the standard deviation of the weighted mean, and the standard deviations of the weighted observations.

SOLUTION By Equation (10.13), the weighted mean for elevation difference is

$$\overline{M} = \frac{18(25.35) + 9(25.41) + 6(25.38) + 3(25.30)}{18 + 9 + 6 + 3} = +25.366 \text{ ft}$$

TABLE 10.2 Route Data for Example 10.5

Route	Length (miles)	ΔElev	w
1	1	+25.35	18
2	2	+25.41	9
3	3	+25.38	6
4	6	+25.30	3

170 WEIGHTS OF OBSERVATIONS

TABLE 10.3 Data for Standard Deviations in Example 10.5

Route	w	v	v²	wv²
1	18	+0.016	0.0002	0.0045
2	9	−0.044	0.0019	0.0176
3	6	−0.014	0.0002	0.0012
4	3	+0.066	0.0043	0.0130
				0.0363

The arithmetic mean for this set of observations is 25.335, but the weighted mean is 25.366. To find the standard deviations for the weighted observations, the data in Table 10.3 are first created.

By Equation (10.20), the standard deviation of unit weight is

$$S_0 = \sqrt{\frac{0.0363}{3}} = \pm 0.11 \text{ ft}$$

By Equation (10.24), the standard deviation of the weighted mean is

$$S_{\bar{M}} = \sqrt{\frac{0.0363}{36(3)}} = \pm 0.018 \text{ ft}$$

By Equation (10.22), the standard deviations for the weighted observations are

$$S_1 = \sqrt{\frac{0.0363}{18(3)}} = \pm 0.026 \text{ ft} \quad S_2 = \sqrt{\frac{0.0363}{9(3)}} = \pm 0.037 \text{ ft}$$

$$S_3 = \sqrt{\frac{0.0363}{6(3)}} = \pm 0.045 \text{ ft} \quad S_4 = \sqrt{\frac{0.0363}{3(3)}} = \pm 0.063 \text{ ft}$$

PROBLEMS

10.1 An angle was measured as 49°27′30″ using an engineer's transit and had a standard deviation of ±30″. It was measured again using a repeating optical theodolite as 49°27′24″ with a standard deviation of ±10″. This angle was measured a third time with a directional theodolite as 49°27′22″ with a standard deviation of ±2″. Calculate the weighted mean of the angle and its standard deviation.

10.2 An angle was measured at four different times, with the following results:

Day	Angle	S
1	136°14'34"	±12.2"
2	136°14'36"	±6.7"
3	136°14'28"	±8.9"
4	136°14'26"	±9.5"

What is the most probable value for the angle and the standard deviation in the mean?

10.3 A distance was measured by pacing as 154 ft with a standard deviation of ±2.5 ft. It was then observed as 153.86 ft with a steel tape having a standard deviation of ±0.05 ft. Finally, it was measured as 153.89 ft with an EDM instrument with a standard deviation of ±0.02 ft. What is the most probable value for the distance and its standard deviation?

10.4 A distance was measured by pacing as 267 ft with a standard deviation of ±3 ft. It was then measured as 268.94 ft with a steel tape and had a standard deviation of ±0.05 ft. Finally, it was measured as 268.99 ft with an EDM. The EDM instrument and reflector setup standard deviations were ±0.005 ft and ±0.01 ft, respectively, and the manufacturer's estimated standard deviation for the EDM instrument is ±(3 mm + 3 ppm). What is the most probable value for the distance and the standard deviation of the weighted mean?

10.5 What standard deviation is computed for each weighted observation of Problem 10.4?

10.6 Compute the standard deviation for the taped observation in Problem 10.4 assuming standard deviations of ±5°F in temperature, ±3 lb in pull, ±0.005 ft in tape length, and ±0.01 ft in reading and marking the tape. The temperature at the time of the observation was 78°F, the calibrated tape length was 99.993 ft, and the field tension was recorded as 20 lb. The cross-sectional area of the tape is 0.004 in^2, its modulus of elasticity is 29,000,000 lb/in^2, its coefficient of thermal expansion is 6.45 × 10^{-5} °F^{-1}, and its weight is 2.5 lb. Assume horizontal taping with full tape lengths for all but the last partial distance with ends-only support.

10.7 Do Problem 10.3 using the standard deviation information for the EDM distance in Problem 10.4 and the tape calibration data in Problem 10.6.

10.8 A zenith angle was measured six times with both faces of a total station. The average direct reading is 88°05′16″ with a standard deviation of ±12.8″. With the reverse face, it was observed as 271°54′32″ with a standard deviation of ±9.6″. What is the most probable value for the zenith angle in the direct face?

10.9 Three crews level to a benchmark following three different routes. The lengths of the routes and the observed differences in elevation are:

Route	ΔElev (ft)	Length (ft)
1	14.80	3200
2	14.87	4800
3	14.83	3900

What is:

(a) the best value for the difference in elevation?
(b) the standard deviation for the weighted elevation difference?
(c) the standard deviation for the weighted observations?

10.10 Find the standard deviation for Problem 10.9.

CHAPTER 11

PRINCIPLES OF LEAST SQUARES

11.1 INTRODUCTION

In surveying, observations must often satisfy established numerical relationships known as *geometric constraints*. As examples, in a closed polygon traverse, horizontal angle and distance measurements should conform to the geometric constraints given in Section 8.4, and in a differential leveling loop, the elevation differences should sum to a given quantity. However, because the geometric constraints meet perfectly rarely, if ever, the data are adjusted.

As discussed in earlier chapters, errors in observations conform to the laws of probability; that is, they follow normal distribution theory. Thus, they should be adjusted in a manner that follows these mathematical laws. Whereas the mean has been used extensively throughout history, the earliest works on least squares started in the late eighteenth century. Its earliest application was primarily for adjusting celestial observations. Laplace first investigated the subject and laid its foundation in 1774. The first published article on the subject, entitled "Méthode des moindres quarrés" (Method of Least Squares), was written in 1805 by Legendre. However, it is well known that although Gauss did not publish until 1809, he developed and used the method extensively as a student at the University of Göttingen beginning in 1794 and thus is given credit for the development of the subject. In this chapter, equations for performing least squares adjustments are developed and their use is illustrated with several examples.

11.2 FUNDAMENTAL PRINCIPLE OF LEAST SQUARES

To develop the principle of least squares, a specific case is considered. Suppose that there are n independent equally weighted measurements, $z_1, z_2, \ldots, z_n$, of the same quantity z, which has a *most probable value* denoted by M. By definition,

$$\begin{aligned} M - z_1 &= v_1 \\ M - z_2 &= v_2 \\ &\vdots \\ M - z_n &= v_n \end{aligned} \tag{11.1}$$

where the v's are the residual errors. Note that residuals behave in a manner similar to errors, and thus they can be used interchangeably in the normal distribution function given by Equation (3.2). Substituting v for x, there results

$$f_x(v) = y = \frac{1}{\sigma\sqrt{2\pi}} e^{-v^2/2\sigma^2} = Ke^{-h^2v^2} \tag{11.2}$$

where $h = 1/\sigma\sqrt{2}$ and $K = h/\sqrt{\pi}$.

As discussed in Chapter 3, probabilities are represented by areas under the normal distribution curve. Thus, the individual probabilities for the occurrence of residuals $v_1, v_2, \ldots, v_n$ are obtained by multiplying their respective ordinates $y_1, y_2, \ldots, y_n$ by some infinitesimally small increment of v, Δv. The following probability statements result:

$$\begin{aligned} P_1 &= y_1\,\Delta v = Ke^{-h^2v_1^2}\,\Delta v \\ P_2 &= y_2\,\Delta v = Ke^{-h^2v_2^2}\,\Delta v \\ &\vdots \\ P_n &= y_n\,\Delta v = Ke^{-h^2v_n^2}\,\Delta v \end{aligned} \tag{11.3}$$

From Equation (3.1), the probability of the simultaneous occurrence of all the residuals v_1 through v_n is the product of the individual probabilities and thus

$$P = (Ke^{-h^2v_1^2}\,\Delta v)(Ke^{-h^2v_2^2}\,\Delta v) \cdots (Ke^{-h^2v_n^2}\,\Delta v) \tag{11.4}$$

Simplifying Equation (11.4) yields

11.2 FUNDAMENTAL PRINCIPLE OF LEAST SQUARES

$$P = K^n(\Delta v)^n e^{-h^2(v_1^2+v_2^2+\cdots+v_n^2)} \tag{11.5}$$

M is the quantity that is to be selected in such a way that it gives the greatest probability of occurrence, or, stated differently, the value of M that maximizes the value of P. Figure 11.1 shows a plot of e^{-x} versus x. From this plot it is readily seen that e^{-x} is maximized by minimizing x, and thus in relation to Equation (11.5), the probability P is maximized when the quantity $v_1^2 + v_2^2 + \cdots + v_n^2$ is minimized. In other words, *to maximize P, the sum of the squares of the residuals must be minimized*. Equation (11.6) expresses the *fundamental principle of least squares*:

$$\sum v^2 = v_1^2 + v_2^2 + \cdots + v_n^2 = \text{minimum} \tag{11.6}$$

This condition states: *The most probable value (MPV) for a quantity obtained from repeated observations of equal weight is the value that renders the sum of the residuals squared a minimum.* From calculus, the minimum value of a function can be found by taking its first derivative and equating the resulting function with zero. That is, the condition stated in Equation (11.6) is enforced by taking the first derivative of the function with respect to the unknown variable M and setting the results equal to zero. Substituting Equation (11.1) into Equation (11.6) yields

$$\sum v^2 = (M - z_1)^2 + (M - z_2)^2 + \cdots + (M - z_n)^2 \tag{11.7}$$

Taking the first derivative of Equation (11.7) with respect to M and setting the resulting equation equal to zero yields

$$\frac{d(\sum v^2)}{dM} = 2(M - z_1)(1) + 2(M - z_2)(1) + \cdots + 2(M - z_n)(1) = 0 \tag{11.8}$$

Now dividing Equation (11.8) by 2 and simplifying yields

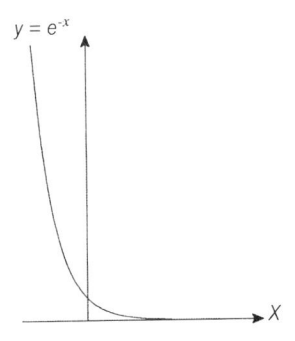

Figure 11.1 Plot of e^{-x}.

$$M - z_1 + M - z_2 + \cdots + M - z_n = 0$$

$$nM = z_1 + z_2 + \cdots + z_n$$

$$M = \frac{z_1 + z_2 + \cdots + z_n}{n} \tag{11.9}$$

In Equation (11.9) the quantity $(z_1 + z_2 + \cdots + z_n)/n$ is the mean of the values observed. *This is proof that when a quantity has been observed independently several times, the MPV is the arithmetic mean.*

11.3 FUNDAMENTAL PRINCIPLE OF WEIGHTED LEAST SQUARES

In Section 11.2, the fundamental principle of a least squares adjustment was developed for observations having equal or unit weights. The more general case of least squares adjustment assumes that the observations have varying degrees of precision, and thus varying weights.

Consider a set of measurements $z_1, z_2, \ldots, z_n$ having relative weights $w_1, w_2, \ldots, w_n$ and residuals $v_1, v_2, \ldots, v_n$. Denote the weighted MPV as M. As in Section 11.2, the residuals are related to the observations through Equations (11.1), and the total probability of their simultaneous occurrence is given by Equation (11.5). However, notice in Equation (11.2) that $h^2 = 1/2\sigma^2$, and since weights are inversely proportional to variances, they are directly proportional to h^2. Thus, Equation (11.5) can be rewritten as

$$P = K^n(\Delta v)^n e^{-(w_1 v_1^2 + w_2 v_2^2 + \cdots + w_n v_n^2)} \tag{11.10}$$

To maximize P in Equation (11.10), the negative exponent must be minimized. To achieve this, the sum of the products of the weights times their respective residuals squared must be minimized. This is is the condition imposed in weighted least squares adjustment. The condition of weighted least squares adjustment in equation form is

$$w_1 v_1^2 + w_2 v_2^2 + \cdots + w_n v_n^2 = \Sigma\, wv^2 \to \text{minimum} \tag{11.11}$$

Substituting the values for the residuals given in Equation (11.1) into Equation (11.11) yields

$$w_1(M - z_1)^2 + w_2(M - z_2)^2 + \cdots + w_n(M - z_n)^2 \to \text{minimum} \tag{11.12}$$

The condition for a weighted least squares adjustment is: *The most probable value for a quantity obtained from repeated observations having various weights is that value which renders the sum of the weight times their respective squared residual a minimum.*

The minimum condition is imposed by differentiating Equation (11.12) with respect to M and setting the resulting equation equal to zero. This yields

$$2w_1(M - z_1)(1) + 2w_2(M - z_2)(1) + \cdots + 2w_n(M - z_n)(1) = 0 \quad (11.13)$$

Dividing Equation (11.13) by 2 and rearranging results in

$$w_1(M - z_1) + w_2(M - z_2) + \cdots + w_n(M - z_n) = 0 \quad (11.14a)$$

Rearranging Equation (11.14a) gives

$$w_1 z_1 + w_2 z_2 + \cdots + w_n z_n = w_1 M + w_2 M + \cdots + w_n M \quad (11.14b)$$

Equation (11.14b) can be written as $\Sigma wz = \Sigma wM$. Thus,

$$M = \frac{\Sigma wz}{\Sigma w} \quad (11.15)$$

Notice that Equation (11.15) is the same as Equation (10.13), which is the formula for computing the weighted mean.

11.4 STOCHASTIC MODEL

The determination of variances, and subsequently the weights of the observations, is known as the *stochastic model* in a least squares adjustment. In Section 11.3 the inclusion of weights in the adjustment was discussed. It is crucial to the adjustment to select a proper stochastic (weighting) model since, as was discussed in Section 10.1, the weight of an observation controls the amount of correction it receives during the adjustment. However, development of the stochastic model is important not only to weighted adjustments. When doing an unweighted adjustment, all observations are assumed to be of equal weight, and thus the stochastic model is created implicitly. The foundations for selecting a proper stochastic model in surveying were established in Chapters 7 to 10. It will be shown in Chapter 21 that failure to select the stochastic model properly will also affect one's ability to isolate blunders in observation sets.

11.5 FUNCTIONAL MODEL

A functional model in adjustment computations is an equation or set of equations that represents or defines an adjustment condition. It must be either known or assumed. If the functional model represents the physical situation adequately, the observational errors can be expected to conform to the normal distribution curve. For example, a well-known functional model states that the sum of angles in a triangle is 180°. This model is adequate if the survey is limited to a small region. However, when the survey covers very large areas, this model does not account for the systematic errors caused by Earth's curvature. In this case, the functional model is inadequate and needs to be modified to include corrections for spherical excess. In traversing, the functional model of plane computations is suitable for smaller surveys, but if the extent of the survey becomes too large, the model must again be changed to account for the systematic errors caused by Earth's curvature. This can be accomplished by transforming the observations into a plane mapping system such as the state plane coordinate system or by using geodetic observation equations. Needless to say, if the model does not fit the physical situation, an incorrect adjustment will result. In Chapter 23 we discuss a three-dimensional geodetic and the systematic errors that must be taken into account in a three-dimensional geodetic adjustment.

There are two basic forms for functional models: the conditional and parametric adjustments. In a *conditional adjustment,* geometric conditions are enforced on the observations and their residuals. Examples of conditional adjustment are: (1) the sum of the angles in a closed polygon is $(n-2)180°$, where n is the number of sides in the polygon; (2) the latitudes and departures of a polygon traverse sum to zero; and (3) the sum of the angles in the horizon equal 360°. A least squares adjustment example using condition equations is given in Section 11.13.

When performing a *parametric adjustment,* observations are expressed in terms of unknown parameters that were never observed directly. For example, the well-known coordinate equations are used to model the angles, directions, and distances observed in a horizontal plane survey. The adjustment yields the most probable values for the coordinates (parameters), which in turn provide the most probable values for the adjusted observations.

The choice of the functional model will determine which quantities or parameters are adjusted. A primary purpose of an adjustment is to ensure that all observations are used to find the most probable values for the unknowns in the model. In least squares adjustments, no matter if conditional or parametric, the geometric checks at the end of the adjustment are satisfied and the same adjusted observations are obtained. In complicated networks, it is often difficult and time consuming to write the equations to express all conditions that must be met for a conditional adjustment. Thus, this book will focus on the parametric adjustment, which generally leads to larger systems of equations but is straightforward in its development and solution and, as a result, is well suited to computers.

The combination of stochastic and functional models results in a mathematical model for the adjustment. The stochastic and functional models must both be correct if the adjustment is to yield the most probable values for the unknown parameters. That is, it is just as important to use a correct stochastic model as it is to use a correct functional model. Improper weighting of observations will result in the unknown parameters being determined incorrectly.

11.6 OBSERVATION EQUATIONS

Equations that relate observed quantities to both observational residuals and independent unknown parameters are called *observation equations*. One equation is written for each observation and for a unique set of unknowns. For a unique solution of unknowns, the number of equations must equal the number of unknowns. Usually, there are more observations (and hence equations) than unknowns, and this permits determination of the most probable values for the unknowns based on the principle of least squares.

11.6.1 Elementary Example of Observation Equation Adjustment

As an example of a least squares adjustment by the observation equation method, consider the following three equations:

$$\begin{aligned}(1)\ & x + y = 3.0 \\ (2)\ & 2x - y = 1.5 \\ (3)\ & x - y = 0.2\end{aligned} \quad (11.16)$$

Equations (11.16) relate the two unknowns, x and y, to the quantities observed (the values on the right side of the equations). One equation is redundant since the values for x and y can be obtained from any two of the three equations. For example, if Equations (1) and (2) are solved, x would equal 1.5 and y would equal 1.5, but if Equations (2) and (3) are solved, x would equal 1.3 and y would equal 1.1, and if Equations (1) and (3) are solved, x would equal 1.6 and y would equal 1.4. Based on the inconsistency of these equations, the observations contain errors. Therefore, new expressions, called *observation equations*, can be rewritten that include residuals. The resulting set of equations is

$$\begin{aligned}(4)\ & x + y - 3.0 = v_1 \\ (5)\ & 2x - y - 1.5 = v_2 \\ (6)\ & x - y - 0.2 = v_3\end{aligned} \quad (11.17)$$

Equations (11.17) relate the unknown parameters to the observations and their errors. Obviously, it is possible to select values of v_1, v_2, and v_3 that will yield the same values for x and y no matter which pair of equations are used. For example, to obtain consistencies through all of the equations, ar-

180 PRINCIPLES OF LEAST SQUARES

bitrarily let $v_1 = 0$, $v_2 = 0$, and $v_3 = -0.2$. In this arbitrary solution, x would equal 1.5 and y would equal 1.5, no matter which pair of equations is solved. This is a consistent solution, however, there are other values for the v's that will produce a smaller sum of squares.

To find the least squares solution for x and y, the residual equations are squared and these squared expressions are added to give a function, $f(x,y)$, that equals Σv^2. Doing this for Equations (11.17) yields

$$f(x,y) = \sum v^2 = (x + y - 3.0)^2 + (2x - y - 1.5)^2 + (x - y - 0.2)^2$$

(11.18)

As discussed previously, to minimize a function, its derivatives must be set equal to zero. Thus, in Equation (11.18), the partial derivatives of Equation (11.18) with respect to each unknown must be taken and set equal to zero. This leads to the two equations

$$\frac{\partial f(x,y)}{\partial x} = 2(x + y - 3.0) + 2(2x - y - 1.5)(\ ) + 2(x - y - 0.2) = 0$$

$$\frac{\partial f(x,y)}{\partial y} = 2(x + y - 3.0) + 2(2x - y - 1.5)(-1) + 2(x - y - 0.2)(-1) = 0$$

(11.19)

Equations (11.19) are called *normal equations*. Simplifying them gives reduced normal equations of

$$6x - 2y - 6.2 = 0$$
$$-2x + 3y - 1.3 = 0$$

(11.20)

Simultaneous solution of Equations (11.20) yields x equal to 1.514 and y equal to 1.442. Substituting these adjusted values into Equations (11.17), numerical values for the three residuals can be computed. Table 11.1 provides a comparison of the arbitrary solution to the least squares solution. The tab-

TABLE 11.1 Comparison of an Arbitrary and a Least Squares Solution

Arbitrary Solution		Least Squares Solution	
$v_1 = 0$	$v_1^2 = 0.0$	$v_1 = -0.044$	$v_1^2 = 0.002$
$v_2 = 0$	$v_2^2 = 0.0$	$v_2 = 0.085$	$v_2^2 = 0.007$
$v_3 = -0.02$	$v_3^2 = 0.04$	$v_3 = -0.128$	$v_3^2 = 0.016$
	0.04		0.025

ulated summations of residuals squared shows that the least squares solution yields the smaller total and thus the better solution. In fact, it is the most probable solution for the unknowns based on the observations.

11.7 SYSTEMATIC FORMULATION OF THE NORMAL EQUATIONS

11.7.1 Equal-Weight Case

In large systems of observation equations, it is helpful to use systematic procedures to formulate the normal equations. In developing these procedures, consider the following generalized system of linear observation equations having variables of $(A, B, C, \ldots, N)$:

$$a_1 A + b_1 B + c_1 C + \cdots + n_1 N = l_1 + v_1$$
$$a_2 A + b_2 B + c_2 C + \cdots + n_2 N = l_2 + v_2 \qquad (11.21)$$
$$\vdots$$
$$a_m A + b_m B + c_m C + \cdots + n_m N = l_m + v_m$$

The squares of the residuals for Equations (11.21) are

$$v_1^2 = (a_1 A + b_1 B + c_1 C + \cdots + n_1 N - l_1)^2$$
$$v_2^2 = (a_2 A + b_2 B + c_2 C + \cdots + n_2 N - l_2)^2 \qquad (11.22)$$
$$\vdots$$
$$v_m^2 = (a_m A + b_m B + c_m C + \cdots + n_m N - l_m)^2$$

Summing Equations (11.22), the function $f(A,B,C, \ldots, N) = \Sigma v^2$ is obtained. This expresses the equal-weight least squares condition as

$$\Sigma v^2 = (a_1 A + b_1 B + c_1 C + \cdots + n_1 N - l_1)^2$$
$$+ (a_2 A + b_2 B + c_2 C + \cdots + n_2 N - l_2)^2$$
$$+ \cdots + (a_m A + b_m B + c_m C + \cdots + n_m N - l_m)^2 \qquad (11.23)$$

According to least squares theory, the minimum for Equation (11.23) is found by setting the partial derivatives of the function with respect to each unknown equal to zero. This results in the normal equations

182 PRINCIPLES OF LEAST SQUARES

$$\frac{\partial \sum v^2}{\partial A} = 2(a_1A + b_1B + c_1C + \cdots + n_1N - l_1)a_1$$
$$+ 2(a_2A + b_2B + c_2C + \cdots + n_2N - l_2)a_2 + \cdots$$
$$+ 2(a_mA + b_mB + c_mC + \cdots + n_mN - l_m)a_m = 0$$

$$\frac{\partial \sum v^2}{\partial B} = 2(a_1A + b_1B + c_1C + \cdots + n_1N - l_1)b_1$$
$$+ 2(a_2A + b_2B + c_2C + \cdots + n_2N - l_2)b_2 + \cdots$$
$$+ 2(a_mA + b_mB + c_mC + \cdots + n_mN - l_m)b_m = 0$$

$$\frac{\partial \sum v^2}{\partial C} = 2(a_1A + b_1B + c_1C + \cdots + n_1N - l_1)c_1 \qquad (11.24)$$
$$+ 2(a_2A + b_2B + c_2C + \cdots + n_2N - l_2)c_2 + \cdots$$
$$+ 2(a_mA + b_mB + c_mC + \cdots + n_mN - l_m)c_m = 0$$

$$\vdots$$

$$\frac{\partial \sum v^2}{\partial N} = 2(a_1A + b_1B + c_1C + \cdots + n_1N - l_1)n_1$$
$$+ 2(a_2A + b_2B + c_2C + \cdots + n_2N - l_2)n_2 + \cdots$$
$$+ 2(a_mA + b_mB + c_mC + \cdots + n_mN - l_m)n_m = 0$$

Dividing each expression by 2 and regrouping the remaining terms in Equation (11.24) results in

$$(a_1^2 + a_2^2 + \cdots + a_m^2)A + (a_1b_1 + a_2b_2 + \cdots + a_mb_m)B$$
$$+ (a_1c_1 + a_2c_2 + \cdots + a_mc_m)C + \cdots + (a_1n_1 + a_2n_2 + \cdots + a_mn_m)N$$
$$- (a_1l_1 + a_2l_2 + \cdots + a_ml_m) = 0$$
$$(b_1a_1 + b_2a_2 + \cdots + b_ma_m)A + (b_1^2 + b_2^2 + \cdots + b_m^2)B$$
$$+ (b_1c_1 + b_2c_2 + \cdots + b_mc_m)C + \cdots + (b_1n_1 + b_2n_2 + \cdots + b_mn_m)N$$
$$- (b_1l_1 + b_2l_2 + \cdots + b_ml_m) = 0 \qquad (11.25)$$

11.7 SYSTEMATIC FORMULATION OF THE NORMAL EQUATIONS

$$(c_1a_1 + c_2a_2 + \cdots + c_ma_m)A + (c_1b_1 + c_2b_2 + \cdots + c_mb_m)B$$
$$+ (c_1^2 + c_2^2 + \cdots + c_m^2)C + \cdots + (c_1n_1 + c_2n_2 + \cdots + c_mn_m)N$$
$$- (c_1l_1 + c_2l_2 + \cdots + c_ml_m) = 0$$
$$\vdots$$
$$(n_1a_1 + n_2a_2 + \cdots + n_ma_m)A + (n_1b_1 + n_2b_2 + \cdots + n_mb_m)B$$
$$+ (n_1c_1 + n_2c_2 + \cdots + n_mc_m)C + \cdots + (n_1^2 + n_2^2 + \cdots + n_m^2)N$$
$$- (n_1l_1 + n_2l_2 + \cdots + n_ml_m) = 0$$

Generalized equations expressing normal Equations (11.25) are now written as

$$\left(\sum a^2\right)A + \left(\sum ab\right)B + \left(\sum ac\right)C + \cdots + \left(\sum an\right)N = \sum al$$
$$\left(\sum ba\right)A + \left(\sum b^2\right)B + \left(\sum bc\right)C + \cdots + \left(\sum bn\right)N = \sum bl \quad (11.26)$$
$$\left(\sum ca\right)A + \left(\sum cb\right)B + \left(\sum c^2\right)C + \cdots + \left(\sum cn\right)N = \sum cl$$
$$\left(\sum na\right)A + \left(\sum nb\right)B + \left(\sum nc\right)C + \cdots + \left(\sum n^2\right)N = \sum nl$$

In Equation (11.26) the a's, b's, c's, ..., n's are the coefficients for the unknowns A, B, C, ..., N; the l values are the observations; and Σ signifies summation from $i = 1$ to m.

11.7.2 Weighted Case

In a manner similar to that of Section 11.7.1, it can be shown that normal equations can be formed systematically for weighted observation equations in the following manner:

$$\sum (wa^2)A + \sum (wab)B + \sum (wac)C + \cdots + \sum (wan)N = \sum wal$$
$$\sum (wba)A + \sum (wb^2)B + \sum (wbc)C + \cdots + \sum (wbn)N = \sum wbl$$
$$\sum (wca)A + \sum (wcb)B + \sum (wc^2)C + \cdots + \sum (wcn)N = \sum wcl$$
$$\vdots$$
$$\sum (wna)A + \sum (wnb)B + \sum (wnc)C + \cdots + \sum (wn^2)N = \sum wnl$$
$$(11.27)$$

In Equation (11.27), w are the weights of the observations, l; the a's, b's, c's, ..., n's are the coefficients for the unknowns A, B, C, ..., N; the l values are the observations; and Σ signifies summation from $i = 1$ to m.

Notice that the terms in Equations (11.27) are the same as those in Equations (11.26) except for the addition of the w's which are the relative weights of the observations. In fact, Equations (11.27) can be thought of as the general set of equations for forming the normal equations, since if the weights are equal, they can all be given a value of 1. In this case they will cancel out of Equations (11.27) to produce the special case given by Equations (11.26).

11.7.3 Advantages of the Systematic Approach

Using the systematic methods just demonstrated, the normal equations can be formed for a set of linear equations without writing the residual equations, compiling their summation equation, or taking partial derivatives with respect to the unknowns. Rather, for any set of linear equations, the normal equations for the least squares solution can be written directly.

11.8 TABULAR FORMATION OF THE NORMAL EQUATIONS

Formulation of normal equations from observation equations can be simplified further by handling Equations (11.26) and (11.27) in a table. In this way, a large quantity of numbers can be manipulated easily. Tabular formulation of the normal equations for the example in Section 11.4.1 is illustrated below. First, Equations (11.17) are made compatible with the generalized form of Equations (11.21). This is shown in Equations (11.28).

$$
\begin{aligned}
(7)\ & x + y = 3.0 + v_1 \\
(8)\ & 2x - y = 1.5 + v_2 \\
(9)\ & x - y = 0.2 + v_3
\end{aligned}
\qquad (11.28)
$$

In Equations (11.28), there are two unknowns, x and y, with different coefficients for each equation. Placing the coefficients and the observations, l's, for each expression of Equation (11.28) into a table, the normal equations are formed systematically. Table 11.2 shows the coefficients, appropriate products, and summations in accordance with Equations (11.26).

After substituting the appropriate values for Σa^2, Σab, Σb^2, Σal, and Σbl from Table 11.2 into Equations (11.26), the normal equations are

$$
\begin{aligned}
6x - 2y &= 6.2 \\
-2x + 3y &= 1.3
\end{aligned}
\qquad (11.29)
$$

TABLE 11.2 Tabular Formation of Normal Equations

Eq.	a	b	l	a^2	ab	b^2	al	bl
7	1	1	3.0	1	1	1	3.0	3.0
8	2	−1	1.5	4	−2	1	3.0	−1.5
9	1	−1	0.2	1	−1	1	0.2	−0.2
				6	−2	3	6.2	1.3

Notice that Equations (11.29) are exactly the same as those obtained in Section 11.6 using the theoretical least squares method. That is, Equations (11.29) match Equations (11.20).

11.9 USING MATRICES TO FORM THE NORMAL EQUATIONS

Note that the number of normal equations in a parametric least squares adjustment is always equal to the number of unknown variables. Often, the system of normal equations becomes quite large. But even when dealing with three unknowns, their solution by hand is time consuming. As a consequence, computers and matrix methods as described in Appendixes A through C are used almost always today. In the following subsections we present the matrix methods used in performing a least squares adjustment.

11.9.1 Equal-Weight Case

To develop the matrix expressions for performing least squares adjustments, an analogy will be made with the systematic procedures demonstrated in Section 11.7. For this development, let a system of observation equations be represented by the matrix notation

$$AX = L + V \qquad (11.30)$$

where

$$A = \begin{bmatrix} a_{11} & a_{12} & \cdots & a_{1n} \\ a_{21} & a_{22} & \cdots & a_{2n} \\ \vdots & \vdots & \vdots & \vdots \\ a_{m1} & a_{m2} & \cdots & a_{mn} \end{bmatrix}$$

$$X = \begin{bmatrix} x_1 \\ x_2 \\ \vdots \\ x_n \end{bmatrix} \quad L = \begin{bmatrix} l_1 \\ l_2 \\ \vdots \\ l_m \end{bmatrix} \quad V = \begin{bmatrix} v_1 \\ v_2 \\ \vdots \\ v_m \end{bmatrix}$$

186 PRINCIPLES OF LEAST SQUARES

Note that the system of observation equations (11.30) is identical to Equations (11.21) except that the unknowns are $x_1, x_2, \ldots, x_n$ instead of $A, B, \ldots, N$, and the coefficients of the unknowns are $a_{11}, a_{12}, \ldots, a_{1n}$ instead of $a_1, b_1, \ldots, n_1$.

Subjecting the foregoing matrices to the manipulations given in the following expression, Equation (11.31) produces the normal equations [i.e., matrix Equation (11.31a) is exactly equivalent to Equations (11.26)]:

$$A^T A X = A^T L \qquad (11.31a)$$

Equation (11.31a) can also be expressed as

$$N X = A^T L \qquad (11.31b)$$

The correspondence between Equations (11.31) and (11.26) becomes clear if the matrices are multiplied and analyzed as follows:

$$A^T A = \begin{bmatrix} a_{11} & a_{21} & \cdots & a_{m1} \\ a_{12} & a_{22} & \cdots & a_{m2} \\ \vdots & \vdots & \vdots & \vdots \\ a_{1n} & a_{2n} & \cdots & a_{mn} \end{bmatrix} \begin{bmatrix} a_{11} & a_{12} & \cdots & a_{1n} \\ a_{21} & a_{22} & \cdots & a_{2n} \\ \vdots & \vdots & \vdots & \vdots \\ a_{m1} & a_{m2} & \cdots & a_{mn} \end{bmatrix} = \begin{bmatrix} n_{11} & n_{12} & \cdots & n_{1n} \\ n_{21} & n_{22} & \cdots & a_{2n} \\ \vdots & \vdots & \vdots & \vdots \\ a_{n1} & a_{n2} & \cdots & a_{nn} \end{bmatrix} = N$$

$$A^T L = \begin{bmatrix} a_{11} & a_{21} & \cdots & a_{m1} \\ a_{12} & a_{22} & \cdots & a_{m2} \\ \vdots & \vdots & \vdots & \vdots \\ a_{1n} & a_{2n} & \cdots & a_{mn} \end{bmatrix} \begin{bmatrix} l_1 \\ l_2 \\ \vdots \\ l_m \end{bmatrix} = \begin{bmatrix} \sum_{i=1}^{m} a_{i1} l_i \\ \sum_{i=1}^{m} a_{i2} l_i \\ \vdots \\ \sum_{i=1}^{m} a_{in} l_i \end{bmatrix}$$

The individual elements of the N matrix can be expressed in the following summation forms:

$$n_{11} = \sum_{i=1}^{m} a_{i1}^2 \qquad n_{12} = \sum_{i=1}^{m} a_{i1} a_{i2} \qquad \cdots \qquad n_{1n} = \sum_{i=1}^{m} a_{i1} a_{in}$$

$$n_{21} = \sum_{i=1}^{m} a_{i2} a_{i1} \qquad n_{22} = \sum_{i=1}^{m} a_{i2}^2 \qquad \cdots \qquad n_{2n} = \sum_{i=1}^{m} a_{i2} a_{in}$$

$$\vdots \qquad\qquad \vdots \qquad\qquad \ddots \qquad \vdots$$

$$n_{n1} = \sum_{i=1}^{m} a_{in} a_{i1} \qquad n_{n2} = \sum_{i=1}^{m} a_{in} a_{i2} \qquad \cdots \qquad n_{nn} = \sum_{i=1}^{m} a_{in}^2$$

11.9 USING MATRICES TO FORM THE NORMAL EQUATIONS

By comparing the summations above with those obtained in Equations (11.26), it should be clear that they are the same. Therefore, it is demonstrated that Equations (11.31a) and (11.31b) produce the normal equations of a least squares adjustment. By inspection, it can also be seen that the N matrix is always symmetric (i.e., $n_{ij} = n_{ji}$).

By employing matrix algebra, the solution of normal equations such as Equation (11.31a) is

$$X = (A^T A)^{-1} A^T L = N^{-1} A^T L \tag{11.32}$$

Example 11.1 To demonstrate this procedure, the problem of Section 11.6 will be solved. Equations (11.28) can be expressed in matrix form as

$$AX = \begin{bmatrix} 1 & 1 \\ 2 & -1 \\ 1 & -1 \end{bmatrix} \begin{bmatrix} x \\ y \end{bmatrix} = \begin{bmatrix} 3.0 \\ 1.5 \\ 0.2 \end{bmatrix} + \begin{bmatrix} v_1 \\ v_2 \\ v_3 \end{bmatrix} = L + V \tag{a}$$

Applying Equation (11.31) to Equation (a) yields

$$A^T A X = NX = \begin{bmatrix} 1 & 2 & 1 \\ 1 & -1 & -1 \end{bmatrix} \begin{bmatrix} 1 & 1 \\ 2 & -1 \\ 1 & -1 \end{bmatrix} \begin{bmatrix} x \\ y \end{bmatrix} = \begin{bmatrix} 6 & -2 \\ -2 & 3 \end{bmatrix} \begin{bmatrix} x \\ y \end{bmatrix} \tag{b}$$

$$A^T L = \begin{bmatrix} 1 & 2 & 1 \\ 1 & -1 & -1 \end{bmatrix} \begin{bmatrix} 3.0 \\ 1.5 \\ 0.2 \end{bmatrix} = \begin{bmatrix} 6.2 \\ 1.3 \end{bmatrix} \tag{c}$$

Finally, the adjusted unknowns, the X matrix, are obtained using the matrix methods of Equation (11.32). This yields

$$X = N^{-1} A^T L = \begin{bmatrix} 6 & -2 \\ -2 & 3 \end{bmatrix}^{-1} \begin{bmatrix} 6.2 \\ 1.3 \end{bmatrix} = \begin{bmatrix} 1.514 \\ 1.442 \end{bmatrix} \tag{d}$$

Notice that the normal equations and the solution in this method are the same as those obtained in Section 11.6.

11.9.2 Weighted Case

A system of weighted linear observation equations can be expressed in matrix notation as

$$WAX = WL + WV \tag{11.33}$$

Using the methods demonstrated in Section 11.9.1, it is possible to show that the normal equations for this weighted system are

$$A^TWAX = A^TWL \qquad (11.34a)$$

Equation (11.34a) can also be expressed as

$$NX = A^TWL \qquad (11.34b)$$

where $N = A^TWA$.

Using matrix algebra, the least squares solution of these weighted normal equations is

$$X = (A^TWA)^{-1}A^TWL = N^{-1}A^TWL \qquad (11.35)$$

In Equation (11.35), W is the weight matrix as defined in Chapter 10.

11.10 LEAST SQUARES SOLUTION OF NONLINEAR SYSTEMS

In Appendix C we discuss a method of solving a nonlinear system of equations using a Taylor series approximation. Following this procedure, the least squares solution for a system of nonlinear equations can be found as follows:

Step 1: Write the first-order Taylor series approximation for each equation.
Step 2: Determine initial approximations for the unknowns in the equations of step 1.
Step 3: Use matrix methods similar to those discussed in Section 11.9 to find the least squares solution for the equations of step 1 (these are corrections to the initial approximations).
Step 4: Apply the corrections to the initial approximations.
Step 5: Repeat steps 1 through 4 until the corrections become sufficiently small.

A system of nonlinear equations that are linearized by a Taylor series approximation can be written as

$$JX = K + V \qquad (11.36)$$

where the *Jacobian matrix* J contains the coefficients of the linearized observation equations. The individual matrices in Equation (11.36) are

11.10 LEAST SQUARES SOLUTION OF NONLINEAR SYSTEMS

$$J = \begin{bmatrix} \frac{\partial F_1}{\partial x_1} & \frac{\partial F_1}{\partial x_2} & \cdots & \frac{\partial F_1}{\partial x_n} \\ \frac{\partial F_2}{\partial x_1} & \frac{\partial F_2}{\partial x_2} & \cdots & \frac{\partial F_2}{\partial x_n} \\ \vdots & \vdots & \vdots & \vdots \\ \frac{\partial F_m}{\partial x_1} & \frac{\partial F_m}{\partial x_2} & \cdots & \frac{\partial F_m}{\partial x_n} \end{bmatrix}$$

$$X = \begin{bmatrix} dx_1 \\ dx_2 \\ \vdots \\ dx_n \end{bmatrix} \quad K = \begin{bmatrix} l_1 - f_1(x_1, x_2, \ldots, x_n) \\ l_2 - f_2(x_1, x_2, \ldots, x_n) \\ \vdots \\ l_m - f_m(x_1, x_2, \ldots, x_n) \end{bmatrix} \quad V = \begin{bmatrix} v_1 \\ v_2 \\ \vdots \\ v_m \end{bmatrix}$$

The vector of least squares corrections in the equally weighted system of Equation (11.36) is given by

$$X = (J^T J)^{-1} J^T K = N^{-1} J^T K \tag{11.37}$$

Similarly, the system of weighted equations is

$$WJX = WK \tag{11.38}$$

and its solution is

$$X = (J^T W J)^{-1} J^T W K = N^{-1} J^T W K \tag{11.39}$$

where W is the weight matrix as defined in Chapter 10. Notice that the least squares solution of a nonlinear system of equations is similar to the linear case. In fact, the only difference is the use of the Jacobian matrix rather than the coefficient matrix and the use of the K matrix rather than the observation matrix, L. Many authors use the same nomenclature for both the linear and nonlinear cases. In these cases, the differences in the two systems of equations are stated implicitly.

Example 11.2 Find the least squares solution for the following system of nonlinear equations:

$$\begin{array}{rl} F: & x + y - 2y^2 = -4 \\ G: & x^2 + y^2 = 8 \\ H: & 3x^2 - y^2 = 7.7 \end{array} \tag{e}$$

190 PRINCIPLES OF LEAST SQUARES

SOLUTION

Step 1: Determine the elements of the *J* matrix by taking partial derivatives of Equation (*e*) with respect to the unknowns *x* and *y*. Then write the first-order Taylor series equations.

$$\frac{\partial F}{\partial x} = 1 \qquad \frac{\partial G}{\partial x} = 2x \qquad \frac{\partial H}{\partial x} = 6x$$

$$\frac{\partial F}{\partial y} = 1 - 4y \qquad \frac{\partial G}{\partial y} = 2y \qquad \frac{\partial H}{\partial y} = -2y$$

$$JX = \begin{bmatrix} 1 & 1 - 4y_0 \\ 2x_0 & 2y_0 \\ 6x_0 & -2y_0 \end{bmatrix} \begin{bmatrix} dx \\ dy \end{bmatrix} = \begin{bmatrix} -4 - F(x_0, y_0) \\ 8 - G(x_0, y_0) \\ 7.7 - H(x_0, y_0) \end{bmatrix} = K \qquad (f)$$

Step 2: Determine initial approximations for the solution of the equations. Initial approximations can be derived by solving any two equations for *x* and *y*. This was done in Section C.3 for the equations for *F* and *G*, and their solution was $x_0 = 2$ and $y_0 = 2$. Using these values, the evaluation of the equations yields

$$F(x_0, y_0) = -4 \qquad G(x_0, y_0) = 8 \qquad H(x_0, y_0) = 8 \qquad (g)$$

Substituting Equations (*g*) into the *K* matrix of Equation (*f*), the *K* matrix becomes

$$K = \begin{bmatrix} -4 - (-4) \\ 8 - 8 \\ 7.7 - 8 \end{bmatrix} = \begin{bmatrix} 0 \\ 0 \\ -0.3 \end{bmatrix}$$

It should not be surprising that the first two rows of the *K* matrix are zero since the initial approximations were determined using these two equations. In successive iterations, these values will change and all terms will become nonzero.

Step 3: Solve the system using Equation (11.37).

$$N = J^\mathsf{T} J = \begin{bmatrix} 1 & 4 & 12 \\ -7 & 4 & -4 \end{bmatrix} \begin{bmatrix} 1 & -7 \\ 4 & 4 \\ 12 & -4 \end{bmatrix} = \begin{bmatrix} 161 & -39 \\ -39 & 81 \end{bmatrix} \qquad (h)$$

$$J^\mathsf{T} K = \begin{bmatrix} 1 & 4 & 12 \\ -7 & 4 & -1 \end{bmatrix} \begin{bmatrix} 0 \\ 0 \\ -0.3 \end{bmatrix} = \begin{bmatrix} -3.6 \\ 1.2 \end{bmatrix}$$

Substituting the matrices of Equation (h) into Equation (11.37), the solution for the first iteration is[1]

$$X = N^{-1}J^{T}K = \begin{bmatrix} -0.02125 \\ 0.00458 \end{bmatrix}$$

Step 4: Apply the corrections to the initial approximations for the first iteration.

$$x_0 = 2.00 - 0.02125 = 1.97875 \qquad y_0 = 2.00 + 0.00458 = 2.00458$$

Step 5: Repeating steps 2 through 4 results in

$$X = N^{-1}J^{T}K = \begin{bmatrix} 157.61806 & -38.75082 \\ -38.75082 & 81.40354 \end{bmatrix}^{-1} \begin{bmatrix} -0.017225 \\ -0.003307 \end{bmatrix} = \begin{bmatrix} -0.00011 \\ -0.00001 \end{bmatrix}$$

$$x = 1.97875 - 0.00011 = 1.97864$$

$$y = 2.00458 - 0.00001 = 2.00457$$

Iterating a third time yields extremely small corrections, and thus the final solution, rounded to the hundredths place, is $x = 1.98$ and $y = 2.00$. Notice that N changed by a relatively small amount from the first iteration to the second iteration. If the initial approximations are close to their final values, this can be expected. Thus, when doing these computations by hand, it is common to use the initial N for each iteration, making it only necessary to recompute $J^{T}K$ between iterations. However, this procedure should be used with caution since if the initial approximations are poor, it will result in an incorrect solution. One should always perform complete computations when doing the solution with the aid of a computer.

11.11 LEAST SQUARES FIT OF POINTS TO A LINE OR CURVE

Frequently in engineering work, it is desirable or necessary to fit a straight line or curve to a set of points with known coordinates. In solving this type

---

[1] Note that although the solution represents more significant figures than can be warranted by the observations, it is important to carry more digits than are desired for the final solution. Failure to carry enough digits can result in a system that will never converge; rather, it may *bounce* above and below the solution, or it may take more iterations, due to these rounding errors. This mistake has been made by many beginning students. The answer should be rounded only after solving the problem.

192 PRINCIPLES OF LEAST SQUARES

of problem, it is first necessary to decide on the appropriate functional model for the data. The decision as to whether to use a straight line, parabola, or some other higher-order curve can generally be made after plotting the data and studying their form or by checking the size of the residuals after the least squares solution with the first line or curve selected.

11.11.1 Fitting Data to a Straight Line

Consider the data illustrated in Figure 11.2. The straight line shown in the figure can be represented by the equation

$$y = mx + b \tag{11.40}$$

In Equation (11.40), x and y are the coordinates of a point, m is the slope of a line, and b is the y intercept at $x = 0$. If the points were truly linear and there were no observational or experimental errors, all coordinates would lie on a straight line. However, this is rarely the case, as shown in Figure 11.2, and thus it is possible that (1) the points contain errors, (2) the functional model is a higher-order curve, or both. If a line is selected as the model for the data, the equation of the best-fitting straight line is found by adding residuals to Equations (11.40). This accounts for the errors shown in the figure. Equations for the four data points A, B, C, and D of Figure 11.2 are rewritten as

$$\begin{aligned} y_A + v_{yA} &= mx_A + b \\ y_B + v_{yB} &= mx_B + b \\ y_C + v_{yC} &= mx_C + b \\ y_D + v_{yD} &= mx_D + b \end{aligned} \tag{11.41}$$

Equations (11.41) contain two unknowns, m and b, with four observations. Their matrix representation is

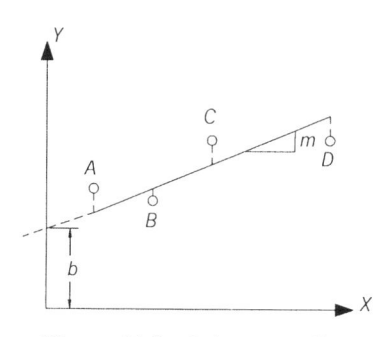

Figure 11.2 Points on a line.

11.11 LEAST SQUARES FIT OF POINTS TO A LINE OR CURVE

$$AX = L + V \tag{11.42}$$

where

$$A = \begin{bmatrix} x_a & 1 \\ x_b & 1 \\ x_c & 1 \\ x_d & 1 \end{bmatrix} \quad X = \begin{bmatrix} m \\ b \end{bmatrix} \quad L = \begin{bmatrix} y_a \\ y_b \\ y_c \\ y_d \end{bmatrix} \quad V = \begin{bmatrix} v_{y_a} \\ v_{y_b} \\ v_{y_c} \\ v_{y_d} \end{bmatrix}$$

Equation (11.42) is solved by the least squares method using Equation (11.32). If some data were more reliable than others, relative weights could be introduced and a weighted least squares solution could be obtained using Equation (11.35).

Example 11.3 Find the best-fit straight line for the following points, whose x and y coordinates are given in parentheses.

A: (3.00, 4.50)   B: (4.25, 4.25)   C: (5.50, 5.50)   D: (8.00, 5.50)

SOLUTION Following Equations (11.41), the four observation equations for the coordinate pairs are

$$3.00m + b = 4.50 + v_a$$
$$4.25m + b = 4.25 + v_b \tag{i}$$
$$5.50m + b = 5.50 + v_c$$
$$8.00m + b = 5.50 + v_d$$

Rewriting Equations (i) into matrix form yields

$$\begin{bmatrix} 3.00 & 1 \\ 4.25 & 1 \\ 5.50 & 1 \\ 8.00 & 1 \end{bmatrix} \begin{bmatrix} m \\ b \end{bmatrix} = \begin{bmatrix} 4.50 \\ 4.25 \\ 5.50 \\ 5.50 \end{bmatrix} + \begin{bmatrix} v_A \\ v_B \\ v_C \\ v_D \end{bmatrix} \tag{j}$$

To form the normal equations, premultiply matrices A and L of Equation (j) by A^T and get

$$\begin{bmatrix} 121.3125 & 20.7500 \\ 20.7500 & 4.0000 \end{bmatrix} \begin{bmatrix} m \\ b \end{bmatrix} = \begin{bmatrix} 105.8125 \\ 19.7500 \end{bmatrix} \tag{k}$$

The solution of Equation (k) is

194 PRINCIPLES OF LEAST SQUARES

$$X = \begin{bmatrix} m \\ b \end{bmatrix} = \begin{bmatrix} 121.3125 & 20.7500 \\ 20.7500 & 4.0000 \end{bmatrix}^{-1} \begin{bmatrix} 105.8125 \\ 19.7500 \end{bmatrix} = \begin{bmatrix} 0.246 \\ 3.663 \end{bmatrix}$$

Thus, the most probable values for m and b to the nearest hundredth are 0.25 and 3.66, respectively. To obtain the residuals, Equation (11.30) is rearranged and solved as

$$V = AX - L = \begin{bmatrix} 3.00 & 1 \\ 4.25 & 1 \\ 5.50 & 1 \\ 8.00 & 1 \end{bmatrix} \begin{bmatrix} 0.246 \\ 3.663 \end{bmatrix} - \begin{bmatrix} 4.50 \\ 4.25 \\ 5.50 \\ 5.50 \end{bmatrix} = \begin{bmatrix} -0.10 \\ 0.46 \\ -0.48 \\ 0.13 \end{bmatrix}$$

11.11.2 Fitting Data to a Parabola

For certain data sets or in special situations, a parabola will fit the situation best. An example would be fitting a vertical curve to an existing roadbed. The general equation of a parabola is

$$Ax^2 + Bx + C = y \qquad (11.43)$$

Again, since the data rarely fit the equation exactly, residuals are introduced. For the data shown in Figure 11.3, the following observation equations can be written:

$$\begin{aligned} Ax_a^2 + Bx_a + C &= y_a + v_a \\ Ax_b^2 + Bx_b + C &= y_b + v_b \\ Ax_c^2 + Bx_c + C &= y_c + v_c \\ Ax_d^2 + Bx_d + C &= y_d + v_d \\ Ax_e^2 + Bx_e + C &= y_e + v_e \end{aligned} \qquad (11.44)$$

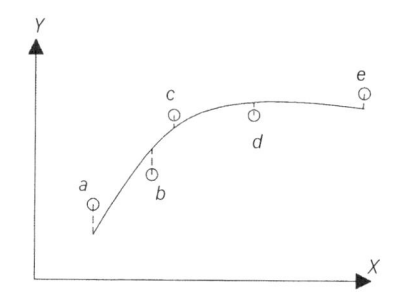

Figure 11.3 Points on a parabolic curve.

Equations (11.44) contain three unknowns, A, B, and C, with five equations. Thus, this represents a redundant system that can be solved using least squares. In terms of the unknown coefficients, Equations (11.44) are linear and can be represented in matrix form as

$$AX = L + V \tag{11.45}$$

Since this is a linear system, it is solved using Equation (11.32). If weights were introduced, Equation (11.35) would be used. The steps taken would be similar to those used in Section 11.11.1.

11.12 CALIBRATION OF AN EDM INSTRUMENT

Calibration of an EDM is necessary to ensure confidence in the distances it measures. In calibrating these devices, if they internally make corrections and reductions for atmospheric conditions, Earth curvature, and slope, it is first necessary to determine if these corrections are made properly. Once these corrections are applied properly, the instruments with their reflectors must be checked to determine their constant and scaling corrections. This is often accomplished using a calibration baseline. The observation equation for an electronically observed distance on a calibration baseline is

$$SD_A + C = D_H - D_A + V_{DH} \tag{11.46}$$

In Equation (11.46), S is a scaling factor for the EDM; C is an instrument–reflector constant; D_H is the horizontal distance observed with all atmospheric and slope corrections applied; D_A is the published horizontal calibrated distance for the baseline; and V_{DH} is the residual error for each observation. This is a linear equation with two unknowns, S and C. Systems of these equations can be solved using Equation (11.31).

Example 11.4 A surveyor wishes to use an instrument–reflector combination that has an unknown constant value. Calibration baseline observations were made carefully, and following the manufacturer's recommendations, the necessary corrections were applied for the atmospheric conditions, Earth curvature, and slope. Use these corrected distances and their published values, listed in Table 11.3, to determine the instrument–reflector constant (C) and scaling factor (S) for the system.

TABLE 11.3 EDM Instrument–Reflector Calibration Data

Distance	D_A (m)	D_H (m)	Distance	D_A (m)	D_H (m)
0–150	149.9975	150.0175	150–0	149.9975	150.0174
0–430	430.0101	430.0302	430–0	430.0101	430.0304
0–1400	1400.003	1400.0223	1400–0	1400.003	1400.0221
150–430	280.0126	280.0327	430–150	280.0126	280.0331
150–1400	1250.0055	1250.0248	1400–150	1250.0055	1250.0257
430–1400	969.9929	970.0119	430–1400	969.9929	970.0125

SOLUTION Following Equation (11.46), the matrix equation for this problem is

$$\begin{bmatrix} 149.9975 & 1 \\ 149.9975 & 1 \\ 430.0101 & 1 \\ 430.0101 & 1 \\ 1400.0030 & 1 \\ 1400.0030 & 1 \\ 280.0126 & 1 \\ 280.0126 & 1 \\ 1250.0055 & 1 \\ 1250.0055 & 1 \\ 969.9929 & 1 \\ 969.9929 & 1 \end{bmatrix} \begin{bmatrix} S \\ C \end{bmatrix} = \begin{bmatrix} 150.0175 - 149.9975 \\ 150.0174 - 149.9975 \\ 430.0302 - 430.0101 \\ 430.0304 - 430.0101 \\ 1400.0223 - 1400.0030 \\ 1400.0221 - 1400.0030 \\ 280.0327 - 280.0126 \\ 280.0331 - 280.0126 \\ 1250.0248 - 1250.0055 \\ 1250.0257 - 1250.0055 \\ 970.0119 - 969.9929 \\ 970.0125 - 969.9929 \end{bmatrix} + V$$

Using Equation (11.32), the solution is $S = -0.0000007$ (-0.7 ppm) and $C = 0.0203$. Thus, the constant value for the instrument–reflector pair is approximately 0.020 m, or 20 mm.

11.13 LEAST SQUARES ADJUSTMENT USING CONDITIONAL EQUATIONS

As stated in Section 11.5, observations can also be adjusted using conditional equations. In this section this form of adjustment is demonstrated by using the condition that the sum of the angles in the horizon at a single station must equal 360°.

Example 11.5 While observing angles at a station, the horizon was closed. The following observations and their standard deviations were obtained:

11.13 LEAST SQUARES ADJUSTMENT USING CONDITIONAL EQUATIONS

No.	Angle	S (″)
a_1	134°38′56″	±6.7
a_2	83°17′35″	±9.9
a_3	142°03′14″	±4.3

What are the most probable values for these observations?

SOLUTION In a conditional adjustment, the most probable set of residuals are found that satisfy a given functional condition. In this case, the condition is that the sum of the three angles is equal to 360°. Since the three angles observed actually sum to 359°59′45″, the angular misclosure is 15″. Thus, errors are present. The following residual equations are written for the observations listed above.

$$v_1 + v_2 + v_3 = 360° - (a_1 + a_2 + a_3) = 15'' \qquad (l)$$

In Equation (*l*), the *a*'s represent the observations and the *v*'s are residuals.

Applying the fundamental condition for a weighted least squares adjustment, the following equation must be minimized:

$$F = w_1 v_1^2 + w_2 v_2^2 + w_3 v_3^2 \qquad (m)$$

where the *w*'s are weights, which are the inverses of the squares of the standard deviations.

Equation (*l*) can be rearranged such that v_3 is expressed as a function of the other two residuals, or

$$v_3 = 15 - (v_1 + v_2) \qquad (n)$$

Substituting Equation (*n*) into Equation (*m*) yields

$$F = w_1 v_1^2 + w_2 v_2^2 + w_3 [15 - (v_1 + v_2)]^2 \qquad (o)$$

Taking the partial derivatives of *F* with respect to both v_1 and v_2, respectively, in Equation (*o*) results in the following two equations:

$$\frac{\partial F}{\partial v_1} = 2w_1 v_1 + 2w_3[15'' - (v_1 + v_2)](-1) = 0 \qquad (p)$$

$$\frac{\partial F}{\partial v_2} = 2w_2 v_2 + 2w_3[15'' - (v_1 + v_2)](-1) = 0$$

Rearranging Equations (*p*) and substituting in the appropriate weights yields the following normal equations:

198 PRINCIPLES OF LEAST SQUARES

$$\left(\frac{1}{6.7^2} + \frac{1}{4.3^2}\right)v_1 + \frac{1}{4.3^2}v_2 = 15\left(\frac{1}{4.3^2}\right)$$

$$\frac{1}{4.3^2}v_1 + \left(\frac{1}{9.9^2} + \frac{1}{4.3^2}\right)v_2 = 15\left(\frac{1}{4.3^2}\right)$$

(q)

Solving Equations (q) for v_1 and v_2 yields

$$v_1 = 4.2''$$

$$v_2 = 9.1''$$

By substituting these residual values into Equation (n), residual v_3 is computed as

$$v_3 = 15'' - (4.2'' + 9.1'') = 1.7''$$

Finally, the adjusted observations are obtained by adding to the observations the residuals that were computed.

No.	Observed Angle	v ('')	Adjusted Angle
a_1	134°38′56″	4.2	134°39′00.2″
a_2	83°17′35″	9.1	83°17′44.1″
a_3	142°03′14″	1.7	142°03′15.7″
			360°00′00.0″

Note that geometric closure has been enforced in the adjusted angles to make their sum exactly 360°. Note also that the angle having the smallest standard deviation received the smallest correction (i.e., its residual is smallest).

11.14 EXAMPLE 11.5 USING OBSERVATION EQUATIONS

Example 11.5 can also be done using observation equations. In this case, the three observations are related to their adjusted values and their residuals by writing observation equations

11.14 EXAMPLE 11.5 USING OBSERVATION EQUATIONS

$$a_1 = 134°38'56'' + v_1$$
$$a_2 = 83°17'35'' + v_2 \qquad (r)$$
$$a_3 = 142°03'14'' + v_3$$

While these equations relate the adjusted observations to their observed values, they cannot be solved in this form. What is needed is the *constraint*,[2] which states that the sum of the three angles equals 360°. This equation is

$$a_1 + a_2 + a_3 = 360° \qquad (s)$$

Rearranging Equation (s) to solve for a_3 yields

$$a_3 = 360° - (a_1 + a_2) \qquad (t)$$

Substituting Equation (t) into Equations (r) produces

$$a_1 = 134°38'56'' + v_1$$
$$a_2 = 83°17'35'' + v_2 \qquad (u)$$
$$360° - (a_1 + a_2) = 142°03'14'' + v_3$$

This is a linear problem with two unknowns, a_1 and a_2. The weighted observation equation solution is obtained by solving Equation (11.35). The appropriate matrices for this problem are

$$A = \begin{bmatrix} 1 & 0 \\ 0 & 1 \\ -1 & -1 \end{bmatrix} \quad W = \begin{bmatrix} \frac{1}{6.7^2} & 0 & 0 \\ 0 & \frac{1}{9.9^2} & 0 \\ 0 & 0 & \frac{1}{4.3^2} \end{bmatrix} \quad L = \begin{bmatrix} 134°38'56'' \\ 83°17'35'' \\ 142°03'14'' - 360° \end{bmatrix}$$

Performing matrix manipulations, the coefficients of the normal equations are

---

[2] Chapter 20 covers the use of constraint equations in least squares adjustment.

$$A^TWA = \begin{bmatrix} 1 & 0 & -1 \\ 0 & 1 & -1 \end{bmatrix} \begin{bmatrix} \frac{1}{6.7^2} & 0 & 0 \\ 0 & \frac{1}{9.9^2} & 0 \\ 0 & 0 & \frac{1}{4.3^2} \end{bmatrix} \begin{bmatrix} 1 & 0 \\ 0 & 1 \\ -1 & 1 \end{bmatrix}$$

$$= \begin{bmatrix} 0.07636 & 0.05408 \\ 0.050408 & 0.06429 \end{bmatrix}$$

$$A^TWL = \begin{bmatrix} 14.7867721 \\ 12.6370848 \end{bmatrix}$$

Finally, X is computed as

$$X = (A^TWA)^{-1}A^TWL = \begin{bmatrix} 134°39'00.2'' \\ 83°17'44.1'' \end{bmatrix}$$

Using Equation (t), it can now be determined that a_3 is $360° - 134°39'00.2'' - 83°17'44.1'' = 142°03'15.7''$. The same result is obtained as in Section 11.13. It is important to note that no matter what method of least squares adjustment is used, if the procedures are performed properly, the same solution will always be obtained. This example involved *constraint equation* (t). This topic is covered in more detail in Chapter 20.

PROBLEMS

11.1 Calculate the most probable values for A and B in the equations below by the method of least squares. Consider the observations to be of equal weight. (Use the tabular method to form normal equations.)
(a) $3A + 2B = 7.80 + v_1$
(b) $2A - 3B = 5.55 + v_2$
(c) $6A - 7B = 8.50 + v_3$

11.2 If observations (a), (b), and (c) in Problem 11.1 have weights of 6, 4, and 3, respectively, solve the equations for the most probable values of A and B using weighted least squares. (Use the tabular method to form normal equations.)

11.3 Repeat Problem 11.1 using matrices.

11.4 Repeat Problem 11.2 using matrices.

11.5 Solve the following nonlinear equations using the least squares method.
(a) $x^2 + 3xy - y^2 = 16.0$
(b) $7x^3 - 3y^2 = 71.7$
(c) $2x - 6xy + 3y^2 = 3.2$

11.6 The following coordinates of points on a line were computed for a block. What are the slope and y intercept of the line? What is the azimuth of the line?

Point	X (ft)	Y (ft)
1	1254.72	2951.76
2	1362.50	3205.13
3	1578.94	3713.80
4	1843.68	4335.92

11.7 What are the most probable values for the three angles observed to close the horizon at station Red. The observed values and their standard deviations are:

Angle	Value	S (")
1	123°32'56"	±2.5
2	110°07'28"	±1.5
3	126°19'44"	±4.9

11.8 Determine the most probable values for the three interior of a triangle that were measured as:

Angle	Value	S (")
1	58°26'48"	±5.1
2	67°06'56"	±4.3
3	54°26'24"	±2.6

11.9 Eight blocks of the Main Street are to be reconstructed. The existing street consists of short, jogging segments as tabulated in the traverse survey data below. Assuming coordinates of $X = 1000.0$ and $Y = 1000.0$ at station A, and that the azimuth of AB is 90°, define a new straight alignment for a reconstructed street passing through this area which best conforms to the present alignment. Give the Y intercept and the azimuth of the new alignment.

Course	Length (ft)	Station	Angle to Right
AB	635.74	B	180°01′26″
BC	364.82	C	179°59′52″
CD	302.15	D	179°48′34″
DE	220.08	E	180°01′28″
EF	617.36	F	179°59′05″
FG	429.04	G	180°01′37″
GH	387.33	H	179°59′56″
HI	234.28		

11.10 Use the ADJUST software to do Problem 11.9.

11.11 The property corners on a single block with an alley are shown as a straight line with a Due East bearing on a recorded plat. During a recent survey, all the lot corners were found, and measurements from station A to each were obtained. The surveyor wishes to determine the possibility of disturbance of the corners by checking their fit to a straight line. A sketch of the situation is shown in Figure P11.11, and the results of the survey are given below. Assuming that station A has coordinates of $X = 5000.00$ and $Y = 5000.00$ and that the coordinates of the backsight station are $X = 5000.10$ and $Y = 5200.00$, determine the best-fitting line for the corners. Give the Y intercept and the bearing of the best-fit line.

Course	Distance (ft)	Angle at A
AB	100.02	90°00′16″
AC	200.12	90°00′08″
AD	300.08	89°59′48″
AE	399.96	90°01′02″
AF	419.94	89°59′48″
AG	519.99	90°00′20″
AH	620.04	89°59′36″
AI	720.08	90°00′06″

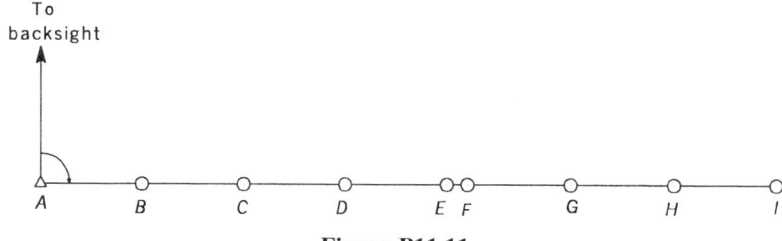

Figure P11.11

11.12 Use the ADJUST software to do Problem 11.11.

11.13 Calculate a best-fit parabola for the following data obtained on a survey of an existing vertical curve, and determine the deviation (residuals) of the road from this best-fit curve. The curve starts at station 10+00 and ends at station 18+00. List the adjusted station elevations and their residuals.

Station	Elevation	Station	Elevation
10+00	51.2	15+00	46.9
11+00	49.5	16+00	47.3
12+00	48.2	17+00	48.3
13+00	47.3	18+00	49.6
14+00	46.8		

11.14 Use the ADJUST software to do Problem 11.13.

11.15 Using a procedure similar to that in Section 11.7.1, derive Equations (11.27).

11.16 Using a procedure similar to that used in Section 11.9.1, show that the matrix operations in Equation (11.34) result in the normal equations for a linear set of weighted observation equations.

11.17 Discuss the importance of selecting the stochastic model when adjusting data.

11.18 The values for three angles in a triangle, observed using a total station and the directional method, are

Angle	Number of Repetitions	Value
A	2	14°25′20″
B	3	58°16′00″
C	6	107°19′10″

The observed lengths of the course are

$$AB = 971.25 \text{ ft} \quad BC = 253.25 \text{ ft} \quad CA = 865.28 \text{ ft}$$

The following estimated errors are assumed for each measurement:

$$\sigma_i = \pm 0.003 \text{ ft} \quad \sigma_t = \pm 0.020 \text{ ft} \quad \sigma_{\text{DIN}} = \pm 2.0''$$

What are the most probable values for the angles? Use the conditional equation method.

11.19 Do Problem 11.18 using observation equations and a constraint as presented in Section 11.13.

11.20 The following data were collected on a calibration baseline. Atmospheric refraction and Earth curvature corrections were made to the measured distances, which are in units of meters. Determine the instrument–reflector constant and any scaling factor.

Distance	D_A	D_H	Distance	D_A	D_H
0–150	149.9104	149.9447	150–0	149.9104	149.9435
0–430	430.001	430.0334	430–0	430.001	430.034
0–1400	1399.9313	1399.9777	1400–0	1399.9313	1399.9519
150–430	280.0906	280.1238	430–150	280.0906	280.123
150–1400	1250.0209	1250.0795	1400–150	1250.0209	1250.0664
430–1400	969.9303	969.9546	1400–430	969.9303	969.9630

11.21 A survey of the centerline of a horizontal metric curve is done to determine the as-built curve specifications. The coordinates for the points along the curve are:

Point	X (ft)	Y (ft)
1	10,006.82	10,007.31
2	10,013.12	10,015.07
3	10,024.01	10,031.83
4	10,032.44	10,049.95
5	10,038.26	10,069.04
6	10,041.39	10,088.83

(a) Using Equation (C.10), compute the most probable values for the radius and center of the circle.

(b) If two points located on the tangents have coordinates of (9987.36, 9987.40) and (10,044.09, 10,119.54), what are the coordinates of the *PC* and *PT* of the curve?

CHAPTER 12

ADJUSTMENT OF LEVEL NETS

12.1 INTRODUCTION

Differential leveling observations are used to determine differences in elevation between stations. As with all observations, these measurements are subject to random errors that can be adjusted using the method of least squares. In this chapter the observation equation method for adjusting differential leveling observations by least squares is developed, and several examples are given to illustrate the adjustment procedures.

12.2 OBSERVATION EQUATION

To apply the method of least squares in leveling adjustments, a prototype observation equation is first written for any elevation difference. Figure 12.1 illustrates the functional relationship for the elevation difference observed between two stations, I and J. The equation is expressed as

$$E_j - E_i = \Delta\text{Elev}_{ij} + v_{\Delta\text{Elev}_{ij}} \tag{12.1}$$

This prototype observation equation relates the unknown elevations of any two stations, I and J, with the differential leveling observation ΔElev_{ij} and its residual $\Delta v_{\text{Elev}_{ij}}$. This equation is fundamental in performing least squares adjustments of differential level nets.

205

206 ADJUSTMENT OF LEVEL NETS

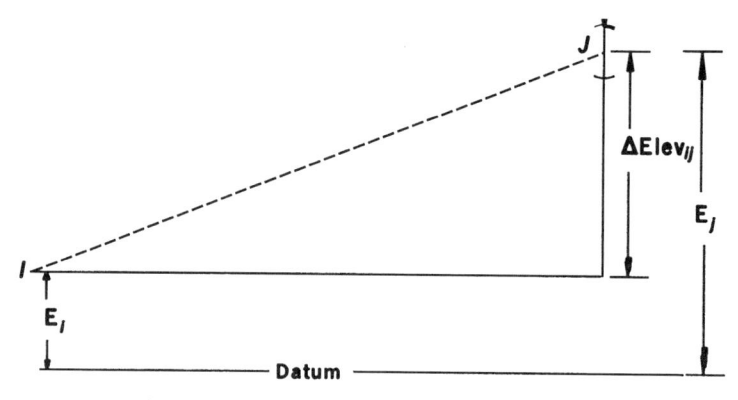

Figure 12.1 Differential leveling observation.

12.3 UNWEIGHTED EXAMPLE

In Figure 12.2, a leveling network and its survey data are shown. Assume that all observations are equal in weight. In this figure, arrows indicate the direction of leveling, and thus for line 1, leveling proceeds from benchmark X to A with an observed elevation difference of $+5.10$ ft. By substituting into prototype equation (12.1), an observation equation is written for each observation in Figure 12.2. The resulting equations are

Line	Observed Elevation Difference
1	5.10
2	2.34
3	−1.25
4	−6.13
5	−0.68
6	−3.00
7	1.70

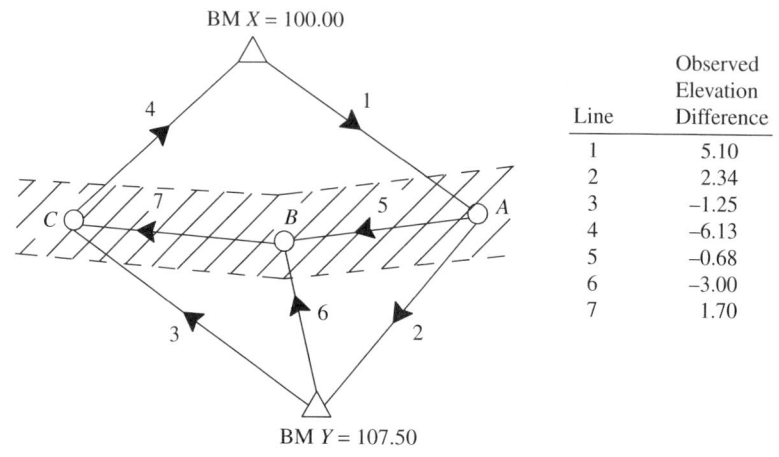

Figure 12.2 Interlocking leveling network.

12.3 UNWEIGHTED EXAMPLE

$$
\begin{aligned}
A & & -\text{BM } X &= 5.10 + v_1 \\
-A & & +\text{BM } Y &= 2.34 + v_2 \\
& C & -\text{BM } Y &= -1.25 + v_3 \\
& -C & +\text{BM } X &= -6.13 + v_4 \quad (12.2)\\
-A \quad B & & &= -0.68 + v_5 \\
B & & -\text{BM } Y &= -3.00 + v_6 \\
-B \quad C & & &= +1.70 + v_7
\end{aligned}
$$

Rearranging so that the known benchmarks are on the right-hand side of the equations and substituting in their appropriate elevations yields

$$
\begin{aligned}
A & & &= +105.10 + v_1 \\
-A & & &= -105.16 + v_2 \\
& C &&= +106.25 + v_3 \\
& -C &&= -106.13 + v_4 \quad (12.3)\\
-A \quad +B & & &= -0.68 + v_5 \\
B & & &= +104.50 + v_6 \\
-B \quad +C & & &= +1.70 + v_7
\end{aligned}
$$

In this example there are three unknowns, A, B, and C. In matrix form, Equations (12.2) are written as

$$AX + B = L + V \quad (12.4a)$$

where

$$A = \begin{bmatrix} 1 & 0 & 0 \\ -1 & 0 & 0 \\ 0 & 0 & 1 \\ 0 & 0 & -1 \\ -1 & 1 & 0 \\ 0 & 1 & 0 \\ 0 & -1 & 1 \end{bmatrix} \quad X = \begin{bmatrix} A \\ B \\ C \end{bmatrix} \quad B = \begin{bmatrix} -100.00 \\ +107.50 \\ -107.50 \\ 100.00 \\ 0 \\ -107.50 \\ 0 \end{bmatrix}$$

ADJUSTMENT OF LEVEL NETS

$$L = \begin{bmatrix} 5.10 \\ 2.34 \\ -1.25 \\ -6.13 \\ -0.68 \\ -3.00 \\ 1.70 \end{bmatrix} \quad V = \begin{bmatrix} v_1 \\ v_2 \\ v_3 \\ v_4 \\ v_5 \\ v_6 \\ v_7 \end{bmatrix}$$

In Equation (12.4a), the B matrix is a vector of the constants (benchmarks) collected from the left side of the equation and L is a collection of elevation differences observed using differential leveling. The right side of Equation (12.3) is equal to $L - B$. It is a collection of the constants in the observation equations and is often referred to as the *constants matrix*, L where L is the difference between the differential leveling observations and constants in B. Since the benchmarks can also be thought of as observations, this combination of benchmarks and differential leveling observations is referred to as L in this book, and Equation (12.4a) is simplified as

$$AX = L + V \qquad (12.4b)$$

Also note in the A matrix that when an unknown does not appear in an equation, its coefficient is zero. Since this is an unweighted example, according to Equation (11.31) the normal equations are

$$A^TA = NX = \begin{bmatrix} 3 & 1 & 0 \\ 1 & 3 & 1 \\ 0 & 1 & 3 \end{bmatrix} \begin{bmatrix} A \\ B \\ C \end{bmatrix} \quad \text{and} \quad A^TL = \begin{bmatrix} 210.94 \\ 102.12 \\ 214.08 \end{bmatrix} \qquad (12.5)$$

Using Equation (11.32), the solution of Equation (12.5) is

$$X = N^{-1}A^TL = \begin{bmatrix} 3 & -1 & 0 \\ -1 & 3 & -1 \\ 0 & -1 & 3 \end{bmatrix}^{-1} \begin{bmatrix} 210.94 \\ 102.12 \\ 214.08 \end{bmatrix}$$

$$= \begin{bmatrix} 0.38095 & 0.14286 & 0.04762 \\ 0.14286 & 0.42857 & 0.14286 \\ 0.04762 & 0.14286 & 0.38095 \end{bmatrix} \begin{bmatrix} 210.94 \\ 102.12 \\ 214.08 \end{bmatrix} = \begin{bmatrix} 105.14 \\ 104.48 \\ 106.19 \end{bmatrix}$$

$$(12.6)$$

From Equation (12.6), the most probable elevations for A, B, and C are 105.14, 104.48, and 106.19, respectively. The rearranged form of Equation (12.4b) is used to compute the residuals as

$$V = AX - L \qquad (12.7)$$

From Equation (12.7), the matrix solution for V is

$$V = \begin{bmatrix} 1 & 0 & 0 \\ -1 & 0 & 0 \\ 0 & 0 & 1 \\ 0 & 0 & -1 \\ -1 & 1 & 0 \\ 0 & 1 & 0 \\ 0 & -1 & 1 \end{bmatrix} \begin{bmatrix} 105.141 \\ 104.483 \\ 106.188 \end{bmatrix} - \begin{bmatrix} 105.10 \\ -105.16 \\ 106.25 \\ -106.13 \\ -0.68 \\ 104.50 \\ 1.70 \end{bmatrix} = \begin{bmatrix} 0.041 \\ 0.019 \\ -0.062 \\ -0.058 \\ 0.022 \\ -0.017 \\ 0.005 \end{bmatrix}$$

12.4 WEIGHTED EXAMPLE

In Section 10.6 it was shown that relative weights for adjusting level lines are inversely proportional to the lengths of the lines:

$$w = \frac{1}{\text{length}} \qquad (12.8)$$

The application of weights to the level circuit's least squares adjustment is illustrated by including the variable line lengths for the unweighted example of Section 12.3. These line lengths for the leveling network of Figure 12.2 and their corresponding relative weights are given in Table 12.1. For convenience, each length is divided into the constant 12, so that integer "relative weights" were obtained. (Note that this is an unnecessary step in the adjustment.) The observation equations are now formed as in Section 12.3, except that in the weighted case, each equation is multiplied by its weight.

$$
\begin{aligned}
w_1(A) &= w_1(+105.10) + w_1 v_1 \\
w_2(-A) &= w_2(-105.16) + w_2 v_2 \\
w_3(C) &= w_3(+106.25) + w_3 v_3 \\
w_4(-C) &= w_4(-106.13) + w_4 v_4 \quad (12.9) \\
w_5(-A+B) &= w_5(-0.68) + w_5 v_5 \\
w_6(+B) &= w_6(+104.50) + w_6 v_6 \\
w_7(-B+C) &= w_7(+1.70) + w_7 v_7
\end{aligned}
$$

After dropping the residual terms in Equation (12.9), they can be written in terms of matrices as

210 ADJUSTMENT OF LEVEL NETS

TABLE 12.1 Weights for the Example in Figure 12.2

Line	Length (miles)	Relative Weights
1	4	12/4 = 3
2	3	12/3 = 4
3	2	12/2 = 6
4	3	12/3 = 4
5	2	12/2 = 6
6	2	12/2 = 6
7	2	12/2 = 6

$$\begin{bmatrix} 3 & 0 & 0 & 0 & 0 & 0 & 0 \\ 0 & 4 & 0 & 0 & 0 & 0 & 0 \\ 0 & 0 & 6 & 0 & 0 & 0 & 0 \\ 0 & 0 & 0 & 4 & 0 & 0 & 0 \\ 0 & 0 & 0 & 0 & 6 & 0 & 0 \\ 0 & 0 & 0 & 0 & 0 & 6 & 0 \\ 0 & 0 & 0 & 0 & 0 & 0 & 6 \end{bmatrix} \begin{bmatrix} 1 & 0 & 0 \\ -1 & 0 & 0 \\ 0 & 0 & 1 \\ 0 & 0 & -1 \\ -1 & 1 & 0 \\ 0 & 1 & 0 \\ 0 & -1 & 0 \end{bmatrix} \begin{bmatrix} A \\ B \\ C \end{bmatrix}$$

$$= \begin{bmatrix} 3 & 0 & 0 & 0 & 0 & 0 & 0 \\ 0 & 4 & 0 & 0 & 0 & 0 & 0 \\ 0 & 0 & 6 & 0 & 0 & 0 & 0 \\ 0 & 0 & 0 & 4 & 0 & 0 & 0 \\ 0 & 0 & 0 & 0 & 6 & 0 & 0 \\ 0 & 0 & 0 & 0 & 0 & 6 & 0 \\ 0 & 0 & 0 & 0 & 0 & 0 & 6 \end{bmatrix} \begin{bmatrix} 105.10 \\ -105.16 \\ 106.25 \\ -106.13 \\ -0.68 \\ 104.50 \\ 1.70 \end{bmatrix}$$

(12.10)

Applying Equation (11.34), we find that the normal equations are

$$(A^TWA)X = NX = A^TWL \tag{12.11}$$

where

$$N = \begin{bmatrix} 1 & -1 & 0 & 0 & -1 & 0 & 0 \\ 0 & 0 & 0 & 0 & 1 & 1 & -1 \\ 0 & 0 & 1 & -1 & 0 & 0 & 1 \end{bmatrix} \begin{bmatrix} 3 & 0 & 0 & 0 & 0 & 0 & 0 \\ 0 & 4 & 0 & 0 & 0 & 0 & 0 \\ 0 & 0 & 6 & 0 & 0 & 0 & 0 \\ 0 & 0 & 0 & 4 & 0 & 0 & 0 \\ 0 & 0 & 0 & 0 & 6 & 0 & 0 \\ 0 & 0 & 0 & 0 & 0 & 6 & 0 \\ 0 & 0 & 0 & 0 & 0 & 0 & 6 \end{bmatrix}$$

$$\begin{bmatrix} 1 & 0 & 0 \\ -1 & 0 & 0 \\ 0 & 0 & 1 \\ 0 & 0 & -1 \\ -1 & 1 & 0 \\ 0 & 1 & 0 \\ 0 & -1 & 1 \end{bmatrix} = \begin{bmatrix} 13 & -6 & 0 \\ -6 & 18 & -6 \\ 0 & -6 & 16 \end{bmatrix}$$

$$A^TWL = \begin{bmatrix} 740.02 \\ 612.72 \\ 1072.22 \end{bmatrix}$$

By using Equation (11.35), the solution for the X matrix is

$$X = N^{-1}(A^TWL) = \begin{bmatrix} 0.0933 & 0.0355 & 0.0133 \\ 0.0355 & 0.0770 & 0.0289 \\ 0.0133 & 0.0289 & 0.0733 \end{bmatrix} \begin{bmatrix} 740.02 \\ 612.72 \\ 1072.22 \end{bmatrix} = \begin{bmatrix} 105.150 \\ 104.489 \\ 106.197 \end{bmatrix} \quad (12.12)$$

Equation (12.7) is now used to compute the residuals as

$$V = AX - L = \begin{bmatrix} 1 & 0 & 0 \\ -1 & 0 & 0 \\ 0 & 0 & 1 \\ 0 & 0 & -1 \\ -1 & 1 & 0 \\ 0 & 1 & 0 \\ 0 & -1 & 1 \end{bmatrix} \begin{bmatrix} 105.150 \\ 104.489 \\ 106.197 \end{bmatrix} - \begin{bmatrix} 105.10 \\ -105.16 \\ 106.25 \\ -106.13 \\ -0.68 \\ 104.50 \\ 1.70 \end{bmatrix} = \begin{bmatrix} 0.050 \\ 0.010 \\ -0.053 \\ -0.067 \\ 0.019 \\ -0.011 \\ 0.008 \end{bmatrix}$$

It should be noted that these adjusted values (X matrix) and residuals (V matrix) differ slightly from those obtained in the unweighted adjustment of Section 12.3. This illustrates the effect of weights on an adjustment. Although the differences in this example are small, for precise level circuits it is both logical and wise to use a weighted adjustment since a correct stochastic model will place the errors back in the observations that probably produced the errors.

12.5 REFERENCE STANDARD DEVIATION

Equation (10.20) expressed the standard deviation for a weighted set of observations as

$$S_0 = \sqrt{\frac{\Sigma wv^2}{n-1}} \quad (12.13)$$

However, Equation (12.13) applies to a multiple set of observations for a single quantity where each observation has a different weight. Often, observations are obtained that involve several unknown parameters that are related functionally like those in Equations (12.3) or (12.9). For these types of observations, the standard deviation in the unweighted case is

$$S_0 = \sqrt{\frac{\Sigma v^2}{m-n}} = \sqrt{\frac{\Sigma v^2}{r}} \quad \text{which in matrix form is} \quad S_0 = \sqrt{\frac{V^T V}{r}} \quad (12.14)$$

In Equation (12.14), Σv^2 is expressed in matrix form as $V_i^T V$, m is the number of observations, and n is the number of unknowns. There are $r = m - n$ redundant measurements or degrees of freedom in the adjustment.

The standard deviation for the weighted case is

$$S_0 = \sqrt{\frac{\Sigma wv^2}{m-n}} = \sqrt{\frac{\Sigma wv^2}{r}} \quad \text{which in matrix form is} \quad S_0 = \sqrt{\frac{V^T W V}{r}} \quad (12.15)$$

where Σwv^2 in matrix form is $V^T W V$.

Since these standard deviations relate to the overall adjustment and not a single quantity, they are referred to as *reference standard deviations*. Computations of the reference standard deviations for both unweighted and weighted examples are illustrated below.

12.5.1 Unweighted Example

In the example of Section 12.3, there are $7 - 3$, or 4, degrees of freedom. Using the residuals given in Equation (12.7) and using Equation (12.14), the reference standard deviation in the unweighted example is

$$S_0 = \sqrt{\frac{(0.041)^2 + (0.019)^2 + (-0.062)^2 + (-0.058)^2 + (0.022)^2 + (-0.017)^2 + (0.005)^2}{7-3}}$$

$$= \pm 0.05 \quad (12.16)$$

This can be computed using the matrix expression in Equation (12.14) as

$$S_0 = \sqrt{\frac{V^T V}{r}}$$

$$= \sqrt{\frac{[0.041 \quad 0.019 \quad -0.062 \quad -0.058 \quad 0.022 \quad -0.017 \quad 0.005] \begin{bmatrix} 0.041 \\ 0.019 \\ -0.062 \\ -0.058 \\ 0.022 \\ -0.017 \\ 0.005 \end{bmatrix}}{}}$$

$$= \sqrt{\frac{0.010}{4}} = \pm 0.05 \quad (12.17)$$

12.5.2 Weighted Example

Notice that the weights are used when computing the reference standard deviation in Equation (12.15). That is, each residual is squared and multiplied by its weight, and thus the reference standard deviation computed using non-matrix methods is

$$S_0 = \sqrt{\frac{3(0.050)^2 + 4(0.010)^2 + 6(-0.053)^2 + 4(-0.067)^2 + 6(0.019)^2 + 6(-0.011)^2 + 6(0.008)^2}{7 - 3}}$$

$$= \sqrt{\frac{0.04598}{4}} = \pm 0.107 \qquad (12.18)$$

It is left as an exercise to verify this result by solving the matrix expression of Equation (12.15).

12.6 ANOTHER WEIGHTED ADJUSTMENT

Example 12.1 The level net shown in Figure 12.3 is observed with the following results (the elevation differences and standard deviations are given in meters, and the elevation of A is 437.596 m):

From	To	ΔElev (m)	σ (m)	From	To	ΔElev (m)	σ (m)
A	B	10.509	0.006	D	A	−7.348	0.003
B	C	5.360	0.004	B	D	−3.167	0.004
C	D	−8.523	0.005	A	C	15.881	0.012

What are the most probable values for the elevations of B, C, and D?

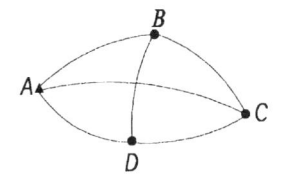

Figure 12.3 Differential leveling network for Example 12.1.

214 ADJUSTMENT OF LEVEL NETS

SOLUTION

Step 1: Write the observation equations without their weights:

$$
\begin{align}
(1) \quad &+B &&= A + 10.509 + v_1 = 448.105 + v_1 \\
(2) \quad &-B + C &&= 5.360 + v_2 \\
(3) \quad & -C + D &&= -8.523 + v_3 \\
(4) \quad & -D &&= -A - 7.348 + v_4 = -444.944 + v_4 \\
(5) \quad &-B + D &&= -3.167 + v_5 \\
(6) \quad & +C &&= A = 15.881 + v_6 = 453.477 + v_6
\end{align}
$$

Step 2: Rewrite observation equations in matrix form $AX = L + V$ as

$$
\begin{bmatrix} 1 & 0 & 0 \\ -1 & 1 & 0 \\ 0 & -1 & 1 \\ 0 & 0 & -1 \\ -1 & 0 & 1 \\ 0 & 1 & 0 \end{bmatrix} \begin{bmatrix} A \\ B \\ C \end{bmatrix} = \begin{bmatrix} 448.105 \\ 5.360 \\ -8.523 \\ -444.944 \\ -3.167 \\ 453.477 \end{bmatrix} + \begin{bmatrix} v_1 \\ v_2 \\ v_3 \\ v_4 \\ v_5 \\ v_6 \end{bmatrix} \quad (12.19)
$$

Step 3: In accordance with Equations (10.4) and (10.6), form the weight matrix as

$$
W = \begin{bmatrix} \dfrac{1}{0.006^2} & 0 & 0 & 0 & 0 & 0 \\ 0 & \dfrac{1}{0.004^2} & 0 & 0 & 0 & 0 \\ 0 & 0 & \dfrac{1}{0.005^2} & 0 & 0 & 0 \\ 0 & 0 & 0 & \dfrac{1}{0.003^2} & 0 & 0 \\ 0 & 0 & 0 & 0 & \dfrac{1}{0.004^2} & 0 \\ 0 & 0 & 0 & 0 & 0 & \dfrac{1}{0.012^2} \end{bmatrix} \quad (12.20)
$$

from which

$$W = \begin{bmatrix} 27{,}778 & 0 & 0 & 0 & 0 & 0 \\ 0 & 62{,}500 & 0 & 0 & 0 & 0 \\ 0 & 0 & 40{,}000 & 0 & 0 & 0 \\ 0 & 0 & 0 & 111{,}111 & 0 & 0 \\ 0 & 0 & 0 & 0 & 62{,}500 & 0 \\ 0 & 0 & 0 & 0 & 0 & 6944 \end{bmatrix} \quad (12.21)$$

Step 4: Compute the normal equations using Equation (11.34):

$$(A^T W A) X = N X = A^T W L \quad (12.22)$$

where

$$N = \begin{bmatrix} 152{,}778 & -62{,}500 & -62{,}500 \\ -62{,}500 & 109{,}444 & -40{,}000 \\ -62{,}500 & -40{,}000 & 213{,}611 \end{bmatrix} \quad X = \begin{bmatrix} B \\ C \\ D \end{bmatrix}$$

$$A^T W L = \begin{bmatrix} 12{,}310{,}298.611 \\ 3{,}825{,}065.833 \\ 48{,}899{,}364.722 \end{bmatrix}$$

Step 5: Solving for the X matrix using Equation (11.35) yields

$$X = \begin{bmatrix} 448.1087 \\ 453.4685 \\ 444.9436 \end{bmatrix} \quad (12.23)$$

Step 6: Compute the residuals using the matrix expression $V = AX - L$:

$$V = \begin{bmatrix} 448.1087 \\ 5.3598 \\ -8.5249 \\ -444.9436 \\ -3.1651 \\ 453.4685 \end{bmatrix} - \begin{bmatrix} 448.105 \\ 5.360 \\ -8.523 \\ -444.944 \\ -3.167 \\ 453.477 \end{bmatrix} = \begin{bmatrix} 0.0037 \\ -0.0002 \\ -0.0019 \\ 0.0004 \\ 0.0019 \\ -0.0085 \end{bmatrix} \quad (12.24)$$

Step 7: Calculate the reference standard deviation for the adjustment using the matrix expression of Equation (12.15):

$$V^TWV = [0.0037 \quad -0.0002 \quad -0.0019 \quad 0.0004 \quad 0.0019 \quad -0.0085]$$

$$W \begin{bmatrix} 0.0037 \\ -0.0002 \\ -0.0019 \\ 0.0004 \\ 0.0019 \\ -0.0085 \end{bmatrix}$$

$$= [1.26976] \tag{12.25}$$

Since the number of system redundancies is the number of observations minus the number of unknowns, $r = 6 - 3 = 3$, and thus

$$S_0 = \sqrt{\frac{1.26976}{3}} = \pm 0.6575 \tag{12.26}$$

Step 8: Tabulate the results showing both the adjusted elevation differences, their residuals, and final adjusted elevations.

From	To	Adjusted ΔElev	Residual	Station	Adjusted Elevation
A	B	10.513	0.004	A	437.596
B	C	5.360	0.000	B	448.109
C	D	-8.525	-0.002	C	453.468
D	A	-7.348	0.000	D	444.944
B	D	-3.165	0.002		
A	C	15.872	-0.009		

PROBLEMS

Note: For problems requiring least squares adjustment, if a computer program is not distinctly specified for use in the problem, it is expected that the least squares algorithm will be solved using the program MATRIX, which is included on the CD supplied with the book.

Figure P12.1

[Leveling network diagram: BM A = 100.00 and BM B = 101.60 at top connecting via lines (1) and (2) to point Y; Y connects via line (3) to X; X connects via lines (4) and (5) to BM C = 108.05 and BM D = 106.07 at bottom.]

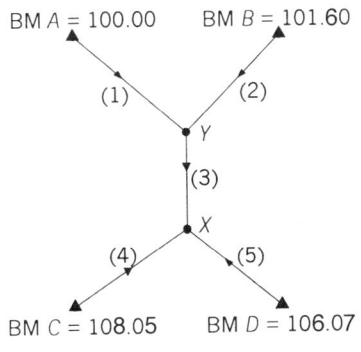

Figure P12.1

12.1 For the leveling network in Figure P12.1, calculate the most probable elevations for X and Y. Use an unweighted least squares adjustment with the observed values given in the accompanying table. Assume units of feet.

Line	ΔElev (ft)
1	+3.68
2	+2.06
3	+2.02
4	−2.37
5	−0.38

12.2 For Problem 12.1, compute the reference standard deviation and tabulate the adjusted observations and their residuals.

12.3 Repeat Problem 12.1 using ADJUST.

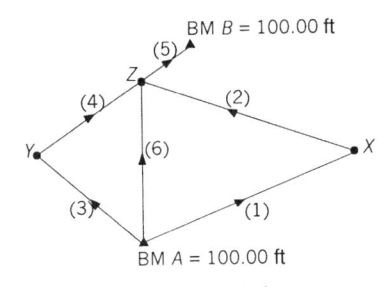

Figure P12.4

[Leveling network diagram with BM B = 100.00 ft at top, BM A = 100.00 ft at bottom, and points X, Y, Z connected by lines (1) through (6).]

12.4 For the leveling network shown in Figure P12.4, calculate the most probable elevations for X, Y, and Z. The observed values and line lengths are given in the table. Apply appropriate weights in the computations.

Line	Length (mi)	ΔElev (ft)
1	3	+1.02
2	3	−0.95
3	1.5	+1.96
4	1.5	−1.99
5	1	−0.04
6	2	+0.05

12.5 For Problem 12.4, compute the reference standard deviation and tabulate the adjusted observations and their residuals.

12.6 Use ADJUST to solve Problem 12.4.

12.7 A line of differential level is run from benchmark Oak (elevation 753.01) to station 13+00 on a proposed alignment. It continued along the alignment to 19+00. Rod readings were taken on stakes at each full station. The circuit then closed on benchmark Bridge, which has an elevation of 772.52 ft. The elevation differences observed are, in order, −3.03, 4.10, 4.03, 7.92, 7.99, −6.00, −6.02, and 2.98 ft. A third tie between benchmark Rock (elevation of 772.39 ft) and station 16+00 is observed as −6.34 ft. What are:

(a) the most probable values for the adjusted elevations?
(b) the reference standard deviation for the adjustment?
(c) the adjusted observations and their residuals?

12.8 Use ADJUST to solve Problem 12.7.

12.9 If the elevation of A is 257.891 m, adjust the following leveling data using the weighted least squares method.

From	To	ΔElev (m)	Distance (km)
A	B	5.666	1
B	C	48.025	4.5
C	D	3.021	6
D	E	−13.987	2.5
E	F	20.677	5
F	G	−32.376	7.6
G	A	−30.973	2.4
A	C	53.700	5.8
C	F	9.634	4.3
F	D	−6.631	3.8

(a) What are the most probable values for the elevations of the stations?
(b) What is the reference standard deviation?
(c) Tabulate the adjusted observations and their residuals.

12.10 Use ADJUST to solve Problem 12.9.

12.11 If the elevation of station 1 is 2395.67 ft, use weighted least squares to adjust the following leveling.

From	To	ΔElev (ft)	Distance (mi)	From	To	ΔElev (ft)	Distance (mi)
1	2	37.17	3.00	2	3	−9.20	3.63
3	4	34.24	1.56	4	5	−10.92	1.98
5	6	−23.12	0.83	6	1	−28.06	0.93
1	7	16.99	1.61	7	3	11.21	1.21
2	7	−19.99	2.91	7	6	10.89	1.41
6	3	−0.04	3.06	3	8	74.93	1.77
8	5	−51.96	2.98	6	8	74.89	8.03
8	4	−41.14	1.08				

(a) What are the most probable values for the elevations for the stations?
(b) What is the adjustment's reference standard deviation?
(c) Tabulate the adjusted observations and their residuals.

12.12 Use ADJUST to solve Problem 12.11.

12.13 Precise procedures were applied with a level that can be read to within ±0.4 mm/1 m. The line of sight was held to within ±3″ of horizontal, and the sight distances were approximately 50 m in length. Use these specifications and Equation (9.20) to compute standard deviations and hence weights. The elevation of A is 100.000 m. Adjust the network by weighted least squares.

From	To	ΔElev (m)	Number of Setups	From	To	ΔElev (m)	Number of Setups
A	B	12.383	16	M	D	−38.238	23
B	C	−16.672	25	C	M	30.338	16
C	D	−7.903	37	M	B	−13.676	38
D	A	12.190	26	A	M	26.058	19

(a) What are the most probable values for the elevations of the stations?

(b) What is the reference standard deviation for the adjustment?

(c) Tabulate the adjusted observations and their residuals.

12.14 Repeat Problem 12.13 using the number of setups for weighting following the procedures discussed in Section 10.6.

12.15 In Problem 12.13, the estimated error in reading the rod is 1.4 mm/km. The bubble sensitivity of the instrument is 12 mm/km and the average sight distances are 50 m. What are:

(a) the estimated standard errors for the observations?

(b) the most probable values for the elevations of the stations?

(c) the reference variance for the adjustment?

(d) Tabulate the adjusted observations and their residuals.

12.16 Demonstrate that $\Sigma v^2 = V^T V$.

12.17 Demonstrate that $\Sigma wv^2 = V^T W V$.

Programming Problems

12.18 Write a program that reads a file of differential leveling observations and writes the matrices A, W, and L in a format suitable for the MATRIX program. Using this package, solve Problem 12.11.

12.19 Write a computational package that reads the matrices A, W, and L, computes the least squares solution for the unknown station elevations, and writes a file of adjusted elevation differences and their residuals. Using this package, solve Problem 12.11.

12.20 Write a computational package that reads a file of differential leveling observations, computes the least squares solution for the adjusted station elevations, and writes a file of adjusted elevation differences and their residuals. Using this package, solve Problem 12.11.

CHAPTER 13

PRECISION OF INDIRECTLY DETERMINED QUANTITIES

13.1 INTRODUCTION

Following an adjustment, it is important to know the estimated errors in both the adjusted observations and the derived quantities. For example, after adjusting a level net as described in Chapter 12, the uncertainties in both adjusted elevation differences and computed benchmark elevations should be determined. In Chapter 5, error propagation formulas were developed for indirectly measured quantities which were functionally related to observed values. In this chapter, error propagation formulas are developed for the quantities computed in a least squares solution.

13.2 DEVELOPMENT OF THE COVARIANCE MATRIX

Consider an adjustment involving weighted observation equations like those in the level circuit example of Section 12.4. The matrix form for the system of weighted observation equation is

$$WAX = WL + WV \tag{13.1}$$

and the least squares solution of the weighted observation equations is given by

$$X = (A^TWA)^{-1}A^TWL \tag{13.2}$$

221

In this equation, X contains the most probable values for the unknowns, whereas the true values are X_{true}. The true values differ from X by some small amount ΔX, such that

$$X + \Delta X = X_{\text{true}} \tag{13.3}$$

where ΔX represents the errors in the adjusted values.

Consider now a small incremental change, ΔL, in the measured values, L, which changes X to its true value, $X + \Delta X$. Then Equation (13.2) becomes

$$X + \Delta X = (A^T W A)^{-1} A^T W (L + \Delta L) \tag{13.4}$$

Expanding Equation (13.4) yields

$$X + \Delta X = (A^T W A)^{-1} A^T W L + (A^T W A)^{-1} A^T W \Delta L \tag{13.5}$$

Note in Equation (13.2) that $X = (A^T W A)^{-1} A^T W L$, and thus subtracting this from Equation (13.5) yields

$$\Delta X = (A^T W A)^{-1} A^T W \Delta L \tag{13.6}$$

Recognizing ΔL as the errors in the observations, Equation (13.6) can be rewritten as

$$\Delta X = (A^T W A)^{-1} A^T W V \tag{13.7}$$

Now let

$$B = (A^T W A)^{-1} A^T W \tag{13.8}$$

then

$$\Delta X = BV \tag{13.9}$$

Multiplying both sides of Equation (13.9) by their transposes results in

$$\Delta X \, \Delta X^T = (BV)(BV)^T \tag{13.10}$$

Applying the matrix property $(BV)^T = V^T B^T$ to Equation (13.10) gives

$$\Delta X \, \Delta X^T = B V V^T B^T \tag{13.11}$$

The expanded left side of Equation (13.11) is

13.2 DEVELOPMENT OF THE COVARIANCE MATRIX

$$\Delta X\, \Delta X^{\mathrm{T}} = \begin{bmatrix} \Delta x_1^2 & \Delta x_1\,\Delta x_2 & \Delta x_1\,\Delta x_3 & \cdots & \Delta x_1\,\Delta x_n \\ \Delta x_2\,\Delta x_1 & \Delta x_2^2 & \Delta x_2\,\Delta x_3 & \cdots & \Delta x_2\,\Delta x_n \\ \Delta x_3\,\Delta x_1 & \Delta x_3\,\Delta x_2 & \Delta x_3^2 & \cdots & \Delta x_3\,\Delta x_n \\ \vdots & \vdots & \vdots & \ddots & \vdots \\ \Delta x_n\,\Delta x_1 & \Delta x_n\,\Delta x_2 & \Delta x_n\,\Delta x_3 & \cdots & \Delta x_n^2 \end{bmatrix} \quad (13.12)$$

Also, the expanded right side of Equation (13.11) is

$$B \begin{bmatrix} v_1^2 & v_1 v_2 & v_1 v_3 & \cdots & v_1 v_m \\ v_2 v_1 & v_2^2 & v_2 v_3 & \cdots & v_2 v_m \\ v_3 v_1 & v_3 v_2 & v_3^2 & \cdots & v_3 v_m \\ \vdots & \vdots & \vdots & \ddots & \vdots \\ v_m v_1 & v_m v_2 & v_m v_3 & \cdots & v_m^2 \end{bmatrix} B^{\mathrm{T}} \quad (13.13)$$

Assume that it is possible to repeat the entire sequence of observations many times, say a times, and that each time a slightly different solution occurs, yielding a different set of X's. Averaging these sets, the left side of Equation (13.11) becomes

$$\frac{1}{a}\sum(\Delta X)(\Delta X)^{\mathrm{T}} = \begin{bmatrix} \dfrac{\sum \Delta x_1^2}{a} & \dfrac{\sum \Delta x_1\,\Delta x_2}{a} & \cdots & \dfrac{\sum \Delta x_1\,\Delta x_n}{a} \\ \dfrac{\sum \Delta x_2\,\Delta x_1}{a} & \dfrac{\sum \Delta x_2^2}{a} & \cdots & \dfrac{\sum \Delta x_2\,\Delta x_n}{a} \\ \vdots & \vdots & \ddots & \vdots \\ \dfrac{\sum \Delta x_n\,\Delta x_1}{a} & \dfrac{\sum \Delta x_n\,\Delta x_2}{a} & \cdots & \dfrac{\sum \Delta x_n^2}{a} \end{bmatrix} \quad (13.14)$$

If a is large, the terms in Equation (13.14) are the variances and covariances as defined in Equation (6.7) and Equation (13.14) can be rewritten as

$$\begin{bmatrix} S_{x_1}^2 & S_{x_1 x_2} & \cdots & S_{x_1 x_n} \\ S_{x_2 x_1} & S_{x_2}^2 & \cdots & S_{x_2 x_n} \\ \vdots & \vdots & \ddots & \vdots \\ S_{x_n x_1} & S_{x_n x_2} & \cdots & S_{x_n}^2 \end{bmatrix} = S_x^2 \quad (13.15)$$

Also, considering a sets of observations, Equation (13.13) becomes

224 PRECISION OF INDIRECTLY DETERMINED QUANTITIES

$$B \begin{bmatrix} \frac{\sum v_1^2}{a} & \frac{\sum v_1 v_2}{a} & \cdots & \frac{\sum v_1 v_m}{a} \\ \frac{\sum v_2 v_1}{a} & \frac{\sum v_2^2}{a} & \cdots & \frac{\sum v_2 v_m}{a} \\ \vdots & \vdots & \ddots & \vdots \\ \frac{\sum v_m v_1}{a} & \frac{\sum v_m v_2}{a} & \cdots & \frac{\sum v_m^2}{a} \end{bmatrix} B^\mathrm{T} \qquad (13.16)$$

Recognizing the diagonal terms as variances of the quantities observed, $s_{l_i}^2$, off-diagonal terms as the covariances, $S_{l_i l_j}^2$, and the fact that the matrix is symmetric, Equation (13.16) can be rewritten as

$$B \begin{bmatrix} S_{l_1}^2 & S_{l_1 l_2} & \cdots & S_{l_1 l_m} \\ S_{l_2 l_1} & S_{l_2}^2 & \cdots & S_{l_2 l_m} \\ \vdots & \vdots & \ddots & \vdots \\ S_{l_m l_1} & S_{l_m l_2} & \cdots & S_{l_m}^2 \end{bmatrix} B^\mathrm{T} \qquad (13.17)$$

In Section 10.1 it was shown that the weight of an observation is inversely proportional to its variance. Also, from Equation (10.5), the variance of an observation of weight w can be expressed in terms of the reference variance as

$$S_i^2 = \frac{S_0^2}{w_i} \qquad (13.18)$$

Recall from Equation (10.3) that $W = Q^{-1} = \sigma_0^2 \Sigma^{-1}$. Therefore, $\Sigma = \sigma_0^2 W^{-1}$, and substituting Equation (13.18) into matrix (13.17) and replacing σ_0 with S_0 yields

$$S_0^2 B W_{ll}^{-1} B^\mathrm{T} \qquad (13.19)$$

Substituting Equation (13.8) into Equation (13.19) gives

$$S_0^2 B W^{-1} B^\mathrm{T} = S_0^2 (A^\mathrm{T} W A)^{-1} A^\mathrm{T} W W^{-1} W^\mathrm{T} A \, [(A^\mathrm{T} W A)^{-1}]^\mathrm{T} \qquad (13.20)$$

Since the matrix of the normal equations is symmetric, it follows that

$$[(A^\mathrm{T} W A)^{-1}]^\mathrm{T} = (A^\mathrm{T} W A)^{-1} \qquad (13.21)$$

Also, since the weight matrix W is symmetric, $W^\mathrm{T} = W$, and thus Equation (13.20) reduces to

$$S_0^2(A^TWA)^{-1}(A^TWA)(A^TWA)^{-1} = S_0^2(A^TWA)^{-1} \qquad (13.22)$$

Equation (13.15) is the left side of Equation (13.11), for which Equation (13.22) is the right. That is,

$$S_x^2 = S_0^2(A^TWA)^{-1} = S_0^2 N^{-1} = S_0^2 Q_{xx} \qquad (13.23)$$

In least squares adjustment, the matrix Q_{xx} of Equation (13.23) is known as the *variance–covariance matrix*, or simply the *covariance matrix*. Diagonal elements of the matrix when multiplied by S_0^2 give variances of the adjusted quantities, and the off-diagonal elements multiplied by S_0^2 yield covariances. From Equation (13.23), the estimated standard deviation S_i for any unknown parameter computed from a system of observation equations is expressed as

$$S_i \doteq S_0 \sqrt{q_{x_i x_i}} \qquad (13.24)$$

where $q_{x_i x_i}$ is the diagonal element (from the *i*th row and *i*th column) of the Q_{xx} matrix, which as noted in Equation (13.23), is equal to the inverse of the matrix of normal equations. Since the normal equation matrix is symmetric, its inverse is also symmetric, and thus the covariance matrix is a symmetric matrix (i.e., element ij = element ji).[1]

13.3 NUMERICAL EXAMPLES

The results of the level net adjustment in Section 12.3 will be used to illustrate the computation of estimated errors for the adjusted unknowns. From Equation (12.6), the N^{-1} matrix, which is also the Q_{xx} matrix, is

$$Q_{xx} = \begin{bmatrix} 0.38095 & 0.14286 & 0.04762 \\ 0.14286 & 0.42857 & 0.14286 \\ 0.04762 & 0.14286 & 0.38095 \end{bmatrix}$$

Also, from Equation (12.17), $S_0 = \pm 0.05$. Now by Equation (13.24), the estimated standard deviations for the unknown benchmark elevations A, B, and C are

---

[1] Note that an estimate of the reference variance, σ_0^2, may be computed using either Equation (12.13) or (12.14). However, it should be remembered that this only gives an estimate of the *a priori* (before the adjustment) value for the reference variance. The validity of this estimate can be checked using a χ^2 test as discussed in Chapter 5. If it is a valid estimate for σ_0^2, the a priori value for the reference variance should be used in the computations discussed in this and subsequent chapters. Thus, if the a priori value for σ_0^2 is known, it should be used when computing the *a posteriori* (after the adjustment) statistics. When weights are determined as $1/\sigma_i^2$, the implicit assumption made is that the a priori value for $\sigma_0^2 = 1$ [see Equations (10.5) and (10.6)].

$$S_A = S_0\sqrt{q_{AA}} = \pm 0.05\sqrt{0.38095} = \pm 0.031 \text{ ft}$$
$$S_B = S_0\sqrt{q_{BB}} = \pm 0.05\sqrt{0.42857} = \pm 0.033 \text{ ft}$$
$$S_C = S_0\sqrt{q_{CC}} = \pm 0.05\sqrt{0.38095} = \pm 0.031 \text{ ft}$$

In the weighted example of Section 12.4, it should be noted that although this is a weighted adjustment, the a priori value for the reference variance is not known because weights were determined as 1/distance and not $1/\sigma_i^2$. From Equation (12.12), the Q_{xx} matrix is

$$Q_{XX} = \begin{bmatrix} 0.0933 & 0.0355 & 0.0133 \\ 0.0355 & 0.0770 & 0.0289 \\ 0.0133 & 0.0289 & 0.0733 \end{bmatrix}$$

Recalling that in Equation (12.18), $S_0 = \pm 0.107$, the estimated errors in the computed elevations of benchmarks A, B, and C are

$$S_A = S_0\sqrt{q_{AA}} = \pm 0.07\sqrt{0.0933} = \pm 0.033 \text{ ft}$$
$$S_B = S_0\sqrt{q_{BB}} = \pm 0.07\sqrt{0.0770} = \pm 0.030 \text{ ft}$$
$$S_C = S_0\sqrt{q_{CC}} = \pm 0.07\sqrt{0.0733} = \pm 0.029 \text{ ft}$$

These standard deviations are at the 68% probability level, and if other percentage errors are desired, these values should be multiplied by their respective t values as discussed in Chapter 3.

13.4 STANDARD DEVIATIONS OF COMPUTED QUANTITIES

In Section 6.1 the generalized law of propagation of variances was developed. Recalled here for convenience, Equation (6.13) was written as

$$\Sigma_{ll} = A\Sigma_{xx}A^T$$

where $\hat{l}$ represents the adjusted observations, Σ_{ll} the covariance matrix of the adjusted observations, Σ_{xx} the covariance matrix of the unknown parameters [i.e., $(A^TWA)^{-1}$], and A, the coefficient matrix. Rearranging Equation (10.2) and using sample statistics, there results $\Sigma_{xx} = S_0^2 Q_{xx}$. Also, from Equation (13.23), $S_x^2 = S_0^2 Q_{xx} = S_0^2(A^TWA)^{-1}$ and thus $\Sigma_{xx} = S_x^2$. Substituting this equality into Equation (a), the estimated standard deviations of the adjusted observations is

13.4 STANDARD DEVIATIONS OF COMPUTED QUANTITIES

$$\Sigma_{ll}^2 = S_l^2 = A\Sigma_{xx}A^T = AS_0^2(A^TWA)^{-1}A^T = S_0^2 AQ_{xx}A^T = S_0^2 Q_{ll} \quad (13.25)$$

where $AQ_{xx}A^T = Q_{ll}$, which is known as the *covariance matrix of the adjusted observations*.

Example 13.1 Consider the linear example in Section 12.3. By Equation (13.25), the estimated standard deviations in the adjusted elevation differences are

$$S_l^2 = 0.050^2 \begin{bmatrix} 1 & 0 & 0 \\ -1 & 0 & 0 \\ 0 & 0 & 1 \\ 0 & 0 & -1 \\ -1 & 1 & 0 \\ 0 & 1 & 0 \\ 0 & -1 & 1 \end{bmatrix} \begin{bmatrix} 0.38095 & 0.14286 & 0.04762 \\ 0.14286 & 0.42857 & 0.14286 \\ 0.04762 & 0.14286 & 0.38095 \end{bmatrix}$$

$$\times \begin{bmatrix} 1 & -1 & 0 & 0 & -1 & 0 & 0 \\ 0 & 0 & 0 & 0 & 1 & 1 & -1 \\ 0 & 0 & 1 & -1 & 0 & 0 & 1 \end{bmatrix} \quad (13.26)$$

Performing the required matrix multiplications in Equation (13.26) yields

$$S_l^2 = 0.050^2$$

$$\times \begin{bmatrix} 0.38095 & -0.38095 & 0.04762 & -0.04762 & -0.23810 & 0.14286 & -0.09524 \\ -0.38095 & 0.38095 & -0.04762 & 0.04762 & 0.23810 & -0.14286 & 0.09524 \\ 0.04762 & -0.04762 & 0.38095 & -0.38095 & 0.09524 & 0.14286 & 0.23810 \\ -0.04762 & 0.04762 & -0.38095 & 0.38095 & -0.09524 & -0.14286 & -0.23810 \\ -0.23810 & 0.23810 & 0.09524 & -0.09524 & 0.52381 & 0.28571 & -0.19048 \\ 0.14286 & -0.14286 & 0.14286 & -0.14286 & 0.28571 & 0.42857 & -0.28571 \\ -0.09524 & 0.09524 & 0.23810 & -0.23810 & -0.19048 & -0.28571 & 0.52381 \end{bmatrix}$$

$$(13.27)$$

The estimated standard deviation of an observation is found by taking the square root of the corresponding diagonal element of the S_l^2 matrix (leveling from A to B). For instance, for the fifth observation, $S_l(5,5)$ applies and the estimated error in the adjusted elevation difference of that observation is

$$S_{\Delta AB} = \pm 0.050\sqrt{0.52381} = \pm 0.036 \text{ ft}$$

An interpretation of the meaning of the value just calculated is that there is a 68% probability that the true value is within the range ± 0.036 ft of the adjusted elevation difference ($l_5 + v_5 = -106.13 + 0.022 = -106.108$). That

228 PRECISION OF INDIRECTLY DETERMINED QUANTITIES

is, the true value lies between -106.072 and -106.144 ft with 68% probability.

Careful examination of the matrix manipulations involved in solving Equation (13.25) for Example 13.1 reveals that the effort can be reduced significantly. In fact, to obtain the estimated standard deviation in the fifth element, only the fifth row of the coefficient matrix, A, which represents the elevation difference between A and B, need be used in the calculations. That row is $[-1 \quad 1 \quad 0]$. Thus, to compute the standard deviation in this observation, the following computations could be made:

$$S^2_{\Delta AB} = 0.050^2 [-1 \quad 1 \quad 0] \begin{bmatrix} 0.38095 & 0.14286 & 0.04762 \\ 0.14286 & 0.42857 & 0.14286 \\ 0.04762 & 0.14286 & 0.38095 \end{bmatrix} \begin{bmatrix} -1 \\ 1 \\ 0 \end{bmatrix}$$

$$= 0.050^2 [0.2389 \quad -0.28571 \quad -0.09524] \begin{bmatrix} -1 \\ 1 \\ 0 \end{bmatrix} \quad (13.28)$$

$$= 0.050^2 [0.52381]$$

$$S_{\Delta AB} = \pm 0.050\sqrt{0.52381} = \pm 0.036 \text{ ft}$$

Note that this shortcut method produces the same value. Furthermore, because of the zero in the third position of this row from the coefficient matrix, the matrix operations in Equation (13.28) could be further reduced to

$$S^2_{\Delta AB} = 0.050^2 [-1 \quad 1] \begin{bmatrix} 0.38095 & 0.14286 \\ 0.14286 & 0.42857 \end{bmatrix} \begin{bmatrix} -1 \\ 1 \end{bmatrix} = 0.050^2 [0.52381]$$

Another use for Equation (13.25) is in the computation of adjusted uncertainties for observations that were never made. For instance, in the example of Section 12.3, the elevation difference between benchmarks X and B was not observed. But from the results of the adjustment, this elevation difference is $104.48 - 100.00 = 4.48$ ft. The estimated error in this difference can be found by writing an observation equation for it (i.e., $B = X + \Delta \text{Elev}_{XB}$). This equation does not involve either A or C, and thus in matrix form this difference would be expressed as

$$[0 \quad 1 \quad 0] \quad (13.29)$$

Using this row matrix in the same procedure as in Equation (13.28) yields

$$S^2_{\Delta XB} = 0.050^2 [0 \quad 1 \quad 0] \begin{bmatrix} 0.38095 & 0.14286 & 0.04762 \\ 0.14286 & 0.42857 & 0.14286 \\ 0.04762 & 0.14286 & 0.38095 \end{bmatrix} \begin{bmatrix} 0 \\ 1 \\ 0 \end{bmatrix}$$

$$= 0.050^2 [0.42857]$$

Hence,

$$S_{\Delta XB} = \pm 0.050 \sqrt{0.42857} = \pm 0.033 \text{ ft}$$

Again, recognizing the presence of the zeros in the row matrix, these computations can be simplified to

$$S^2_{\Delta XB} = 0.050^1 [1][0.42857][1] = 0.050^2 [0.42857]$$

The method illustrated above of eliminating unnecessary matrix computations is formally known as *matrix partitioning*.

Computing uncertainties of quantities that were not actually observed has application in many areas. For example, suppose that in a triangulation adjustment, the x and y coordinates of stations A and B are calculated and the covariance matrix exists. Equation (13.25) could be applied to determine the estimated error in the length of line AB calculated from the adjusted coordinates of A and B. This is accomplished by relating the length AB to the unknown parameters as

$$\overline{AB} = \sqrt{(X_b - X_a)^2 + (Y_b - Y_a)^2} \qquad (13.30)$$

This subject is discussed further in Chapter 15.

An important observation that should be made about the Q_{ll} and Q_{xx} matrices is that only the coefficient matrix, A, is used in their formation. Since the A matrix contains coefficients that express the relationships of the unknowns to each other, it depends only on the *geometry* of the problem. The only other term in Equation (13.25) is the reference variance, and that depends on the quality of the measurements. These are important concepts that will be revisited in Chapter 21 when simulation of surveying networks is discussed.

PROBLEMS

For each problem, calculate the estimated errors for the adjusted benchmark elevations.

13.1 The reference variance of an adjustment is 0.89. The covariance matrix and unknown parameter matrix are

$$Q_{xx} = \begin{bmatrix} 0.5486 & 0.1864 & 0.0937 \\ 0.1864 & 0.4987 & -0.1678 \\ 0.0937 & -0.1678 & 0.8439 \end{bmatrix} \quad X = \begin{bmatrix} A \\ B \\ C \end{bmatrix}$$

What is the estimated error in the adjusted value for:
(a) A?
(b) B?
(c) C?

13.2 In Problem 13.1, the adjustment had nine degrees of freedom.
(a) Did the adjustment pass the χ^2 test at a 95% confidence level?
(b) Assuming it passed the χ^2 test in part (a), what are the estimated errors in the adjusted parameters?

For Problems 13.3 to 13.8, determine the estimated errors in the adjusted elevations.

13.3 Problem 12.1
13.4 Problem 12.4
13.5 Problem 12.7
13.6 Problem 12.9
13.7 Problem 12.11
13.8 Problem 12.13

For each problem, calculate the estimated errors for the adjusted elevation differences.

13.9 Problem 12.1
13.10 Problem 12.4
13.11 Problem 12.7
13.12 Problem 12.9
13.13 Problem 12.11
13.14 Calculate the adjusted length $\overline{AD}$ and its estimated error given Figure P13.14 and observational data below (assume equal weights):

$$l_1 = 100.01 \quad l_2 = 200.00 \quad l_3 = 300.02$$
$$l_4 = 99.94 \quad l_5 = 200.02 \quad l_6 = 299.98$$

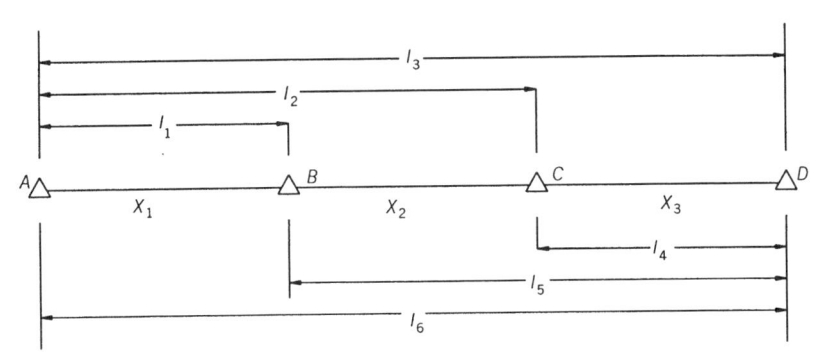

Figure P13.14

13.15 Use Figure P13.15 and the data below to answer the following questions.

Elevation of BM A = 263.453 m

Obs	From	To	ΔElev (m)	σ (m)
1	BM A	V	25.102	±0.018
2	BM B	V	−6.287	±0.019
3	V	X	10.987	±0.016
4	V	Y	24.606	±0.021
5	BM B	Y	17.993	±0.017
6	BM A	X	36.085	±0.021
7	Y	X	−13.295	±0.018

Elevation of BM B = 294.837 m

Obs	From	To	ΔElev (m)	σ (m)
8	Y	Z	−20.732	±0.022
9	W	Z	18.455	±0.022
10	V	W	−14.896	±0.021
11	BM A	W	10.218	±0.017
12	BM B	X	4.693	±0.020
13	W	X	25.883	±0.018
14	X	Z	−7.456	±0.020

What is:
(a) the most probable elevation for each of stations V, W, X, Y, and Z?
(b) the estimated error in each elevation?
(c) the estimated error in each adjusted observation?
(d) the estimated error in the elevation difference from benchmark A to station Z?

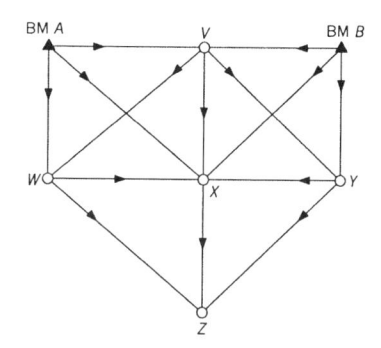

Figure P13.15

13.16 Do a χ^2 test in Problem 13.15. What observation might contain a blunder?

13.17 Repeat Problem 13.15 without observations 3, 4, and 10.

13.18 Repeat Problem 13.15 without observations 4, 8, 9, and 12.

13.19 Use ADJUST to do Problems 13.15, 13.17, and 13.18. Explain any differences in the adjustment results.

Programming Problems

13.20 Adapt the program developed in Problem 12.17 to compute and tabulate the adjusted:

(a) elevations and their estimated errors.

(b) elevation differences and their estimated errors.

13.21 Adapt the program developed in Problem 12.18 to compute and tabulate the adjusted:

(a) elevations and their estimated errors.

(b) elevation differences and their estimated errors.

CHAPTER 14

ADJUSTMENT OF HORIZONTAL SURVEYS: TRILATERATION

14.1 INTRODUCTION

Horizontal surveys are performed for the purpose of determining precise relative horizontal positions of points. They have traditionally been accomplished by trilateration, triangulation, and traverse. These traditional types of surveys involve making distance, direction, and angle observations. As with all types of surveys, errors will occur in making these measurements, and thus they must be analyzed and if acceptable, adjusted. In the following three chapters, procedures are described for adjusting trilateration, triangulation, and traverse surveys, in that order.

In recent years, the *global positioning system* (GPS) has gradually been replacing these traditional procedures for conducting precise horizontal control surveys. In fact, GPS not only yields horizontal positions, but it gives ellipsoidal heights as well. Thus, GPS provides three-dimensional surveys. Again as with all observations, GPS observations contain errors and must be adjusted. In Chapter 17 we discuss the subject of GPS surveying in more detail and illustrate methods for adjusting networks surveyed by this procedure.

Horizontal surveys, especially those covering a large extent, must account for the systematic effects of the Earth's curvature. One way this can be accomplished is to do the computations using coordinates from a mathematically rigorous map projection system such as the state plane system or a local plane coordinate system that accounts rigorously for the Earth's curvature. Map projection coordinate systems care presented in Appendix F. In the following chapters, methods are developed for adjusting horizontal surveys using parametric equations that are based on plane coordinates. In Chapter 23, a three-dimensional geodetic network adjustment is developed for traditional surveying observations, including differential leveling, slant distances, and vertical angles.

It should be noted that if state plane coordinates are used, the numbers are usually rather large. Consequently, when they are used in mathematical computations, errors due to round-off and truncation can occur. This can be prevented by translating the origin of the coordinates prior to adjustment, a process that simply involves subtracting a constant value from all coordinates. Then after the adjustment is finished, the true origin is restored by adding the constants to the adjusted values. This procedure is demonstrated with the following example.

Example 14.1 Assume that the NAD 83 state plane coordinates of three control stations to be used in a horizontal survey adjustment are as given below. Translate the origin.

Point	Easting (m)	Northing (m)
A	698,257.171	172,068.220
B	698,734.839	171,312.344
C	698,866.717	170,696.617

SOLUTION

Step 1: Many surveyors prefer to work in feet, and some jobs require it. Thus, in this step the eastings and northings, respectively, are converted to X and Y values in feet by multiplying by 3.28083333. This is the factor for converting meters to U.S. survey feet and is based on there being exactly 39.37 inches per meter. After making the multiplications, the coordinates in feet are as follows:

Point	X (ft)	Y (ft)
A	2,290,865.40	564,527.15
B	2,292,432.55	562,047.25
C	2,292,865.22	560,027.15

Step 2: To reduce the sizes of these numbers, an X constant is subtracted from each X coordinate and a Y constant is subtracted from each Y coordinate. For convenience, these constants are usually rounded to the nearest thousandth and are normally selected to give the smallest possible coordinates without producing negative values. In this instance, 2,290,000 ft and 560,000 ft are used as the X and Y constants, respectively. Subtracting these values from the coordinates yields

Point	X' (ft)	Y' (ft)
A	865.40	4527.15
B	2432.55	2047.25
C	2865.22	27.15

These X' and Y' coordinates can then be used in the adjustment. After the adjustment is complete, the coordinates are translated back to their state plane values by reversing the steps described above, that is, by adding 2,290,000 ft to all adjusted X coordinates, and adding 560,000 ft to all adjusted Y coordinates. If desired, they can be converted back to meters also.

In the horizontal adjustment problems solved later in this the book, either translated state plane coordinates or local plane coordinates are used. In this chapter we concentrate on adjusting trilateration surveys, those involving only horizontal distance observations. This method of conducting horizontal surveys became common with the introduction of EDM instruments that enable accurate distance observations to be made rapidly and economically. Trilateration is still possible using today's modern total station instruments, but as noted, the procedure is now giving way to traversing and GPS surveys.

14.2 DISTANCE OBSERVATION EQUATION

In adjusting trilateration surveys using the parametric least squares method, observation equations are written that relate the observed quantities and their inherent random errors to the most probable values for the x and y coordinates (the parameters) of the stations involved. Referring to Figure 14.1, the following distance equation can be written for any observation IJ:

$$l_{ij} + v_{l_{ij}} = \sqrt{(x_j - x_i)^2 + (y_j - y_i)^2} \qquad (14.1)$$

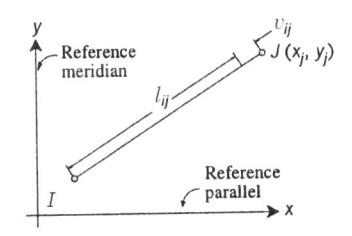

Figure 14.1 Observation of a distance.

In Equation (14.1), l_{ij} is the observed distance of a line between stations I and J, $v_{l_{ij}}$ the residual in the observation l_{ij}, x_i and y_i the most probable coordinate values for station I, and x_j and y_j the most probable coordinate values for station J. Equation (14.1) is a nonlinear function involving the unknown variables x_i, y_i, x_j, and y_j that can be rewritten as

$$F(x_i, y_i, x_j, y_j) = l_{ij} + v_{l_{ij}} \tag{14.2}$$

where

$$F(x_i, y_i, x_j, y_j) = \sqrt{(x_j - x_i)^2 + (y_j - y_i)^2}$$

As discussed in Section 11.10, a system of nonlinear equations such as Equation (14.2) can be linearized and solved using a first-order Taylor series approximation. The linearized form of Equation (14.2) is

$$F(x_i, y_i, x_j, y_j) = F(x_{i_0}, y_{i_0}, x_{j_0}, y_{j_0}) + \left(\frac{\partial F}{\partial x_i}\right)_0 dx_i + \left(\frac{\partial F}{\partial y_i}\right)_0 dy_i$$
$$+ \left(\frac{\partial F}{\partial x_j}\right)_0 dx_j + \left(\frac{\partial F}{\partial y_j}\right)_0 dy_j \tag{14.3}$$

where $(\partial F/\partial x_i)_0$, $(\partial F/\partial y_i)_0$, $(\partial F/\partial x_j)_0$, and $(\partial F/\partial y_j)_0$ are the partial derivatives of F with respect to x_i, y_i, x_j, and y_j, respectively, evaluated with the approximate coordinate values x_{i_0}, y_{i_0}, x_{j_0}, and y_{j_0}; x_i, y_i, x_j, and y_j are the unknown parameters; and dx_i, dy_i, dx_j, and dy_j are the corrections to the approximation coordinate values such that

$$x_i = x_{i_0} + dx_i \qquad y_i = y_{i_0} + dy_i$$
$$x_j = x_{j_0} + dx_j \qquad y_j = y_{j_0} + dy_j \tag{14.4}$$

The evaluation of partial derivatives is straightforward and will be illustrated with $\partial F/\partial x_i$. Equation (14.2) can be rewritten as

$$F(x_i, y_i, x_j, y_j) = [(x_j - x_i)^2 + (y_j - y_i)^2]^{1/2} \tag{14.5}$$

Taking the derivative of Equation (14.5) with respect to x_i yields

$$\frac{\partial F}{\partial x_i} = \frac{1}{2}[(x_j - x_i)^2 + (y_j - y_i)^2]^{1/2}[2(x_j - x_i)(-1)] \tag{14.6}$$

Simplifying Equation (14.6) yields

$$\frac{\partial F}{\partial x_i} = \frac{-(x_j - x_i)}{\sqrt{(x_j - x_i)^2 + (y_j - y_i)^2}} = \frac{x_i - x_j}{IJ} \tag{14.7}$$

Employing the same procedure, the remaining partial derivatives are

$$\frac{\partial F}{\partial y_i} = \frac{y_i - y_j}{IJ} \qquad \frac{\partial F}{\partial x_j} = \frac{x_j - x_i}{IJ} \qquad \frac{\partial F}{\partial y_j} = \frac{y_j - y_i}{IJ} \tag{14.8}$$

If Equations (14.7) and (14.8) are substituted into Equation (14.3) and the results substituted into Equation (14.2), the following prototype linearized distance observation equation is obtained:

$$\left(\frac{x_i - x_j}{IJ}\right)_0 dx_i + \left(\frac{y_i - y_j}{IJ}\right)_0 dy_i + \left(\frac{x_j - x_i}{IJ}\right)_0 dx_j + \left(\frac{y_j - y_i}{IJ}\right)_0 dy_j$$
$$= k_{l_{ij}} + v_{l_{ij}} \tag{14.9}$$

where $(\cdot)_0$ is evaluated at the approximate parameter values, $k_{l_{ij}} = l_{ij} - IJ_0$, and

$$IJ_0 = F(x_{i_0}, y_{i_0}, x_{j_0}, y_{j_0}) = \sqrt{(x_{j_0} - x_{i_0})^2 + (y_{j_0} - y_{i_0})^2}$$

14.3 TRILATERATION ADJUSTMENT EXAMPLE

Even though the geometric figures used in trilateration are many and varied, they are equally adaptable to the observation equation method in a parametric adjustment. Consider the example shown in Figure 14.2, where the distances are observed from three stations with known coordinates to a common unknown station U. Since the unknown station has two unknown coordinates and there are three observations, this results in one redundant observation. That is, the coordinates of station U could be determined using any two of

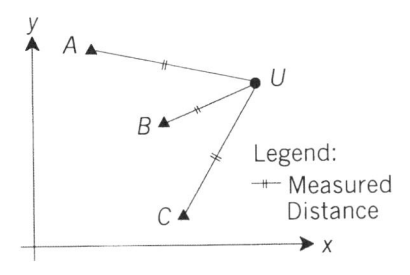

Figure 14.2 Trilateration example.

238 ADJUSTMENT OF HORIZONTAL SURVEYS: TRILATERATION

the three observations. But all three observations can be used simultaneously and adjusted by the method of least squares to determine the most probable value for the coordinates of the station.

The observation equations are developed by substituting into prototype equation (14.9). For example, the equation for distance AU is formed by interchanging subscript I with A and subscript J with U in Equation (14.9). In a similar fashion, an equation can be created for each line observed using the following subscript substitutions:

I	J
A	U
B	U
C	U

When one end of the observed line is a control station, its coordinates are fixed, and thus those terms can be dropped in prototype equation (14.9).[1] This can be thought of as setting the dx and dy corrections for the control station equal to zero. In this example, station U always takes the position of J in the prototype equation, and thus only the coefficients corresponding to dx_j and dy_j are used. Using the appropriate substitutions, the following three linearized observation equations result.

$$\frac{x_{u0} - x_a}{AU_0} dx_u + \frac{y_{u0} - y_a}{AU_0} dy_u = (l_{AU} - AU_0) + v_{l_{AU}}$$

$$\frac{x_{u0} - x_b}{BU_0} dx_u + \frac{y_{u0} - y_b}{BU_0} dy_u = (l_{BU} - AU_0) + v_{l_{BU}} \quad (14.10)$$

$$\frac{x_{u0} - x_c}{CU_0} dx_u + \frac{y_{u0} - y_c}{CU_0} dy_u = (l_{CU} - CU_0) + v_{l_{CU}}$$

In Equation (14.10),

$$AU_0 = \sqrt{(x_{u0} - x_a)^2 + (y_{u0} - y_a)^2} \qquad BU_0 = \sqrt{(x_{u0} - x_b)^2 + (y_{u0} - y_b)^2}$$

$$CU_0 = \sqrt{(x_{u0} - x_c)^2 + (y_{u0} - y_c)^2}$$

l_{au}, l_{bu}, and l_{cu} are the observed distances with residuals v; and x_{u0} and y_{u0} are initial coordinate approximations for station U. Equations (14.10) can be expressed in matrix form as

---

[1] The method of dropping control station coordinates from the adjustment, known as *elimination of constraints,* is covered in Chapter 20.

14.3 TRILATERATION ADJUSTMENT EXAMPLE

$$JX = K + V \qquad (14.11)$$

where J is the Jacobian matrix of partial derivatives, X the matrix or unknown corrections dx_u and dy_u, K the matrix of constants (i.e., the observed lengths, minus their corresponding lengths computed from the initial approximate coordinates), and V the residual matrix. Equation (14.11) in expanded form is

$$\begin{bmatrix} \dfrac{x_{u_0} - x_a}{AU_0} & \dfrac{y_{u_0} - y_a}{AU_0} \\ \dfrac{x_{u_0} - x_b}{BU_0} & \dfrac{y_{u_0} - y_b}{BU_0} \\ \dfrac{x_{u_0} - x_c}{CU_0} & \dfrac{y_{u_0} - y_c}{CU_0} \end{bmatrix} \begin{bmatrix} dx_u \\ dy_u \end{bmatrix} = \begin{bmatrix} l_{AU} - AU_0 \\ l_{BU} - BU_0 \\ l_{CU} - CU_0 \end{bmatrix} + \begin{bmatrix} v_{l_{AU}} \\ v_{l_{BU}} \\ v_{l_{CU}} \end{bmatrix} \qquad (14.12)$$

The Jacobian matrix can be formed systematically using the following steps.

Step 1: Head each column with an unknown value.
Step 2: Create a row for every observation.
Step 3: Substitute in the appropriate coefficient corresponding to the column into each row.

If this procedure is followed for this problem, the Jacobian matrix is

$$\begin{bmatrix} & & dx_u & dy_u \\ AU & | & \dfrac{\partial F}{\partial x_u} & \dfrac{\partial F}{\partial y_u} \\ BU & | & \dfrac{\partial F}{\partial x_u} & \dfrac{\partial F}{\partial y_u} \\ CU & | & \dfrac{\partial F}{\partial x_u} & \dfrac{\partial F}{\partial y_u} \end{bmatrix}$$

Once Equation (14.12) is created, the corrections of dx_u and dy_u, and thus the most probable coordinate values, x_u and y_u, can be computed using Equation (11.37). Of course, to obtain the final adjusted values, the solution must be iterated, as discussed in Section 11.10.

Example 14.2 To clarify the computational procedure, a numerical example for Figure 14.2 is presented. Suppose that the observed distances l_{AU}, l_{BU}, and l_{CU} are 6049.00, 4736.83, and 5446.49 ft, respectively, and the control stations have coordinates in units of feet of

240 ADJUSTMENT OF HORIZONTAL SURVEYS: TRILATERATION

$$x_a = 865.40 \qquad x_b = 2432.55 \qquad x_c = 2865.22$$

$$y_a = 4527.15 \qquad y_b = 2047.25 \qquad y_c = 27.15$$

(Note that these are the translated coordinates obtained in Example 14.1.) Compute the most probable coordinates for station U.

SOLUTION Perform the first iteration.

Step 1: Calculate approximate coordinates for station U.

(a) Calculate azimuth AB from the coordinate values of stations A and B.

$$Az_{AB} = \tan^{-1} \frac{x_b - x_a}{y_b - y_a} + 180°$$

$$= \tan^{-1} \frac{2432.55 - 865.40}{2047.25 - 4527.15} + 180°$$

$$= 147°42'34''$$

(b) Calculate the distance between stations A and B from their coordinate values.

$$AB = \sqrt{(x_b - x_a)^2 + (x_b - x_a)^2}$$

$$= \sqrt{(2432.55 - 865.20)^2 + (2047.25 - 4527.15)^2}$$

$$= 2933.58 \text{ ft}$$

(c) Calculate azimuth AU_0 using the cosine law in triangle AUB:

$$c^2 = a^2 + b^2 - 2ab \cos C$$

$$\cos(\angle UAB) = \frac{6049.00^2 + 2933.58^2 - 4736.83^2}{2(6049.00)(2933.58)}$$

$$\angle UAB = 50°06'50''$$

$$Az_{AU_0} = 147°42'34'' - 50°06'50'' = 97°35'44''$$

(d) Calculate the coordinates for station U.

$$x_{u_0} = 865.40 + 6049.00 \sin 97°35'44'' = 6861.325 \text{ ft}$$

$$y_{u_0} = 4527.15 + 6049.00 \cos 97°35'44'' = 3727.596 \text{ ft}$$

14.3 TRILATERATION ADJUSTMENT EXAMPLE

Step 2: Calculate AU_0, BU_0, and CU_0. For this first iteration, AU_0 and BU_0 are exactly equal to their respective observed distances since x_{u0} and y_{u0} were calculated using these quantities. Thus,

$$AU_0 = 6049.00 \qquad BU_0 = 4736.83$$

$$CU_0 = \sqrt{(6861.325 - 2865.22)^2 + (3727.596 - 27.15)^2} = 5446.298 \text{ ft}$$

Step 3: Formulate the matrices.
(a) The elements of the Jacobian matrix in Equation (14.12) are[2]

$$j_{11} = \frac{6861.325 - 865.40}{6049.00} = 0.991 \qquad j_{12} = \frac{3727.596 - 4527.15}{6049.00} = -0.132$$

$$j_{21} = \frac{6861.325 - 2432.55}{4736.83} = 0.935 \qquad j_{22} = \frac{3727.596 - 2047.25}{4736.83} = 0.355$$

$$j_{31} = \frac{6861.325 - 2865.22}{5446.298} = 0.734 \qquad j_{32} = \frac{3727.596 - 27.15}{5446.298} = 0.679$$

(b) The elements of the K matrix in Equation (14.12) are

$$k_1 = 6049.00 - 6049.00 = 0.000$$

$$k_2 = 4736.83 - 4736.83 = 0.000$$

$$k_3 = 5446.49 - 5446.298 = 0.192$$

Step 4: The matrix solution using Equation (11.37) is

$$X = (J^T J)^{-1} J^T K$$

$$J^T J = \begin{bmatrix} 0.991 & 0.935 & 0.734 \\ -0.132 & 0.355 & 0.679 \end{bmatrix} \begin{bmatrix} 0.991 & -0.132 \\ 0.935 & 0.355 \\ 0.735 & 0.679 \end{bmatrix} = \begin{bmatrix} 2.395 & 0.699 \\ 0.699 & 0.605 \end{bmatrix}$$

$$(J^T J)^{-1} = \frac{1}{0.960} \begin{bmatrix} 0.605 & -0.699 \\ -0.699 & 2.395 \end{bmatrix}$$

---

[2] Note that the denominators in the coefficients of step 3a are distances computed from the approximate coordinates. Only the distances computed for the first iteration will match the measured distances exactly. Do not use measured distances for the denominators of these coefficients.

$$J^TK = \begin{bmatrix} 0.991 & 0.935 & 0.734 \\ -0.132 & 0.355 & 0.679 \end{bmatrix} \begin{bmatrix} 0.000 \\ 0.000 \\ 0.192 \end{bmatrix} = \begin{bmatrix} 0.141 \\ 0.130 \end{bmatrix}$$

$$X = \frac{1}{0.960} \begin{bmatrix} 0.605 & -0.699 \\ -0.699 & 2.395 \end{bmatrix} \begin{bmatrix} 0.141 \\ 0.130 \end{bmatrix} = \begin{bmatrix} -0.006 \\ 0.222 \end{bmatrix}$$

The revised coordinates of U are

$$x_u = 6861.325 - 0.006 = 6861.319$$

$$y_u = 3727.596 + 0.222 = 3727.818$$

Now perform the second iteration.

Step 1: Calculate AU_0, BU_0, and CU_0.

$$AU_0 = \sqrt{(6861.319 - 865.40)^2 + (3727.818 - 4527.15)^2} = 6048.965 \text{ ft}$$
$$BU_0 = \sqrt{(6861.319 - 2432.55)^2 + (3727.818 - 2047.25)^2} = 4736.909 \text{ ft}$$
$$CU_0 = \sqrt{(6861.319 - 2865.22)^2 + (3727.818 - 27.15)^2} = 5446.444 \text{ ft}$$

Notice that these computed distances no longer match their observed counterparts.

Step 2: Formulate the matrices. With these minor changes in the lengths, the J matrix (to three places) does not change, and thus $(J^TJ)^{-1}$ does not change either. However, the K matrix does change, as shown by the following computations.

$$k_1 = 6049.00 - 6048.965 = 0.035$$

$$k_2 = 4736.83 - 4736.909 = -0.079$$

$$k_3 = 5446.49 - 5446.444 = 0.046$$

Step 3: Matrix solution

$$J^TK = \begin{bmatrix} 0.991 & 0.935 & 0.734 \\ -0.132 & 0.355 & 0.679 \end{bmatrix} \begin{bmatrix} 0.035 \\ -0.079 \\ 0.046 \end{bmatrix} = \begin{bmatrix} -0.005 \\ -0.001 \end{bmatrix}$$

$$X = \frac{1}{0.960} \begin{bmatrix} 0.605 & -0.699 \\ -0.699 & 2.395 \end{bmatrix} \begin{bmatrix} -0.005 \\ -0.001 \end{bmatrix} = \begin{bmatrix} -0.002 \\ 0.001 \end{bmatrix}$$

The revised coordinates of U are

$$x_u = 6861.319 - 0.002 = 6861.317$$

$$y_u = 3727.818 + 0.001 = 3727.819$$

Satisfactory convergence is shown by the very small corrections in the second iteration. This problem has also been solved using the program AD-JUST. Values computed include the most probable coordinates for station U, their standard deviations, the adjusted lengths of the observed distances, their residuals and standard deviations, and the reference variance and reference standard deviation. These are tabulated as shown below.

```
*****************
Adjusted stations
*****************
Station           X               Y              Sx          Sy
=================================================================
   U           6,861.32        3,727.82         0.078       0.154


******************************
Adjusted Distance Observations
******************************
Station      Station
Occupied     Sighted      Distance          V           S
=================================================================
   A            U         6,048.96        -0.037       0.090
   B            U         4,736.91         0.077       0.060
   C            U         5,446.44        -0.047       0.085
```

$$\text{Adjustment Statistics}$$
$$S_0^2 = 0.00954$$
$$S_0 = \pm 0.10$$

14.4 FORMULATION OF A GENERALIZED COEFFICIENT MATRIX FOR A MORE COMPLEX NETWORK

In the trilaterated network of Figure 14.3, all lines were observed. Assume that stations A and C are control stations. For this network, there are 10 observations and eight unknowns. Stations A and C can be fixed by giving the terms dx_a, dy_a, dx_c, and dy_c zero coefficients, which effectively drops these

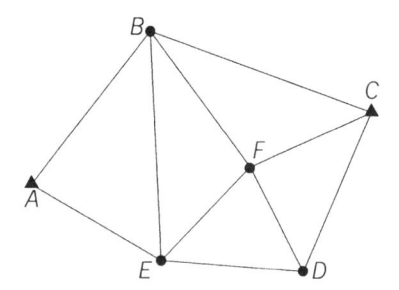

244 ADJUSTMENT OF HORIZONTAL SURVEYS: TRILATERATION

Figure 14.3 Trilateration network.

terms from the solution. The coefficient matrix formulated from prototype equation (14.9) has nonzero elements, as indicated in Table 14.1. In this table the appropriate coefficient from Equation (14.9) is indicated by its corresponding unknown terms of dx_i, dy_i, dx_j, or dy_j.

14.5 COMPUTER SOLUTION OF A TRILATERATED QUADRILATERAL

The quadrilateral shown in Figure 14.4 was adjusted using the software MATRIX. In this problem, points Bucky and Badger are control stations whose coordinates are held fixed. The five distances observed are:

Line	Distance (ft)
Badger–Wisconsin	5870.302
Badger–Campus	7297.588
Wisconsin–Campus	3616.434
Wisconsin–Bucky	5742.878
Campus–Bucky	5123.760

TABLE 14.1 Structure of the Normal Matrix for the Complex Network in Figure 14.3

Distance, IJ	dx_b	dy_b	dx_d	dy_d	dx_e	dy_e	dx_f	dy_f
AB	dx_j	dy_j	0	0	0	0	0	0
AE	0	0	0	0	dx_j	dy_j	0	0
BC	dx_i	dy_i	0	0	0	0	0	0
BF	dx_i	dy_i	0	0	0	0	dx_j	dy_j
BE	dx_i	dy_i	0	0	dx_j	dy_j	0	0
CD	0	0	dx_j	dy_j	0	0	0	0
CF	0	0	0	0	0	0	dx_j	dy_j
DF	0	0	dx_i	dy_i	0	0	dx_j	dy_j
DE	0	0	dx_i	dy_i	dx_j	dy_j	0	0
EF	0	0	0	0	dx_i	dy_i	dx_j	dy_j

14.5 COMPUTER SOLUTION OF A TRILATERATED QUADRILATERAL

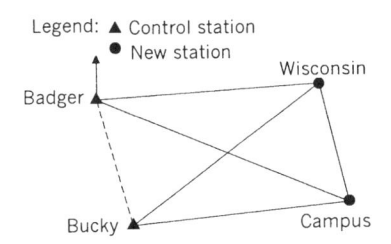

Figure 14.4 Quadrilateral network.

The state plane control coordinates in units of feet for station Badger are $x = 2,410,000.000$ and $y = 390,000.000$, and for Bucky are $x = 2,411,820.000$ and $y = 386,881.222$.

Step 1: To solve this problem, approximate coordinates are first computed for stations Wisconsin and Campus. This is done following the procedures used in Section 14.3, with the resulting initial approximations being

Wisconsin: $x = 2,415,776.819$ $y = 391,043.461$

Campus: $x = 2,416,898.227$ $y = 387,602.294$

Step 2: Following prototype equation (14.9) and the procedures outlined in Section 14.4, a table of coefficients is established. For the sake of brevity in Table 14.2, the following station assignments were made: Badger = 1, Bucky = 2, Wisconsin = 3, and Campus = 4.

After forming the J matrix, the K matrix is computed. This is done in a manner similar to step 3 of the first iteration in Example 14.2. The matrices

TABLE 14.2 Structure of the Coefficient or J Matrix for the Example in Figure 14.4

	$dx_{\text{Wisconsin}}$	$dy_{\text{Wisconsin}}$	dx_{Campus}	dy_{Campus}
Badger–Wisconsin 1–3	$\dfrac{x_{30}-x_1}{(1-3)_0}$	$\dfrac{y_{30}-y_1}{(1-3)_0}$	0	0
Badger–Campus 1–4	0	0	$\dfrac{x_{40}-x_1}{(1-4)_0}$	$\dfrac{y_{40}-y_1}{(1-4)_0}$
Wisconsin–Campus 3–4	$\dfrac{x_{30}-x_{40}}{(3-4)_0}$	$\dfrac{y_{30}-y_{40}}{(3-4)_0}$	$\dfrac{x_{40}-x_{30}}{(3-4)_0}$	$\dfrac{y_{40}-y_{30}}{(3-4)_0}$
Wisconsin–Bucky 3–2	$\dfrac{x_{30}-x_{20}}{(3-2)_0}$	$\dfrac{y_{30}-y_{20}}{(3-2)_0}$	0	0
Campus–Bucky 4–2	0	0	$\dfrac{x_{40}-x_{20}}{(4-2)_0}$	$\dfrac{y_{40}-y_{20}}{(4-2)_0}$

246 ADJUSTMENT OF HORIZONTAL SURVEYS: TRILATERATION

were entered into a file following the formats listed in the Help file for program MATRIX. Following are the input data, matrices for the first and last iterations of this three-iteration solution, and the final results tabulated.

```
*********************************************
Initial approximations for unknown stations
*********************************************
Station              X                Y
========================================
Wisconsin   2,415,776.819    391,043.461
   Campus   2,416,898.227    387,602.294

Control Stations
~~~~~~~~~~~~
Station              X                Y
========================================
  Badger    2,410,000.000    390,000.000
   Bucky    2,411,820.000    386,881.222

*********************
Distance Observations
*********************
Occupied    Sighted    Distance
================================
   Badger   Wisconsin   5,870.302
   Badger   Campus      7,297.588
Wisconsin   Campus      3,616.434
Wisconsin   Bucky       5,742.878
   Campus   Bucky       5,123.760
```

First Iteration Matrices

```
J Dim: 5x4                                              K Dim: 5x1   X Dim 4x1
=================================================     ==========   =========
   0.98408   0.17775   0.00000   0.00000               -0.00026    0.084751
   0.00000   0.00000   0.94457  -0.32832               -5.46135   -0.165221
  -0.30984   0.95079   0.30984  -0.95079               -2.84579   -5.531445
   0.68900   0.72477   0.00000   0.00000               -0.00021    0.959315
   0.00000   0.00000   0.99007   0.14058               -5.40507   =========
=================================================     ==========

JtJ Dim: 4x4
==========================================
   1.539122    0.379687   -0.096003    0.294595
   0.379687    1.460878    0.294595   -0.903997
  -0.096003    0.294595    1.968448   -0.465525
   0.294595   -0.903997   -0.465525    1.031552
==========================================
```

```
Inv(N) Dim: 4x4
==========================================
  1.198436  -1.160169  -0.099979  -1.404084
 -1.160169   2.635174   0.194272   2.728324
 -0.099979   0.194272   0.583337   0.462054
 -1.404084   2.728324   0.462054   3.969873
==========================================
```

Final Iteration

```
J Dim: 5x4                                       K Dim:5x1    X Dim 4x1
=====================================            ==========   =========
  0.98408  0.17772  0.00000   0.00000             -0.05468     0.000627
  0.00000  0.00000  0.94453  -0.32843              0.07901    -0.001286
 -0.30853  0.95121  0.30853  -0.95121             -0.03675    -0.000040
  0.68902  0.72474  0.00000   0.00000              0.06164     0.001814
  0.00000  0.00000  0.99002   0.14092             -0.06393    =========
=====================================            ==========
```

```
JtJ Dim: 4x4
==========================================
  1.538352   0.380777  -0.095191   0.293479
  0.380777   1.461648   0.293479  -0.904809
 -0.095191   0.293479   1.967465  -0.464182
  0.293479  -0.904809  -0.464182   1.032535
==========================================
```

```
Qxx = Inv(N) Dim: 4x4
==========================================
  1.198574  -1.160249  -0.099772  -1.402250
 -1.160249   2.634937   0.193956   2.725964
 -0.099772   0.193956   0.583150   0.460480
 -1.402250   2.725964   0.460480   3.962823
==========================================
```

```
Qll = J Qxx Jt Dim: 5x5
=====================================================
  0.838103   0.233921  -0.108806   0.182506  -0.189263
  0.233921   0.662015   0.157210  -0.263698   0.273460
 -0.108806   0.157210   0.926875   0.122656  -0.127197
  0.182506  -0.263698   0.122656   0.794261   0.213356
 -0.189263   0.273460  -0.127197   0.213356   0.778746
=====================================================
```

248 ADJUSTMENT OF HORIZONTAL SURVEYS: TRILATERATION

```
*****************
Adjusted stations
*****************
Station         X               Y              Sx       Sy
===========================================================
Wisconsin   2,415,776.904   391,043.294    0.1488   0.2206
   Campus   2,416,892.696   387,603.255    0.1038   0.2705


******************************
Adjusted Distance Observations
******************************
 Occupied    Sighted    Distance       V         S
===========================================================
   Badger  Wisconsin   5,870.357     0.055    0.1244
   Badger     Campus   7,297.509    -0.079    0.1106
Wisconsin     Campus   3,616.471     0.037    0.1308
Wisconsin      Bucky   5,742.816    -0.062    0.1211
   Campus      Bucky   5,123.824     0.064    0.1199

-----Reference Standard Deviation = ±0.135905-----
      Iterations » 3
```

Notes

1. As noted earlier, it is important that observed distances not be used in the denominator of the coefficients matrix, *J*. This is not only theoretically incorrect but can cause slight differences in the final solution, or even worse, it can cause the system to diverge from any solution! Always compute distances based on the current approximate coordinates.
2. The final portion of the output lists the adjusted *x* and *y* coordinates of the stations, the reference standard deviation, the standard deviations of the adjusted coordinates, the adjusted line lengths, and their residuals.
3. The Q_{xx} matrix was listed on the last iteration only. It is needed for calculating the estimated errors of the adjusted coordinates using Equation (14.24) and is also necessary for calculating error ellipses. The subject of error ellipses is discussed in Chapter 19.

14.6 ITERATION TERMINATION

When programming a nonlinear least squares adjustment, some criteria must be established to determine the appropriate point at which to stop the iteration process. Since it is possible to have a set of data that has no solution, it is also important to determine when that condition occurs. In this section we describe three methods commonly used to indicate the appropriate time end the iteration process.

14.6.1 Method of Maximum Iterations

The simplest procedure of iteration termination involves limiting the number of iterations to a predetermined maximum. The risk with this method is that if this maximum is too low, a solution may not be reached at the conclusion of the process, and if it is too high, time is wasted on unnecessary iterations. Although this method does not assure convergence, it can prevent the adjustment from continuing indefinitely, which could occur if the solution diverges. When good initial approximations are supplied for the unknown parameters, a limit of 10 iterations should be well beyond what is required for a solution since the least squares method converges quadratically.

14.6.2 Maximum Correction

This method was used in earlier examples. It involves monitoring the absolute size of the corrections. When all corrections become negligibly small, the iteration process is stopped. The term *negligible* is relative. For example, if distances are observed to the nearest foot, it would be foolish to assume that the size of the corrections will become less than some small fraction of a foot. Generally, *negligible* is interpreted as a correction that is less than one-half the least count of the smallest unit of measure. For instance, if all distances are observed to the nearest 0.01 ft, it would be appropriate to assume convergence when the absolute size of all corrections is less than 0.005 ft. Although the solution may continue to converge with continued iterations, the work to get these corrections is not warranted based on the precision of the observations.

14.6.3 Monitoring the Adjustment's Reference Variance

The best method for determining convergence involves monitoring the reference variance and its changes between iterations. Since the least squares method converges quadratically, the iteration process should definitely be stopped if the reference variance increases. An increasing reference variance suggests a diverging solution, which happens when one of two things has occurred: (1) a large blunder exists in the data set and no solution is possible, or (2) the maximum correction size is less than the precision of the observations. In the second case, the best solution for the given data set has already been reached, and when another iteration is attempted, the solution will converge, only to diverge on the next iteration. This *apparent bouncing* in the solution is caused by convergence limits being too stringent for the quality of the data.

By monitoring the reference variance, convergence and divergence can be detected. Convergence is assumed when the change in the reference variance falls below some predefined percentage. Convergence can generally be assumed when the change in the reference variance is less than 1% between iterations. If the size of the reference variance increases, the solution is diverging and the iteration process should be stopped. It should be noted that monitoring changes in the reference variance will always show convergence or divergence in the solution, and thus it is better than any method discussed previously. However, all methods should be used in concert when doing an adjustment.

PROBLEMS

Note: For problems requiring least squares adjustment, if a computer program is not distinctly specified for use in the problem, it is expected that the least squares algorithm will be solved using the program MATRIX, which is included on the CD supplied with the book.

14.1 Given the following observed values for the lines in Figure 14.2:

$AU = 2828.83$ ft $\quad BU = 2031.55$ ft $\quad CU = 2549.83$ ft

the control coordinates of A, B, and C are:

Station	x (ft)	y (ft)
A	1418.17	4747.14
B	2434.53	3504.91
C	3234.86	2105.56

What are the most probable values for the adjusted coordinates of station U?

14.2 Do a weighted least squares adjustment using the data in Problem 14.1 with weights base on the following observational errors.

$AU = \pm 0.015$ ft $\quad BU = 0.011$ ft $\quad CU = 0.012$ ft

(a) What are the most probable values for the adjusted coordinates of station U?
(b) What is the reference standard deviation of unit weight?

(c) What are the estimated standard deviations of the adjusted coordinates?

(d) Tabulate the adjusted distances, their residuals, and the standard deviations.

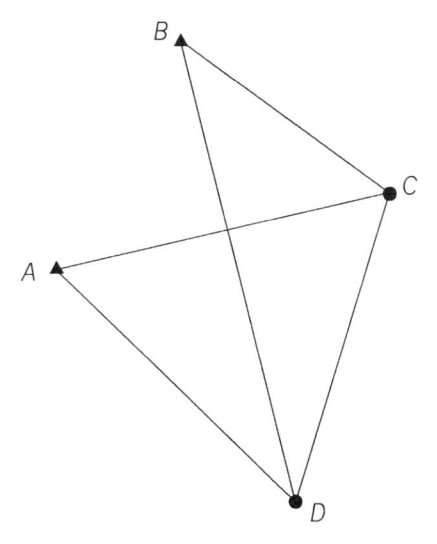

Figure 14.3

14.3 Do a least squares adjustment for the following values observed for the lines in Figure P14.3.

$$AC = 2190.04 \text{ ft} \qquad AD = 3397.25 \text{ ft} \qquad BC = 2710.38 \text{ ft}$$

$$BD = 2250.05 \text{ ft} \qquad CD = 2198.45 \text{ ft}$$

In the adjustment, hold the coordinates of stations A and B (in units of feet) of

$$x_a = 1423.08 \text{ ft} \quad \text{and} \quad y_a = 4796.24 \text{ ft}$$

$$x_b = 1776.60 \text{ ft} \quad \text{and} \quad y_b = 2773.32 \text{ ft}$$

(a) What are the most probable values for the adjusted coordinates of stations C and D?

(b) What is the reference standard deviation of unit weight?

252 ADJUSTMENT OF HORIZONTAL SURVEYS: TRILATERATION

(c) What are the estimated standard deviations of the adjusted coordinates?
(d) Tabulate the adjusted distances, their residuals, and the standard deviations.

14.4 Repeat Problem 14.3 using a weighted least squares adjustment, where the distance standard deviations are

$$DA = \pm 0.016 \text{ ft} \quad BC = \pm 0.0.018 \text{ ft} \quad BD = \pm 0.0.017 \text{ ft}$$

$$AC = \pm 0.0.016 \text{ ft} \quad CD = \pm 0.016 \text{ ft}$$

14.5 Use the ADJUST software to do Problems 14.3 and 14.4. Explain any differences in the adjustments.

14.6 Using the trilaterated Figure 14.3 and the data below, compute the most probable station coordinates and their standard deviations.

Initial approximations of stations

Station	Easting (m)	Northing (m)
B	12,349.500	14,708.750
D	17,927.677	11,399.956
E	13,674.750	10,195.970
F	14,696.838	12,292.118

Control stations

Station	Easting (m)	Northing (m)
A	10,487.220	11,547.206
C	16,723.691	14,258.338

Distance observations

Station Occupied	Station Sighted	Distance (m)	σ (m)
A	B	3669.240	0.013
B	C	4397.254	0.015
C	D	3101.625	0.012
D	E	4420.055	0.015
E	A	3462.076	0.013
B	E	4703.319	0.016
B	F	3369.030	0.012
F	E	2332.063	0.010
F	C	2823.857	0.011
F	D	3351.737	0.012

14.7 Using the station coordinates and trilateration data given below, find:
(a) the most probable coordinates for station E.
(b) the reference standard deviation of unit weight.
(c) the standard deviations of the adjusted coordinates.
(d) the adjusted distances, their residuals, and the standard deviations.

PROBLEMS

Control stations

Station	Easting (m)	Northing (m)
A	100,643.154	38,213.066
B	101,093.916	67,422.484
C	137,515.536	67,061.874
D	139,408.739	37,544.403

Initial approximations

Station	Easting (m)	Northing (m)
E	119,665.336	53,809.452

Distance observations

From	To	Distance (m)	S (m)
A	E	24,598.543	0.074
B	E	23,026.189	0.069
C	E	22,231.945	0.067
D	E	25,613.764	0.077

14.8 Using the station coordinates and trilateration data given below, find:
(a) the most probable coordinates for the unknown stations.
(b) the reference standard deviation of unit weight.
(c) the standard deviations of the adjusted coordinates.
(d) the adjusted distances, their residuals, and the standard deviations.

Control station

Station	X (ft)	Y (ft)
A	92,890.04	28,566.74
D	93,971.87	80,314.29

Initial approximations

Station	X (ft)	Y (ft)
B	93,611.26	47,408.62
C	93,881.71	64,955.36
E	111,191.00	38,032.76
F	110,109.17	57,145.10
G	110,019.02	73,102.09
H	131,475.32	28,837.20
I	130,213.18	46,777.56
J	129,311.66	64,717.91
K	128,590.44	79,142.31

Distance observations

From	To	Distance (ft)	S (ft)
A	B	18,855.74	0.06
B	C	17,548.79	0.05
C	D	15,359.17	0.05
D	G	17,593.38	0.05
C	G	18,077.20	0.06
C	F	18,009.22	0.06
B	F	19,156.82	0.06

B	E	19,923.71	0.06
A	E	20,604.19	0.06
H	E	22,271.36	0.07
I	E	20,935.94	0.06
H	I	17,984.75	0.06
I	J	17,962.99	0.06
J	K	14,442.41	0.05
I	F	22,619.85	0.07
J	F	20,641.79	0.06
J	G	21,035.82	0.06
K	G	19,529.02	0.06
E	F	19,142.85	0.06
F	G	15,957.22	0.05

14.9 Use the ADJUST software to do Problem 14.5.

14.10 Use the ADJUST software to do Problem 14.8.

14.11 Describe the methods used to detect convergence in a nonlinear least squares adjustment and the advantages and disadvantages of each.

Programming Problems

14.12 Create a computational program that computes the distance, coefficients, and $k_{l_{ij}}$ in Equation (14.9) between stations I and J given their initial coordinate values. Use this spreadsheet to determine the matrix values necessary for solving Problem 14.5.

14.13 Create a computational program that reads a data file containing station coordinates and distances and generates the J, W, and K matrices, which can be used by the MATRIX program. Demonstrate that this program works by using the data of Problem 14.5.

14.14 Create a computational program that reads a file containing the J, W, and K matrices and finds the most probable value for the station coordinates, the reference standard deviation, and the standard deviations of the station coordinates. Demonstrate that this program works by solving Problem 14.5.

14.15 Create a computational program that reads a file containing control station coordinates, initial approximations of unknown stations, and distance observations. The program should generate the appropriate matrices for a least squares adjustment, do the adjustment, and print out the final adjusted coordinates, their standard deviations, the final adjusted distances, their residuals, and the standard deviations in the adjusted distances. Demonstrate that this program works by solving Problem 14.5.

CHAPTER 15

ADJUSTMENT OF HORIZONTAL SURVEYS: TRIANGULATION

15.1 INTRODUCTION

Prior to the development of electronic distance measuring equipment and the global positioning system, triangulation was the preferred method for extending horizontal control over long distances. The positions of widely spaced stations were computed from measured angles and a minimal number of measured distances called *baselines*. This method was used extensively by the U.S. Coast and Geodetic Survey in extending much of the national network. Triangulation is still used by many surveyors in establishing horizontal control, although surveys that combine trilateration (distance observations) with triangulation (angle observations) are more common. In this chapter, methods are described for adjusting triangulation networks using least squares.

A least squares triangulation adjustment can use condition equations or observation equations written in terms of either azimuths or angles. In this chapter the observation equation method is presented. The procedure involves a parametric adjustment where the parameters are coordinates in a plane rectangular system such as state plane coordinates. In the examples, the specific types of triangulations known as intersections, resections, and quadrilaterals are adjusted.

15.2 AZIMUTH OBSERVATION EQUATION

The azimuth equation in parametric form is

$$\text{azimuth} = \alpha + C \qquad (15.1)$$

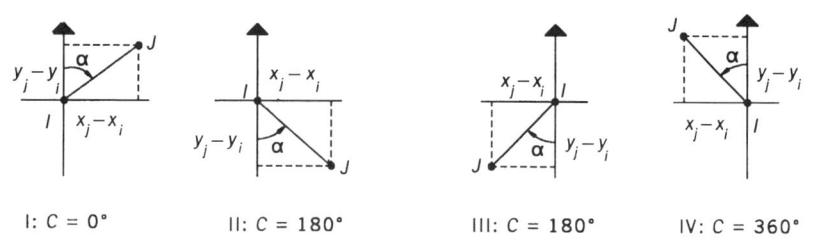

Figure 15.1 Relationship between the azimuth and the computed angle, α.

I: $C = 0°$ II: $C = 180°$ III: $C = 180°$ IV: $C = 360°$

where $\alpha = \tan^{-1}[(x_j - x_i)/(y_j - y_i)]$; x_i and y_i are the coordinates of the occupied station I; x_j and y_j are the coordinates of the sighted station J; and C is a constant that depends on the quadrant in which point J lies, as shown in Figure 15.1.

From the figure, Table 15.1 can be constructed, which relates the algebraic sign of the computed angle α in Equation (15.1) to the value of C and the value of the azimuth.

15.2.1 Linearization of the Azimuth Observation Equation

Referring to Equation (15.1), the complete observation equation for an observed azimuth of line IJ is

$$\tan^{-1}\frac{x_j - x_i}{y_j - y_i} + C = \text{Az}_{ij} + v_{\text{Az}_{ij}} \tag{15.2}$$

where Az_{ij} is the observed azimuth, $v_{\text{Az}_{ij}}$ the residual in the observed azimuth, x_i and y_i the most probable values for the coordinates of station I, x_j and y_j the most probable values for the coordinates of station J, and C a constant with a value based on Table 15.1. Equation (15.2) is a nonlinear function involving variables x_i, y_i, x_j, and y_j, that can be rewritten as

$$F(x_i, y_i, x_j, y_j) = \text{Az}_{ij} + v_{\text{Az}_{ij}} \tag{15.3}$$

where

TABLE 15.1 Relationship between the Quadrant, C, and the Azimuth of the Line

Quadrant	Sign($x_j - x_i$)	Sign($y_j - y_i$)	Sign α	C	Azimuth
I	+	+	+	0	α
II	+	−	−	180°	$\alpha + 180°$
III	−	−	+	180°	$\alpha + 180°$
IV	−	+	−	360°	$\alpha + 360°$

15.2 AZIMUTH OBSERVATION EQUATION

$$F(x_i, y_i, x_j, y_j) = \tan^{-1} \frac{x_j - x_i}{y_j - y_i} + C$$

As discussed in Section 11.10, nonlinear equations such as (15.3) can be linearized and solved using a first-order Taylor series approximation. The linearized form of Equation (15.3) is

$$F(x_i, y_i, x_j, y_j) = F(x_i, y_i, x_j, y_j)_0 + \left(\frac{\partial F}{\partial x_i}\right)_0 dx_i + \left(\frac{\partial F}{\partial y_i}\right)_0 dy_i + \left(\frac{\partial F}{\partial x_j}\right)_0 dx_j$$

$$+ \left(\frac{\partial F}{\partial y_j}\right)_0 dy_j \qquad (15.4)$$

where $(\partial F/\partial x_i)_0$, $(\partial F/\partial y_i)_0$, $(\partial F/\partial x_j)_0$, and $(\partial F/\partial y_j)_0$ are the partial derivatives of F with respect to x_i, y_i, x_j, and y_j that are evaluated at the initial approximations x_{i_0}, y_{i_0}, x_{j_0}, and y_{j_0}, and dx_i, dy_i, dx_j, and dy_j are the corrections applied to the initial approximations after each iteration such that

$$x_i = x_{i_0} + dx_i \quad y_i = y_{i_0} + dy_i \quad x_j = x_{j_0} + dx_{ji} \quad y_j = y_{j_0} + dy_i \quad (15.5)$$

To determine the partial derivatives of Equation (15.4) requires the prototype equation for the derivative of $\tan^{-1} u$ with respect to x, which is

$$\frac{d}{dx} \tan^{-1} u = \frac{1}{1 + u^2} \frac{du}{dx} \qquad (15.6)$$

Using Equation (15.6), the procedure for determining the $\partial F/\partial x_i$ is demonstrated as follows:

$$\frac{\partial F}{\partial x_i} = \frac{1}{1 + [(x_j - x_i)/(y_j - y_i)]^2} \frac{-1}{y_j - y_i}$$

$$= \frac{-1(y_j - y_i)}{(x_j - x_i)^2 + (y_j - y_i)^2} \qquad (15.7)$$

$$= \frac{y_i - y_j}{IJ^2}$$

By employing the same procedure, the remaining partial derivatives are

$$\frac{\partial F}{\partial y_i} = \frac{x_j - x_i}{IJ^2} \quad \frac{\partial F}{\partial x_j} = \frac{x_j - y_i}{IJ^2} \quad \frac{\partial F}{\partial y_j} = \frac{x_i - x_j}{IJ^2} \qquad (15.8)$$

where $IJ^2 = (x_j - x_i)^2 + (y_j - y_i)^2$.

258 ADJUSTMENT OF HORIZONTAL SURVEYS: TRIANGULATION

If Equations (15.7) and (15.8) are substituted into Equation (15.4) and the results then substituted into Equation (15.3), the following prototype azimuth equation is obtained:

$$\left(\frac{y_i - y_j}{IJ^2}\right)_0 dx_i + \left(\frac{x_j - x_i}{IJ^2}\right)_0 dy_i + \left(\frac{y_j - y_i}{IJ^2}\right)_0 dx_j + \left(\frac{x_i - x_j}{IJ^2}\right)_0 dy_j$$
$$= k_{Az_{ij}} + v_{Az_{ij}} \tag{15.9}$$

Both

$$k_{Az_{ij}} = Az_{ij} - \left[\tan^{-1}\left(\frac{x_j - x_i}{y_j - y_i}\right)_0 + C\right] \quad \text{and} \quad IJ^2 = (x_j - x_i)_0^2 + (y_j - y_i)_0^2$$

are evaluated using the approximate coordinate values of the unknown parameters.

15.3 ANGLE OBSERVATION EQUATION

Figure 15.2 illustrates the geometry for an angle observation. In the figure, B is the backsight station, F the foresight station, and I the instrument station. As shown in the figure, an angle observation equation can be written as the difference between two azimuth observations, and thus for clockwise angles,

$$\angle BIF = Az_{IF} - Az_{IB} = \tan^{-1}\frac{x_f - x_i}{y_f - y_i} - \tan^{-1}\frac{x_b - x_i}{y_b - y_i} + D = \theta_{bif} + v_{\theta_{bif}}$$
$$\tag{15.10}$$

where θ_{bif} is the observed clockwise angle, $v_{\theta_{bif}}$ the residual in the observed angle, x_b and y_b the most probable values for the coordinates of the backsighted station B, x_i and y_i the most probable values for the coordinates of

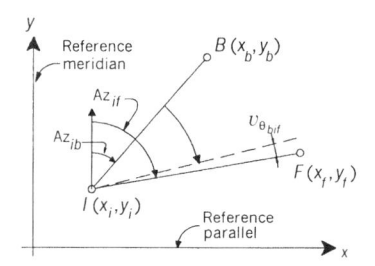

Figure 15.2 Relationship between an angle and two azimuths.

15.3 ANGLE OBSERVATION EQUATION

the instrument station I, x_f and y_f the most probable values for the coordinates of the foresighted station F, and D a constant that depends on the quadrants in which the backsight and foresight occur. This term can be computed as the difference between the C terms from Equation (15.1) as applied to the backsight and foresight azimuths; that is,

$$D = C_{if} - C_{ib}$$

Equation (15.10) is a nonlinear function of x_b, y_b, x_i, y_i, x_f, and y_f that can be rewritten as

$$F(x_b, y_b, x_i, y_i, x_f, y_f) = \theta_{bif} + v_{\theta_{bif}} \qquad (15.11)$$

where

$$F(x_b, y_b, x_i, y_i, x_f, y_f) = \tan^{-1} \frac{x_f - x_i}{y_f - y_i} - \tan^{-1} \frac{x_b - x_i}{y_b - y_i} + D$$

Equation (15.11) expressed as a linearized first-order Taylor series expansion is

$$F(x_b, y_b, x_i, y_i, x_f, y_f) = F(x_b, y_b, x_i, y_i, x_f, y_f)_0 + \left(\frac{\partial F}{\partial x_b}\right)_0 dx_b + \left(\frac{\partial F}{\partial y_b}\right)_0 dy_b$$
$$+ \left(\frac{\partial F}{\partial x_i}\right)_0 dx_i + \left(\frac{\partial F}{\partial y_i}\right)_0 dy_i + \left(\frac{\partial F}{\partial x_f}\right)_0 dx_f + \left(\frac{\partial F}{\partial y_f}\right)_0 dy_f$$
$$(15.12)$$

where $\partial F/\partial x_b$, $\partial F/\partial y_b$, $\partial F/\partial x_i$, $\partial F/\partial y_i$, $\partial F/\partial x_f$, and $\partial F/\partial y_f$ are the partial derivatives of F with respect to x_b, y_b, x_i, y_i, x_f, and y_f, respectively.

Evaluating partial derivatives of the function F and substituting into Equation (15.12), then substituting into Equation (15.11), results in the following equation:

$$\left(\frac{y_i - y_b}{IB^2}\right)_0 dx_b + \left(\frac{x_b - x_i}{IB^2}\right)_0 dy_b + \left(\frac{y_b - y_i}{IB^2} - \frac{y_f - y_i}{IF^2}\right)_0 dx_i$$
$$+ \left(\frac{x_i - x_b}{IB^2} - \frac{x_i - x_f}{IF^2}\right)_0 dy_i + \left(\frac{y_f - y_i}{IF^2}\right)_0 dx_f + \left(\frac{x_i - x_f}{IF^2}\right)_0 dy_f \quad (15.13)$$
$$= k_{\theta_{bif}} + v_{\theta_{bif}}$$

where

$$k_{\theta_{bif}} = \theta_{bif} - \theta_{bifo} \qquad \theta_{bifo} = \tan^{-1}\left(\frac{x_f - x_i}{y_f - y_i}\right)_0 - \tan^{-1}\left(\frac{x_b - x_i}{y_b - y_i}\right)_0 + D$$

$$IB^2 = (x_b - x_i)^2 + (y_b - y_i)^2 \qquad IF^2 = (x_f - x_i)^2 + (y_f - y_i)^2$$

are evaluated at the approximate values for the unknowns.

In formulating the angle observation equation, remember that I is always assigned to the instrument station, B to the backsight station, and F to the foresight station. This station designation must be followed strictly in employing prototype equation (15.13), as demonstrated in the numerical examples that follow.

15.4 ADJUSTMENT OF INTERSECTIONS

When an unknown station is visible from two or more existing control stations, the angle intersection method is one of the simplest and sometimes most practical ways for determining the horizontal position of a station. For a unique computation, the method requires observation of at least two horizontal angles from two control points. For example, angles θ_1, and θ_2 observed from control stations R and S in Figure 15.3, will enable a unique computation for the position of station U. If additional control is available, computations for the unknown station's position can be strengthened by observing redundant angles such as angles θ_3 and θ_4 in Figure 15.3 and applying the method of least squares. In that case, for each of the four independent angles, a linearized observation equation can be written in terms of the two unknown coordinates, x_u and y_u.

Example 15.1 Using the method of least squares, compute the most probable coordinates of station U in Figure 15.3 by the least squares intersection procedure. The following unweighted horizontal angles were observed from control stations R, S, and T:

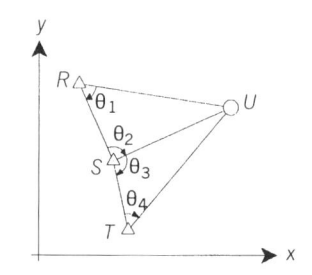

Figure 15.3 Intersection example.

15.4 ADJUSTMENT OF INTERSECTIONS

$\theta_1 = 50°06'50''$ $\quad$ $\theta_2 = 101°30'47''$ $\quad$ $\theta_3 = 98°41'17''$ $\quad$ $\theta_4 = 59°17'01''$

The coordinates for the control stations R, S, and T are

$$x_r = 865.40 \quad x_s = 2432.55 \quad x_t = 2865.22$$

$$y_r = 4527.15 \quad y_s = 2047.25 \quad y_t = 27.15$$

SOLUTION

Step 1: Determine initial approximations for the coordinates of station U.

(a) Using the coordinates of stations R and S, the distance RS is computed as

$$RS = \sqrt{(2432.55 - 865.40)^2 + (4527.15 - 2047.25)^2} = 2933.58 \text{ ft}$$

(b) From the coordinates of stations R and S, the azimuth of the line between R and S can be determined using Equation (15.2). Then the initial azimuth of line RU can be computed by subtracting θ_1 from the azimuth of line RS:

$$Az_{RS} = \tan^{-1} \frac{x_s - x_r}{y_s - y_r} + C = \tan^{-1} \frac{865.40 - 2432.55}{4527.15 - 2047.25} + 180°$$

$$= 147° + 42'34''$$

$$Az_{RU_0} = 147°42'34'' - 50°06'50'' = 97°35'44''$$

(c) Using the sine law with triangle RUS, an initial length for RU_0 can be calculated as

$$RU_0 = \frac{RS \sin \theta_2}{\sin(180° - \theta_1 - \theta_2)} = \frac{2933.58 \sin(100°30'47'')}{\sin(28°27'23'')} = 6049.00 \text{ ft}$$

(d) Using the azimuth and distance for RU_0 computed in steps 1(b) and 1(c), initial coordinates for station U are computed as

$$x_{u_0} = x_r + RU_0 \sin Az_{RU_0} = 865.40 + 6049.00 \sin(97°35'44'')$$

$$= 6861.35$$

$$y_{u_0} = y_r + RU_0 \cos Az_{RU_0} = 865.40 + 6049.00 \cos(97°35'44'')$$

$$= 3727.59$$

262 ADJUSTMENT OF HORIZONTAL SURVEYS: TRIANGULATION

(e) Using the appropriate coordinates, the initial distances for SU and TU are calculated as

$$SU_0 = \sqrt{(6861.35 - 2432.55)^2 + (3727.59 - 2047.25)^2}$$
$$= 4736.83 \text{ ft}$$
$$TU_0 = \sqrt{(6861.35 - 2865.22)^2 + (3727.59 - 27.15)^2}$$
$$= 5446.29 \text{ ft}$$

Step 2: Formulate the linearized equations. As in the trilateration adjustment, control station coordinates are held fixed during the adjustment by assigning zeros to their dx and dy values. Thus, these terms drop out of prototype equation (15.13). In forming the observation equations, b, i, and f are assigned to the backsight, instrument, and foresight stations, respectively, for each angle. For example, with angle θ_1, B, I, and F are replaced by U, R, and S, respectively. By combining the station substitutions shown in Table 15.2 with prototype equation (15.13), the following observation equations are written for the four observed angles.

$$\left(\frac{y_r - y_u}{RU^2}\right)_0 dx_u + \left(\frac{x_u - x_r}{RU^2}\right)_0 dy_u$$

$$= \theta_1 - \left[\tan^{-1}\frac{x_s - x_r}{y_s - y_r} - \tan^{-1}\left(\frac{x_u - x_r}{y_u - y_r}\right)_0 + 0°\right] + v_1$$

$$\left(\frac{y_u - y_s}{SU^2}\right)_0 dx_u + \left(\frac{x_s - x_u}{SU^2}\right)_0 dy_u$$

$$= \theta_2 - \left[\tan^{-1}\left(\frac{x_u - x_s}{y_u - y_s}\right)_0 - \tan^{-1}\frac{x_r - x_s}{y_r - y_s} + 0°\right] + v_2$$

$$\left(\frac{y_s - y_u}{SU^2}\right)_0 dx_u + \left(\frac{x_u - x_s}{SU^2}\right)_0 dy_u$$

$$= \theta_3 - \left[\tan^{-1}\frac{x_t - x_s}{y_t - y_s} - \tan^{-1}\left(\frac{x_u - x_s}{y_u - y_s}\right)_0 + 180°\right] + v_3$$

$$\left(\frac{y_u - y_t}{TU^2}\right)_0 dx_u + \left(\frac{x_t - x_u}{TU^2}\right)_0 dy_u$$

$$= \theta_4 - \left[\tan^{-1}\left(\frac{x_u - x_t}{y_u - y_t}\right)_0 - \tan^{-1}\frac{x_s - x_t}{y_s - y_t} + 0°\right] + v_4$$

(15.14)

15.4 ADJUSTMENT OF INTERSECTIONS

TABLE 15.2 Substitutions

Angle	B	I	F
θ_1	U	R	S
θ_2	R	S	U
θ_3	U	S	T
θ_4	S	T	U

Substituting the appropriate values into Equations (15.14) and multiplying the left side of the equations by ρ to achieve unit consistency,[1] the following J and K matrices are formed:

$$J = \rho \begin{bmatrix} \dfrac{4527.15 - 3727.59}{6049.00^2} & \dfrac{6861.35 - 865.40}{6049.00^2} \\ \dfrac{3727.59 - 2047.25}{4736.83^2} & \dfrac{2432.55 - 6861.35}{4736.83^2} \\ \dfrac{2047.25 - 3727.59}{4736.83^2} & \dfrac{6861.35 - 2432.55}{4736.83^2} \\ \dfrac{3727.59 - 27.15}{5446.29^2} & \dfrac{2865.22 - 6861.35}{5446.29^2} \end{bmatrix} = \begin{bmatrix} 4.507 & 33.800 \\ 15.447 & -40.713 \\ -15.447 & 40.713 \\ 25.732 & -27.788 \end{bmatrix}$$

$$K = \begin{bmatrix} 50°06'50'' - \left(\tan^{-1} \dfrac{2432.55 - 865.40}{2047.25 - 4527.15} - \tan^{-1} \dfrac{6861.35 - 865.40}{3727.59 - 4527.15} + 0° \right) \\ 101°30'47'' - \left(\tan^{-1} \dfrac{6861.35 - 2432.55}{3727.59 - 2047.25} - \tan^{-1} \dfrac{865.40 - 2432.55}{4527.15 - 2047.25} + 0° \right) \\ 98°41'17'' - \left(\tan^{-1} \dfrac{2865.22 - 2432.55}{27.15 - 2047.25} - \tan^{-1} \dfrac{6861.35 - 2432.55}{3727.59 - 2047.25} + 180° \right) \\ 59°17'01'' - \left(\tan^{-1} \dfrac{6861.35 - 2865.22}{3727.59 - 27.15} - \tan^{-1} \dfrac{2432.55 - 2865.22}{2047.25 - 27.15} + 0° \right) \end{bmatrix}$$

$$= \begin{bmatrix} 0.00'' \\ 0.00'' \\ -0.69'' \\ -20.23'' \end{bmatrix}$$

---

[1] For these observations to be dimensionally consistent, the elements of the K and V matrices must be in radian measure, or in other words, the coefficients of the K and J elements must be in the same units. Since it is most common to work in the sexagesimal system, and since the magnitudes of the angle residuals are generally in the range of seconds, the units of the equations are converted to seconds by (1) multiplying the coefficients in the equation by ρ, which is the number of seconds per radian, or 206,264.8"/rad, and (2) computing the K elements in units of seconds.

264 ADJUSTMENT OF HORIZONTAL SURVEYS: TRIANGULATION

Notice that the initial coordinates for x_{u_0} and y_{u_0} were calculated using θ_1 and θ_2, and thus their *K*-matrix values are zero for the first iteration. These values will change in subsequent iterations.

Step 3: Matrix solution. The least squares solution is found by applying Equation (11.37).

$$J^TJ = \begin{bmatrix} 1159.7 & -1820.5 \\ -1820.5 & 5229.7 \end{bmatrix}$$

$$Q_{xx} = (J^TJ)^{-1} = \begin{bmatrix} 0.001901 & 0.000662 \\ 0.000662 & 0.000422 \end{bmatrix}$$

$$J^TK = \begin{bmatrix} -509.9 \\ 534.1 \end{bmatrix}$$

$$X = (J^TJ)^{-1}(J^TK) = \begin{bmatrix} 0.001901 & 0.000662 \\ 0.000662 & 0.000422 \end{bmatrix} \begin{bmatrix} -509.9 \\ 534.1 \end{bmatrix} = \begin{bmatrix} dx_u \\ dy_u \end{bmatrix}$$

$dx_u = -0.62$ ft   and $dy_u = -0.11$ ft

Step 4: Add the corrections to the initial coordinates for station *U*:

$$x_u = x_{u_0} + dx_u = 6861.35 - 0.62 = 6860.73$$
$$y_u = y_{u_0} + dy_u = 3727.59 - 0.11 = 3727.48$$
(15.15)

Step 5: Repeat steps 2 through 4 until negligible corrections occur. The next iteration produced negligible corrections, and thus Equations (15.15) produced the final adjusted coordinates for station *U*.

Step 6: Compute post-adjustment statistics. The residuals for the angles are

$$V = JX - K = \begin{bmatrix} 4.507 & 33.80 \\ 15.447 & -40.713 \\ -15.447 & 40.713 \\ 25.732 & -27.788 \end{bmatrix} \begin{bmatrix} -0.62 \\ -0.11 \end{bmatrix} - \begin{bmatrix} 0.00'' \\ 0.00'' \\ -0.69'' \\ -20.23'' \end{bmatrix}$$

$$= \begin{bmatrix} -6.5'' \\ -5.1'' \\ 5.8'' \\ 7.3'' \end{bmatrix}$$

The reference variance (standard deviation of unit weight) for the adjustment is computed using Equation (12.14) as

$$V^\mathrm{T}V = \begin{bmatrix} -6.5 & -5.1 & 5.8 & 7.3 \end{bmatrix} \begin{bmatrix} -6.5 \\ -5.1 \\ 5.8 \\ 7.3 \end{bmatrix} = [155.2]$$

$$S_0 = \sqrt{\frac{V^\mathrm{T}V}{m-n}} = \sqrt{\frac{155.2}{4-2}} = \pm 8.8''$$

The estimated errors for the adjusted coordinates of station U, given by Equation (13.24), are

$$S_{x_u} = S_0\sqrt{Q_{x_u x_u}} = \pm 8.8\sqrt{0.001901} = \pm 0.38 \text{ ft}$$

$$S_{y_u} = S_0\sqrt{Q_{y_u y_u}} = \pm 8.8\sqrt{0.000422} = \pm 0.18 \text{ ft}$$

The estimated error in the position of station U is given by

$$S_u = \sqrt{S_x^2 + S_y^2} = \sqrt{0.38^2 + 0.18^2} = \pm 0.42 \text{ ft}$$

15.5 ADJUSTMENT OF RESECTIONS

Resection is a method used for determining the unknown horizontal position of an occupied station by observing a minimum of two horizontal angles to a minimum of three stations whose horizontal coordinates are known. If more than three stations are available, redundant observations are obtained and the position of the unknown occupied station can be computed using the least squares method. Like intersection, resection is suitable for locating an occasional station and is especially well adapted over inaccessible terrain. This method is commonly used for orienting total station instruments in locations favorable for staking projects by radiation using coordinates.

Consider the resection position computation for the occupied station U of Figure 15.4 having observed the three horizontal angles shown between stations P, Q, R, and S whose positions are known. To determine the position of station U, two angles could be observed. The third angle provides a check and allows a least squares solution for computing the coordinates of station U.

Using prototype equation (15.13), a linearized observation equation can be written for each angle. In this problem, the vertex station is occupied and is the only unknown station. Thus, all coefficients in the Jacobian matrix follow the form used for the coefficients of dx_i and dy_i in prototype equation (15.13).

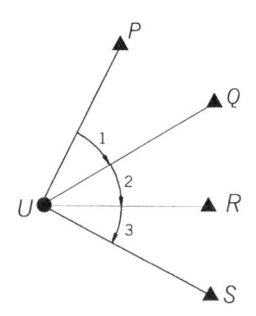

266 ADJUSTMENT OF HORIZONTAL SURVEYS: TRIANGULATION

Figure 15.4 Resection example.

The method of least squares yields corrections, dx_u and dy_u, which gives the most probable coordinate values for station U.

15.5.1 Computing Initial Approximations in the Resection Problem

In Figure 15.4 only two angles are necessary to determine the coordinates of station U. Using stations P, Q, R, and U, a procedure to find the station U's approximate coordinate values is

Step 1: Let

$$\angle QPU + \angle URQ = G = 360° - (\angle 1 + \angle 2 + \angle RQP) \quad (15.16)$$

Step 2: Using the sine law with triangle PQU yields

$$\frac{QU}{\sin \angle QPU} = \frac{PQ}{\sin \angle 1} \quad (a)$$

and with triangle URQ,

$$\frac{QU}{\sin \angle URQ} = \frac{QR}{\sin \angle 2} \quad (b)$$

Step 3: Solving Equations (*a*) and (*b*) for QU and setting the resulting equations equal to each other gives

$$\frac{PQ \sin \angle PQU}{\sin \angle 1} = \frac{QR \sin \angle URQ}{\sin \angle 2} \quad (c)$$

Step 4: From Equation (*c*), let H be defined as

$$H = \frac{\sin \angle QPU}{\sin \angle URQ} = \frac{QR \sin \angle 1}{PQ \sin \angle 2} \quad (15.17)$$

15.5 ADJUSTMENT OF RESECTIONS

Step 5: From Equation (15.16),

$$\angle QPU = G - \angle URQ \tag{d}$$

Step 6: Solving Equation (15.17) for the $\sin\angle QPU$, and substituting Equation (d) into the result gives

$$\sin(G - \angle URQ) = H \sin\angle URQ \tag{e}$$

Step 7: From trigonometry

$$\sin(\alpha - \beta) = \sin\alpha \cos\beta - \cos\alpha \sin\beta$$

Applying this relationship to Equation (e) yields

$$\sin G - \angle URQ = \sin G \cos\angle URQ - \cos G \sin\angle URQ \tag{f}$$

$$\sin G - \angle URQ = H \sin \angle URQ \tag{g}$$

Step 8: Dividing Equation (g) by $\cos \angle URQ$ and rearranging yields

$$\sin G = \tan\angle URQ [H + \cos(G)] \tag{h}$$

Step 9: Solving Equation (h) for $\angle URQ$ gives

$$\angle URQ = \tan^{-1} \frac{\sin G}{H + \cos G} \tag{15.18}$$

Step 10: From Figure 15.4,

$$\angle RQU = 180° - (\angle 2 + \angle URQ) \tag{15.19}$$

Step 11: Again applying the sine law yields

$$RU = \frac{QR \sin \angle RQU}{\sin \angle 2} \tag{15.20}$$

Step 12: Finally, the initial coordinates for station U are

$$x_u = x_r + RU \sin(Az_{RQ} - \angle URQ) \tag{15.21}$$
$$y_u = y_r + RU \cos(Az_{RQ} - \angle URQ)$$

Example 15.2 The following data are obtained for Figure 15.4. Control stations P, Q, R, and S have the following (x,y) coordinates: P (1303.599, 1458.615), Q (1636.436, 1310.468), R (1503.395, 888.362), and S (1506.262, 785.061). The observed values for angles 1, 2, and 3 with standard deviations are as follows:

Backsight	Occupied	Foresight	Angle	S (″)
P	U	Q	30°29′33″	5
Q	U	R	38°30′31″	6
R	U	S	10°29′57″	6

What are the most probable coordinates of station U?

SOLUTION Using the procedures described in Section 15.5.1, the initial approximations for the coordinates of station U are:

(a) From Equation (15.10),

$$\angle RQP = Az_{PQ} - Az_{QR} = 293°59'38.4'' - 197°29'38.4''$$
$$= 96°30'00.0''$$

(b) Substituting the appropriate angular values into Equation (15.16) gives

$$G = 360° - (30°29'33'' + 38°30'31'' + 96°30'00.0'') = 194°29'56''$$

(c) Substituting the appropriate station coordinates into Equation (14.1) yields

$$PQ = 364.318 \quad \text{and} \quad QR = 442.576$$

(d) Substituting the appropriate values into Equation (15.17) yields H as

$$H = \frac{442.576 \sin(30°29'33'')}{364.318 \sin(38°30'31'')} = 0.990027302$$

(e) Substituting previously determined G and H into Equation (15.18), $\angle URQ$ is computed as

$$\angle URQ = \tan^{-1} \frac{\sin(194°29'56'')}{0.990027302 + \cos(194°29'56'')} + 180°$$
$$= -85°00'22'' + 180° = 94°59'36.3''$$

15.5 ADJUSTMENT OF RESECTIONS

(f) Substituting the value of $\angle URQ$ into Equation (15.19), $\angle RQU$ is determined to be

$$\angle RQU = 180° - (38°30'31'' + 94°59'36.3'') = 46°29'52.7''$$

(g) From Equation (15.20), RU is

$$RU = \frac{442.576 \sin(46°29'52.7'')}{\sin(38°30'31'')} = 515.589$$

(h) Using Equation (15.1), the azimuth of RQ is

$$Az_{RQ} = \tan^{-1} \frac{1636.436 - 1503.395}{1310.468 - 888.362} + 0° = 17°29'38.4''$$

(i) From Figure 15.4, Az_{RU} is computed as

$$Az_{RQ} = 197°29'38.4'' - 180° = 17°29'38.4''$$

$$Az_{RU} = Az_{RQ} - \angle URQ = 360° + 17°29'38.4'' - 94°59'36.3''$$
$$= 282°30'02.2''$$

(j) Using Equation (15.21), the coordinates for station U are

$$x_u = 1503.395 + 515.589 \sin Az_{RU} = 1000.03$$

$$y_u = 888.362 + 515.589 \cos Az_{RU} = 999.96$$

For this problem, using prototype equation (15.13), the J and K matrices are

$$J = \rho \begin{bmatrix} \left(\dfrac{y_p - y_u}{UP^2} - \dfrac{y_q - y_u}{UQ^2}\right)_0 & \left(\dfrac{x_u - x_p}{UP^2} - \dfrac{x_u - x_q}{UQ^2}\right)_0 \\ \left(\dfrac{y_q - y_u}{UQ} - \dfrac{y_r - y_u}{UR^2}\right)_0 & \left(\dfrac{x_u - x_q}{UQ^2} - \dfrac{x_u - x_r}{UR^2}\right)_0 \\ \left(\dfrac{y_r - y_u}{UR^2} - \dfrac{y_s - y_u}{US^2}\right)_0 & \left(\dfrac{x_u - x_r}{UR^2} - \dfrac{x_u - x_s}{US1^2}\right)_0 \end{bmatrix}$$

$$K = \begin{bmatrix} (\angle 1 - \angle 1_0)'' \\ (\angle 2 - \angle 2_0)'' \\ (\angle 3 - \angle 3_0)'' \end{bmatrix}$$

270 ADJUSTMENT OF HORIZONTAL SURVEYS: TRIANGULATION

Also, the weight matrix W is a diagonal matrix composed of the inverses of the variances of the angles observed, or

$$W = \begin{bmatrix} \frac{1}{5^2} & 0 & 0 \\ 0 & \frac{1}{6^2} & 0 \\ 0 & 0 & \frac{1}{6^2} \end{bmatrix}$$

Using the data given for the problem together with the initial approximations computed, numerical values for the matrices were calculated and the adjustment performed using the program ADJUST. The following results were obtained after two iterations. The reader is encouraged to adjust these example problems using both the MATRIX and ADJUST programs supplied.

ITERATION 1

```
J MATRIX                          K MATRIX     X MATRIX
======================         ========     ========
184.993596    54.807717        -0.203359    -0.031107
214.320813   128.785353        -0.159052     0.065296
 59.963802   -45.336838        -6.792817
```

ITERATION 2

```
J MATRIX                          K MATRIX     X MATRIX
======================         ========     ========
185.018081    54.771738         1.974063    0.000008
214.329904   128.728773        -1.899346    0.000004
 59.943758   -45.340316        -1.967421
```

```
INVERSE MATRIX
=======================
 0.00116318  -0.00200050
-0.00200050   0.00500943
```

Adjusted stations

```
Station     X           Y           Sx        Sy
=============================================
   U     999.999    1,000.025     0.0206    0.0427
```

Adjusted Angle Observations

Station Backsighted	Station Occupied	Station Foresighted	Angle	V	S (")
P	U	Q	30° 29' 31"	-2.0"	2.3
Q	U	R	38° 30' 33"	1.9"	3.1
R	U	S	10° 29' 59"	2.0"	3.0

Redundancies = 1

Reference Variance = 0.3636

Reference So = ±0.60

15.6 ADJUSTMENT OF TRIANGULATED QUADRILATERALS

The quadrilateral is the basic figure for triangulation. Procedures like those used for adjusting intersections and resections are also used to adjust this figure. In fact, the parametric adjustment using the observation equation method can be applied to any triangulated geometric figure, regardless of its shape.

The procedure for adjusting a quadrilateral consists of first using a minimum number of the observed angles to solve the triangles, and computing initial values for the unknown coordinates. Corrections to these initial coordinates are then calculated by applying the method of least squares. The procedure is iterated until the solution converges. This yields the most probable coordinate values. A statistical analysis of the results is then made. The following example illustrates the procedure.

Example 15.3 The following observations are supplied for Figure 15.5. Adjust this figure by the method of unweighted least squares. The observed angles are as follows:

1 = 42°35'29.0°   3 = 79°54'42.1"   5 = 21°29'23.9"   7 = 31°20'45.8"
2 = 87°35'10.6"   4 = 18°28'22.4"   6 = 39°01'35.4"   8 = 39°34'27.9"

The fixed coordinates are

x_A = 9270.33 y_A = 8448.90 x_D = 15,610.58 y_D = 8568.75

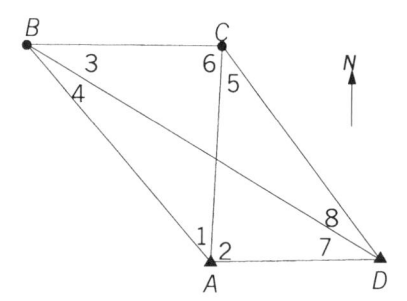

272 ADJUSTMENT OF HORIZONTAL SURVEYS: TRIANGULATION

Figure 15.5 Quadrilateral.

SOLUTION The coordinates of stations B and C are to be computed in this adjustment. The Jacobian matrix has the form shown in Table 15.3. The subscripts b, i, and f of the dx's and dy's in the table indicate whether stations B and C are the backsight, instrument, or foresight station in Equation (15.13), respectively. In developing the coefficient matrix, of course, the appropriate station coordinate substitutions must be made to obtain each coefficient.

A computer program has been used to form the matrices and solve the problem. In the program, the angles were entered in the order of 1 through 8. The X matrix has the form

$$X = \begin{bmatrix} dx_b \\ dy_b \\ dx_c \\ dy_c \end{bmatrix}$$

The following self-explanatory computer listing gives the solution for this example. As shown, one iteration was satisfactory to achieve convergence, since the second iteration produced negligible corrections. Residuals, adjusted

TABLE 15.3 Structure of the Coefficient or J Matrix in Example 15.3

	Unknowns			
Angle	dx_b	dy_b	dx_c	dy_c
1	$dx(b)$	$dy(b)$	$dx(f)$	$dy(f)$
2	0	0	$dx(b)$	$dy(b)$
3	$dx(i)$	$dy(i)$	$dx(b)$	$dy(b)$
4	$dx(i)$	$dy(i)$	0	0
5	0	0	$dx(i)$	$dy(i)$
6	$dx(f)$	$dy(f)$	$dx(i)$	$dy(i)$
7	$dx(f)$	$dy(f)$	0	0
8	$dy(b)$	$dy(b)$	$dx(f)$	$dy(f)$

coordinates, their estimated errors, and adjusted angles are tabulated at the end of the listing.

```
***********************************************
Initial approximations for unknown stations
***********************************************
Station         X              Y
===============================
    B      2,403.600      16,275.400
    C      9,649.800      24,803.500

Control Stations
~~~~~~~~~~~~~
Station         X              Y
===============================
    A      9,270.330       8,448.900
    D     15,610.580       8,568.750


*******************
Angle Observations
*******************
   Station    Station     Station
 Backsighted  Occupied  Foresighted            Angle
==================================================
      B          A           C       42° 35' 29.0"
      C          A           D       87° 35' 10.6"
      C          B           D       79° 54' 42.1"
      D          B           A       18° 28' 22.4"
      D          C           A       21° 29' 23.9"
      A          C           B       39° 01' 35.4"
      A          D           B       31° 20' 45.8"
      B          D           C       39° 34' 27.9"
```

Iteration 1

```
J Matrix                                                K MATRIX     X MATRIX
-------------------------------------------------       ---------    -----------
-14.891521   -13.065362    12.605250    -0.292475      -1.811949    1  -0.011149
  0.000000     0.000000   -12.605250     0.292475      -5.801621    2   0.049461
 20.844399    -0.283839   -14.045867    11.934565       3.508571    3   0.061882
  8.092990     1.414636     0.000000     0.000000       1.396963    4   0.036935
  0.000000     0.000000     1.409396    -4.403165      -1.833544
-14.045867    11.934565     1.440617   -11.642090       5.806415
  6.798531    11.650726     0.000000     0.000000      -5.983393
 -6.798531   -11.650726    11.195854     4.110690       1.818557
```

274 ADJUSTMENT OF HORIZONTAL SURVEYS: TRIANGULATION

Iteration 2

```
J Matrix                                                        K MATRIX    X MATRIX
-------------------------------------------------------------   ---------   -----------
-14.891488   -13.065272    12.605219   -0.292521   -2.100998   1  0.000000
  0.000000     0.000000   -12.605219    0.292521   -5.032381   2 -0.000000
 20.844296    -0.283922   -14.045752   11.934605    4.183396   3  0.000000
  8.092944     1.414588     0.000000    0.000000    1.417225   4 -0.000001
  0.000000     0.000000     1.409357   -4.403162   -1.758129
-14.045752    11.934605     1.440533  -11.642083    5.400377
  6.798544    11.650683     0.000000    0.000000   -6.483846
 -6.798544   -11.650683    11.195862    4.110641    1.474357
```

INVERSE MATRIX
```
-------------------------------
 0.00700   -0.00497    0.00160   -0.01082
-0.00497    0.00762    0.00148    0.01138
 0.00160    0.00148    0.00378    0.00073
-0.01082    0.01138    0.00073    0.02365
```

****
Adjusted stations
****

Station	X	Y	Sx	Sy
B	2,403.589	16,275.449	0.4690	0.4895
C	9,649.862	24,803.537	0.3447	0.8622

****
Adjusted Angle Observations
****

Station Backsighted	Station Occupied	Station Foresighted	Angle	V	S
B	A	C	42° 35' 31.1"	2.10"	3.65
C	A	D	87° 35' 15.6"	5.03"	4.33
C	B	D	79° 54' 37.9"	-4.18"	4.29
D	B	A	18° 28' 21.0"	-1.42"	3.36
D	C	A	21° 29' 25.7"	1.76"	3.79
A	C	B	39° 01' 30.0"	-5.40"	4.37
A	D	B	31° 20' 52.3"	6.48"	4.24
B	D	C	39° 34' 26.4"	-1.47"	3.54

****
Adjustment Statistics
****

```
              Iterations  = 2
           Redundancies   = 4
       Reference Variance = 31.42936404
           Reference So   = ±5.6062
              Convergence!
```

PROBLEMS

15.1 Given the following observations and control station coordinates to accompany Figure P15.1, what are the most probable coordinates for station E using an unweighted least squares adjustment?

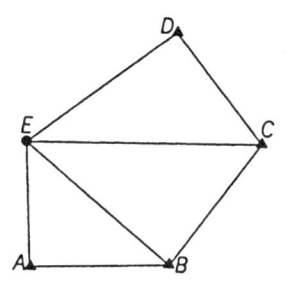

Figure P15.1

Control stations

Station	X (ft)	Y (ft)
A	10,000.00	10,000.00
B	11,498.58	10,065.32
C	12,432.17	11,346.19
D	11,490.57	12,468.51

Angle observations

Backsight, b	Occupied, i	Foresight, f	Angle	S (″)
E	A	B	90°59′57″	5.3
A	B	E	40°26′02″	4.7
E	B	C	88°08′55″	4.9
B	C	E	52°45′02″	4.7
E	C	D	51°09′55″	4.8
C	D	E	93°13′14″	5.0

15.2 Repeat Problem 15.1 using a weighted least squares adjustment with weights of $1/S^2$ for each angle. What are:

276 ADJUSTMENT OF HORIZONTAL SURVEYS: TRIANGULATION

(a) the most probable coordinates for station E?
(b) the reference standard deviation of unit weight?
(c) the standard deviations in the adjusted coordinates for station E?
(d) the adjusted angles, their residuals, and the standard deviations?

15.3 Given the following observed angles and control coordinates for the resection problem of Figure 15.4:

$$1 = 49°47'03" \qquad 2 = 33°21'55" \qquad 3 = 47°58'53"$$

Assuming equally weighted angles, what are the most probable coordinates for station U?

Control stations

Station	X (m)	Y (m)
P	2423.077	3890.344
Q	3627.660	3602.291
R	3941.898	2728.314
S	3099.018	1858.429

15.4 If the estimated standard deviations for the angles in Problem 15.3 are $S_1 = \pm 3.1"$, $S_2 = \pm 3.0"$, and $S_3 = \pm 3.1"$, what are:
(a) the most probable coordinates for station U?
(b) the reference standard deviation of unit weight?
(c) the standard deviations in the adjusted coordinates of station U?
(d) the adjusted angles, their residuals, and the standard deviations?

15.5 Given the following control coordinates and observed angles for an intersection problem:

Control stations

Station	X (m)	Y (m)
A	100,643.154	38,213.066
B	101,093.916	67,422.484
C	137,515.536	67,061.874
D	139,408.739	37,491.846

Angle observations

Backsight	Occupied	Foresight	Angle	S (")
D	A	E	319°39'50"	5.0
A	B	E	305°21'17"	5.0
B	C	E	322°50'35"	5.0
C	D	E	313°10'22"	5.0

What are:
(a) the most probable coordinates for station E?
(b) the reference standard deviation of unit weight?
(c) the standard deviations in the adjusted coordinates of station E?
(d) the adjusted angles, their residuals, and the standard deviations?

15.6 The following control station coordinates, observed angles, and standard deviations apply to the quadrilateral in Figure 15.5.

Control stations

Station	X (ft)	Y (ft)
A	2546.64	1940.26
D	4707.04	1952.54

Initial approximations

Station	X (ft)	Y (ft)
B	2243.86	3969.72
C	4351.06	4010.64

Angle observations

Backsight	Occupied	Foresight	Angle	S (″)
B	A	C	49°33′30″	4.2
C	A	D	48°35′54″	4.2
C	B	D	40°25′44″	4.2
D	B	A	42°11′56″	4.2
D	C	A	50°53′07″	4.2
A	C	B	47°48′47″	4.2
A	D	B	39°38′34″	4.2
B	D	C	40°52′20″	4.2

Do a weighted adjustment using the standard deviations to calculate weights. What are:
(a) the most probable coordinates for stations B and C?
(b) the reference standard deviation of unit weight?
(c) the standard deviations in the adjusted coordinates for stations B and C?
(d) the adjusted angles, their residuals, and the standard deviations?

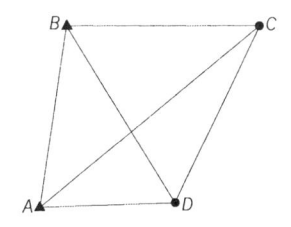

Figure P15.7

15.7 For Figure P15.7 and the following observations, perform a weighted least squares adjustment.
(a) Station coordinate values and standard deviations.
(b) Angles, their residuals, and the standard deviations.

Control stations

Station	X (m)	Y (m)
A	114,241.071	91,294.643
B	116,607.143	108,392.857

Initial approximations

Station	X (m)	Y (m)
C	135,982.143	107,857.143
D	131,567.500	90,669.643

Angle observations

Backsight	Occupied	Foresight	Angle	S (″)
B	A	C	44°49′15.4″	2.0
C	A	D	39°21′58.0″	2.0
C	B	D	48°14′48.9″	2.0
D	B	A	48°02′49.6″	2.0
D	C	A	38°17′38.0″	2.0
A	C	B	38°53′03.9″	2.0
A	D	B	47°45′56.8″	2.0
B	D	C	54°34′26.1″	2.0

15.8 Do Problem 15.7 using an unweighted least squares adjustment. Compare and discuss any differences or similarities between these results and those obtained in Problem 15.7.

15.9 The following observations were obtained for the triangulation chain shown in Figure P15.9.

Control stations

Station	X (m)	Y (m)
A	103,482.143	86,919.643
B	118,303.571	86,919.643
G	104,196.429	112,589.286
H	118,080.357	112,767.857

Initial approximations

Station	X (m)	Y (m)
C	103,616.071	96,116.071
D	117,991.071	95,580.357
E	104,375.000	104,196.429
F	118,169.643	104,598.214

Angle observations

B	I	F	Angle	S (″)	B	I	F	Angle	S (″)
C	A	D	58°19′52″	3	D	A	B	30°49′56″	3
A	B	C	32°03′11″	3	C	B	D	55°52′51″	3
D	C	B	29°55′01″	3	B	C	A	58°46′53″	3
B	D	A	61°14′02″	3	A	D	C	32°58′06″	3
E	C	F	54°24′00″	3	F	C	D	32°22′05″	3

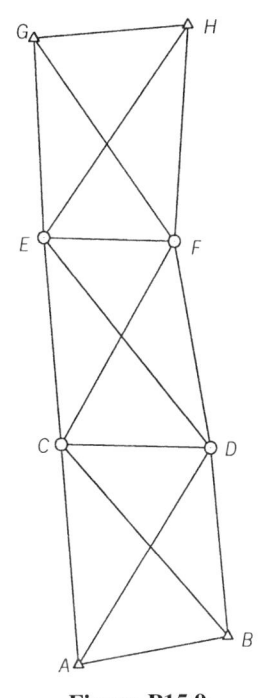

C	D	E	30°11′27°	3	E	D	F	58°48′32″	3
D	E	C	63°02′21″	3	F	E	D	33°59′36″	3
D	F	C	58°37′50″	3	C	F	E	28°34′00″	3
H	E	F	30°21′08″	3	G	E	H	59°11′48″	3
E	F	G	31°25′55°	3	G	F	H	59°36′31″	3
H	G	F	30°30′01″	3	F	G	E	59°01′04″	3
F	H	E	58°37′08″	3	E	H	G	31°17′11″	3

Figure P15.9

Use ADJUST to perform a weighted least squares adjustment. Tabulate the final adjusted:

(a) station coordinates and their standard deviations.

(b) angles, their residuals, and the standard deviations.

15.10 Repeat Problem 15.9 using an unweighted least squares adjustment. Compare and discuss any differences or similarities between these results and those obtained in Problem 15.9. Use the program ADJUST in computing the adjustment.

15.11 Using the control coordinates from Problem 14.3 and the following angle observations, compute the least squares solution and tabulate the final adjusted:

(a) station coordinates and their standard deviations.

(b) angles, their residuals, and the standard deviations.

Angle observations

Backsight	Occupied	Foresight	Angle	S (″)
B	A	C	280°41′06″	5.2
C	A	D	39°21′53″	5.1
C	B	D	51°36′16″	5.2
D	B	A	255°50′03″	5.2
D	C	A	101°27′17″	5.2
A	C	B	311°52′38″	5.2
A	D	B	324°07′04″	5.1
B	D	C	75°03′50″	5.2

15.12 The following observations were obtained for a triangulation chain.

Control stations

Station	X (ft)	Y (ft)
A	92,890.04	28,566.74
D	93,971.87	80,314.29

Initial approximations

Station	X (ft)	Y (ft)
B	93,611.26	47,408.62
C	93,881.71	64,955.36
E	111,191.00	38,032.76
F	110,109.17	57,145.10
G	110,019.02	73,102.09
H	131,475.32	28,837.20
I	130,213.18	46,777.56
J	129,311.66	64,717.91
K	128,590.44	79,142.31

Angle observations

B	I	F	Angle	S (″)	B	I	F	Angle	S (″)
B	A	E	60°27′28″	2.2	E	B	A	64°07′06″	2.1
F	B	E	58°37′14″	2.1	C	B	F	58°34′09″	2.1
F	C	B	65°10′51″	2.3	G	C	F	52°29′14″	2.0
D	C	G	62°52′42″	2.1	G	D	C	66°08′08″	2.2
D	G	K	137°46′57″	2.4	K	G	J	41°30′18″	2.7
J	G	F	66°11′15″	2.1	F	G	C	63°32′13″	2.1
C	G	D	50°59′11″	2.3	B	F	C	56°14′56″	2.6
C	F	G	63°58′29″	2.1	G	F	J	68°48′05″	2.1
J	F	I	48°48′05″	2.3	I	F	E	59°28′49″	2.4
E	F	B	62°41′29″	2.2	A	E	B	55°25′19″	2.3
B	E	F	58°41′13″	2.5	F	E	I	68°33′06″	2.0
I	E	H	49°04′27″	2.1	H	E	A	128°15′52″	2.6
E	H	I	61°35′24″	2.5	H	I	E	69°20′10″	2.0
E	I	F	51°58′06″	2.1	F	I	J	59°50′35″	2.2
I	J	F	71°21′12″	2.2	F	J	G	45°00′39″	2.1
G	J	K	63°38′57″	2.2	J	K	G	74°50′46″	2.3

Use ADJUST to perform a weighted least squares adjustment. Tabulate the final adjusted:

(a) station coordinates and their standard deviations.

(b) angles, their residuals, and the standard deviations.

15.13 Do Problem 15.12 using an unweighted least squares adjustment. Compare and discuss any differences or similarities between these results and those obtained in Problem 15.12. Use the program ADJUST in computing the adjustment.

Use the ADJUST software to do the following problems.

15.14 Problem 15.2

15.15 Problem 15.4

15.16 Problem 15.5

15.17 Problem 15.6

15.18 Problem 15.9

Programming Problems

15.19 Write a computational program that computes the coefficients for prototype equations (15.9) and (15.13) and their k values given the coordinates of the appropriate stations. Use this program to determine the matrix values necessary to do Problem 15.6.

15.20 Prepare a computational program that reads a file of station coordinates, observed angles, and their standard deviations and then:

(a) writes the data to a file in a formatted fashion.

(b) computes the J, K, and W matrices.

(c) writes the matrices to a file that is compatible with the MATRIX program.

(d) test this program with Problem 15.6.

15.21 Write a computational program that reads a file containing the J, K, and W matrices and then:

(a) writes these matrices in a formatted fashion.

(b) performs one iteration of either a weighted or unweighted least squares adjustment of Problem 15.6.

(c) writes the matrices used to compute the solution and the corrections to the station coordinates in a formatted fashion.

15.22 Write a computational program that reads a file of station coordinates, observed angles, and their standard deviations and then:

(a) writes the data to a file in a formatted fashion.
(b) computes the J, K, and W matrices.
(c) performs either a relative or equal weight least squares adjustment of Problem 15.6.
(d) writes the matrices used to compute the solution and tabulates the final adjusted station coordinates and their estimated errors and the adjusted angles, together with their residuals and their estimated errors.

15.23 Prepare a computational program that solves the resection problem. Use this program to compute the initial approximations for Problem 15.3.

CHAPTER 16

ADJUSTMENT OF HORIZONTAL SURVEYS: TRAVERSES AND NETWORKS

16.1 INTRODUCTION TO TRAVERSE ADJUSTMENTS

Of the many methods that exist for traverse adjustment, the characteristic that distinguishes the method of least squares from other methods is that distance, angle, and direction observations are adjusted simultaneously. Furthermore, the adjusted observations not only satisfy all geometrical conditions for the traverse but provide the most probable values for the given data set. Additionally, the observations can be rigorously weighted based on their estimated errors and adjusted accordingly. Given these facts, together with the computational power now provided by computers, it is hard to justify not using least squares for all traverse adjustment work.

In this chapter we describe methods for making traverse adjustments by least squares. As was the case in triangulation adjustments, traverses can be adjusted by least squares using either observation equations or conditional equations. Again, because of the relative ease with which the equations can be written and solved, the parametric observation equation approach is discussed.

16.2 OBSERVATION EQUATIONS

When adjusting a traverse using parametric equations, an observation equation is written for each distance, direction, or angle. The necessary linearized observation equations developed previously are recalled in the following equations.

284   ADJUSTMENT OF HORIZONTAL SURVEYS: TRAVERSES AND NETWORKS

Distance observation equation:

$$\left(\frac{x_i - x_j}{IJ}\right)_0 dx_i + \left(\frac{y_i - y_j}{IJ}\right)_0 dy_i + \left(\frac{x_j - x_i}{IJ}\right)_0 dx_j + \left(\frac{y_j - y_i}{IJ}\right)_0 dy_j$$
$$= k_{l_{ij}} + v_{l_{ij}} \tag{16.1}$$

Angle observation equation:

$$\left(\frac{y_i - y_b}{IB^2}\right)_0 dx_b + \left(\frac{x_b - x_i}{IB^2}\right)_0 dy_b + \left(\frac{y_b - y_i}{IB^2} - \frac{y_f - y_i}{IF^2}\right)_0 dx_i$$
$$+ \left(\frac{x_i - x_b}{IB^2} - \frac{x_i - x_f}{IF^2}\right)_0 dy_i + \left(\frac{y_f - y_i}{IF^2}\right)_0 dx_f + \left(\frac{x_i - x_f}{IF^2}\right)_0 dy_f \tag{16.2}$$
$$= k_{\theta_{bif}} + v_{\theta_{bif}}$$

Azimuth observation equation:

$$\left(\frac{y_i - y_j}{IJ^2}\right)_0 dx_i + \left(\frac{x_j - x_i}{IJ^2}\right)_0 dy_i + \left(\frac{y_j - y_i}{IJ^2}\right)_0 dx_j + \left(\frac{x_i - x_j}{IJ^2}\right)_0 dy_j$$
$$= k_{Az_{ij}} + v_{Az_{ij}} \tag{16.3}$$

The reader should refer to Chapters 14 and 15 to review the specific notation for these equations. As demonstrated with the examples that follow, the azimuth equation may or may not be used in traverse adjustments.

16.3 REDUNDANT EQUATIONS

As noted earlier, one observation equation can be written for each angle, distance, or direction observed in a closed traverse. Thus, if there are n sides in the traverse, there are n distances and $n + 1$ angles, assuming that one angle exists for orientation of the traverse. For example, each closed traverse in Figure 16.1 has four sides, four distances, and five angles. Each traverse also has three points whose positions are unknown, and each point introduces two unknown coordinates into the solution. Thus, there is a maximum of $2(n - 1)$ unknowns for any closed traverse. From the foregoing, no matter the number of sides, there will always be a minimum of $r = (n + n + 1) - 2(n - 1) = 3$ redundant equations for any closed traverse. That is, every closed traverse that is fixed in space both positionally and rotationally has a minimum of three redundant equations.

Figure 16.1 (a) Polygon and (b) link traverses.

16.4 NUMERICAL EXAMPLE

Example 16.1 To illustrate a least squares traverse adjustment, the simple link traverse shown in Figure 16.2 will be used. The observational data are:

Distance (ft)	Angle
$RU = 200.00 \pm 0.05$	$\theta_1 = 240°00' \pm 30''$
$US = 100.00 \pm 0.08$	$\theta_2 = 150°00' \pm 30''$
	$\theta_3 = 240°01' \pm 30''$

SOLUTION

Step 1: Calculate initial approximations for the unknown station coordinates.

$$x_{u_0} = 1000.00 + 200.00 \sin(60°) = 1173.20 \text{ ft}$$

$$y_{u_0} = 1000.00 + 200.00 \cos(60°) = 1100.00 \text{ ft}$$

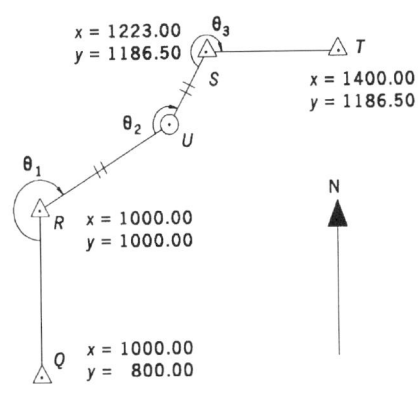

Figure 16.2 Simple link traverse.

ADJUSTMENT OF HORIZONTAL SURVEYS: TRAVERSES AND NETWORKS

Step 2: Formulate the X and K matrices. The traverse in this problem contains only one unknown station with two unknown coordinates. The elements of the X matrix thus consist of the dx_u and dy_u terms. They are the unknown corrections to be applied to the initial approximations for the coordinates of station U. The values in the K matrix are derived by subtracting computed quantities, based on initial coordinates, from their respective observed quantities. Note that since the first and third observations were used to compute initial approximations for station U, their K-matrix values will be zero in the first iteration.

$$X = \begin{bmatrix} dx_u \\ dy_u \end{bmatrix} \quad K = \begin{bmatrix} k_{l_{RU}} \\ k_{l_{US}} \\ k_{\theta_1} \\ k_{\theta_2} \\ k_{\theta_3} \end{bmatrix} = \begin{bmatrix} 200.00 \text{ ft} - 200.00 \text{ ft} \\ 100.00 \text{ ft} - 99.81 \text{ ft} \\ 240°00'00'' - 240°00'00'' \\ 150°00'00'' - 149°55'51'' \\ 240°01'00'' - 240°04'12'' \end{bmatrix} = \begin{bmatrix} 0.00 \text{ ft} \\ 0.019 \text{ ft} \\ 0'' \\ 249'' \\ -192'' \end{bmatrix}$$

Step 3: Calculate the Jacobian matrix. Since the observation equations are nonlinear, the Jacobian matrix must be formed to obtain the solution. The J matrix is formed using prototype equations (16.1) for distances and (16.2) for angles. As explained in Section 15.4, since the units of the K matrix that relate to the angles are in seconds, the angle coefficients of the J matrix must be multiplied by ρ to also obtain units of seconds.

In developing the J matrix using prototype equations (16.1) and (16.2), subscript substitutions were as shown in Table 16.1. Substitutions of numerical values and computation of the J matrix follow.

$$J = \begin{bmatrix} \dfrac{1173.20 - 1000.00}{200.00} & \dfrac{1100.00 - 1000.00}{200.00} \\ \dfrac{1173.20 - 1223.00}{99.81} & \dfrac{1100.00 - 1186.50}{99.81} \\ \dfrac{1100.00 - 1000.00}{200.00^2}\rho & \dfrac{1000.00 - 1173.20}{200.00^2}\rho \\ \left(\dfrac{1000.00 - 1100.00}{200.00^2} - \dfrac{1186.50 - 1100.00}{99.81^2}\right)\rho & \left(\dfrac{1173.20 - 1000.00}{200.00^2} - \dfrac{1173.20 - 1223.00}{99.81^2}\right)\rho \\ \dfrac{1186.50 - 1100.00}{99.81^2}\rho & \dfrac{1173.50 - 1223.00}{99.81^2}\rho \end{bmatrix}$$

TABLE 16.1 Subscript Substitution

Observation	Subscript Substitution
Length RU	$R = i, U = j$
Length US	$U = i, S = j$
Angle θ_1	$Q = b, R = i, U = f$
Angle θ_2	$R = b, U = i, S = f$
Angle θ_3	$U = b, S = i, T = f$

$$J = \begin{bmatrix} 0.866 & 0.500 \\ -0.499 & -0.867 \\ 515.7 & -893.2 \\ -2306.6 & 1924.2 \\ 1709.9 & -1031.1 \end{bmatrix}$$

Step 4: Formulate the W matrix. The fact that distance and angle observations have differing observational units and are combined in an adjustment is resolved by using relative weights that are based on observational variances in accordance with Equation (10.6). This weighting makes the observation equations dimensionally consistent. If weights are not used in traverse adjustments (i.e., equal weights are assumed), the least squares problem will generally either give unreliable results or result in a system of equations that has no solution. Since weights influence the correction size that each observation will receive, it is extremely important to use variances that correspond closely to the observational errors. The error propagation procedures discussed in Chapter 7 aid in the determination of the estimated errors. Repeating Equation (10.6), the distance and angle weights for this problem are

$$\text{distances: } w_{l_{IJ}} = \frac{1}{S_{l_{IJ}}^2} \quad \text{and} \quad \text{angles: } w_{\theta_{bif}} = \frac{1}{S_{\theta_{bif}}^2} \qquad (16.4)$$

Again, the units of the weight matrix must match those of the J and K matrices. That is, the angular weights must be in the same units of measure (seconds) as the counterparts in the other two matrices. Based on the estimated errors in the observations, the W matrix, which is diagonal, is

$$W = \begin{bmatrix} \frac{1}{0.05^2} & & & & & (\text{zeros}) \\ & \frac{1}{0.08^2} & & & & \\ & & \frac{1}{30^2} & & & \\ & & & \frac{1}{30^2} & & \\ & & & & \frac{1}{30^2} & \\ (\text{zeros}) & & & & & \frac{1}{30^2} \end{bmatrix}$$

$$= \begin{bmatrix} 400.00 & & & & & (\text{zeros}) \\ & 156.2 & & & & \\ & & 0.0011 & & & \\ & & & 0.0011 & & \\ (\text{zeros}) & & & & & 0.0011 \end{bmatrix}$$

Step 5: Solve the matrix system. This problem is iterative and was solved according to Equation (11.39) using the program MATRIX. (Output from the solution follows.) The first iteration yielded the following corrections to the initial coordinates.

$$dx_u = -0.11 \text{ ft}$$
$$dy_u = -0.01 \text{ ft}$$

Note that a second iteration produced zeros for dx_u and dy_u. The reader is encouraged to use the MATRIX or ADJUST program to duplicate these results.

Step 6: Compute the a posteriori adjustment statistics. Also from the program MATRIX, the residuals and reference standard deviation are

$$v_{ru} = -0.11 \text{ ft} \quad v_{us} = -0.12 \text{ ft}$$
$$v_{\theta 1} = -49'' \quad v_{\theta 2} = -17'' \quad v_{\theta 3} = 6'' \quad S_0 = \pm 1.82$$

A χ^2 test was used as discussed in Section 5.4 to see if the a posteriori reference variance differed significantly from its a priori value of 1.[1] The test revealed that there was no statistically significant difference between the a posteriori value of $(1.82)^2$ and its a priori value of 1 at a 99% confidence level, and thus the a priori value should be used for the reference variance when computing the standard deviations of the coordinates. By applying Equation (13.24), the estimated errors in the adjusted coordinates are

$$S_{xU} = 1.00\sqrt{0.00053} = \pm 0.023 \text{ ft}$$
$$S_{yU} = 1.00\sqrt{0.000838} = \pm 0.029 \text{ ft}$$

Following are the results from the program ADJUST.

```
*******************************************
Initial approximations for unknown stations
*******************************************
Station   Northing   Easting
==============================
   U      1,100.00   1,173.20
```

---

[1] Since weights are calculated using the formula $w_i = \sigma_0^2/\sigma_i^2$, using weights of $1/\sigma_i^2$ implies an a priori value of 1 for the reference variance (see Chapter 10).

16.4 NUMERICAL EXAMPLE

```
Control Stations
~~~~~~~~~~~~~
Station   Easting    Northing
=============================
    Q   1,000.00      800.00
    R   1,000.00    1,000.00
    S   1,223.00    1,186.50
    T   1,400.00    1,186.50

Distance Observations
~~~~~~~~~~~~~~~~~~~
 Station  Station
Occupied  Sighted  Distance      σ
==================================
    R        U      200.00     0.050
    U        S      100.00     0.080

Angle Observations
~~~~~~~~~~~~~~~~
    Station      Station      Station
  Backsighted   Occupied    Foresighted       Angle         σ
=============================================================
       Q            R             U       240° 00' 00"    30"
       R            U             S       150° 00' 00"    30"
       U            S             T       240° 01' 00"    30"
```

First Iteration Matrices

```
J Dim: 5x2                            K Dim: 5x1    X Dim: 2x1
======================                ==========    ==========
     0.86602       0.50001             0.00440        -0.11
    -0.49894      -0.86664             0.18873        -0.01
   515.68471    -893.16591             2.62001      ==========
 -2306.62893    1924.25287           249.36438
  1790.94422   -1031.08696          -191.98440
======================                ==========

W Dim: 5x5
=================================================
 400.00000    0.00000    0.00000    0.00000    0.00000
   0.00000  156.25000    0.00000    0.00000    0.00000
   0.00000    0.00000    0.00111    0.00000    0.00000
   0.00000    0.00000    0.00000    0.00111    0.00000
   0.00000    0.00000    0.00000    0.00000    0.00111
=================================================
```

290 ADJUSTMENT OF HORIZONTAL SURVEYS: TRAVERSES AND NETWORKS

```
N Dim: 2x2                              Qxx Dim: 2x2
=========================               ==================
 10109.947301   -7254.506002              0.000530   0.000601
 -7254.506002    6399.173533              0.000601   0.000838
=========================               ==================
```

Final Iteration
```
J Dim: 5x2                              K Dim: 5x1    X Dim: 2x1
=========================               ==========    ==========
     0.86591        0.50020                0.10723       0.0000
    -0.49972       -0.86619                0.12203       0.0000
   516.14929     -893.51028               48.62499    ==========
 -2304.96717     1925.52297               17.26820
  1788.81788    -1032.01269               -5.89319
=========================               ==========
```

```
N Dim: 2x2                              Qxx Dim: 2x2
=========================               ==================
 10093.552221  -7254.153057               0.000532   0.000602
 -7254.153057   6407.367420               0.000602   0.000838
=========================               ==================
```

```
J Qxx Jt Dim: 5x5
============================================================
  0.001130  -0.001195   -0.447052    0.055151    0.391901
 -0.001195   0.001282    0.510776   -0.161934   -0.348843
 -0.447052   0.510776  255.118765  -235.593233  -19.525532
  0.055151  -0.161934 -235.593233   586.956593 -351.363360
  0.391901  -0.348843  -19.525532  -351.363360  370.888892
============================================================
```

```
*****************
Adjusted stations
*****************

Station   Northing    Easting     S-N     S-E
==============================================
      U   1,099.99   1,173.09    0.029   0.023
```

```
*******************************
Adjusted Distance Observations
*******************************
 Station    Station
Occupied    Sighted    Distance        V        S
========================================
    R          U        199.89      -0.11    0.061
    U          S         99.88      -0.12    0.065

***************************
Adjusted Angle Observations
***************************
  Station    Station      Station
 Backsight   Occupied    Foresight        Angle         V         S
===================================================
    Q           R            U       239° 59' 11"    -49"     29.0"
    R           U            S       149° 59' 43"    -17"     44.1"
    S           S            T       240° 01' 06"      6"     35.0"

-----Reference Standard Deviation = ±1.82-----
     Iterations » 2
```

16.5 MINIMUM AMOUNT OF CONTROL

All adjustments require some form of control, and failure to supply a sufficient amount will result in an indeterminate solution. A traverse requires a minimum of one control station to fix it in position and one line of known direction to fix it in angular orientation. When a traverse has the minimum amount of control, it is said to be *minimally constrained*. It is not possible to adjust a traverse without this minimum. If minimal constraint is not available, necessary control values can be assumed and the computational process carried out in arbitrary space. This enables the observed data to be tested for blunders and errors. In Chapter 21 we discuss minimally constrained adjustments.

A *free network adjustment* involves using a *pseudoinverse* to solve systems that have less than the minimum amount of control. This material is beyond the scope of this book. Readers interested in this subject should consult Bjerhammar (1973) or White (1987) in the bibliography.

16.6 ADJUSTMENT OF NETWORKS

With the introduction of an EDM instrument, and particularly the total station, the speed and reliability of making angle and distance observations have increased greatly. This has led to observational systems that do not conform to the basic systems of trilateration, triangulation, or traverse. For example, it is common to collect more than the minimum observations at a station during a horizontal control survey. This creates what is called a *complex network,* referred to more commonly as a *network.* The least squares solution of a network is similar to that of a traverse. That is, observation equations are written for each observation using the prototype equations given in Section 16.2. Coordinate corrections are found using Equation (11.39) and a posteriori error analysis is carried out.

Example 16.2 A network survey was conducted for the project shown in Figure 16.3. Station Q has control coordinates of (1000.00, 1000.00) and the azimuth of line QR is $0°06'24.5''$ with an estimated error of $\pm 0.001''$. The observations and their estimated errors are listed in Table 16.2. Adjust this survey by the method of least squares.

SOLUTION Using standard traverse coordinate computational methods, the initial approximations for station coordinates (x,y) were determined to be

R: (1003.07, 2640.00) S: (2323.07, 2638.46) T: (2661.74, 1096.08)

Under each station heading in the observation columns, a letter representing the appropriate prototype equation dx and dy coefficient appears. For example, for the first distance QR, station Q is substituted for i in prototype equation (16.1) and station R replaces j. For the first angle, observed at Q from R to S, station R takes on the subscript b, Q becomes i, and S is substituted for f in prototype equation (16.2).

Table 16.3 shows the structure of the coefficient matrix for this adjustment and indicates by subscripts where the nonzero values occur. In this table, the column headings are the elements of the unknown X matrix dx_r, dy_r, dx_s, dy_s, dx_t, and dy_t. Note that since station Q is a control station, its corrections are set to zero and thus dx_q and dy_q are not included in the adjustment. Note also in this table that the elements which have been left blank are zeros.

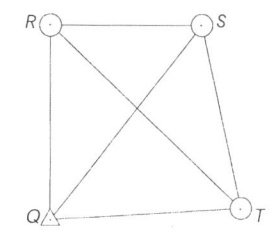

Figure 16.3 Horizontal network.

16.6 ADJUSTMENT OF NETWORKS

TABLE 16.2 Data for Example 16.2

Distance observations

Occupied, i	Sighted, j	Distance$_{ij}$ (ft)	S (ft)
Q	R	1640.016	0.026
R	S	1320.001	0.024
S	T	1579.123	0.025
T	Q	1664.524	0.026
Q	S	2105.962	0.029
R	T	2266.035	0.030

Angle observations

Backsight, b	Instrument, i	Foresight, f	Angle	S (″)
R	Q	S	38°48′50.7″	4.0
S	Q	T	47°46′12.4″	4.0
T	Q	R	273°24′56.5″	4.4
Q	R	S	269°57′33.4″	4.7
R	S	T	257°32′56.8″	4.7
S	T	Q	279°04′31.2″	4.5
S	R	T	42°52′51.0″	4.3
S	R	Q	90°02′26.7″	4.5
Q	S	R	51°08′45.0″	4.3
T	S	Q	51°18′16.2″	4.0
Q	T	R	46°15′02.0″	4.0
R	T	S	34°40′05.7″	4.0

Azimuth observations

From, i	To, j	Azimuth	S (″)
Q	R	0°06′24.5″	0.001

To fix the orientation of the network, the direction of course QR is included as an observation, but with a very small estimated error, ±0.001″. The last row of Table 16.3 shows the inclusion of this constrained observation using prototype equation (16.3). Since for azimuth QR only the foresight station, R, is an unknown, only coefficients for the foresight station j are included in the coefficient matrix.

Below are the necessary matrices for the first iteration when doing the weighted least squares solution of the problem. Note that the numbers have been truncated to five decimal places for publication purposes only. Following these initial matrices, the results of the adjustment are listed, as determined with program ADJUST.

294 ADJUSTMENT OF HORIZONTAL SURVEYS: TRAVERSES AND NETWORKS

TABLE 16.3 Format for Coefficient Matrix J of Example 16.4

Observation	dx_r	dy_r	dx_s	dy_s	dx_t	dy_t
QR	j	j				
RS	i	i	j	j		
ST			i	i	j	j
TQ					i	i
QS			j	j		
RT	i	i			j	j
∠RQS	b	b	f	f		
∠SQT			b	b	f	f
∠TQR	f	f			b	b
∠QRS	i	i	f	f		
∠RST	b	b	i	i	f	f
∠STQ			b	b	i	i
∠SRT	i	i	b	b	f	f
∠SRQ	i	i	b	b		
∠QSR	f	f	i	i		
∠TSQ			i	i	b	b
∠QTR	f	f			i	i
∠RTS	b	b	f	f	i	i
Az QR	j	j				

$$J = \begin{bmatrix}
0.00187 & 1.00000 & 0.00000 & 0.00000 & 0.00000 & 0.00000 \\
-1.00000 & 0.00117 & 1.00000 & -0.00117 & 0.00000 & 0.00000 \\
0.00000 & 0.00000 & -0.21447 & 0.97673 & 0.21447 & -0.97673 \\
0.00000 & 0.00000 & 0.00000 & 0.00000 & 0.99833 & 0.05772 \\
0.00000 & 0.00000 & 0.62825 & 0.77801 & 0.00000 & 0.00000 \\
-0.73197 & 0.68133 & 0.00000 & 0.00000 & 0.73197 & -0.68133 \\
-125.77078 & 0.23544 & 76.20105 & -61.53298 & 0.00000 & 0.00000 \\
0.00000 & 0.00000 & -76.20105 & 61.53298 & 7.15291 & -123.71223 \\
125.77078 & -0.23544 & 0.00000 & 0.00000 & -7.15291 & 123.71223 \\
-125.58848 & 156.49644 & -0.18230 & -156.26100 & 0.00000 & 0.00000 \\
-0.18230 & -156.26100 & 127.76269 & 184.27463 & -127.58038 & -28.01362 \\
0.00000 & 0.00000 & -127.58038 & -28.01362 & 134.73329 & -95.69861 \\
61.83602 & -89.63324 & 0.18230 & 156.26100 & -62.01833 & -66.62776 \\
125.58848 & -156.49644 & 0.18230 & 156.26100 & 0.00000 & 0.00000 \\
0.18230 & 156.26100 & -76.38335 & -94.72803 & 0.00000 & 0.00000 \\
0.00000 & 0.00000 & -51.37934 & -89.54660 & 127.58038 & 28.01362 \\
62.01833 & 66.62776 & 0.00000 & 0.00000 & -69.17123 & 57.08446 \\
-62.01833 & -66.62776 & 127.58038 & 28.01362 & -65.56206 & 38.61414 \\
125.770798 & -0.23544 & 0.00000 & 0.00000 & 0.00000 & 0.00000
\end{bmatrix}$$

16.6 ADJUSTMENT OF NETWORKS

The weight matrix is

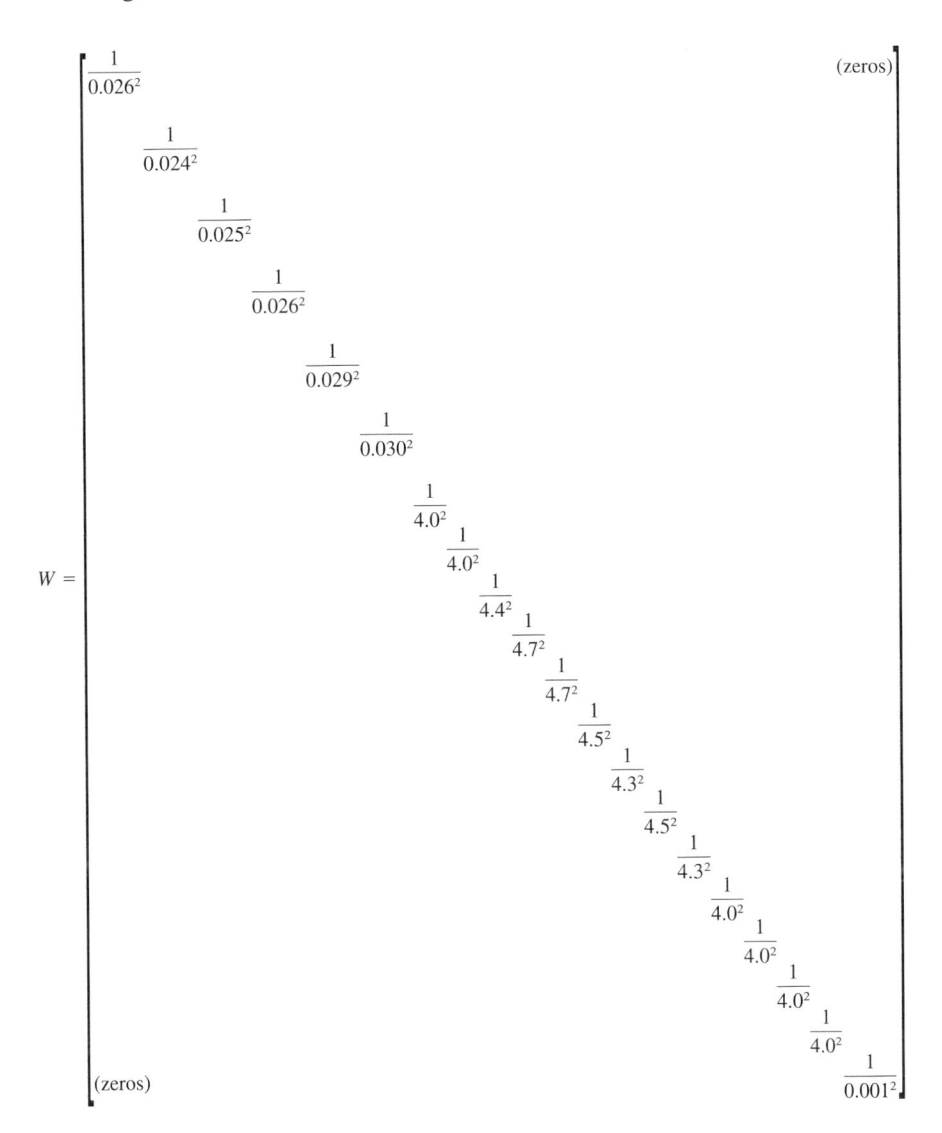

$$W = \begin{bmatrix} \frac{1}{0.026^2} & & & & & & & & & & & & & & & & & & & & & & \text{(zeros)} \\ & \frac{1}{0.024^2} & & & & & & & & & & & & & & & & & & & & & \\ & & \frac{1}{0.025^2} & & & & & & & & & & & & & & & & & & & & \\ & & & \frac{1}{0.026^2} & & & & & & & & & & & & & & & & & & & \\ & & & & \frac{1}{0.029^2} & & & & & & & & & & & & & & & & & & \\ & & & & & \frac{1}{0.030^2} & & & & & & & & & & & & & & & & & \\ & & & & & & \frac{1}{4.0^2} & & & & & & & & & & & & & & & & \\ & & & & & & & \frac{1}{4.0^2} & & & & & & & & & & & & & & & \\ & & & & & & & & \frac{1}{4.4^2} & & & & & & & & & & & & & & \\ & & & & & & & & & \frac{1}{4.7^2} & & & & & & & & & & & & & \\ & & & & & & & & & & \frac{1}{4.7^2} & & & & & & & & & & & & \\ & & & & & & & & & & & \frac{1}{4.5^2} & & & & & & & & & & & \\ & & & & & & & & & & & & \frac{1}{4.3^2} & & & & & & & & & & \\ & & & & & & & & & & & & & \frac{1}{4.5^2} & & & & & & & & & \\ & & & & & & & & & & & & & & \frac{1}{4.3^2} & & & & & & & & \\ & & & & & & & & & & & & & & & \frac{1}{4.0^2} & & & & & & & \\ & & & & & & & & & & & & & & & & \frac{1}{4.0^2} & & & & & & \\ & & & & & & & & & & & & & & & & & \frac{1}{4.0^2} & & & & & \\ & & & & & & & & & & & & & & & & & & \frac{1}{4.0^2} & & & & \\ \text{(zeros)} & & & & & & & & & & & & & & & & & & & \frac{1}{0.001^2} \end{bmatrix}$$

296 ADJUSTMENT OF HORIZONTAL SURVEYS: TRAVERSES AND NETWORKS

The K matrix is

$$K = \begin{bmatrix} 0.0031 \\ -0.0099 \\ -0.0229 \\ -0.0007 \\ -0.0053 \\ -0.0196 \\ -0.0090 \\ -0.5988 \\ 0.2077 \\ -2.3832 \\ 1.4834 \\ -1.4080 \\ -1.0668 \\ 2.4832 \\ -0.0742 \\ -3.4092 \\ -22.1423 \\ 2.4502 \\ -0.3572 \\ 0.0000 \end{bmatrix}$$

Following is a summary of the results from ADJUST.

```
Number of Control Stations      » 1
Number of Unknown Stations      » 3
Number of Distance observations » 6
Number of Angle observations    » 12
Number of Azimuth observations  » 1


*******************************************
Initial approximations for unknown stations
*******************************************
Station        X           Y
===========================
      R    1,003.06    2,640.01
      S    2,323.07    2,638.47
      T    2,661.75    1,096.07
```

16.6 ADJUSTMENT OF NETWORKS

```
Control Stations
~~~~~~~~~~~~
Station           X          Y
===========================
      Q    1,000.00  1,000.00

*********************
Distance Observations
*********************
 Station   Station
Occupied   Sighted   Distance      S
==================================
      Q       R    1,640.016    0.026
      R       S    1,320.001    0.024
      S       T    1,579.123    0.025
      T       Q    1,664.524    0.026
      Q       S    2,105.962    0.029
      R       T    2,266.035    0.030

******************
Angle Observations
******************
   Station      Station     Station
 Backsighted   Occupied   Foresighted            Angle         S
==============================================================
      R           Q             S      38° 48' 50.7"    4.0"
      S           Q             T      47° 46' 12.4"    4.0"
      T           Q             R     273° 24' 56.5"    4.4"
      Q           R             S     269° 57' 33.4"    4.7"
      R           S             T     257° 32' 56.8"    4.7"
      S           T             Q     279° 04' 31.2"    4.5"
      S           R             T      42° 52' 51.0"    4.3"
      S           R             Q      90° 02' 26.7"    4.5"
      Q           S             R      51° 08' 45.0"    4.3"
      T           S             Q      51° 18' 16.2"    4.0"
      Q           T             R      46° 15' 02.0"    4.0"
      R           T             S      34° 40' 05.7"    4.0"

********************
Azimuth Observations
********************
 Station   Station
Occupied   Sighted      Azimuth       S
=====================================
      Q       R     0° 06' 24.5    0.0"
```

Iteration 1

```
K MATRIX    X MATRIX
~~~~~~      ~~~~~~
  0.0031   -0.002906
 -0.0099   -0.035262
 -0.0229   -0.021858
 -0.0007    0.004793
 -0.0053    0.003996
 -0.0196   -0.014381
 -0.0090
 -0.5988
  0.2077
 -2.3832
  1.4834
 -1.4080
 -1.0668
  2.4832
 -0.0742
 -3.4092
-22.1423
  2.4502
 -0.3572
```

Iteration 2

```
K MATRIX    X MATRIX
~~~~~~      ~~~~~~
  0.0384    0.000000
 -0.0176   -0.000000
 -0.0155   -0.000000
 -0.0039    0.000000
  0.0087   -0.000000
 -0.0104    0.000000
 -2.0763
 -0.6962
  2.3725
 -0.6444
 -0.5048
 -3.3233
 -1.3435
  0.7444
  3.7319
 -5.2271
-18.5154
  0.7387
  0.0000
```

16.6 ADJUSTMENT OF NETWORKS

```
INVERSE MATRIX
~~~~~~~~~~
0.00000000  0.00000047   0.00000003   0.00000034  0.00000005   0.00000019
0.00000047  0.00025290   0.00001780   0.00018378  0.00002767   0.00010155
0.00000003  0.00001780   0.00023696  -0.00004687  0.00006675  -0.00008552
0.00000034  0.00018378  -0.00004687   0.00032490  0.00010511   0.00022492
0.00000005  0.00002767   0.00006675   0.00010511  0.00027128   0.00011190
0.00000019  0.00010155  -0.00008552   0.00022492  0.00011190   0.00038959
```

```
****************
Adjusted stations
****************
Station          X          Y        Sx      Sy
================================================
   R      1,003.06   2,639.97    0.000   0.016
   S      2,323.07   2,638.45    0.015   0.018
   T      2,661.75   1,096.06    0.016   0.020
```

```
*******************************
Adjusted Distance Observations
*******************************
Station   Station
Occupied  Sighted   Distance         V         S
=================================================
    Q        R     1,639.978   -0.0384    0.0159
    R        S     1,320.019    0.0176    0.0154
    S        T     1,579.138    0.0155    0.0158
    T        Q     1,664.528    0.0039    0.0169
    Q        S     2,105.953   -0.0087    0.0156
    R        T     2,266.045    0.0104    0.0163
```

```
**************************
Adjusted Angle Observations
**************************
  Station    Station    Station
Backsighted  Occupied  Foresighted      Angle           V       S
==================================================================
     R          Q           S     38° 48' 52.8"    2.08"    1.75
     S          Q           T     47° 46' 13.1"    0.70"    1.95
     T          Q           R    273° 24' 54.1"   -2.37"    2.40
     Q          R           S    269° 57' 34.0"    0.64"    2.26
     R          S           T    257° 32' 57.3"    0.50"    2.50
     S          T           Q    279° 04' 34.5"    3.32"    2.33
     S          R           T     42° 52' 52.3"    1.34"    1.82
     S          R           Q     90° 02' 26.0"   -0.74"    2.26
     Q          S           R     51° 08' 41.3"   -3.73"    1.98
     T          S           Q     51° 18' 21.4"    5.23"    2.04
```

```
             Q           T            R    46° 15' 20.5"   18.52   1.82
             R           T            S    34° 40' 05.0"   -0.74"  1.72

****************************
Adjusted Azimuth Observations
****************************
Station     Station
Occupied    Sighted       Azimuth        V         S
==========================================================
      Q            R    0° 06' 24.5"    0.00"    0.00"

            ***************************************
                        Adjustment Statistics
            ***************************************
                        Iterations   = 2
                        Redundancies = 13

                    Reference Variance = 2.20
                      Reference So     = ±1.5
            Passed X² test at 99.0% significance level!
                       X² lower value = 3.57
                       X² upper value = 29.82
       The a priori value of 1 used in computations involving
                       the reference variance.
                              Convergence!
```

16.7 χ^2 TEST: GOODNESS OF FIT

At the completion of a least-squares adjustment, the significance of the computed reference variance, S_0^2, can be checked statistically. This check is often referred to as a *goodness-of-fit test* since the computation of S_0^2 is based on Σv^2. That is, as the residuals become larger, so will the reference variance computed, and thus the model computed deviates more from the values observed. However, the size of the residuals is not the only contributing factor to the size of the reference variance in a weighted adjustment. The stochastic model also plays a role in the size of this value. Thus, when a χ^2 test indicates that the null hypothesis should be rejected, it may be due to a blunder in the data or an incorrect decision by the operator in selecting the stochastic model for the adjustment. In Chapters 21 and 25 these matters are discussed in greater detail. For now, the reference variance of the adjustment of Example 16.2 will be checked.

In Example 16.2 there are 13 degrees of freedom and the computed reference variance, S_0^2, is 2.2. In Chapter 10 it was shown that the a priori value

TABLE 16.4 Two-Tailed χ^2 Test on S_0^2

$$H_0: S^2 = 1$$

$$H_a: S^2 \neq 1$$

Test statistic:

$$\chi^2 = \frac{\nu S^2}{\sigma^2} = \frac{13(2.2)}{1} = 28.6$$

Rejection region:

$$28.6 = \chi^2 > \chi^2_{0.005,13} = 29.82$$

$$28.6 = \chi^2 < \chi^2_{0.995,13} = 3.565$$

for the reference variance was 1. A check can now be made to compare the computed value for the reference variance against its a priori value using a two-tailed χ^2 test. For this adjustment, a significance level of 0.01 was selected. The procedures for doing the test were outlined in Section 5.4, and the results for this example are shown in Table 16.4. Since $\alpha/2$ is 0.005 and the adjustment had 13 redundant observations, the critical χ^2 value from the table is 29.82. Now it can be seen that the χ^2 value computed is less that the tabular value, and thus the test fails to reject the null hypothesis, H_0. The value of 1 for S_0^2 can and should be used when computing the standard deviations for the station coordinates and observations since the computed value is only an estimate.

PROBLEMS

Note: For problems requiring least squares adjustment, if a computer program is not distinctly specified for use in the problem, it is expected that the least squares algorithm will be solved using the program MATRIX, which is included on the CD supplied with the book.

16.1 For the link traverse shown in Figure P16.1, assume that the distance and angle standard deviations are ±0.027 ft and ±5″, respectively. Using the control below, adjust the data given in the figure using weighted least squares. The control station coordinates in units of feet are

A: $x = 944.79$ $y = 756.17$ C: $x = 6125.48$ $y = 1032.90$

Mk1: $x = 991.31$ $y = 667.65$ Mk2: $x = 6225.391$ $y = 1037.109$

(a) What is the reference standard deviation, S_0?
(b) List the adjusted coordinates of station B and give the standard deviations.
(c) Tabulate the adjusted observations, the residuals, and the standard deviations.
(d) List the inverted normal matrix used in the last iteration.

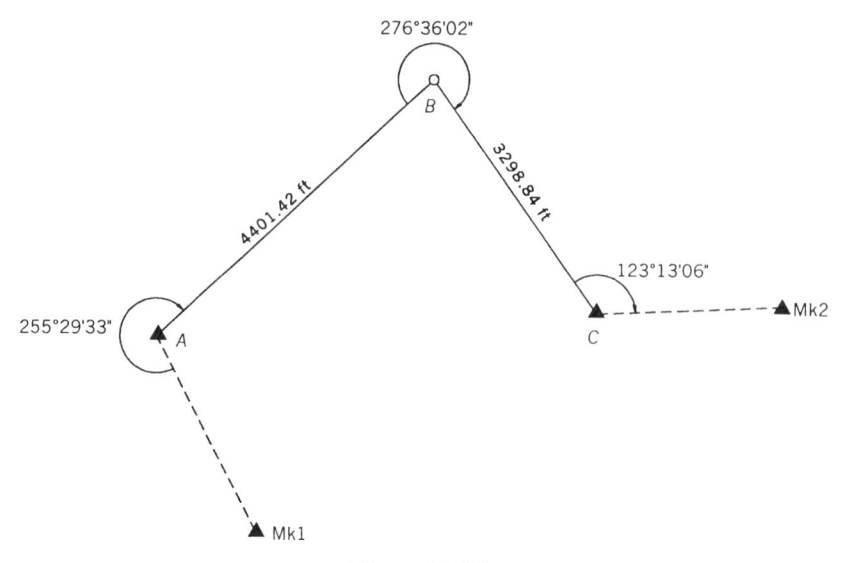

Figure P16.1

16.2 Adjust by the method of least squares the closed traverse in Figure P16.2. The data are given below.
(a) What is the reference standard deviation, S_0?
(b) List the adjusted coordinates of the unknown stations and the standard deviations.
(c) Tabulate the adjusted observations, the residuals, and the standard deviations.
(d) List the inverted normal matrix used in the last iteration.

Observed angles

Angle	Value	S (")
XAB	62°38′55.4″	5.6
BAC	56°18′41.9″	5.3
CBA	74°24′19.2″	5.4
ACB	49°16′55.9″	5.3

Observed distances

Course	Distance (ft)	S (ft)
AB	1398.82	0.020
BC	1535.70	0.021
CA	1777.73	0.022

PROBLEMS 303

Control stations

Station	X (ft)	Y (ft)
X	1490.18	2063.39
A	1964.28	1107.14

Unknown stations

Station	X (ft)	Y (ft)
B	2791.96	2234.82
C	3740.18	1026.78

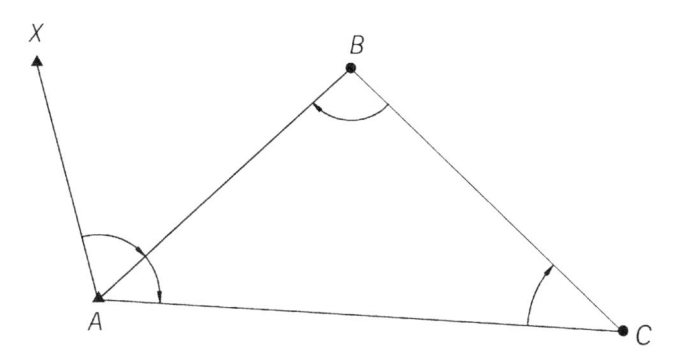

Figure P16.2

16.3 Adjust the network shown in Figure P16.3 by the method of least squares. The data are listed below.
(a) What is the reference standard deviation, S_0?
(b) List the adjusted coordinates of the unknown stations and the standard deviations.
(c) Tabulate the adjusted observations, the residuals, and the standard deviations.
(d) List the inverted normal matrix used in the last iteration.

Control station

Station	X (m)	Y (m)
A	1776.596	2162.848

Unknown stations

Station	X (m)	Y (m)
B	5339.61	2082.65
C	5660.39	6103.93
D	2211.95	6126.84

Distance observations

Course	Distance (m)	S (m)
AB	3563.905	0.013
BC	4034.021	0.014
CD	3448.534	0.013
DA	3987.823	0.014
AC	5533.150	0.018

Angle observations

Angle	Value	S (")
DAC	38°18'44"	4.0
CAB	46°42'38"	4.0
ABC	93°16'18"	4.0
BCA	40°01'11"	4.0
ACD	45°47'57"	4.0
DCA	314°12'00"	4.0

304 ADJUSTMENT OF HORIZONTAL SURVEYS: TRAVERSES AND NETWORKS

The azimuth of line AB is $91°17'19.9'' \pm 0.001''$.

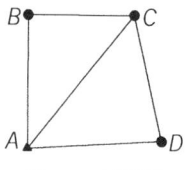

Figure P16.3

16.4 Perform a weighted least squares adjustment using the data given in Problem 15.4 and the additional distances given below.
   (a) What is the reference standard deviation, S_0?
   (b) List the adjusted coordinates of the unknown station and the standard deviations.
   (c) Tabulate the adjusted observations, the residuals, and the standard deviations.
   (d) List the inverted normal matrix used in the last iteration.

Course	Distance (ft)	S (ft)
PU	1214.44	0.021
QU	1605.03	0.021
RU	1629.19	0.021
SU	1137.33	0.021

16.5 Do a weighted least squares adjustment using the data given in Problems 14.7 and 15.5.
   (a) What is the reference standard deviation, S_0?
   (b) List the adjusted coordinates of the unknown station and the standard deviations.
   (c) Tabulate the adjusted observations, the residuals, and the standard deviations.
   (d) List the inverted normal matrix used in the last iteration.

16.6 Using the program ADJUST, do a weighted least squares adjustment using the data given in Problem 15.7 with the additional distances given below.
   (a) What is the reference standard deviation, S_0?
   (b) List the adjusted coordinates of the unknown stations and the standard deviations.
   (c) Tabulate the adjusted observations, the residuals, and the standard deviations.
   (d) List the inverted normal matrix used in the last iteration.

Course	Distance (m)	S (m)
AD	17,337.708	0.087
AC	27,331.345	0.137
BD	23,193.186	0.116
BC	19,382.380	0.097
CD	17,745.364	0.089

16.7 Using the program ADJUST, do a weighted least squares adjustment using the data given in Problem 15.9 with the additional distances given below.
(a) What is the reference standard deviation, S_0?
(b) List the adjusted coordinates of the unknown stations and the standard deviations.
(c) Tabulate the adjusted observations, the residuals, and the standard deviations.
(d) List the inverted normal matrix used in the last iteration.

Course	Distance (m)	S (m)
AC	9197.385	0.028
AD	16,897.138	0.051
BC	17,329.131	0.052
BD	8666.341	0.026
CD	14,384.926	0.043
CE	8115.898	0.025
CF	16,845.056	0.051
DE	16,113.175	0.049
DF	9019.629	0.027
EF	13,800.459	0.042
EG	8394.759	0.026
EH	16,164.944	0.049
FG	16,096.755	0.048
FH	8170.129	0.025

16.8 Using the Program ADJUST, do a weighted least squares adjustment using the data given in Problems 14.4 and 15.11.
(a) What is the reference standard deviation, S_0?
(b) List the adjusted coordinates of the unknown station and the standard deviations.
(c) Tabulate the adjusted observations, the residuals, and the standard deviations.
(d) List the inverted normal matrix used in the last iteration.

16.9 Using the program ADJUST, do a weighted least squares adjustment using the data given below.
   (a) What is the reference standard deviation, S_0?
   (b) List the adjusted coordinates of the unknown stations and the standard deviations.
   (c) Tabulate the adjusted observations, the residuals, and the standard deviations.
   (d) List the inverted normal matrix used in the last iteration.

Control station

Station	X (ft)	Y (ft)
A	108,250.29	33,692.06

Unknown stations

Station	X (ft)	Y (ft)
B	104,352.50	54,913.38
C	106,951.03	75,528.38
D	155,543.53	75,701.62
E	160,220.88	57,165.44
F	154,763.88	57,165.44
G	131,436.82	54,645.29
H	129,558.23	61,487.41

Distance observations

Course	Distance (ft)	S (ft)
AB	21,576.31	0.066
BC	20,778.13	0.063
CD	48,592.81	0.146
DE	19,117.21	0.059
EF	5457.00	0.020
FA	52,101.00	0.157
AG	31,251.45	0.094
GH	7095.33	0.024
HD	29,618.90	0.090

Angle observations

Angle	Value	S (")
BAG	58°18′15″	3.0
ABC	197°35′31″	3.0
BCD	262°36′41″	3.0
HDC	28°28′29″	3.0
DHG	103°19′34″	3.0
HGA	243°14′58″	3.0
EDH	75°28′59″	3.0
DEF	284°09′44″	3.0
EFA	153°13′19″	3.0
GAF	15°19′32″	3.0

16.10 Using the program ADJUST, do a weighted least squares adjustment using the data given in Problem 15.12 and the distances listed below.
   (a) What is the reference standard deviation, S_0?
   (b) List the adjusted coordinates of the unknown stations and the standard deviations.
   (c) Tabulate the adjusted observations, the residuals, and the standard deviations.
   (d) List the inverted normal matrix used in the last iteration.

Distance observations

Course	Distance (ft)	S (ft)	Course	Distance (ft)	S (ft)
AB	18,855.64	0.06	BE	19,923.80	0.06
AE	20,604.01	0.06	BC	17,548.84	0.05
EH	22,271.40	0.07	CF	18,009.19	0.06
EI	20,935.95	0.06	FG	15,957.20	0.05
EF	19,142.86	0.06	CD	15,359.20	0.05
BF	19,156.67	0.06	CG	18,077.08	0.06
DG	17,593.29	0.05	IJ	17,962.96	0.06
IF	22,619.87	0.07	HI	17,984.70	0.06
FJ	20,641.65	0.06	JG	21,035.70	0.06
JK	14,442.37	0.05	KG	19,528.95	0.06

16.11 Using the program ADJUST, do a weighted least squares adjustment using the following.
  (a) What is the reference standard deviation, S_0?
  (b) List the adjusted coordinates of the unknown stations and the standard deviations.
  (c) Tabulate the adjusted observations, the residuals, and the standard deviations.

Control stations

Station	X (ft)	Y (ft)
A	5,545.96	5504.56
E	11,238.72	7535.81

Unknown stations

Station	X (ft)	Y (ft)
B	9949.16	6031.81
C	5660.12	8909.83
D	9343.18	9642.46
F	8848.38	6617.78
G	7368.43	7154.46
H	6255.96	6624.33

Distance observations

Course	Distance (ft)	S (ft)	Course	Distance (ft)	S (ft)
AB	4434.66	0.020	AH	1325.89	0.015
BE	1981.15	0.016	HG	1232.33	0.015
ED	2833.91	0.017	GC	2623.49	0.016
DC	3989.03	0.019	GD	3177.97	0.017
CA	3407.18	0.018	DF	3064.98	0.017
FH	2592.35	0.016	FB	1247.03	0.015
BD	3661.14	0.018			

Angle observations

Stations	Angle	S (")	Stations	Angle	S (")
CAH	34°27'25"	3.4	HAB	50°47'41"	3.4
ABF	34°51'20"	3.5	ABE	137°26'38"	3.2
FBD	52°27'02"	3.5	DBE	50°08'16"	3.2
EDB	32°27'10"	3.1	BED	97°24'33"	3.3
BDG	47°54'56"	3.1	DCG	52°39'33"	3.1
GCA	45°50'58"	3.2	AHG	212°17'20"	3.8
GHF	25°28'45"	3.5	FHA	122°13'54"	3.5
HGC	67°24'11"	3.5	DGF	71°27'44"	3.3
FGH	134°48'45"	3.7	GFB	188°10'24"	3.7
BFH	152°07'08"	3.5	HFG	19°42'31"	3.4
CGD	86°19'10"	3.2	GDC	41°00'56"	3.1

For Problems 16.11 through 16.15, does the reference variance computed for the adjustment pass the χ^2 test at a level of significance of 0.05?

16.12 Example 16.1

16.13 Problem 16.1

16.14 Problem 16.2

16.15 Problem 16.3

16.16 Problem 16.4

Programming Problems

16.17 Write a computational program that reads a file of station coordinates and observations and then:

(a) writes the data to a file in a formatted fashion.

(b) computes the J, K, and W matrices.

(c) writes the matrices to a file that is compatible with the MATRIX program.

(d) Demonstrate this program with Problem 16.6.

16.18 Write a program that reads a file containing the J, K, and W matrices and then:

(a) writes these matrices in a formatted fashion.

(b) performs one iteration of Problem 16.6.

(c) writes the matrices used to compute the solution, and tabulates the corrections to the station coordinates in a formatted fashion.

16.19 Write a program that reads a file of station coordinates and observations and then:

- (a) writes the data to a file in a formatted fashion.
- (b) computes the J, K, and W matrices.
- (c) performs a weighted least squares adjustment of Problem 16.6.
- (d) writes the matrices used in computations in a formatted fashion to a file.
- (e) computes the final adjusted station coordinates, their estimated errors, the adjusted observations, their residuals, and their estimated errors, and writes them to a file in a formatted fashion.

16.20 Develop a computational program that creates the coefficient, weight, and constant matrices for a network. Write the matrices to a file in a format usable by the MATRIX program supplied with this book. Demonstrate its use with Problem 16.6.

CHAPTER 17

ADJUSTMENT OF GPS NETWORKS

17.1 INTRODUCTION

For the past five decades, NASA and the U.S. military have been engaged in a space research program to develop a precise positioning and navigation system. The first-generation system, called *TRANSIT,* used six satellites and was based on the *Doppler principle.* TRANSIT was made available for commercial use in 1967, and shortly thereafter its use in surveying began. The establishment of a worldwide network of control stations was among its earliest and most valuable applications. Point positioning using TRANSIT required very lengthy observing sessions, and its accuracy was at the 1-m level. Thus, in surveying it was suitable only for control work on networks consisting of widely spaced points. It was not satisfactory for everyday surveying applications such as traversing or engineering layout.

Encouraged by the success of TRANSIT, a new research program was developed that ultimately led to the creation of the NAVSTAR Global Positioning System (GPS). This second-generation positioning and navigation system utilizes a constellation of 24 orbiting satellites. The accuracy of GPS was improved substantially over that of the TRANSIT system, and the disadvantage of lengthy observing sessions was also eliminated. Although developed for military applications, civilians, including surveyors, also found uses for the GPS system.

Since its introduction, GPS has been used extensively. It is reliable, efficient, and capable of yielding extremely high accuracies. GPS observations can be taken day or night and in any weather conditions. A significant advantage of GPS is that visibility between surveyed points is not necessary. Thus, the time-consuming process of clearing lines of sight is avoided. Although most of the earliest applications of GPS were in control work, improvements have now made the system convenient and practical for use in virtually every type of survey, including property surveys, topographic mapping, and construction staking.

In this chapter we provide a brief introduction to GPS surveying. We explain the basic measurements involved in the system, discuss the errors in those measurements, describe the nature of the adjustments needed to account for those errors, and give the procedures for making adjustments of networks surveyed using GPS. An example problem is given to demonstrate the procedures.

17.2 GPS OBSERVATIONS

Fundamentally, the global positioning system operates by observing distances from receivers located on ground stations of unknown locations to orbiting GPS satellites whose positions are known precisely. Thus, conceptually, GPS surveying is similar to conventional resection, in which distances are observed with an EDM instrument from an unknown station to several control points. (The conventional resection procedure was discussed in Chapter 15 and illustrated in Example 15.2.) Of course, there are some differences between GPS position determination and conventional resection. Among them is the process of observing distances and the fact that the control stations used in GPS work are satellites.

All of the 24 satellites in the GPS constellation orbit Earth at nominal altitudes of 20,200 km. Each satellite continuously broadcasts unique electronic signals on two *carrier frequencies*. These carriers are modulated with *pseudorandom noise* (PRN) *codes*. The PRN codes consist of unique sequences of binary values (zeros and ones) that are superimposed on the carriers. These codes appear to be random, but in fact they are generated according to a known mathematical algorithm. The frequencies of the carriers and PRN codes are controlled very precisely at known values.

Distances are determined in GPS surveying by taking observations on these transmitted satellite signals. Two different observational procedures are used: positioning by pseudoranging, and positioning by carrier-phase measurements. *Pseudoranging* involves determining distances (ranges) between satellites and receivers by observing precisely the time it takes transmitted signals to travel from satellites to ground receivers. This is done by determining changes in the PRN codes that occur during the time it takes signals to travel from the satellite transmitter to the antenna of the receiver. Then from the known frequency of the PRN codes, very precise travel times are determined. With the velocity and travel times of the signals known, the pseudoranges can be computed. Finally, based on these ranges, the positions of the ground stations can be calculated. Because pseudoranging is based on observing PRN codes, this GPS observation technique is also often referred to as the *code measurement procedure.*

In the *carrier-phase procedure,* the quantities observed are phase changes that occur as a result of the carrier wave traveling from the satellites to the receivers. The principle is similar to the phase-shift method employed by electronic distance-measuring instruments. However, a major difference is that the satellites are moving, so that the signals cannot be returned to the transmitters for "true" phase-shift measurements. Instead, the phase shifts must be observed at the receivers. But to make true phase-shift observations, the clocks in the satellites and receivers would have to be perfectly synchronized, which of course cannot be achieved. To overcome this timing problem and to eliminate other errors in the system, *differencing* techniques (taking differences between phase observations) are used. Various differencing procedures can be applied. *Single differencing* is achieved by simultaneously observing two satellites with one receiver. Single differencing eliminates satellite clock biases. *Double differencing* (subtracting the results of single differences from two receivers) eliminates receiver clock biases and other systematic errors.

Another problem in making carrier-phase measurements is that only the phase shift of the last cycle of the carrier wave is observed, and the number of full cycles in the travel distance is unknown. (In EDM work this problem is overcome by progressively transmitting longer wavelengths and observing their phase shifts.) Again, because the satellites are moving, this cannot be done in GPS work. However, by extending the differencing technique to what is called *triple differencing,* this ambiguity in the number of cycles cancels out of the solution. Triple differencing consists of differencing the results of two double differences and thus involves making observations at two different times to two satellites from two stations.

In practice, when surveys are done by observing carrier phases, four or more satellites are observed simultaneously using two or more receivers located on ground stations. Also, the observations are repeated many times. This produces a very large number of redundant observations, from which many difference combinations can be computed.

Of the two GPS observing procedures, pseudoranging yields a somewhat lower order of accuracy, but it is preferred for navigation use because it gives instantaneous point positions of satisfactory accuracy. The carrier-phase technique produces a higher order of accuracy and is therefore the choice for high-precision surveying applications. Adjustment of carrier-phase GPS observations is the subject of this chapter.

The differencing techniques used in carrier-phase observations, described briefly above, do not yield positions directly for the points occupied by receivers. Rather, *baselines* (vector distances between stations) are determined. These baselines are actually computed in terms of their coordinate difference components ΔX, ΔY, and ΔZ. These coordinate differences are reported in the reference three-dimensional rectangular coordinate system described in Section 17.4.

To use the GPS carrier-phase procedure in surveying, at least two receivers located on separate stations must be operated simultaneously. For example, assume that two stations A and B were occupied for an observing session, that station A is a control point, and that station B is a point of unknown position. The session would yield coordinate differences ΔX_{AB}, ΔY_{AB}, and ΔZ_{AB} between stations A and B. The X,Y,Z coordinates of station B can then be obtained by adding the baseline components to the coordinates of A as

$$X_B = X_A + \Delta X_{AB}$$
$$Y_B = Y_A + \Delta Y_{AB} \qquad (17.1)$$
$$Z_B = Z_A + \Delta Z_{AB}$$

Because carrier-phase observations do not yield point positions directly, but rather, give baseline components, this method of GPS surveying is referred to as *relative positioning*. In practice, often more than two receivers are used simultaneously in relative positioning, which enables more than one baseline to be determined during each observing session. Also, after the first observing session, additional points are interconnected in the survey by moving the receivers to nearby stations. In this procedure, at least one receiver is left on one of the previously occupied stations. By employing this technique, a network of interconnected points can be created. Figure 17.1 illustrates an example of a GPS network. In this figure, stations A and B are control stations,

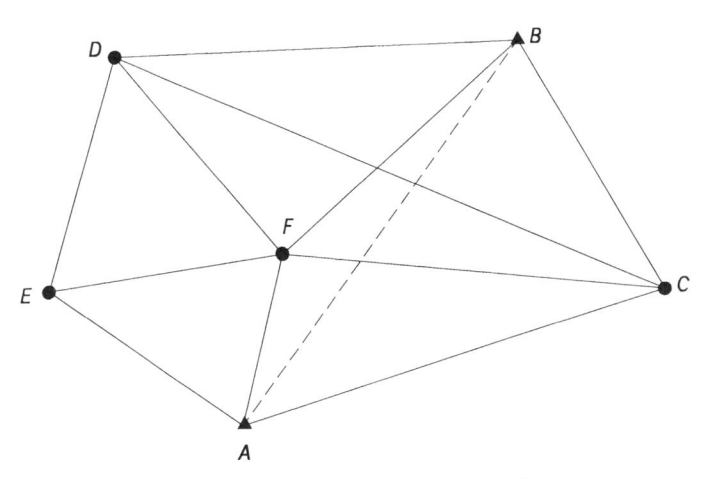

Figure 17.1 GPS survey network.

and stations *C, D, E,* and *F* are points of unknown position. Creation of such networks is a common procedure employed in GPS relative positioning work.

17.3 GPS ERRORS AND THE NEED FOR ADJUSTMENT

As in all types of surveying observations, GPS observations contain errors. The principal sources of these errors are (1) orbital errors in the satellite, (2) signal transmission timing errors due to atmospheric conditions, (3) receiver errors, (4) multipath errors (signals being reflected so that they travel indirect routes from satellite to receiver), and (5) miscentering errors of the receiver antenna over the ground station and receiver height-measuring errors. To account for these and other errors, and to increase the precisions of point position, GPS observations are very carefully made according to strict specifications, and redundant observations are taken. The fact that errors are present in the observations makes it necessary to analyze the measurements for acceptance or rejection. Also, because redundant observations have been made, they must be adjusted so that all observed values are consistent.

In GPS surveying work where the observations are made using carrier-phase observations, there are two stages where least squares adjustment is applied. The first is in processing the redundant observations to obtain the adjusted baseline components (ΔX, ΔY, ΔZ), and the second is in adjusting networks of stations wherein the baseline components have been observed. The latter adjustment is discussed in more detail later in the chapter.

17.4 REFERENCE COORDINATE SYSTEMS FOR GPS OBSERVATIONS

In GPS surveying, three different reference coordinate systems are involved. First, the satellite positions at the instants of their observation are given in a space-related X_s, Y_s, Z_s three-dimensional rectangular coordinate system. This coordinate system is illustrated in Figure 17.2. In the figure, the elliptical orbit of a satellite is shown. It has one of its two foci at *G*, Earth's center of gravity. Two points, *perigee* (point where the satellite is closest to *G*) and *apogee* (point where the satellite is farthest from *G*), define the *line of apsides*. This line, which also passes through the two foci, is the X_s axis of the satellite reference coordinate system. The origin of the system is at *G*, the Y_s axis is in the mean orbital plane, and Z_s is perpendicular to the X_s–Y_s plane. Because a satellite varies only slightly from its mean orbital plane, values of Z_s are small. For each specific instant of time that a given satellite is observed, its coordinates are calculated in its unique X_s, Y_s, Z_s system.

In processing GPS observations, all X_s, Y_s, and Z_s coordinates that were computed for satellite observations are converted to a common Earth-related X_e, Y_e, Z_e three-dimensional geocentric coordinate system. This Earth-centered, Earth-fixed coordinate system, illustrated in Figure 17.3, is also commonly

17.4 REFERENCE COORDINATE SYSTEMS FOR GPS OBSERVATIONS   315

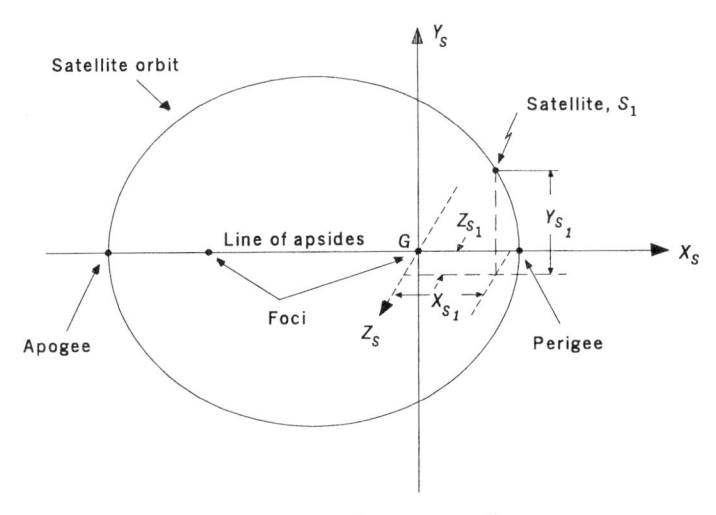

Figure 17.2 Satellite reference coordinate system.

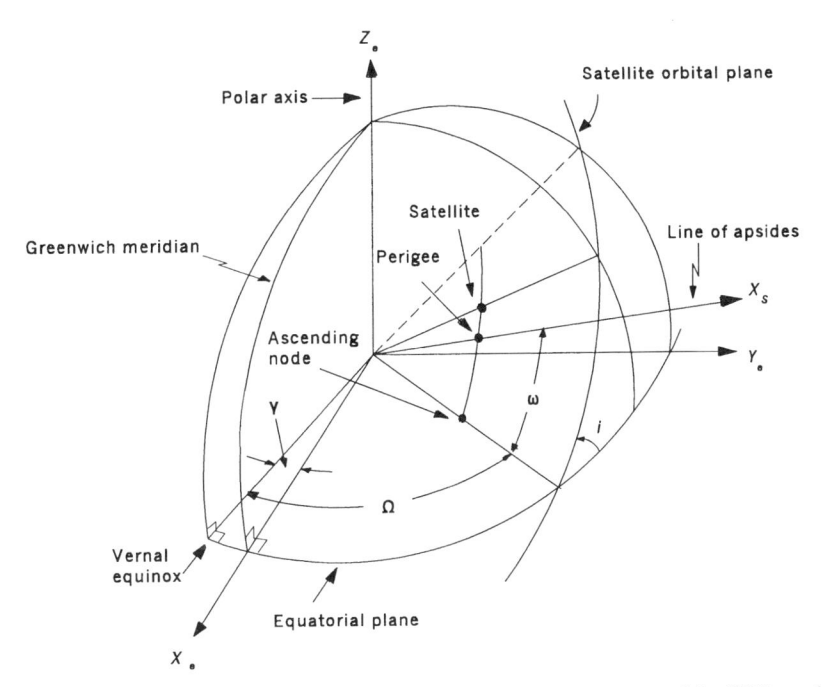

Figure 17.3 Earth-related three-dimensional coordinate system used in GPS carrier-phase differencing computations.

called the *terrestrial geocentric system,* or simply the *geocentric system.* It is in this system that the baseline components are computed based on the differencing of observed carrier phase measurements. The origin of this coordinate system is at Earth's gravitational center. The Z_e axis coincides with Earth's *Conventional Terrestrial Pole* (CTP) axis, the X_e–Y_e plane is perpendicular to the Z_e axis, and the X_e axis passes through the Greenwich Meridian. To convert coordinates from the space-related (X_s, Y_s, Z_s) system to the Earth-centered, Earth-related (X_e, Y_e, Z_e) geocentric system, six parameters are needed. These are (a) the *inclination angle i* (the angle between the orbital plane and Earth's equatorial plane); (b) the *argument of perigee* ω (the angle observed in the orbital plane between the equator and the line of apsides); (c) *right ascension of the ascending node* Ω (the angle observed in the plane of Earth's equator from the vernal equinox to the line of intersection between the orbital and equatorial planes); (d) the *Greenwich hour angle of the vernal equinox* γ (the angle observed in the equatorial plane from the Greenwich meridian to the vernal equinox); (e) the semimajor axis of the orbital ellipse, a; and (f) the eccentricity, e, of the orbital ellipse. The first four parameters are illustrated in Figure 17.3. For any satellite at any instant of time, these four parameters are available. Software provided by GPS equipment manufacturers computes the X_s, Y_s, and Z_s coordinates of satellites at their instants of observation, and it also transforms these coordinates into the X_e, Y_e, Z_e geocentric coordinate system used for computing the baseline components.

For the results of the baseline computations to be useful to local surveyors, the X_e, Y_e, and Z_e coordinates must be converted to geodetic coordinates of latitude, longitude, and height. The geodetic coordinate system is illustrated in Figure 17.4, where the parameters are symbolized by ϕ, λ, and h, respectively. Geodetic coordinates are referenced to the World Geodetic System of 1984, which employs the WGS 84 ellipsoid. The center of this ellipsoid is oriented at Earth's gravitational center, and for most practical purposes it is the same as the GRS 80 ellipsoid used for NAD 83. From latitude and longitude, state plane coordinates (which are more convenient for use by local surveyors) can be computed.

It is important to note that geodetic heights are not *orthometric heights* (elevations referred to the geoid). To convert geodetic heights to orthometric heights, the *geoid heights* (vertical distances between the ellipsoid and geoid) must be subtracted from geodetic heights.

17.5 CONVERTING BETWEEN THE TERRESTRIAL AND GEODETIC COORDINATE SYSTEMS

GPS networks must include at least one control point, but more are preferable. The geodetic coordinates of these control points will normally be given from a previous GPS survey. Prior to processing carrier-phase observations to obtain adjusted baselines for a network, the coordinates of the control stations

17.5 CONVERTING BETWEEN THE TERRESTRIAL AND GEODETIC COORDINATE SYSTEMS

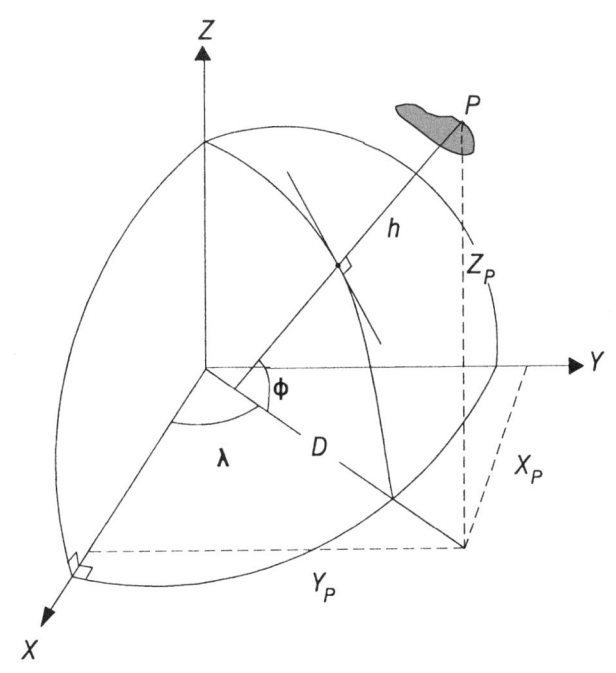

Figure 17.4 Geocentric coordinates (with the Earth-related X_e, Y_e, Z_e geocentric coordinate system superimposed).

in the network must be converted from their geodetic values into the Earth-centered, Earth-related X_e, Y_e, Z_e geocentric system. The equations for making these conversions are

$$X = (N + h) \cos \phi \cos \lambda \tag{17.2}$$

$$Y = (N + h) \cos \phi \sin \lambda \tag{17.3}$$

$$Z = [N(1 - e^2) + h] \sin \phi \tag{17.4}$$

In the equations above, h is the geodetic height of the point, ϕ the geodetic latitude of the point, and λ the geodetic longitude of the point. Also, e is eccentricity for the ellipsoid, which is computed as

$$e^2 = 2f - f^2 \tag{17.5a}$$

or

$$e^2 = \frac{a^2 - b^2}{a^2} \tag{17.5b}$$

where f is the flattening factor of the ellipsoid; a and b are the semimajor and semiminor axes, respectively, of the ellipsoid[1]; and N is the normal to the ellipsoid at the point, which is computed as

$$N = \frac{a}{\sqrt{1 - e^2 \sin^2 \phi}} \tag{17.6}$$

Example 17.1 Control stations A and B of the GPS network of Figure 17.1 have the following NAD83 geodetic coordinates:

$\phi_A = 43°15'46.2890''$ $\quad \lambda_A = -89°59'42.1640''$ $\quad h_A = 1382.618$ m

$\phi_B = 43°23'46.3626''$ $\quad \lambda_B = -89°54'00.7570''$ $\quad h_B = 1235.457$ m

Compute their X_e, Y_e, and Z_e geocentric coordinates.

SOLUTION *For station A:* By Equation (17.5a),

$$e^2 = \frac{2}{298.257223563} - \left(\frac{1}{298.257223563}\right)^2 = 0.006694379990$$

By Equation (17.6),

$$N = \frac{6{,}378{,}137}{\sqrt{1 - e^2 \sin^2(43°15'46.2890'')}} = 6{,}388{,}188.252 \text{ m}$$

By Equation (17.2),

$X_A = (6{,}388{,}188.252 + 1382.618) \cos(43°15'46.2890'') \cos(-89°59'42.1640'')$

$\quad = 402.3509$ m

By Equation (17.3),

$Y_A = (6{,}388{,}188.252 + 1382.618) \cos(43°15'46.2890'') \sin(-89°59'42.1640'')$

$\quad = -4{,}652{,}995.3011$ m

---

[1] The WGS 84 ellipsoid is used, whose a, b, and f values are 6,378,137.0 m, 6,356,752.3142 m, and 1/298.257223563, respectively.

17.5 CONVERTING BETWEEN THE TERRESTRIAL AND GEODETIC COORDINATE SYSTEMS 319

By Equation (17.4),

$$Z_A = [6{,}388{,}188.252(1 - e^2) + 1382.618]\sin(43°15'46.2890'')$$
$$= 4{,}349{,}760.7775 \text{ m}$$

For station B: Following the same procedure as above, the geocentric coordinates for station B are

$$X_B = 8086.0318 \text{ m} \qquad Y_B = -4{,}642{,}712.8474 \text{ m}$$

$$Z_B = 4{,}360{,}439.0833 \text{ m}$$

After completing the network adjustment, it is necessary to convert all X_e, Y_e, and Y_e geocentric coordinates to their geodetic values for use by local surveyors. This conversion process follows these steps (refer to Figure 17.4):

Step 1: Determine the longitude, λ, from

$$\lambda = \tan^{-1}\frac{Y}{X} \tag{17.7}$$

Step 2: Compute D from

$$D = \sqrt{X^2 + Y^2} \tag{17.8}$$

Step 3: Calculate an approximate latitude value ϕ_0 from

$$\phi_0 = \tan^{-1}\frac{Z}{D(1 - e^2)} \tag{17.9}$$

Step 4: Compute an approximate ellipsoid normal value N_0 from

$$N_0 = \frac{a}{\sqrt{1 - e^2 \sin^2 \phi_0}} \tag{17.10}$$

Step 5: Calculate an improved value for latitude ϕ_0 from

$$\phi_0 = \tan^{-1}\frac{Z + e^2 N_0 \sin \phi_0}{D} \tag{17.11}$$

Step 6: Use the value of ϕ_0 computed in step 5, and return to step 4. Iterate steps 4 and 5 until there is negligible change in ϕ_0. Using the values from the last iteration for N_0 and ϕ_0, the value for h is now computed[2] as

$$h = \frac{D}{\cos \phi_0} - N_0 \tag{17.12}$$

Example 17.2 Assume that the final adjusted coordinates for station C of the network of Figure 17.4 were

$$X_C = 12{,}046.5808 \text{ m} \qquad Y_C = -4{,}649{,}394.0826 \text{ m}$$

$$Z_C = 4{,}353{,}160.0634 \text{ m}$$

Compute the NAD83 geodetic coordinates for station C.

SOLUTION By Equation (17.7),

$$\lambda = \tan^{-1} \frac{-4{,}649{,}394.0826}{12{,}046.5808} = -89°51'05.5691''$$

By Equation (17.8),

$$D = \sqrt{(12{,}046.5808)^2 + (-4{,}649{,}394.0826)^2} = 4{,}649{,}409.6889 \text{ m}$$

Using Equation (17.9), the initial value for ϕ_0 is

$$\phi_0 = \tan^{-1} \frac{4{,}353{,}160.0634}{D(1 - e^2)} = 43°18'26.2228''$$

The first iteration for N_0 and ϕ_0 is

$$N_0 = \frac{6{,}378{,}137.0}{1 - e^2 \sin^2(43°18'26.22280'')} = 6{,}388{,}204.8545 \text{ m}$$

$$\phi_0 = \tan^{-1} \frac{4{,}353{,}160.0634 + e^2 (6{,}388{,}204.8545) \sin(43°18'26.22280'')}{D}$$

$$= 43°18'26.1035''$$

The next iteration produced the final values as

---

[2] Equation (17.12) is numerically stable for values of ϕ less than 45°. For values of ϕ greater than 45°, use the equation $h = (Z/\sin \phi_0) - N_0(1 - e^2)$.

$$N_0 = 6{,}388{,}204.8421 \qquad \phi_0 = 43°18'26.1030''$$

Using Equation (17.12), the elevation of station C is

$$h = \frac{D}{\cos(43°18'26.1030'')} - 6{,}388{,}204.8421 = 1103.101 \text{ m}$$

A computer program included with the software package ADJUST will make these coordinate conversions, both from geodetic to geocentric and from geocentric to geodetic. These computations are also demonstrated in a Mathcad worksheet on the CD that accompanies this book.

17.6 APPLICATION OF LEAST SQUARES IN PROCESSING GPS DATA

Least squares adjustment is used at two different stages in processing GPS carrier-phase measurements. First, it is applied in the adjustment that yields baseline components between stations from the redundant carrier-phase observations. Recall that in this procedure, differencing techniques are employed to compensate for errors in the system and to resolve the cycle ambiguities. In the solution, observation equations are written that contain the differences in coordinates between stations as parameters. The reference coordinate system for this adjustment is the X_e, Y_e, Z_e geocentric system. A highly redundant system of equations is obtained because, as described earlier, a minimum of four (and often more) satellites are tracked simultaneously using at least two (and often more) receivers. Furthermore, many repeat observations are taken. This system of equations is solved by least squares to obtain the most probable ΔX, ΔY, and ΔZ components of the baseline vectors. The development of these observation equations is beyond the scope of this book, and thus their solution by least squares is also not covered herein.[3]

Software furnished by manufacturers of GPS receivers will process observed phase changes to form the differencing observation equations, perform the least squares adjustment, and output the adjusted baseline vector components. The software will also output the covariance matrix, which expresses the correlation between the ΔX, ΔY, and ΔZ components of each baseline. The software is proprietary and thus cannot be included herein.

The second stage where least squares is employed in processing GPS observations is in adjusting baseline vector components in networks. This adjustment is made after the least squares adjustment of the carrier-phase

---

[3] Readers interested in studying these observation equations should consult *GPS Theory and Practice* (Hoffman-Wellenhof et al., 2001) or *GPS Satellite Surveying* (Leick, 2004) (see the bibliography).

322 ADJUSTMENT OF GPS NETWORKS

observations is completed. It is also done in the X_e, Y_e, Z_e geocentric coordinate system. In network adjustments, the goal is to make all X coordinates (and all X-coordinate differences) consistent throughout the figure. The same objective applies for all Y and Z coordinates. As an example, consider the GPS network shown in Figure 17.1. It consists of two control stations and four stations whose coordinates are to be determined. A summary of the baseline observations obtained from the least squares adjustment of carrier-phase measurements for this figure is given in Table 17.1. The covariance matrix elements that are listed in the table are used for weighting the observations. These are discussed in Section 17.8 but for the moment can be ignored.

A network adjustment of Figure 17.1 should yield adjusted X coordinates for the stations (and adjusted coordinate differences between stations) that are all mutually consistent. Specifically for this network, the adjusted X coordinate of station C should be obtained by adding ΔX_{AC} to the X coordinate of station A; and the same value should be obtained by adding ΔX_{BC} to the X coordinate of station B, or by adding ΔX_{DC} to the X coordinate of station D, and so on. Equivalent conditions should exist for the Y and Z coordinates. Note that these conditions do not exist for the data of Table 17.1, which contains the unadjusted baseline measurements. The procedure of adjusting GPS networks is described in detail in Section 17.8 and an example is given.

17.7 NETWORK PREADJUSTMENT DATA ANALYSIS

Prior to adjusting GPS networks, a series of procedures should be followed to analyze the data for internal consistency and to eliminate possible blunders. No control points are needed for these analyses. Depending on the actual observations taken and the network geometry, these procedures may consist of analyzing (1) differences between fixed and observed baseline components, (2) differences between repeated observations of the same baseline components, and (3) loop closures. After making these analyses, a minimally constrained adjustment is usually performed that will help isolate any blunders that may have escaped the first set of analyses. Procedures for making these analyses are described in the following subsections.

17.7.1 Analysis of Fixed Baseline Measurements

GPS job specifications often require that baseline observations be taken between fixed control stations. The benefit of making these observations is to verify the accuracy of both the GPS observational system and the control being held fixed. Obviously, the smaller the discrepancies between observed and known baseline lengths, the better. If the discrepancies are too large to be tolerated, the conditions causing them must be investigated before proceeding. Note that in the data of Table 17.1, one fixed baseline (between control points A and B) was observed. Table 17.2 gives the data for comparing

TABLE 17.1 Baseline Data Observed for the Network of Figure 17.1

(1)	(2)	(3)	(4)	(5)	(6)	(7)	(8)	(9)	(10)	(11)
From	To	ΔX	ΔY	ΔZ	\multicolumn{6}{c}{Covariance Matrix Elements}					
A	C	11,644.2232	3,601.2165	3,399.2550	9.884E-4	−9.580E-6	9.520E-6	9.377E-4	−9.520E-6	9.827E-4
A	E	−5,321.7164	3,634.0754	3,173.6652	2.158E-4	−2.100E-6	2.160E-6	1.919E-4	−2.100E-6	2.005E-4
B	C	3,960.5442	−6,681.2467	−7,279.0148	2.305E-4	−2.230E-6	2.070E-6	2.546E-4	−2.230E-6	2.252E-4
B	D	−11,167.6076	−394.5204	−907.9593	2.700E-4	−2.750E-6	2.850E-6	2.721E-4	−2.720E-6	2.670E-4
D	C	15,128.1647	−6,286.7054	−6,371.0583	1.461E-4	−1.430E-6	1.340E-6	1.614E-4	−1.440E-6	1.308E-4
D	E	−1,837.7459	−6,253.8534	−6,596.6697	1.231E-4	−1.190E-6	1.220E-6	1.277E-4	−1.210E-6	1.283E-4
F	A	−1,116.4523	−4,596.1610	−4355.8962	7.475E-5	−7.900E-7	8.800E-7	6.593E-5	−8.100E-7	7.616E-5
F	C	10,527.7852	−994.9377	−956.6246	2.567E-4	−2.250E-6	2.400E-6	2.163E-4	−2.270E-6	2.397E-4
F	E	−6,438.1364	−962.0694	−1,182.2305	9.442E-5	−9.200E-7	1.040E-6	9.959E-5	−8.900E-7	8.826E-5
F	D	−4,600.3787	5,291.7785	5,414.4311	9.330E-5	−9.900E-7	9.000E-7	9.875E-5	−9.900E-7	1.204E-4
F	B	6,567.2311	5,686.2926	6,322.3917	6.643E-5	−6.500E-7	6.900E-7	7.465E-5	−6.400E-7	6.048E-5
B	F	−6,567.2310	−5,686.3033	−6,322.3807	5.512E-5	−6.300E-7	6.100E-7	7.472E-5	−6.300E-7	6.629E-5
A	F	1,116.6883	4,596.4550	4,355.3008	6.619E-5	−8.000E-7	9.000E-7	8.108E-5	−8.200E-7	9.376E-5
A^a	B	7,683.6883	10,282.4550	10,678.3008	7.2397E-4	−7.280E-6	7.520E-6	6.762E-4	−7.290E-6	7.310E-4

[a] Fixed baseline used only for checking, but not included in adjustment.

TABLE 17.2 Comparisons of Measured and Fixed Baseline Components

(1) Component	(2) Measured (m)	(3) Fixed (m)	(4) Difference (m)	(5) ppm[a]
ΔX	7,683.6883	7,683.6809	0.0074	0.44
ΔY	10,282.4550	10,282.4537	0.0013	0.08
ΔZ	10,678.3008	10,678.3058	0.0050	0.30

[a] The total baseline length used in computing these ppm values was 16,697 m, which was derived from the square root of the sum of the squares of ΔX, ΔY, and ΔZ values.

the observed and fixed baseline components. The observed values are listed in column (2), and the fixed components are given in column (3). To compute the fixed values, X_e, Y_e, and Z_e, geocentric coordinates of the two control stations are first determined from their geodetic coordinates according to procedures discussed in Section 17.5. Then the ΔX, ΔY, and ΔZ differences between the X_e, Y_e, and Z_e coordinates for the two control stations are determined. Differences (in meters) between the observed and fixed baseline components are given in column (4). Finally, the differences, expressed in parts per million (ppm), are listed in column (5). These ppm values are obtained by dividing column (4) differences by their corresponding total baseline lengths and multiplying by 1,000,000.

17.7.2 Analysis of Repeat Baseline Measurements

Another procedure employed in evaluating the consistency of the data observed and in weeding out blunders is to make repeat observations of certain baselines. These repeat observations are taken in different sessions and the results compared. For example, in the data of Table 17.1, baselines *AF* and *BF* were repeated. Table 17.3 gives comparisons of these observations using the procedure that was used in Table 17.2. Again, the ppm values listed in column (5) use the total baseline lengths in the denominator, which are com-

TABLE 17.3 Comparison of Repeat Baseline Measurements

Component	First Observation	Second Observation	Difference (m)	ppm
ΔX_{AF}	1116.4577	−1116.4523	0.0054	0.84
ΔY_{AF}	4596.1553	−4596.1610	0.0057	0.88
ΔZ_{AF}	4355.9141	−4355.9062	0.0079	1.23
ΔX_{BF}	−6567.2310	6567.2311	0.0001	0.01
ΔY_{BF}	−5686.3033	5686.2926	0.0107	1.00
ΔZ_{BF}	−6322.3807	6322.3917	0.0110	1.02

puted from the square root of the sum of the squares of the measured baseline components.

The Federal Geodetic Control Subcommittee (FGCS) has developed the document *Geometric Geodetic Accuracy Standards and Specifications for Using GPS Relative Positioning Techniques.* It is intended to serve as a guideline for planning, executing, and classifying geodetic surveys performed by GPS relative positioning methods. This document may be consulted to determine whether or not the ppm values of column (5) are acceptable for the required order of accuracy for the survey. Besides ppm requirements, the FGCS guidelines specify other criteria that must be met for the various orders of accuracy in connection with repeat baseline observations. It is wise to perform repeat observations at the end of each day to check the repeatability of the software, hardware, and field procedures.

17.7.3 Analysis of Loop Closures

GPS networks will typically consist of many interconnected closed loops. For example, in the network of Figure 17.1, a closed loop is formed by points *ACBDEA*. Similarly, *ACFA, CFBC, BDFB,* and so on, are other closed loops. For each closed loop, the algebraic sum of the ΔX components should equal zero. The same condition should exist for the ΔY and ΔZ components. These loop misclosure conditions are very similar to the leveling loop misclosures imposed in differential leveling and latitude and departure misclosures imposed in closed-polygon traverses. An unusually large misclosure within any loop will indicate that either a blunder or a large random error exists in one (or more) of the baselines of the loop.

To compute loop misclosures, the baseline components are simply added algebraically for the loop chosen. For example, the misclosure in X components for loop *ACBDEA* would be computed as

$$cx = \Delta X_{AC} + \Delta X_{CB} + \Delta X_{BD} + \Delta X_{DE} + \Delta X_{EA} \qquad (17.13)$$

where cx is the loop misclosure in X coordinates. Similar equations apply for computing misclosures in Y and Z coordinates.

Substituting numerical values into Equation (17.13), the misclosure in X coordinates for loop *ACBDEA* is

$$cx = 11{,}644.2232 - 3960.5442 - 11{,}167.6076 - 1837.7459 + 5321.7164$$
$$= 0.0419 \text{ m}$$

Similarly, misclosures in Y and Z coordinates for that loop are

$$cy = 3601.2165 + 6681.2467 - 394.5204 - 6253.8534 - 3634.0754$$
$$= 0.0140 \text{ m}$$
$$cz = 3399.2550 + 7279.0148 - 907.9593 - 6596.6697 - 3173.6652$$
$$= -0.0244 \text{ m}$$

For evaluation purposes, loop misclosures are expressed in terms of the ratios of resultant misclosures to the total loop lengths. They are given in ppm. For any loop, the resultant misclosure is the square root of the sum of the squares of its cx, cy, and cz values, and for loop *ACBDEA* the resultant is 0.0505 m. The total length of a loop is computed by summing its legs, each leg being computed from the square root of the sum of the squares of its observed ΔX, ΔY, and ΔZ values. For loop *ACBDEA*, the total loop length is 50,967 m, and the misclosure ppm ratio is therefore (0.0505/50,967) × 1,000,000 = 0.99 ppm. Again, these ppm ratios can be compared against values given in the FGCS guidelines to determine if they are acceptable for the order of accuracy of the survey. As was the case with repeat baseline observations, the FGCS guidelines also specify other criteria that must be met in loop analyses besides the ppm values.

For any network, enough loop closures should be computed so that every baseline is included within at least one loop. This should expose any large blunders that exist. If a blunder does exist, its location can often be determined through additional loop-closure analyses. For example, assume that the misclosure of loop *ACDEA* discloses the presence of a blunder. By also computing the misclosures of loops *AFCA*, *CFDC*, *DFED*, and *EFAE*, the baseline containing the blunder can often be detected. In this example, if a large misclosure were found in loop DFED and all other loops appeared to be blunder free, the blunder would be in line *DE*.

A computer program included within the software package ADJUST will make these loop-closure computations. Documentation on the use of this program is given in its help file.

17.7.4 Minimally Constrained Adjustment

Prior to making the final adjustment of baseline observations in a network, a minimally constrained least squares adjustment is usually performed. In this adjustment, sometimes called a *free adjustment,* any station in the network may be held fixed with arbitrary coordinates. All other stations in the network are therefore free to adjust as necessary to accommodate the baseline observations and network geometry. The residuals that result from this adjustment are strictly related to the baseline observations and not to faulty control coordinates. These residuals are examined and, from them, blunders that may have gone undetected through the first set of analyses can be found and eliminated.

17.8 LEAST SQUARES ADJUSTMENT OF GPS NETWORKS

As noted earlier, because GPS networks contain redundant observations, they must be adjusted to make all coordinate differences consistent. In applying least squares to the problem of adjusting baselines in GPS networks, observation equations are written that relate station coordinates to the coordinate differences observed and their residual errors. To illustrate this procedure, consider the example of Figure 16.1. For line AC of this figure, an observation equation can be written for each baseline component observed as

$$X_C = X_A + \Delta X_{AC} + v_{X_{AC}}$$
$$Y_C = Y_A + \Delta Y_{AC} + v_{Y_{AC}} \quad (17.14)$$
$$Z_C = Z_A + \Delta Z_{AC} + v_{Z_{AC}}$$

Similarly, the observation equations for the baseline components of line CD are

$$X_D = X_C + \Delta X_{CD} + v_{X_{CD}}$$
$$Y_D = Y_C + \Delta Y_{CD} + v_{Y_{CD}} \quad (17.15)$$
$$Z_D = Z_C + \Delta Z_{CD} + v_{Z_{CD}}$$

Observation equations of the foregoing form would be written for all measured baselines in any figure. For Figure 17.1, a total of 13 baselines were observed, so the number of observation equations that can be developed is 39. Also, stations C, D, E, and F each have three unknown coordinates, for a total of 12 unknowns in the problem. Thus, there are $39 - 12 = 27$ redundant observations in the network. The 39 observation equations can be expressed in matrix form as

$$AX = L + V \quad (17.16)$$

If the observation equations for adjusting the network of Figure 17.1 are written in the order in which the observations are listed in Table 17.1, the A, X, L, and V matrices would be

328 ADJUSTMENT OF GPS NETWORKS

$$A = \begin{bmatrix} 1&0&0&0&0&0&0&0&0&0&0&0 \\ 0&1&0&0&0&0&0&0&0&0&0&0 \\ 0&0&1&0&0&0&0&0&0&0&0&0 \\ 0&0&0&1&0&0&0&0&0&0&0&0 \\ 0&0&0&0&1&0&0&0&0&0&0&0 \\ 0&0&0&0&0&1&0&0&0&0&0&0 \\ 1&0&0&0&0&0&0&0&0&0&0&0 \\ 0&1&0&0&0&0&0&0&0&0&0&0 \\ 0&0&1&0&0&0&0&0&0&0&0&0 \\ 0&0&0&0&0&0&1&0&0&0&0&0 \\ 0&0&0&0&0&0&0&1&0&0&0&0 \\ 0&0&0&0&0&0&0&0&1&0&0&0 \\ 1&0&0&0&0&0&-1&0&0&0&0&0 \\ 0&1&0&0&0&0&0&-1&0&0&0&0 \\ 0&0&1&0&0&0&0&0&-1&0&0&0 \\ 0&0&0&1&0&0&-1&0&0&0&0&0 \\ 0&0&0&0&1&0&0&-1&0&0&0&0 \\ 0&0&0&0&0&1&0&0&-1&0&0&0 \\ 0&0&0&0&0&0&0&0&0&-1&0&0 \\ 0&0&0&0&0&0&0&0&0&0&-1&0 \\ 0&0&0&0&0&0&0&0&0&0&0&-1 \\ 1&0&0&0&0&0&0&0&-1&0&0&0 \\ 0&1&0&0&0&0&0&0&0&-1&0&0 \\ 0&0&1&0&0&0&0&0&0&0&-1&0 \\ 0&0&0&1&0&0&0&0&-1&0&0&0 \\ 0&0&0&0&1&0&0&0&0&-1&0&0 \\ 0&0&0&0&0&1&0&0&0&0&-1&0 \\ &&&&&\vdots&&&&&& \\ 0&0&0&0&0&0&0&0&0&1&0&0 \\ 0&0&0&0&0&0&0&0&0&0&1&0 \\ 0&0&0&0&0&0&0&0&0&0&0&1 \end{bmatrix} \quad X = \begin{bmatrix} X_C \\ Y_C \\ Z_C \\ X_E \\ Y_E \\ Z_E \\ X_D \\ Y_D \\ Z_D \\ X_F \\ Y_F \\ Z_F \end{bmatrix} \quad L = \begin{bmatrix} 12046.5741 \\ -4649394.0846 \\ 4353160.0325 \\ -4919.3655 \\ -4649361.2257 \\ 4352934.4427 \\ 12046.5760 \\ -4649394.0941 \\ 4353160.0685 \\ -3081.5758 \\ -46443107.3678 \\ 4359531.1240 \\ 15128.1647 \\ -6286.7054 \\ -6371.0583 \\ -1837.7459 \\ -6253.8534 \\ -6596.6697 \\ -1518.8032 \\ 4648399.1401 \\ -4354116.6737 \\ 10527.7852 \\ -994.9377 \\ -956.6246 \\ -6438.1364 \\ -962.0694 \\ -1182.2305 \\ \vdots \\ 1518.8086 \\ -4648399.1458 \\ 4354116.6916 \end{bmatrix} \quad V = \begin{bmatrix} v_{X_{AC}} \\ v_{Y_{AC}} \\ v_{Z_{AC}} \\ v_{X_{AE}} \\ v_{Y_{AE}} \\ v_{Z_{AE}} \\ v_{X_{BC}} \\ v_{Y_{BC}} \\ v_{Z_{BC}} \\ v_{X_{BD}} \\ v_{Y_{BD}} \\ v_{Z_{BD}} \\ v_{X_{DC}} \\ v_{Y_{DC}} \\ v_{Z_{DC}} \\ v_{X_{DE}} \\ v_{Y_{DE}} \\ v_{Z_{DE}} \\ v_{X_{FA}} \\ v_{Y_{FA}} \\ v_{Z_{FA}} \\ v_{X_{FC}} \\ v_{Y_{FC}} \\ v_{Z_{FC}} \\ v_{X_{FE}} \\ v_{Y_{FE}} \\ v_{Z_{FE}} \\ \vdots \\ v_{X_{AF}} \\ v_{Y_{AF}} \\ v_{Z_{AF}} \end{bmatrix}$$

The numerical values of the elements of the L matrix are determined by rearranging the observation equations. Its first three elements are for the ΔX, ΔY, and ΔZ baseline components of line AC, respectively. Those elements are calculated as follows:

$$L_X = X_A + \Delta X_{AC}$$
$$L_Y = Y_A + \Delta Y_{AC} \tag{17.17}$$
$$L_Z = Z_A + \Delta Z_{AC}$$

The other elements of the L matrix are formed in the same manner as demonstrated for baseline AC. However, before numerical values for the L-matrix elements can be obtained, the X_e, Y_e, and Z_e geocentric coordinates of all control points in the network must be computed. This is done by following the procedures described in Section 17.5 and demonstrated by Example 17.1. That example problem provided the X_e, Y_e, and Z_e coordinates of control points A and B of Figure 17.1, which are used to compute the elements of the L matrix given above.

17.8 LEAST SQUARES ADJUSTMENT OF GPS NETWORKS

Note that the observation equations for GPS network adjustment are linear and that the only nonzero elements of the A matrix are either 1 and -1. This is the same type of matrix that was developed in adjusting level nets by least squares. In fact, GPS network adjustments are performed in the very same manner as level net adjustments, with the exception of the weights. In GPS relative positioning, the three observed baseline components are correlated. Therefore, a covariance matrix of dimensions 3×3 is derived for each baseline as a product of the least squares adjustment of the carrier-phase measurements. This covariance matrix is used to weight the observations in the network adjustment in accordance with Equation (10.4). The weight matrix for any GPS network is therefore a block-diagonal type, with an individual 3×3 matrix for each baseline observed on the diagonal. When more than two receivers are used, additional 3×3 matrices are created in the off-diagonal region of the matrix to provide the correlation that exists between baselines observed simultaneously. All other elements of the matrix are zeros.

The covariances for the observations in Table 17.1 are given in columns (6) through (11). Only the six upper-triangular elements of the 3×3 covariance matrix for each observation are listed. This gives complete weighting information, however, because the covariance matrix is symmetrical. Columns 6 through 11 list σ_x^2, σ_{xy}, σ_{xz}, σ_y^2, σ_{y2}, and σ_z^2, respectively. Thus, the full 3×3 covariance matrix for baseline AC is

$$\Sigma_{AC} = \begin{bmatrix} 9.884 \times 10^{-4} & -9.580 \times 10^{-6} & 9.520 \times 10^{-6} \\ -9.580 \times 10^{-4} & 9.377 \times 10^{-6} & -9.520 \times 10^{-6} \\ 9.520 \times 10^{-4} & -9.520 \times 10^{-6} & 9.827 \times 10^{-6} \end{bmatrix}$$

The complete weight matrix for the example network of Figure 17.1 has dimensions of 39×39. After inverting the full matrix and multiplying by the a priori estimate for the reference variance, S_0^2, in accordance with Equation (10.4), the weight matrix for the network of Figure 17.1 is (note that S_0^2 is taken as 1.0 for this computation and that no correlation between baselines is included):

$$W = \begin{bmatrix}
1011.8 & 10.2 & -9.7 & 0 & 0 & 0 & 0 & 0 & 0 & 0 & 0 & 0 \\
10.2 & 1066.6 & 10.2 & 0 & 0 & 0 & 0 & 0 & 0 & 0 & 0 & 0 \\
-9.7 & 10.2 & 1017.7 & 0 & 0 & 0 & 0 & 0 & 0 & 0 & 0 & 0 \\
0 & 0 & 0 & 4634.5 & 50.2 & -49.4 & 0 & 0 & 0 & 0 & 0 & 0 \\
0 & 0 & 0 & 50.2 & 5209.7 & 54.0 & 0 & 0 & 0 & 0 & 0 & 0 \\
0 & 0 & 0 & -49.4 & 54.0 & 4988.1 & 0 & 0 & 0 & 0 & 0 & 0 \\
0 & 0 & 0 & 0 & 0 & 0 & 4339.1 & 37.7 & -39.5 & 0 & 0 & 0 \\
0 & 0 & 0 & 0 & 0 & 0 & 37.7 & 3927.8 & 38.5 & 0 & 0 & 0 \\
0 & 0 & 0 & 0 & 0 & 0 & -39.5 & 38.5 & 4441.0 & 0 & 0 & 0 \\
 & & & & & & & & \ddots & & & \\
0 & 0 & 0 & 0 & 0 & 0 & 0 & 0 & 0 & 1511.8 & 147.7 & -143.8 \\
0 & 0 & 0 & 0 & 0 & 0 & 0 & 0 & 0 & 147.7 & 12336.0 & 106.5 \\
0 & 0 & 0 & 0 & 0 & 0 & 0 & 0 & 0 & -143.8 & 106.5 & 10667.8
\end{bmatrix}$$

330 ADJUSTMENT OF GPS NETWORKS

The system of observation equations (17.8) is solved by least squares using Equation (11.35). This yields the most probable values for the coordinates of the unknown stations. The complete output for the example of Figure 17.1 obtained using the program ADJUST follows.

```
*****************
Control stations
*****************
Station             X                  Y                    Z
==========================================================
   A           402.35087       -4652995.30109       4349760.77753
   B          8086.03178       -4642712.84739       4360439.08326


*****************
Distance Vectors
*****************
From To      ΔX           ΔY           ΔZ       Covariance matrix elements
=================================================================================
 A   C   11644.2232    3601.2165    3399.2550  9.884E-4 -9.580E-6 9.520E-6 9.377E-4 -9.520E-6 9.827E-4
 A   E   -5321.7164    3634.0754    3173.6652  2.158E-4 -2.100E-6 2.160E-6 1.919E-4 -2.100E-6 2.005E-4
 B   C    3960.5442   -6681.2467   -7279.0148  2.305E-4 -2.230E-6 2.070E-6 2.546E-4 -2.230E-6 2.252E-4
 B   D  -11167.6076    -394.5204    -907.9593  2.700E-4 -2.750E-6 2.850E-6 2.721E-4 -2.720E-6 2.670E-4
 D   C   15128.1647   -6286.7054   -6371.0583  1.461E-4 -1.430E-6 1.340E-6 1.614E-4 -1.440E-6 1.308E-4
 D   E   -1837.7459   -6253.8534   -6596.6697  1.231E-4 -1.190E-6 1.220E-6 1.277E-4 -1.210E-6 1.283E-4
 F   A   -1116.4523   -4596.1610   -4355.8962  7.475E-5 -7.900E-7 8.800E-7 6.593E-5 -8.100E-7 7.616E-5
 F   C   10527.7852    -994.9377    -956.6246  2.567E-4 -2.250E-6 2.400E-6 2.163E-4 -2.270E-6 2.397E-4
 F   E   -6438.1364    -962.0694   -1182.2305  9.442E-5 -9.200E-7 1.040E-6 9.959E-5 -8.900E-7 8.826E-5
 F   D   -4600.3787    5291.7785    5414.4311  9.330E-5 -9.900E-7 9.000E-7 9.875E-5 -9.900E-7 1.204E-4
 F   B    6567.2311    5686.2926    6322.3917  6.643E-5 -6.500E-7 6.900E-7 7.465E-5 -6.400E-7 6.048E-5
 B   F   -6567.2310   -5686.3033   -6322.3807  5.512E-5 -6.300E-7 6.100E-7 7.472E-5 -6.300E-7 6.629E-5
 A   F    1116.4577    4596.1553    4355.9141  6.619E-5 -8.000E-7 9.000E-7 8.108E-5 -8.200E-7 9.376E-5


Normal Matrix
==========================================================
 16093.0    148.0   -157.3        0        0        0   -6845.9     -60.0      69.4  -3896.2     -40.1      38.6
   148.0  15811.5    159.7        0        0        0     -60.0   -6195.0     -67.6    -40.1   -4622.2     -43.4
  -157.3    159.7  17273.4        0        0        0      69.4    -67.6   -7643.2     38.6     -43.4   -4171.5
       0        0        0  23352.1    221.9   -249.8   -8124.3     -75.0      76.5 -10593.3     -96.8     123.8
       0        0        0    221.9  23084.9    227.3     -75.0   -7832.2     -73.2    -96.8  -10043.0    -100.1
       0        0        0   -249.8    227.3  24116.4      76.5    -73.2   -7795.7    123.8    -100.1  -11332.6
 -6845.9    -60.0     69.4  -8124.3    -75.0     76.5   29393.8     278.7   -264.4 -10720.0    -106.7      79.2
   -60.0  -6195.0    -67.6    -75.0  -7832.2    -73.2     278.7   27831.6     260.2   -106.7  -10128.5     -82.5
    69.4   -67.6  -7643.2     76.5    -73.2   -7795.7    -264.4     260.2   27487.5     79.2     -82.5   -8303.5
 -3896.2    -40.1     38.6 -10593.3    -96.8    123.8  -10720.0   -106.7      79.2  86904.9     830.9    -874.3
   -40.1  -4622.2    -43.4    -96.8 -10043.0   -100.1    -106.7  -10128.5     -82.5    830.9   79084.9     758.1
    38.6   -43.4  -4171.5    123.8   -100.1 -11332.6      79.2     -82.5   -8303.5   -874.3     758.1   79234.9


Constant Matrix
==========================================================
      -227790228.2336
    -23050461170.3104
     23480815458.7631
       -554038059.5699
    -24047087640.5196
```

17.8 LEAST SQUARES ADJUSTMENT OF GPS NETWORKS 331

```
   21397654262.6187
     -491968929.7795
  -16764436256.9406
   16302821193.7660
    -5314817963.4907
 -250088821081.7488
  238833986695.9468
```

X Matrix
=============
```
    12046.5808
 -4649394.0826
  4353160.0634
    -4919.3391
 -4649361.2199
  4352934.4534
    -3081.5831
 -4643107.3692
  4359531.1220
     1518.8012
 -4648399.1453
  4354116.6894
```

```
        Degrees of Freedom = 27
        Reference Variance = 0.6135
              Reference So = ±0.78
```

```
**********************
Adjusted Distance Vectors
**********************
```

From	To	ΔX	ΔY	ΔZ	Vx	Vy	Vz
A	C	11644.2232	3601.2165	3399.2550	0.00669	0.00203	0.03082
A	E	-5321.7164	3634.0754	3173.6652	0.02645	0.00582	0.01068
B	C	3960.5442	-6681.2467	-7279.0148	0.00478	0.01153	-0.00511
B	D	-11167.6076	-394.5204	-907.9593	-0.00731	-0.00136	-0.00194
D	C	15128.1647	-6286.7054	-6371.0583	-0.00081	-0.00801	-0.00037
D	E	-1837.7459	-6253.8534	-6596.6697	-0.01005	0.00268	0.00109
F	A	-1116.4523	-4596.1610	-4355.8962	0.00198	0.00524	-0.01563
F	C	10527.7852	-994.9377	-956.6246	-0.00563	0.00047	-0.00140
F	E	-6438.1364	-962.0694	-1182.2305	-0.00387	-0.00514	-0.00545
F	D	-4600.3787	5291.7785	5414.4311	-0.00561	-0.00232	0.00156
F	B	6567.2311	5686.2926	6322.3917	-0.00051	0.00534	0.00220
B	F	-6567.2310	-5686.3033	-6322.3807	0.00041	0.00536	-0.01320
A	F	1116.4577	4596.1553	4355.9141	-0.00738	0.00046	-0.00227

Advanced Statistical Values

From	To	±S	Slope Dist	Prec
A	C	0.0116	12,653.537	1,089,000
A	E	0.0100	7,183.255	717,000
B	C	0.0116	10,644.669	916,000
B	D	0.0097	11,211.408	1,158,000
D	C	0.0118	17,577.670	1,484,000
D	E	0.0107	9,273.836	868,000
F	A	0.0053	6,430.014	1,214,000
F	C	0.0115	10,617.871	921,000
F	E	0.0095	6,616.111	696,000
F	D	0.0092	8,859.036	964,000
F	B	0.0053	10,744.076	2,029,000
B	F	0.0053	10,744.076	2,029,000
A	F	0.0053	6,430.014	1,214,000

Adjusted Coordinates

Station	X	Y	Z	Sx	Sy	Sz
A	402.35087	-4,652,995.30109	4,349,760.77753			
B	8,086.03178	-4,642,712.84739	4,360,439.08326			
C	12,046.58076	-4,649,394.08256	4,353,160.06335	0.0067	0.0068	0.0066
E	-4,919.33908	-4,649,361.21987	4,352,934.45341	0.0058	0.0058	0.0057
D	-3,081.58313	-4,643,107.36915	4,359,531.12202	0.0055	0.0056	0.0057
F	1,518.80119	-4,648,399.14533	4,354,116.68936	0.0030	0.0031	0.0031

PROBLEMS

Note: For problems requiring least squares adjustment, if a computer program is not distinctly specified for use in the problem, it is expected that the least squares algorithm will be solved using the program MATRIX, which is included on the CD supplied with the book.

17.1 Using the WGS 84 ellipsoid parameters, convert the following geodetic coordinates to geocentric coordinates for these points.
  (a) latitude: 40°59′16.2541″ N
     longitude: 75°59′57.0024″ W
     height: 164.248 m

(b) latitude: 41°15'53.0534" N
longitude: 90°02'36.7203" W
height: 229.085 m

(c) latitude: 44°57'45.3603" N
longitude: 66°12'56.2437" W
height: 254.362 m

(d) latitude: 33°58'06.8409" N
longitude: 122°27'42.0462" W
height: 364.248 m

17.2 Using the WGS 84 ellipsoid parameters, convert the following geocentric coordinates (in meters) to geodetic coordinates for these points.

(a) $X = -426,125.836 \quad Y = -5,472,467.695 \quad Z = 3,237,961.360$
(b) $X = -2,623,877.827 \quad Y = -3,664,128.366 \quad Z = 4,498,233.251$
(c) $X = -11,190.917 \quad Y = -4,469,623.638 \quad Z = 4,534,918.934$
(d) $X = 2,051,484.893 \quad Y = -5,188,627.173 \quad Z = 3,080,194.963$

17.3 Given the following GPS observations and geocentric control station coordinates to Figure P17.3, what are the most probable coordinates for stations B and C using a weighted least squares adjustment? (All data were collected with only two receivers.)

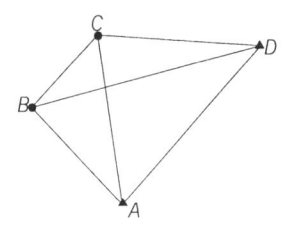

Figure P17.3

Control stations

Station	X (m)	Y (m)	Z (m)
D	1,177,425.88739	−4,674,386.55849	4,162,989.78649
A	1,178,680.69374	−4,673,056.15318	4,164,169.65655

The vector covariance matrices for the ΔX, ΔY, and ΔZ values (in meters) given are as follows. For baseline AB:

$\Delta X = -825.5585 \quad 0.00002199 \quad 0.00000030 \quad 0.00000030$

$\Delta Y = 492.7369 \quad \quad \quad \quad \quad 0.00002806 \quad -0.00000030$

$\Delta Z = 788.9732 \quad \quad \quad \quad \quad \quad \quad \quad \quad 0.00003640$

For baseline *BC*:

$\Delta X = 606.2113$ 0.00003096   −0.00000029   0.00000040
$\Delta Y = 558.8905$ 0.00002709   −0.00000029
$\Delta Z = 546.7241$ 0.00002591

For baseline *CD*:

$\Delta X = 1474.1569$ 0.00004127   −0.00000045   0.00000053
$\Delta Y = 278.7786$ 0.00004315   −0.00000045
$\Delta Z = -155.8336$ 0.00005811

For baseline *AC*:

$\Delta X = -219.3510$ 0.00002440   −0.00000020   0.00000019
$\Delta Y = 1051.6348$ 0.00001700   −0.00000019
$\Delta Z = 1335.6877$ 0.00002352

For baseline *DB*:

$\Delta X = -2080.3644$ 0.00003589   −0.00000034   0.00000036
$\Delta Y = -837.6605$ 0.00002658   −0.00000033
$\Delta Z = -390.9075$ 0.00002982

17.4 Given the following GPS observations and geocentric control station coordinates to accompany Figure P17.4, what are the most probable coordinates for stations *B* and *C* using a weighted least squares adjustment? (All data were collected with only two receivers.)

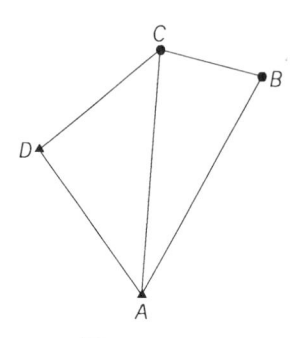

Figure P17.4

Control stations

Station	X (m)	Y (m)	Z (m)
A	593,898.8877	−4,856,214.5456	4,078,710.7059
D	593,319.2704	−4,855,416.0310	4,079,738.3059

The vector covariance matrices for the ΔX, ΔY, and ΔZ values (in meters) given are as follows. For baseline AB:

$\Delta X = 678.034 \quad 5.098\text{E-}6 \quad -1.400\text{E-}5 \quad 6.928\text{E-}6$

$\Delta Y = 1206.714 \qquad\qquad\quad 7.440\text{E-}5 \quad -3.445\text{E-}5$

$\Delta Z = 1325.735 \qquad\qquad\qquad\qquad\qquad\quad 2.018\text{E-}5$

For baseline BC:

$\Delta X = -579.895 \quad 3.404\text{-E6} \quad 2.057\text{E-}6 \quad -3.036\text{E-}7$

$\Delta Y = 145.342 \qquad\qquad\quad 2.015\text{E-}5 \quad -1.147\text{E-}5$

$\Delta Z = 254.820 \qquad\qquad\qquad\qquad\qquad\quad 1.873\text{E-}5$

For baseline AC:

$\Delta X = 98.138 \quad 6.518\text{E-}6 \quad -1.163\text{E-}7 \quad -3.811\text{E-}7$

$\Delta Y = 1352.039 \qquad\qquad\quad 3.844\text{E-}5 \quad -1.297\text{E-}5$

$\Delta Z = 1580.564 \qquad\qquad\qquad\qquad\qquad\quad 2.925\text{E-}5$

For baseline DC:

$\Delta X = 677.758 \quad 9.347\text{E-}6 \quad -1.427\text{E-}5 \quad 8.776\text{E-}6$

$\Delta Y = 553.527 \qquad\qquad\quad 2.954\text{E-}5 \quad -1.853\text{E-}5$

$\Delta Z = 552.978 \qquad\qquad\qquad\qquad\qquad\quad 1.470\text{E-}5$

For baseline DC:

$\Delta X = 677.756 \quad 9.170\text{E-}6 \quad -1.415\text{E-}5 \quad 8.570\text{E-}6$

$\Delta Y = 553.533 \qquad\qquad\quad 3.010\text{E-}5 \quad -1.862\text{E-}5$

$\Delta Z = 552.975 \qquad\qquad\qquad\qquad\qquad\quad 1.460\text{E-}5$

17.5 Given the following GPS observations and geocentric control station coordinates to accompany Figure P17.5, what are the most probable coordinates for station E using a weighted least squares adjustment? (All data were collected with only two receivers.)

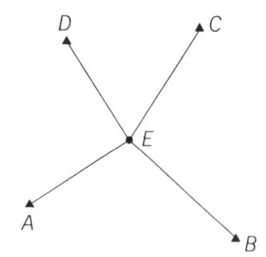

Figure P17.5

Control stations

Station	X (m)	Y (m)	Z (m)
A	−1,683,429.825	−4,369,532.522	4,390,283.745
B	−1,524,701.610	−4,230,122.822	4,511,075.501
C	−1,480,308.035	−4,472,815.181	4,287,476.008
D	−1,725,386.928	−4,436,015.964	4,234,036.124

The vector covariance matrices for the ΔX, ΔY, and ΔZ values (in meters) given are as follows. For baseline AE:

$\Delta X = $ 94,208.555 0.00001287 −0.00000016 −0.00000019

$\Delta Y = -61,902.843$ 0.00001621 −0.00000016

$\Delta Z = -24,740.272$ 0.00001538

For baseline BE:

$\Delta X = -64,519.667$ 0.00003017 −0.00000026 0.00000021

$\Delta Y = -77,506.853$ 0.00002834 −0.00000025

$\Delta Z = -96,051.488$ 0.00002561

For baseline CE:

$\Delta X = -108,913.237$ 0.00008656 −0.00000081 −0.00000087

$\Delta Y = $ 165,185.492   0.00007882 −0.00000080

$\Delta Z = $ 127,548.005   0.00008647

For baseline DE:

$$\Delta X = 136{,}165.650 \quad 0.00005893 \quad -0.00000066 \quad -0.00000059$$
$$\Delta Y = 128{,}386.277 \quad\quad\quad\quad\quad\quad 0.00006707 \quad -0.00000064$$
$$\Delta Z = 180{,}987.895 \quad\quad\quad\quad\quad\quad\quad\quad\quad\quad\quad 0.00005225$$

For baseline EA:

$$\Delta X = -94{,}208.554 \quad 0.00002284 \quad 0.00000036 \quad -0.00000042$$
$$\Delta Y = -61{,}902.851 \quad\quad\quad\quad\quad\quad 0.00003826 \quad -0.00000035$$
$$\Delta Z = -24{,}740.277 \quad\quad\quad\quad\quad\quad\quad\quad\quad\quad\quad 0.00003227$$

For baseline EB:

$$\Delta X = 64{,}519.650 \quad 0.00008244 \quad 0.00000081 \quad -0.00000077$$
$$\Delta Y = 77{,}506.866 \quad\quad\quad\quad\quad\quad 0.00007737 \quad -0.00000081$$
$$\Delta Z = 96{,}051.486 \quad\quad\quad\quad\quad\quad\quad\quad\quad\quad\quad 0.00008483$$

For baseline EC:

$$\Delta X = 108{,}913.236 \quad 0.00002784 \quad -0.00000036 \quad 0.00000038$$
$$\Delta Y = -165{,}185.494 \quad\quad\quad\quad\quad\quad 0.00003396 \quad -0.00000035$$
$$\Delta Z = -127{,}547.991 \quad\quad\quad\quad\quad\quad\quad\quad\quad\quad\quad 0.00002621$$

For baseline ED:

$$\Delta X = -136{,}165.658 \quad 0.00003024 \quad -0.00000037 \quad 0.00000031$$
$$\Delta Y = -128{,}386.282 \quad\quad\quad\quad\quad\quad 0.00003940 \quad -0.00000036$$
$$\Delta Z = -180{,}987.888 \quad\quad\quad\quad\quad\quad\quad\quad\quad\quad\quad 0.00003904$$

17.6 Given the following GPS observations and geocentric control station coordinates to accompany Figure P17.6, what are the most probable coordinates for stations B, D, and E using a weighted least squares adjustment? (All data were collected with only two receivers.)

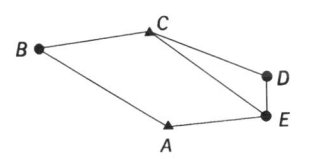

Figure P17.6

338 ADJUSTMENT OF GPS NETWORKS

Control stations

Station	X (m)	Y (m)	Z (m)
A	−1,439,383.018	−5,325,949.910	3,190,645.563
C	−1,454,936.177	−5,240,453.494	3,321,529.500

The vector covariance matrices for the ΔX, ΔY, and ΔZ values (in meters) given are as follows. For baseline AB:

$\Delta X = -118{,}616.114 \quad$ 8.145E-4 $\quad$ −7.870E-6 $\quad$ 7.810E-6

$\Delta Y = 71{,}775.010 \quad\quad\quad\quad\quad$ 7.685E-4 $\quad$ −7.820E-6

$\Delta Z = 62{,}170.130 \quad\quad\quad\quad\quad\quad\quad\quad\quad$ 8.093E-4

For baseline BC:

$\Delta X = 103{,}062.915 \quad$ 8.521E-4 $\quad$ −8.410E-6 $\quad$ 8.520E-6

$\Delta Y = 13{,}721.432 \quad\quad\quad\quad\quad$ 8.040E-4 $\quad$ −8.400E-6

$\Delta Z = 68{,}713.770 \quad\quad\quad\quad\quad\quad\quad\quad\quad$ 8.214E-4

For baseline CD:

$\Delta X = 106{,}488.952 \quad$ 7.998E-4 $\quad$ −7.850E-6 $\quad$ 7.560E-6

$\Delta Y = -41{,}961.364 \quad\quad\quad\quad\quad$ 8.443E-4 $\quad$ −7.860E-6

$\Delta Z = -21{,}442.604 \quad\quad\quad\quad\quad\quad\quad\quad\quad$ 7.900E-4

For baseline DE:

$\Delta X = \phantom{-00{,}00}-7.715 \quad$ 3.547E-4 $\quad$ −3.600E-6 $\quad$ 3.720E-6

$\Delta Y = -35{,}616.922 \quad\quad\quad\quad\quad$ 3.570E-4 $\quad$ −3.570E-6

$\Delta Z = -57{,}297.941 \quad\quad\quad\quad\quad\quad\quad\quad\quad$ 3.512E-4

For baseline EA:

$\Delta X = -90{,}928.118 \quad$ 8.460E-4 $\quad$ −8.380E-6 $\quad$ 8.160E-6

$\Delta Y = -7{,}918.120 \quad\quad\quad\quad\quad$ 8.824E-4 $\quad$ −8.420E-6

$\Delta Z = -52{,}143.439 \quad\quad\quad\quad\quad\quad\quad\quad\quad$ 8.088E-4

For baseline CE:

$$\Delta X = 106{,}481.283 \quad 7.341\text{E-}4 \quad -7.250\text{E-}6 \quad 7.320\text{E-}6$$
$$\Delta Y = -77{,}578.306 \quad\quad\quad\quad\quad 7.453\text{E-}4 \quad -7.290\text{E-}6$$
$$\Delta Z = -78{,}740.573 \quad\quad\quad\quad\quad\quad\quad\quad\quad 7.467\text{E-}4$$

17.7 Given the following GPS observations and geocentric control station coordinates to accompany Figure P17.7, what are the most probable coordinates for stations B, D, E, and F using a weighted least squares adjustment? (All data were collected with only two receivers.)

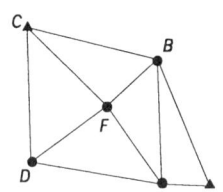

Figure P17.7

Control stations

Station	X (m)	Y (m)	Z (m)
A	−1,612,062.639	−4,384,804.866	4,330,846.142
B	−1,613,505.053	−4,383,572.785	4,331,494.264

The vector covariance matrices for the ΔX, ΔY, and ΔZ values (in meters) given are as follows. For baseline AB:

$$\Delta X = -410.891 \quad 7.064\text{E-}5 \quad -6.500\text{E-}7 \quad 6.400\text{E-}7$$
$$\Delta Y = 979.896 \quad\quad\quad\quad\quad 6.389\text{E-}5 \quad -6.400\text{E-}7$$
$$\Delta Z = 915.452 \quad\quad\quad\quad\quad\quad\quad\quad\quad 6.209\text{E-}5$$

For baseline BC:

$$\Delta X = -1031.538 \quad 1.287\text{E-}5 \quad -1.600\text{E-}7 \quad 1.900\text{E-}7$$
$$\Delta Y = 252.184 \quad\quad\quad\quad\quad 1.621\text{E-}5 \quad -1.600\text{E-}7$$
$$\Delta Z = -267.337 \quad\quad\quad\quad\quad\quad\quad\quad\quad 1.538\text{E-}5$$

For baseline CD:

$$\Delta X = 23.227 \quad 1.220\text{E-}5 \quad -9.000\text{E-}8 \quad 7.000\text{E-}8$$
$$\Delta Y = -1035.622 \quad\quad\quad\quad 1.104\text{E-}5 \quad -9.000\text{E-}8$$
$$\Delta Z = -722.122 \quad\quad\quad\quad\quad\quad\quad\quad\quad 9.370\text{E-}6$$

340 ADJUSTMENT OF GPS NETWORKS

For baseline *DE*:

$\Delta X = 1039.772 \quad 5.335\text{E-}5 \quad -4.900\text{E-}7 \quad 5.400\text{E-}7$
$\Delta Y = -178.623 \quad\quad\quad\quad\quad 4.731\text{E-}5 \quad -4.800\text{E-}7$
$\Delta Z = \quad -3.753 \quad\quad\quad\quad\quad\quad\quad\quad\quad\quad 5.328\text{E-}5$

For baseline *EF*:

$\Delta X = -434.125 \quad 7.528\text{E-}5 \quad -8.300\text{E-}7 \quad 7.500\text{E-}7$
$\Delta Y = \quad 603.788 \quad\quad\quad\quad\quad 8.445\text{E-}5 \quad -8.100\text{E-}7$
$\Delta Z = \quad 566.518 \quad\quad\quad\quad\quad\quad\quad\quad\quad\quad 6.771\text{E-}5$

For baseline *EB*:

$\Delta X = -31.465 \quad 3.340\text{E-}5 \quad -4.900\text{E-}7 \quad 5.600\text{E-}7$
$\Delta Y = \quad 962.058 \quad\quad\quad\quad\quad 5.163\text{E-}5 \quad -4.800\text{E-}7$
$\Delta Z = \quad 993.212 \quad\quad\quad\quad\quad\quad\quad\quad\quad\quad 4.463\text{E-}5$

For baseline *FA*:

$\Delta X = 1845.068 \quad 9.490\text{E-}6 \quad -9.000\text{E-}8 \quad 8.000\text{E-}8$
$\Delta Y = -873.794 \quad\quad\quad\quad\quad 7.820\text{E-}6 \quad -9.000\text{E-}8$
$\Delta Z = -221.422 \quad\quad\quad\quad\quad\quad\quad\quad\quad\quad 1.031\text{E-}5$

For baseline *FB*:

$\Delta X = 402.650 \quad 1.073\text{E-}5 \quad -1.600\text{E-}7 \quad 1.800\text{E-}7$
$\Delta Y = 358.278 \quad\quad\quad\quad\quad 1.465\text{E-}5 \quad -1.600\text{E-}7$
$\Delta Z = 426.706 \quad\quad\quad\quad\quad\quad\quad\quad\quad\quad 9.730\text{E-}6$

For baseline *FC*:

$\Delta X = -628.888 \quad 5.624\text{E-}5 \quad -6.600\text{E-}7 \quad 5.700\text{E-}7$
$\Delta Y = \quad 610.467 \quad\quad\quad\quad\quad 6.850\text{E-}5 \quad -6.300\text{E-}7$
$\Delta Z = \quad 159.360 \quad\quad\quad\quad\quad\quad\quad\quad\quad\quad 6.803\text{E-}5$

For baseline *FD*:

$$\Delta X = -605.648 \quad 8.914\text{E-}5 \quad -8.100\text{E-}7 \quad 8.200\text{E-}7$$
$$\Delta Y = -425.139 \quad\quad\quad\quad\quad\quad 8.164\text{E-}5 \quad -8.100\text{E-}7$$
$$\Delta Z = -562.763 \quad\quad\quad\quad\quad\quad\quad\quad\quad\quad\quad 7.680\text{E-}5$$

17.8 Given the following GPS observations and geocentric control station coordinates to accompany Figure P17.8, what are the most probable coordinates for stations B, D, E, and F using a weighted least squares adjustment? (All data were collected with only two receivers.)

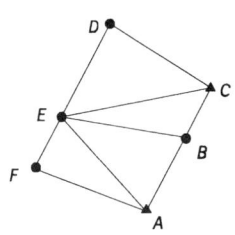

Figure P17.8

Control stations

Station	X (m)	Y (m)	Z (m)
A	−2,413,963.823	−4,395,420.994	3,930,059.456
C	−2,413,073.302	−4,393,796.994	3,932,699.132

The vector covariance matrices for the ΔX, ΔY, and ΔZ values (in meters) given are as follows. For baseline AB:

$$\Delta X = 535.100 \quad 4.950\text{E-}6 \quad -9.000\text{E-}8 \quad 7.000\text{E-}8$$
$$\Delta Y = 974.318 \quad\quad\quad\quad\quad\quad 7.690\text{E-}6 \quad -9.000\text{E-}8$$
$$\Delta Z = 1173.264 \quad\quad\quad\quad\quad\quad\quad\quad\quad\quad 8.090\text{E-}6$$

For baseline BC:

$$\Delta X = 355.412 \quad 5.885\text{E-}5 \quad -6.300\text{E-}7 \quad 7.400\text{E-}7$$
$$\Delta Y = 649.680 \quad\quad\quad\quad\quad\quad 7.168\text{E-}5 \quad -6.500\text{E-}7$$
$$\Delta Z = 1466.409 \quad\quad\quad\quad\quad\quad\quad\quad\quad\quad 6.650\text{E-}5$$

For baseline CD:

$\Delta X = -1368.545 \quad 6.640\text{E-}6 \quad -4.000\text{E-}8 \quad 7.000\text{E-}8$
$\Delta Y = 854.284 \phantom{\quad 6.640\text{E-}6} \quad 4.310\text{E-}6 \quad -4.000\text{E-}8$
$\Delta Z = -71.080 \phantom{\quad 6.640\text{E-}6 \quad -4.000\text{E-}8} \quad 2.740\text{E-}6$

For baseline *DE*:

$\Delta X = -671.715 \quad 1.997\text{E-}5 \quad -2.500\text{E-}7 \quad 2.500\text{E-}7$
$\Delta Y = -1220.263 \phantom{\quad 1.997\text{E-}5} \quad 2.171\text{E-}5 \quad -2.400\text{E-}7$
$\Delta Z = -951.343 \phantom{\quad 1.997\text{E-}5 \quad -2.500\text{E-}7} \quad 3.081\text{E-}5$

For baseline *EF*:

$\Delta X = -374.515 \quad 4.876\text{E-}5 \quad -3.600\text{E-}7 \quad 3.400\text{E-}7$
$\Delta Y = -679.553 \phantom{\quad 4.876\text{E-}5} \quad 2.710\text{E-}5 \quad -3.700\text{E-}7$
$\Delta Z = -1439.338 \phantom{\quad 4.876\text{E-}5 \quad -3.600\text{E-}7} \quad 3.806\text{E-}5$

For baseline *EA*:

$\Delta X = 1149.724 \quad 8.840\text{E-}5 \quad -8.000\text{E}7 \quad 8.300\text{E-}7$
$\Delta Y = -1258.018 \phantom{\quad 8.840\text{E-}5} \quad 7.925\text{E-}5 \quad -8.200\text{E-}7$
$\Delta Z = -1617.250 \phantom{\quad 8.840\text{E-}5 \quad -8.000\text{E}7} \quad 6.486\text{E-}5$

For baseline *EB*:

$\Delta X = 1684.833 \quad 1.861\text{E-}5 \quad -1.600\text{E-}7 \quad 2.000\text{E-}7$
$\Delta Y = -283.698 \phantom{\quad 1.861\text{E-}5} \quad 1.695\text{E-}5 \quad -1.600\text{E-}7$
$\Delta Z = -443.990 \phantom{\quad 1.861\text{E-}5 \quad -1.600\text{E-}7} \quad 1.048\text{E-}5$

For baseline *EC*:

$\Delta X = 2040.254 \quad 6.966\text{E-}5 \quad -6.400\text{E-}7 \quad 7.300\text{E-}7$
$\Delta Y = 365.991 \phantom{\quad 6.966\text{E-}5} \quad 5.665\text{E-}5 \quad -6.300\text{E-}7$
$\Delta Z = 1022.430 \phantom{\quad 6.966\text{E-}5 \quad -6.400\text{E-}7} \quad 7.158\text{E-}5$

For baseline *FA*:

$\Delta X = 1524.252 \quad 2.948\text{E-}5 \quad -3.500\text{E-}7 \quad 3.300\text{E-}7$
$\Delta Y = -578.473 \phantom{\quad 2.948\text{E-}5} \quad 3.380\text{E-}5 \quad -3.500\text{E-}7$
$\Delta Z = -177.914 \phantom{\quad 2.948\text{E-}5 \quad -3.500\text{E-}7} \quad 4.048\text{E-}5$

Given the data in each problem and using the procedure discussed in Section 17.7.2, analyze the repeated baselines.

17.9 Problem 17.4.

17.10 Problem 17.5.

Given the data in each problem and using the procedures discussed in Section 17.7.3, analyze the closures of the loops.

17.11 Problem 17.3, loops *ABCDA, ABCA, ACDA,* and *BCDB*

17.12 Problem 17.4, loops *ABCDA, ACBA,* and *ADCA*

17.13 Problem 17.6, loops *ABCDEA* and *ABCEA*

17.14 Problem 17.7, loops *ABEA, DEFD, BFCB, CDFC,* and *ABCDEA*

17.15 Problem 17.8, loops *ABFA, BFEB,* and *BCDEB*

Use program ADJUST to do each problem.

17.16 Problem 17.6

17.17 Problem 17.7

17.18 Problem 17.8

17.19 Problem 17.14

17.20 Problem 17.15

Programming Problems

17.21 Write a computational package that reads a file of station coordinates and GPS baselines and then
  (a) writes the data to a file in a formatted fashion.
  (b) computes the A, L, and W matrices.
  (c) writes the matrices to a file that is compatible with the MATRIX program.
  (d) Demonstrate this program with Problem 17.8.

17.22 Write a computational package that reads a file containing the A, L, and W matrices and then:
  (a) writes these matrices in a formatted fashion.
  (b) performs a weighted least squares adjustment.
  (c) writes the matrices used to compute the solution and tabulates the station coordinates in a formatted fashion.
  (d) Demonstrate this program with Problem 17.8.

17.23 Write a computational package that reads a file of station coordinates and GPS baselines and then:

(a) writes the data to a file in a formatted fashion.
(b) computes the A, L, and W matrices.
(c) performs a weighted least squares adjustment.
(d) writes the matrices used in computations in a formatted fashion to a file.
(e) computes the final station coordinates, their estimated errors, the adjusted baseline vectors, their residuals, and their estimated errors, and writes them to a file in a formatted fashion.
(f) Demonstrate this program with Problem 17.8.

CHAPTER 18

COORDINATE TRANSFORMATIONS

18.1 INTRODUCTION

The transformation of points from one coordinate system to another is a common problem encountered in surveying and mapping. For instance, a surveyor who works initially in an assumed coordinate system on a project may find it necessary to transfer the coordinates to the state plane coordinate system. In GPS surveying and in the field of photogrammetry, coordinate transformations are used extensively. Since the inception of the North American Datum of 1983 (NAD 83), many land surveyors, management agencies, state departments of transportation, and others have been struggling with the problem of converting their multitudes of stations defined in the 1927 datum (NAD 27) to the 1983 datum. Although several mathematical models have been developed to make these conversions, all involve some form of coordinate transformation. This chapter covers the introductory procedures of using least squares to compute several well-known and often used transformations. More rigorous procedures, which employ the general least squares procedure, are given in Chapter 22.

18.2 TWO-DIMENSIONAL CONFORMAL COORDINATE TRANSFORMATION

The two-dimensional conformal coordinate transformation, also known as the *four-parameter similarity transformation*, has the characteristic that true shape is retained after transformation. It is typically used in surveying when converting separate surveys into a common reference coordinate system. This transformation is a three-step process that involves:

346 COORDINATE TRANSFORMATIONS

1. *Scaling* to create equal dimensions in the two coordinate systems
2. *Rotation* to make the reference axes of the two systems parallel
3. *Translations* to create a common origin for the two coordinate systems

The scaling and rotation are each defined by one parameter. The translations involve two parameters. Thus, there are a total of four parameters in this transformation. The transformation requires a minimum of two points, called *control points*, that are common to both systems. With the minimum of two points, the four parameters of the transformation can be determined uniquely. If more than two control points are available, a least squares adjustment is possible. After determining the values of the transformation parameters, any points in the original system can be transformed.

18.3 EQUATION DEVELOPMENT

Figure 18.1(*a*) and (*b*) illustrate two independent coordinate systems. In these systems, three common control points, *A*, *B*, and *C*, exist (i.e., their coordinates are known in both systems). Points 1 through 4 have their coordinates known only in the *xy* system of Figure 18.1(*b*). The problem is to determine their *XY* coordinates in the system of Figure 18.1(*a*). The necessary equations are developed as follows.

Step 1: Scaling. To make line lengths as defined by the *xy* coordinate system equal to their lengths in the *XY* system, it is necessary to multiply *xy* coordinates by a scale factor, *S*. Thus, the scaled coordinates x' and y' are

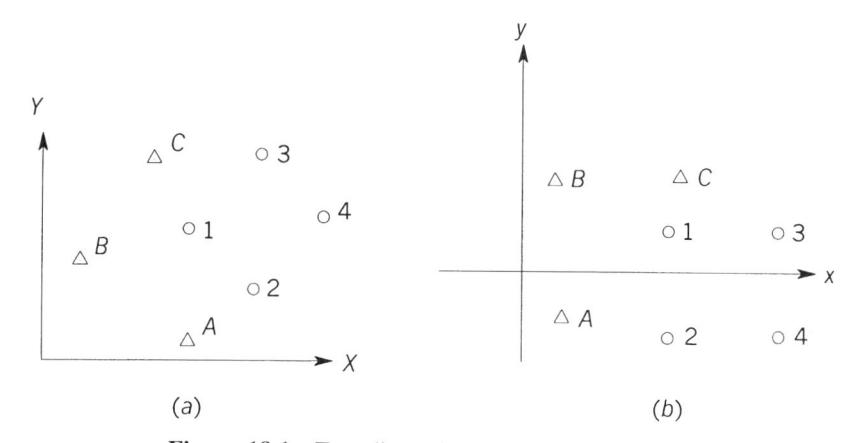

Figure 18.1 Two-dimensional coordinate systems.

$$x' = S\,x$$
$$y' = S\,y \qquad (18.1)$$

Step 2: Rotation. In Figure 18.2, the *XY* coordinate system has been superimposed on the scaled *x'y'* system. The rotation angle, θ, is shown between the *y'* and *Y* axes. To analyze the effects of this rotation, an *X'Y'* system was constructed parallel to the *XY* system such that its origin is common with that of the *x'y'* system. Expressions that give the (*X'*,*Y'*) rotated coordinates for any point (such as point 4 shown) in terms of its *x'y'* coordinates are

$$X' = x' \cos\theta - y' \sin\theta$$
$$Y' = x' \sin\theta + y' \cos\theta \qquad (18.2)$$

Step 3: Translation. To finally arrive at *XY* coordinates for a point, it is necessary to translate the origin of the *X'Y'* system to the origin of the *XY* system. Referring to Figure 18.2, it can be seen that this translation is accomplished by adding translation factors as follows:

$$X = X' + T_X \quad \text{and} \quad Y = Y' + T_Y \qquad (18.3)$$

If Equations (18.1), (18.2), and (18.3) are combined, a single set of equations results that transform the points of Figure 18.1(*b*) directly into Figure 18.1(*a*) as

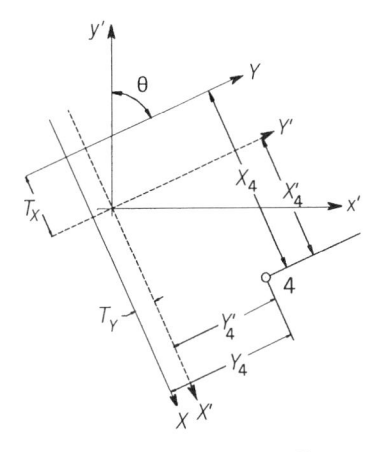

$$X = (S\cos\theta)x - (S\sin\theta)y + T_X$$
$$Y = (S\sin\theta)x + (S\cos\theta)y + T_Y \qquad (18.4)$$

Figure 18.2 Superimposed coordinate systems.

348 COORDINATE TRANSFORMATIONS

Now let $S \cos \theta = a$, $S \sin \theta = b$, $T_X = c$, and $T_Y = d$ and add residuals to make redundant equations consistent. Then Equations (18.4) can be written as

$$ax - by + c = X + v_X$$
$$ay + bx + d = Y + v_Y \tag{18.5}$$

18.4 APPLICATION OF LEAST SQUARES

Equations (18.5) represent the basic observation equations for a two-dimensional conformal coordinate transformation that have four unknowns: a, b, c, and d. The four unknowns embody the transformation parameters S, θ, T_x, and T_y. Since two equations can be written for every control point, only two control points are needed for a unique solution. When more than two are present, a redundant system exists for which a least squares solution can be found. As an example, consider the equations that could be written for the situation illustrated in Figure 18.1. There are three control points, A, B, and C, and thus the following six equations can be written:

$$ax_a - by_a + c = X_A + v_{X_A}$$
$$ay_a + bx_a + d = Y_A + v_{Y_A}$$
$$ax_b - by_b + c = X_B + v_{X_B}$$
$$ay_b + bx_b + d = Y_B + v_{Y_B} \tag{18.6}$$
$$ax_c - by_c + c = X_C + v_{X_C}$$
$$ay_c + bx_c + d = Y_C + v_{Y_C}$$

Equations (18.6) can be expressed in matrix form as

$$AX = L + V \tag{18.7}$$

where

$$A = \begin{bmatrix} x_a & -y_a & 1 & 0 \\ y_a & x_a & 0 & 1 \\ x_b & -y_b & 1 & 0 \\ y_b & x_b & 0 & 1 \\ x_c & -y_c & 1 & 0 \\ y_c & x_c & 0 & 1 \end{bmatrix} \quad X = \begin{bmatrix} a \\ b \\ c \\ d \end{bmatrix} \quad L = \begin{bmatrix} X_A \\ Y_A \\ X_B \\ Y_B \\ X_C \\ Y_C \end{bmatrix} \quad V = \begin{bmatrix} v_{X_A} \\ v_{Y_A} \\ v_{X_B} \\ v_{Y_B} \\ v_{X_C} \\ v_{Y_C} \end{bmatrix}$$

18.4 APPLICATION OF LEAST SQUARES

The redundant system is solved using Equation (11.32). Having obtained the most probable values for the coefficients from the least squares solution, the XY coordinates of any additional points whose coordinates are known in the xy system can then be obtained by applying Equations (18.5) (where the residuals are now considered to be zeros).

After the adjustment, the scale factor S and rotation angle θ can be computed with the following equations:

$$\theta = \tan^{-1}\frac{b}{a}$$
$$S = \frac{a}{\cos\theta} \tag{18.8}$$

Example 18.1 A survey conducted in an arbitrary xy coordinate system produced station coordinates for A, B, and C as well as for stations 1 through 4. Stations A, B, and C also have known state plane coordinates, labeled E and N. It is required to derive the state plane coordinates of stations 1 through 4. Table 18.1 is a tabulation of the arbitrary coordinates and state plane coordinates.

SOLUTION A computer listing from program ADJUST is given below for the problem. The output includes the input data, the coordinates of transformed points, the transformation coefficients, and their estimated errors. Note that the program formed the A and L matrices in accordance with Equation (18.7). After obtaining the solution using Equation (11.32), the program solved Equation (18.8) to obtain the rotation angle and scale factor of the transformation. A complete solution for this example is given in the Mathcad worksheet on the CD that accompanies this book.

TABLE 18.1 Data for Example 18.1

Point	E	N	x	y
A	1,049,422.40	51,089.20	121.622	−128.066
B	1,049,413.95	49,659.30	141.228	187.718
C	1,049,244.95	49,884.95	175.802	135.728
1			174.148	−120.262
2			513.520	−192.130
3			754.444	−67.706
4			972.788	120.994

Two Dimensional Conformal Coordinate Transformation

```
ax − by + Tx = X + VX
bx + ay + Ty = Y + VY
```

	A matrix			L matrix
121.622	128.066	1.000	0.000	1049422.400
−128.066	121.622	0.000	1.000	51089.200
141.228	−187.718	1.000	0.000	1049413.950
187.718	141.228	0.000	1.000	49659.300
175.802	−135.728	1.000	0.000	1049244.950
135.728	175.802	0.000	1.000	49884.950

Transformed Control Points

POINT	X	Y	VX	VY
A	1,049,422.400	51,089.200	−0.004	0.029
B	1,049,413.950	49,659.300	−0.101	0.077
C	1,049,244.950	49,884.950	0.105	−0.106

Transformation Parameters:
```
       a  =    −4.51249 ± 0.00058
       b  =    −0.25371 ± 0.00058
      Tx  = 1050003.715 ± 0.123
      Ty  =   50542.131 ± 0.123

        Rotation = 183° 13′ 05.0″
           Scale = 4.51962
Adjustment's Reference Variance = 0.0195
```

Transformed Points

POINT	X	Y	±σx	±σy
1	1,049,187.361	51,040.629	0.135	0.135
2	1,047,637.713	51,278.829	0.271	0.271
3	1,046,582.113	50,656.241	0.368	0.368
4	1,045,644.713	49,749.336	0.484	0.484

18.5 TWO-DIMENSIONAL AFFINE COORDINATE TRANSFORMATION

The two-dimensional affine coordinate transformation, also known as the *six-parameter transformation*, is a slight variation from the two-dimensional con-

18.5 TWO-DIMENSIONAL AFFINE COORDINATE TRANSFORMATION

formal transformation. In the affine transformation there is the additional allowance for two different scale factors; one in the x direction and the other in the y direction. This transformation is commonly used in photogrammetry for interior orientation. That is, it is used to transform photo coordinates from an arbitrary measurement photo coordinate system to a camera fiducial system and to account for differential shrinkages that occur in the x and y directions. As in the conformal transformation, the affine transformation also applies two translations of the origin, and a rotation about the origin, plus a small non-orthogonality correction between the x and y axes. This results in a total of six unknowns. The mathematical model for the affine transformation is

$$ax + by + c = X + V_X$$
$$dx + ey + f = Y + V_Y \qquad (18.9)$$

These equations are linear and can be solved uniquely when three control points exist (i.e., points whose coordinates are known in the both systems). This is because for each point, an equation set in the form of Equations (18.9) can be written, and three points yield six equations involving six unknowns. If more than three control points are available, a least squares solution can be obtained. Assume, for example, that four common points (1, 2, 3, and 4) exist. Then the equation system would be

$$ax_1 + by_1 + c = X_1 + V_{X1}$$
$$dx_1 + ey_1 + f = Y_1 + V_{Y1}$$
$$ax_2 + by_2 + c = X_2 + V_{X2}$$
$$dx_2 + ey_2 + f = Y_2 + V_{Y2}$$
$$ax_3 + by_3 + c = X_3 + V_{X3}$$
$$dx_3 + ey_3 + f = Y_3 + V_{Y3}$$
$$ax_4 + dy_4 + c = X_4 + V_{X4}$$
$$dx_4 + ey_4 + f = Y_4 + V_{Y4} \qquad (18.10)$$

In matrix notation, Equations (18.10) are expressed as $AX = L + V$, where

$$\begin{bmatrix} x_1 & y_1 & 1 & 0 & 0 & 0 \\ 0 & 0 & 0 & x_1 & y_1 & 1 \\ x_2 & y_2 & 1 & 0 & 0 & 0 \\ 0 & 0 & 0 & x_2 & y_2 & 1 \\ x_3 & y_3 & 1 & 0 & 0 & 0 \\ 0 & 0 & 0 & x_3 & y_3 & 1 \\ x_4 & y_4 & 1 & 0 & 0 & 0 \\ 0 & 0 & 0 & x_4 & y_4 & 1 \end{bmatrix} \begin{bmatrix} a \\ b \\ c \\ d \\ e \\ f \end{bmatrix} = \begin{bmatrix} X_1 \\ Y_1 \\ X_2 \\ Y_2 \\ X_3 \\ Y_3 \\ X_4 \\ Y_4 \end{bmatrix} + \begin{bmatrix} v_{X_1} \\ v_{Y_1} \\ v_{X_2} \\ v_{Y_2} \\ v_{X_3} \\ v_{Y_3} \\ v_{X_4} \\ v_{Y_4} \end{bmatrix} \qquad (18.11)$$

352 COORDINATE TRANSFORMATIONS

The most probable values for the unknown parameters are computed using least squares equation (11.32). They are then used to transfer the remaining points from the *xy* coordinate system to the *XY* coordinate system.

Example 18.2 Photo coordinates, which have been measured using a digitizer, must be transformed into the camera's fiducial coordinate system. The four fiducial points and the additional points were measured in the digitizer's *xy* coordinate system and are listed in Table 18.2 together with the known camera *XY* fiducial coordinates.

SOLUTION A self-explanatory computer solution from the program ADJUST that yields the least squares solution for an affine transformation is shown below. The complete solution for this example is given in the Mathcad worksheet on the CD that accompanies this book.

```
Two Dimensional Affine Coordinate Transformation
-----------------------------------------------------
ax + by + c = X + VX
dx + ey + f = Y + VY
```

		A matrix				L matrix
0.764	5.960	1.000	0.000	0.000	0.000	-113.000
0.000	0.000	0.000	0.764	5.960	1.000	-113.000
5.062	10.541	1.000	0.000	0.000	0.000	0.001
0.000	0.000	0.000	5.062	10.541	1.000	0.001
9.663	6.243	1.000	0.000	0.000	0.000	112.998
0.000	0.000	0.000	9.663	6.243	1.000	112.998
5.350	1.654	1.000	0.000	0.000	0.000	0.001
0.000	0.000	0.000	5.350	1.654	1.000	0.001

TABLE 18.2 Coordinates of Points for Example 18.2

	X	Y	x	y	σx	σy
1	-113.000	0.003	0.764	5.960	0.104	0.112
3	0.001	112.993	5.062	10.541	0.096	0.120
5	112.998	0.003	9.663	6.243	0.112	0.088
7	0.001	-112.999	5.350	1.654	0.096	0.104
306			1.746	9.354		
307			5.329	9.463		

Transformed Control Points

POINT	X	Y	VX	VY
1	−113.000	0.003	0.101	0.049
3	0.001	112.993	−0.086	−0.057
5	112.998	0.003	0.117	0.030
7	0.001	−112.999	−0.086	−0.043

Transformation Parameters:

$$a = 25.37152 \pm 0.02532$$
$$b = 0.82220 \pm 0.02256$$
$$c = -137.183 \pm 0.203$$
$$d = -0.80994 \pm 0.02335$$
$$e = 25.40166 \pm 0.02622$$
$$f = -150.723 \pm 0.216$$

Adjustment's Reference Variance = 2.1828

Transformed Points

POINT	X	Y	$\pm \sigma x$	$\pm \sigma y$
1	−112.899	0.052	0.132	0.141
3	−0.085	112.936	0.125	0.147
5	113.115	0.033	0.139	0.118
7	−0.085	−113.042	0.125	0.134
306	−85.193	85.470	0.134	0.154
307	5.803	85.337	0.107	0.123

18.6 TWO-DIMENSIONAL PROJECTIVE COORDINATE TRANSFORMATION

The two-dimensional projective coordinate transformation is also known as the *eight-parameter transformation*. It is appropriate to use when one two-dimensional coordinate system is projected onto another nonparallel system. This transformation is commonly used in photogrammetry and it can also be used to transform NAD 27 coordinates into the NAD 83 system. In their final form, the two-dimensional projective coordinate transformation equations are

COORDINATE TRANSFORMATIONS

$$X = \frac{a_1 x + b_1 y + c}{a_3 x + b_3 y + 1}$$

$$Y = \frac{a_2 x + b_2 y + c}{a_3 x + b_3 y + 1}$$

(18.12)

Upon inspection, it can be seen that these equations are similar to the affine transformation. In fact, if a_3 and b_3 were equal to zero, these equations are the affine transformation. With eight unknowns, this transformation requires a minimum of four control points (points having coordinates known in both systems). If there are more than four control points, the least squares solution can be used. Since these are nonlinear equations, they must be linearized and solved using Equation (11.37) or (11.39). The linearized form of these equations is

$$\begin{bmatrix} \left(\frac{\partial X}{\partial a_1}\right)_0 & \left(\frac{\partial X}{\partial b_1}\right)_0 & \left(\frac{\partial X}{\partial c_1}\right)_0 & 0 & 0 & 0 & \left(\frac{\partial X}{\partial a_3}\right)_0 & \left(\frac{\partial X}{\partial b_3}\right)_0 \\ 0 & 0 & 0 & \left(\frac{\partial x}{\partial a_2}\right)_0 & \left(\frac{\partial X}{\partial b_2}\right)_0 & \left(\frac{\partial X}{\partial c_2}\right)_0 & \left(\frac{\partial X}{\partial a_3}\right)_0 & \left(\frac{\partial X}{\partial b_3}\right)_0 \end{bmatrix} \begin{bmatrix} da_1 \\ db_1 \\ dc_1 \\ da_2 \\ db_2 \\ dc_2 \\ da_3 \\ db_3 \end{bmatrix}$$

$$= \begin{bmatrix} X - X_0 \\ Y - Y_0 \end{bmatrix}$$

(18.13)

where

$$\frac{\partial X}{\partial a_1} = \frac{x}{a_3 x + b_3 + 1} \quad \frac{\partial X}{\partial b_1} = \frac{y}{a_3 x + b_3 + 1} \quad \frac{\partial X}{\partial c_1} = \frac{1}{a_3 x + b_3 + 1}$$

$$\frac{\partial Y}{\partial a_2} = \frac{x}{a_3 x + b_3 + 1} \quad \frac{\partial Y}{\partial b_2} = \frac{y}{a_3 x + b_3 + 1} \quad \frac{\partial Y}{\partial c_2} = \frac{1}{a_3 x + b_3 + 1}$$

$$\frac{\partial X}{\partial a_3} = -\frac{a_1 x + b_1 y + c_1}{(a_3 x + b_3 + 1)^2} x \quad \frac{\partial X}{\partial b_3} = -\frac{a_1 x + b_1 y + c_1}{(a_3 x + b_3 + 1)^2} y$$

$$\frac{\partial Y}{\partial a_3} = -\frac{a_2 x + b_2 y + c_2}{(a_3 x + b_3 + 1)^2} x \quad \frac{\partial Y}{\partial b_3} = -\frac{a_2 x + b_2 y + c_2}{(a_3 x + b_3 + 1)^2} y$$

For each control point, a set of equations of the form of Equation (18.13) can be written. A redundant system of equations can be solved by least squares to yield the eight unknown parameters. With these values, the re-

18.6 TWO-DIMENSIONAL PROJECTIVE COORDINATE TRANSFORMATION

maining points in the xy coordinate system are transformed into the XY system.

Example 18.3 Given the data in Table 18.3, determine the best-fit projective transformation parameters and use them to transform the remaining points into the XY coordinate system.

Program ADJUST was used to solve this problem and the results follow. The complete solution for this example is given in the Mathcad worksheet on the CD that accompanies this book.

```
Two Dimensional Projective Coordinate Transformation of
File
-------------------------------------------------------
a1x + b1y + c1
---------------   = X + VX
a3x + b3y + 1


a2x + b2y + c2
---------------   = Y + VY
a3x + b3y + 1


Transformation Parameters:
                  a1 =   25.00274 ± 0.01538
                  b1 =    0.80064 ± 0.01896
                  c1 = -134.715   ± 0.377
                  a2 =   -8.00771 ± 0.00954
                  b2 =   24.99811 ± 0.01350
                  c2 = -149.815   ± 0.398
                  a3 =    0.00400 ± 0.00001
                  b3 =    0.00200 ± 0.00001
```

TABLE 18.3 Data for Example 18.3

Point	X	Y	x	y	σ_x	σ_y
1	1420.407	895.362	90.0	90.0	0.3	0.3
2	895.887	351.398	50.0	40.0	0.3	0.3
3	−944.926	641.434	−30.0	20.0	0.3	0.3
4	968.084	−1384.138	50.0	−40.0	0.3	0.3
5	1993.262	−2367.511	110.0	−80.0	0.3	0.3
6	−3382.284	3487.762	−100.0	80.0	0.3	0.3
7			−60.0	20.0	0.3	0.3
8			−100.0	−100.0	0.3	0.3

356 COORDINATE TRANSFORMATIONS

```
Adjustment's Reference Variance = 3.8888
          Number of Iterations = 2
```

Transformed Control Points

POINT	X	Y	VX	VY
1	1,420.165	895.444	−0.242	0.082
2	896.316	351.296	0.429	−0.102
3	−944.323	641.710	0.603	0.276
4	967.345	−1,384.079	−0.739	0.059
5	1,993.461	−2,367.676	0.199	−0.165
6	−3,382.534	3,487.612	−0.250	−0.150

Transformed Points

POINT	X	Y	±σx	±σy
1	1,420.165	895.444	0.511	0.549
2	896.316	351.296	0.465	0.458
3	−944.323	641.710	0.439	0.438
4	967.345	−1,384.079	0.360	0.388
5	1,993.461	−2,367.676	0.482	0.494
6	−3,382.534	3,487.612	0.558	0.563
7	−2,023.678	1,038.310	1.717	0.602
8	−6,794.740	−4,626.976	51.225	34.647

18.7 THREE-DIMENSIONAL CONFORMAL COORDINATE TRANSFORMATION

The three-dimensional conformal coordinate transformation is also known as the *seven-parameter similarity transformation*. It transfers points from one three-dimensional coordinate system to another. It is applied in the process of reducing data from GPS surveys and is also used extensively in photogrammetry. The three-dimensional conformal coordinate transformation involves seven parameters, three rotations, three translations, and one scale factor. The rotation matrix is developed from three consecutive two-dimensional rotations about the x, y, and z axes, respectively. Given in sequence, these are as follows.

In Figure 18.3, the rotation θ_1 about the x axis expressed in matrix form is

18.7 THREE-DIMENSIONAL CONFORMAL COORDINATE TRANSFORMATION

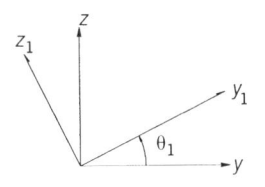

Figure 18.3 θ_1 rotation.

$$X_1 = R_1 X_0 \tag{a}$$

where

$$X_1 = \begin{bmatrix} x_1 \\ y_1 \\ z_1 \end{bmatrix} \quad R_1 = \begin{bmatrix} 1 & 0 & 0 \\ 0 & \cos\theta_1 & \sin\theta_1 \\ 0 & -\sin\theta_1 & \cos\theta_1 \end{bmatrix} \quad X_0 = \begin{bmatrix} x \\ y \\ z \end{bmatrix}$$

In Figure 18.4, the rotation θ_2 about the y axis expressed in matrix form is

$$X_2 = R_2 X_1 \tag{b}$$

where

$$X_2 = \begin{bmatrix} x_2 \\ y_2 \\ z_2 \end{bmatrix} \quad \text{and} \quad R_2 = \begin{bmatrix} \cos\theta_2 & 0 & -\sin\theta_2 \\ 0 & 1 & 0 \\ \sin\theta_2 & 0 & \cos\theta_2 \end{bmatrix}$$

In Figure 18.5, the rotation θ_3 about the z axis expressed in matrix form is

$$X = R_3 X_2 \tag{c}$$

where

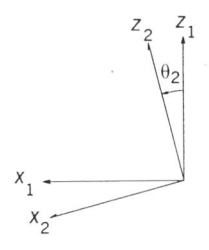

Figure 18.4 θ_2 rotation.

COORDINATE TRANSFORMATIONS

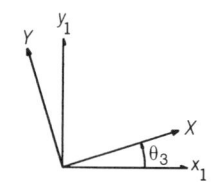

Figure 18.5 θ_3 rotation.

$$X = \begin{bmatrix} X \\ Y \\ Z \end{bmatrix} \quad \text{and} \quad R_3 = \begin{bmatrix} \cos\theta_3 & \sin\theta_3 & 0 \\ -\sin\theta_3 & \cos\theta_3 & 0 \\ 0 & 0 & 1 \end{bmatrix}$$

Substituting Equation (a) into (b) and in turn into (c) yields

$$X = R_3 R_2 R_1 X_0 = R X_0 \tag{d}$$

When multiplied together, the three matrices R_3, R_2, and R_1 in Equation (d) develop a single rotation matrix R for the transformation whose individual elements are

$$R = \begin{bmatrix} r_{11} & r_{12} & r_{13} \\ r_{21} & r_{23} & r_{23} \\ r_{31} & r_{32} & r_{33} \end{bmatrix} \tag{18.14}$$

where

$r_{11} = \cos\theta_2 \cos\theta_3$

$r_{12} = \sin\theta_1 \sin\theta_2 \cos\theta_3 + \cos\theta_1 \sin\theta_3$

$r_{13} = -\cos\theta_1 \sin\theta_2 \cos\theta_3 + \sin\theta_1 \sin\theta_3$

$r_{21} = -\cos\theta_2 \sin\theta_3$

$r_{22} = -\sin\theta_1 \sin\theta_2 \sin\theta_3 + \cos\theta_1 \cos\theta_3$

$r_{23} = \cos\theta_1 \sin\theta_2 \sin\theta_3 + \sin\theta_1 \cos\theta_3$

$r_{31} = \sin\theta_2$

$r_{32} = -\sin\theta_1 \cos\theta_2$

$r_{33} = \cos\theta_1 \cos\theta_2$

Since the rotation matrix is orthogonal, it has the property that its inverse is equal to its transpose. Using this property and multiplying the terms of the matrix X by a scale factor, S, and adding translations factors T_x, T_y, and T_z to

18.7 THREE-DIMENSIONAL CONFORMAL COORDINATE TRANSFORMATION

translate to a common origin yields the following mathematical model for the transformation:

$$X = S(r_{11}x + r_{21}y + r_{31}z) + Tx$$
$$Y = S(r_{12}x + r_{22}y + r_{32}z) + Ty \qquad (18.15)$$
$$Z = S(r_{13}x + r_{23}y + r_{33}z) + Tz$$

Equations (18.15) involve seven unknowns (S, θ_1, θ_2, θ_3, Tx, Ty, Tz). For a unique solution, seven equations must be written. This requires a minimum of two control stations with known XY coordinates and also xy coordinates, plus three stations with known Z and z coordinates. If there are more than the minimum number of control points, a least-squares solution can be used. Equations (18.15) are nonlinear in their unknowns and thus must be linearized for a solution. The following linearized equations can be written for each point as

$$\begin{bmatrix} \left(\dfrac{\partial X}{\partial S}\right)_0 & 0 & \left(\dfrac{\partial X}{\partial \theta_2}\right)_0 & \left(\dfrac{\partial X}{\partial \theta_3}\right)_0 & 1 & 0 & 0 \\ \left(\dfrac{\partial Y}{\partial S}\right)_0 & \left(\dfrac{\partial Y}{\partial \theta_1}\right)_0 & \left(\dfrac{\partial Y}{\partial \theta_2}\right)_0 & \left(\dfrac{\partial Y}{\partial \theta_3}\right)_0 & 0 & 1 & 0 \\ \left(\dfrac{\partial Z}{\partial S}\right)_0 & \left(\dfrac{\partial Z}{\partial \theta_1}\right)_0 & \left(\dfrac{\partial Z}{\partial \theta_2}\right)_0 & \left(\dfrac{\partial Z}{\partial \theta_3}\right)_0 & 0 & 0 & 1 \end{bmatrix} \begin{bmatrix} dS \\ d\theta_1 \\ d\theta_2 \\ d\theta_3 \\ dTx \\ dTy \\ dTz \end{bmatrix} = \begin{bmatrix} X - X_0 \\ Y - Y_0 \\ Z - Z_0 \end{bmatrix}$$

(18.16)

where

$$\dfrac{\partial X}{\partial S} = r_{11}x + r_{21}y + r_{31}z \qquad \dfrac{\partial Y}{\partial S} = r_{12}x + r_{22}y + r_{32}z \qquad \dfrac{\partial Z}{\partial S} = r_{13}x + r_{23}y + r_{33}z$$

$$\dfrac{\partial Y}{\partial \theta_1} = -S[r_{13}x + r_{23}y + r_{33}z] \qquad \dfrac{\partial Z}{\partial \theta_1} = S[r_{12}x + r_{22}y + r_{32}z]$$

$$\dfrac{\partial X}{\partial \theta_2} = S(-x \sin \theta_2 \cos \theta_3 + y \sin \theta_2 \sin \theta_3 + z \cos \theta_2)$$

$$\dfrac{\partial Y}{\partial \theta_2} = S(x \sin \theta_1 \cos \theta_2 \cos \theta_3 - y \sin \theta_1 \cos \theta_2 \sin \theta_3 + z \sin \theta_1 \sin \theta_2)$$

$$\dfrac{\partial Z}{\partial \theta_2} = S(-x \cos \theta_1 \cos \theta_2 \cos \theta_3 + y \cos \theta_1 \cos \theta_2 \sin \theta_3 - z \cos \theta_1 \sin \theta_2)$$

$$\dfrac{\partial X}{\partial \theta_3} = S(r_{21}x - r_{11}y) \qquad \dfrac{\partial Y}{\partial \theta_3} = S(r_{22}x - r_{12}y) \qquad \dfrac{\partial Z}{\partial \theta_3} = S(r_{23}x - r_{13}y)$$

360 COORDINATE TRANSFORMATIONS

Example 18.4 The three-dimensional *xyz* coordinates were measured for six points. Four of these points (1, 2, 3, and 4) were control points whose coordinates were also known in the *XYZ* control system. The data are shown in Table 18.4. Compute the parameters of a three-dimensional conformal coordinate transformation and use them to transform points 5 and 6 in the *XYZ* system.

SOLUTION The results from the program ADJUST are presented below.

```
3D Coordinate Transformation
------------------------------------------------------------
                                                              K
                        J matrix                          matrix
      0.000    102.452   1284.788  1.000  0.000  0.000   -206.164  -0.000
    -51.103     -7.815   -195.197  0.000  1.000  0.000  -1355.718   0.000
  -1287.912    195.697      4.553  0.000  0.000  1.000     53.794   0.000
      0.000    118.747   1418.158  1.000  0.000  0.000    761.082  -0.000
    -62.063     28.850    723.004  0.000  1.000  0.000  -1496.689  -0.000
  -1421.832   -722.441     42.501  0.000  0.000  1.000     65.331  -0.000
      0.000    129.863   1706.020  1.000  0.000  0.000  -1530.174   0.060
    -61.683    -58.003  -1451.826  0.000  1.000  0.000  -1799.945   0.209
  -1709.922   1452.485    -41.580  0.000  0.000  1.000     64.931   0.000
      0.000    204.044   1842.981  1.000  0.000  0.000    -50.417   0.033
   -130.341     -1.911    -46.604  0.000  1.000  0.000  -1947.124  -0.053
  -1849.740     47.857     15.851  0.000  0.000  1.000    137.203   0.043

X matrix
~~~~~~~~~~~~~
 -0.0000347107
 -0.0000103312
 -0.0001056763
  0.1953458986
 -0.0209088384
 -0.0400969773
 -0.0000257795

Measured Points
------------------------------------------------------------
NAME       x           y           z         Sx      Sy      Sz
------------------------------------------------------------
  1    1094.883     820.085    109.821    0.007   0.008   0.005
  2     503.891    1598.698    117.685    0.011   0.008   0.009
  3    2349.343     207.658    151.387    0.006   0.005   0.007
  4    1395.320    1348.853    215.261    0.005   0.008   0.009
```

TABLE 18.4 Data for a Three-Dimensional Conformal Coordinate Transformation

Point	X	Y	Z	$x \pm S_x$	$y \pm S_y$	$z \pm S_z$
1	10,037.81	5262.09	772.04	1094.883 ± 0.007	820.085 ± 0.008	109.821 ± 0.005
2	10,956.68	5128.17	783.00	503.891 ± 0.011	1598.698 ± 0.008	117.685 ± 0.009
3	8,780.08	4840.29	782.62	2349.343 ± 0.006	207.658 ± 0.005	151.387 ± 0.007
4	10,185.80	4700.21	851.32	1395.320 ± 0.005	1348.853 ± 0.008	215.261 ± 0.009
5				265.346 ± 0.005	1003.470 ± 0.007	78.609 ± 0.003
6				784.081 ± 0.006	512.683 ± 0.008	139.551 ± 0.008

CONTROL POINTS

NAME	X	VX	Y	VY	Z	VZ
1	10037.810	0.064	5262.090	0.037	772.040	0.001
2	10956.680	0.025	5128.170	−0.057	783.000	0.011
3	8780.080	−0.007	4840.290	−0.028	782.620	0.007
4	10185.800	−0.033	4700.210	0.091	851.320	−0.024

Transformation Coefficients

```
        Scale =    0.94996 +/- 0.00004
        X-rot =    2°17'05.3" +/- 0° 00' 30.1"
        Y-rot =   -0°33'02.8" +/- 0° 00' 09.7"
        Z-rot =  224°32'10.9" +/- 0° 00' 06.9"
           Tx =   10233.858 +/- 0.065
           Ty =    6549.981 +/- 0.071
           Tz =     720.897 +/- 0.213
```

Reference Standard Deviation: 8.663
Degrees of Freedom: 5
Iterations: 2

Transformed Coordinates

NAME	X	Sx	Y	Sy	Z	Sz
1	10037.874	0.032	5262.127	0.034	772.041	0.040
2	10956.705	0.053	5128.113	0.052	783.011	0.056
3	8780.073	0.049	4840.262	0.041	782.627	0.057
4	10185.767	0.032	4700.301	0.037	851.296	0.067
5	10722.020	0.053	5691.221	0.053	766.068	0.088
6	10043.246	0.040	5675.898	0.042	816.867	0.092

Note that in this adjustment, with four control points available having X, Y, and Z coordinates, 12 equations could be written, three for each point. With seven unknown parameters, this gave 12 − 7 = 5 degrees of freedom in the solution. The complete solution for this example is given in the Mathcad worksheet on the CD that accompanies this book.

18.8 STATISTICALLY VALID PARAMETERS

Besides the coordinate transformations described in preceding sections, it is possible to develop numerous others. For example, polynomial equations of

18.8 STATISTICALLY VALID PARAMETERS

various degrees could be used to transform data. As additional terms are added to a polynomial, the resulting equation will force better fits on any given data set. However, caution should be exercised when doing this since the resulting transformation parameters may not be statistically significant.

As an example, when using a two-dimensional conformal coordinate transformation with a data set having four control points, nonzero residuals would be expected. However, if a projective transformation were used, this data set would yield a unique solution, and thus the residuals would be zero. Is the projective a more appropriate transformation for this data set? Is this truly a better fit? Guidance in the answers to these questions can be obtained by checking the statistical validity of the parameters.

The adjusted parameters divided by their standard deviations represent a t statistic with ν degrees of freedom. If a parameter is to be judged as statistically different from zero, and thus significant, the t value computed (the test statistic) must be greater than $t_{\alpha/2,\nu}$. Simply stated, the test statistic is

$$t = \frac{|\text{parameter}|}{S} \qquad (18.17)$$

For example, in the adjustment in Example 18.2, the following computed t-values are found:

Parameter	S	t-Value
$a = 25.37152$	± 0.02532	1002
$b = 0.82220$	± 0.02256	36.4
$c = -137.183$	± 0.203	675.8
$d = -0.80994$	± 0.02335	34.7
$e = 25.40166$	± 0.02622	968.8
$f = -150.723$	± 0.216	697.8

In this problem there were eight equations involving six unknowns and thus 2 degrees of freedom. From the t-distribution table (Table D.3), $t_{0.025,2} = 4.303$. Because all t values computed are greater than 4.303, each parameter is significantly different from zero at a 95% level of confidence. From the adjustment results of Example 18.3, the t values computed are listed below.

Parameter	Value	S	t-Value
a_1	25.00274	0.01538	1626
b_1	0.80064	0.01896	42.3
c_1	-134.715	0.377	357.3
a_2	-8.00771	0.00954	839.4
b_2	24.99811	0.01350	1851.7
c_2	-149.815	0.398	376.4
a_3	0.00400	0.00001	400
b_3	0.00200	0.00002	100

364 COORDINATE TRANSFORMATIONS

This adjustment has eight unknown parameters and 12 observations. From the t-distribution table (Table D.3), $t_{0.025,4} = 2.776$. By comparing the tabular t value against each computed value, all parameters are again significantly different from zero at a 95% confidence level. This is true for a_3 and b_3 even though they seem relatively small at 0.004 and 0.002, respectively.

Using this statistical technique, a check can be made to determine when the projective transformation is appropriate since it defaults to an affine transformation when a_3 and b_3 are both statistically equal to zero. Similarly, if the confidence intervals at a selected probability level of the two-dimensional conformal coordinate transformation contain two of the parameters from the affine transformation, the computed values of the affine transformation are statistically equal to those from the conformal transformation. Thus, if the interval for a from the conformal transformation contains both a and e from the affine transformation, there is no statistical difference between these parameters. This must also be true for b from the conformal transformation when compared to absolute values of b and d from the affine transformation. Note that a negative sign is part of the conformal coordinate transformation, and thus b and d are generally opposite in signs. If both of these conditions exist, the conformal transformation is the more appropriate adjustment to use for the data given. One must always be sure that a minimum number of unknown parameters are used to solve any problem.

PROBLEMS

Note: For problems requiring least squares adjustment, if a computer program is not distinctly specified for use in the problem, it is expected that the least squares algorithm will be solved using the program MATRIX, which is on within the CD supplied with the book.

18.1 Points A, B, and C have their coordinates known in both an XY and an xy system. Points D, E, F, and G have their coordinates known only in the xy system. These coordinates are shown in the table below. Using a two-dimensional conformal coordinate transformation, determine:
   (a) the transformation parameters.
   (b) the most probable coordinates for D, E, F, and G in the XY coordinate system.
   (c) the rotation angle and scale factor.

Point	X	Y	x	y
A	603,462.638	390,601.450	1221.41	1032.09
B	604,490.074	390,987.136	4607.15	1046.11
C	604,314.613	391,263.879	4200.12	2946.39
D			3975.00	1314.29
E			3585.71	2114.28
F			2767.86	1621.43
G			2596.43	2692.86

18.2 Using a two-dimensional conformal coordinate transformation and the data listed below, determine:
  (a) the transformation parameters.
  (b) the most probable coordinates for 9, 10, 11, and 12 in the XY coordinate system.
  (c) the rotation angle and scale factor.

	Observed		Control	
Point	x	y	X	Y
1	−4.209 ± 0.008	−6.052 ± 0.009	−106.004	−105.901
2	14.094 ± 0.012	9.241 ± 0.010	105.992	106.155
3	−2.699 ± 0.009	10.728 ± 0.007	−105.967	105.939
4	12.558 ± 0.013	−7.563 ± 0.009	105.697	−105.991
5	−3.930 ± 0.005	2.375 ± 0.006	−112.004	−0.024
6	13.805 ± 0.006	0.780 ± 0.011	111.940	−0.108
7	5.743 ± 0.008	10.462 ± 0.005	0.066	112.087
8	4.146 ± 0.009	−7.288 ± 0.003	−0.006	−111.991
9	5.584 ± 0.008	6.493 ± 0.004		
10	9.809 ± 0.010	−8.467 ± 0.009		
11	−4.987 ± 0.006	0.673 ± 0.007		
12	−0.583 ± 0.004	−5.809 ± 0.005		

18.3 Do Problem 18.2 using an unweighted least squares adjustment.

18.4 Do parts (a) and (b) in Problem 18.2 using a two-dimensional affine coordinate transformation.

18.5 Do parts (a) and (b) in Problem 18.2 using a two-dimensional projective coordinate transformation.

18.6 Determine the appropriate two-dimensional transformation for Problem 18.2 at a 0.01 level of significance.

18.7 Using a two-dimensional affine coordinate transformation and the following data, determine:

(a) the transformation parameters.

(b) the most probable XY coordinates for points 9 to 12.

Point	x	y	X	Y
1	−83.485 ± 0.005	1.221 ± 0.007	−113.000	0.003
2	−101.331 ± 0.006	56.123 ± 0.010	−105.962	105.598
3	−43.818 ± 0.011	38.462 ± 0.012	0.001	112.993
4	16.737 ± 0.015	13.140 ± 0.013	105.998	105.996
5	42.412 ± 0.006	−44.813 ± 0.009	112.884	−0.002
6	60.360 ± 0.010	−99.889 ± 0.008	105.889	−105.934
7	2.788 ± 0.006	−82.065 ± 0.012	−0.001	−112.986
8	−57.735 ± 0.003	−56.556 ± 0.005	−105.887	−105.628
9	−63.048 ± 0.008	−89.056 ± 0.008		
10	45.103 ± 0.007	32.887 ± 0.006		
11	−7.809 ± 0.004	98.773 ± 0.010		
12	57.309 ± 0.008	−17.509 ± 0.009		

18.8 Do Problem 18.7 using an unweighted least squares adjustment.

18.9 Do Problem 18.7 using a two-dimensional projective coordinate transformation.

18.10 For the data of Problem 18.7, which two-dimensional transformation is most appropriate, and why? Use a 0.01 level of significance.

18.11 Determine the appropriate two-dimensional coordinate transformation for the following data at a 0.01 level of significance.

Point	X (m)	Y (m)	x (mm)	y (mm)	σ_x	σ_y
1	2181.578	2053.274	89.748	91.009	0.019	0.020
2	1145.486	809.022	49.942	39.960	0.016	0.021
3	−855.426	383.977	−29.467	20.415	0.028	0.028
4	1087.225	−1193.347	50.164	−40.127	0.028	0.028
5	2540.778	−2245.477	109.599	−80.310	0.018	0.021
6	−2595.242	1926.548	−100.971	79.824	0.026	0.022

18.12 Using a weighted three-dimensional conformal coordinate transformation, determine the transformation parameters for the following data set.

Point	X (m)	Y (m)	Z (m)	x (mm)	y (mm)	z (mm)	σ_x	σ_y	σ_z
1	8948.16	6678.50	756.51	1094.97	810.09	804.73	0.080	0.084	0.153
2	8813.93	5755.23	831.67	508.31	1595.68	901.78	0.080	0.060	0.069
3	8512.60	7937.11	803.11	2356.23	197.07	834.47	0.097	0.177	0.202
4	8351.02	6483.62	863.24	1395.18	1397.64	925.96	0.043	0.161	0.120

18.13 Do Problem 18.12 using an unweighted least squares adjustment.

18.14 Using a weighted three-dimensional conformal coordinate transformation and the follow set of data:

(a) determine the transformation parameters.

(b) Compute the *XYZ* coordinates for points 7 to 10.

Control points

Point	X (m)	Y (m)	Z (m)
1	9770.192	16944.028	1235.280
2	16371.750	14998.190	1407.694
3	5417.336	265.432	—
4	27668.765	26963.937	—
5	—	—	1325.885
6	—	—	1070.226

Measured points

Point	x (m)	y (m)	z (m)
1	9845.049 ± 0.015	16,911.947 ± 0.015	1057.242 ± 0.025
2	16,441.006 ± 0.015	14,941.872 ± 0.015	1169.148 ± 0.025
3	5433.174 ± 0.015	250.766 ± 0.0115	1476.572 ± 0.025
4	27781.044 ± 0.015	26864.597 ± 0.015	861.956 ± 0.025
5	8543.224 ± 0.015	22,014.402 ± 0.015	1139.204 ± 0.025
6	4140.096 ± 0.015	24,618.211 ± 0.015	918.253 ± 0.025
7	23,125.031 ± 0.015	4672.275 ± 0.015	1351.655 ± 0.025
8	4893.721 ± 0.015	12,668.887 ± 0.015	1679.184 ± 0.025
9	19967.763 ± 0.015	1603.499 ± 0.015	1210.986 ± 0.025
10	2569.022 ± 0.015	14,610.600 ± 0.015	1359.663 ± 0.025

18.15 Do Problem 11.19, and determine whether the derived constant and scale factor are statistically significant at a 0.01 level of significance.

Use the program ADJUST to do each problem.

18.16 Problem 18.6

18.17 Problem 18.10

18.18 Problem 18.14

Programming Problems

Develop a computational program that calculates the coefficient and constants matrix for each transformation.

18.19 A two-dimensional conformal coordinate transformation

18.20 A two-dimensional affine coordinate transformation

18.21 A two-dimensional projective coordinate transformation

18.22 A three-dimensional conformal coordinate transformation

CHAPTER 19

ERROR ELLIPSE

19.1 INTRODUCTION

As discussed previously, after completing a least squares adjustment, the estimated standard deviations in the coordinates of an adjusted station can be calculated from covariance matrix elements. These standard deviations provide error estimates in the reference axes directions. In graphical representation, they are half the dimensions of a *standard error rectangle* centered on each station. The standard error rectangle has dimensions of $2S_x$ by $2S_y$ as illustrated for station *B* in Figure 19.1, but this is not a complete representation of the error at the station.

Simple deductive reasoning can be used to show the basic problem. Assume in Figure 19.1 that the *XY* coordinates of station *A* have been computed from the observations of distance *AB* and azimuth Az_{AB} that is approximately 30°. Further assume that the observed azimuth has no error at all but that the distance has a large error, say ±2 ft. From Figure 19.1 it should then be readily apparent that the largest uncertainty in the station's position would not lie in either cardinal direction. That is, neither S_x nor S_y represents the largest positional uncertainty for the station. Rather, the largest uncertainty would be collinear with line *AB* and approximately equal to the estimated error in the distance. In fact, this is what happens.

In the usual case, the position of a station is uncertain in both direction and distance, and the estimated error of the adjusted station involves the errors of two jointly distributed variables, the *x* and *y* coordinates. Thus, the positional error at a station follows a *bivariate normal distribution*. The general shape of this distribution for a station is shown in Figure 19.2. In this figure, the three-dimensional *surface plot* [Figure 19.2(*a*)] of a bivariate normal dis-

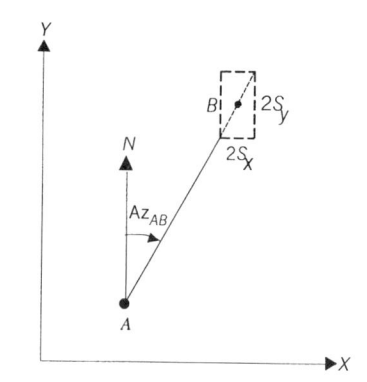

Figure 19.1 Standard error rectangle.

tribution curve is shown along with its *contour plot* [Figure 19.2(*b*)]. Note that the ellipses shown in the *xy* plane of Figure 19.2(*b*) can be obtained by passing planes of varying levels through Figure 19.2(*a*) parallel to the *xy* plane. The volume of the region inside the intersection of any plane with the surface of Figure 19.2(*a*) represents the probability level of the ellipse. The orthogonal projection of the surface plot of Figure 19.2(*a*) onto the *xz* plane would give the normal distribution curve of the *x* coordinate, from which S_x is obtained. Similarly, its orthogonal projection onto the *yz* plane would give the normal distribution in the *y* coordinate from which S_y is obtained.

To fully describe the estimated error of a station, it is only necessary to show the orientation and lengths of the semiaxes of the error ellipse. A detailed diagram of an error ellipse is shown in Figure 19.3. In this figure, the *standard error ellipse* of a station is shown (i.e., one whose arcs are tangent

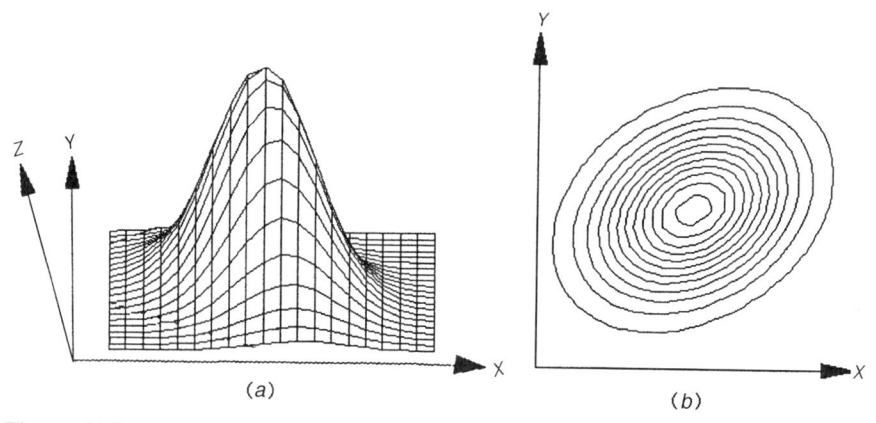

Figure 19.2 (*a*) Three-dimensional view and (*b*) contour plot of a bivariate normal distribution.

19.2 COMPUTATION OF ELLIPSE ORIENTATION AND SEMIAXES

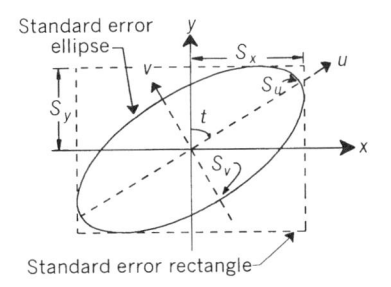

Figure 19.3 Standard error ellipse.

to the sides of the standard error rectangle). The orientation of the ellipse depends on the t angle, which fixes the directions of the auxiliary, orthogonal uv axes along which the ellipse axes lie. The u axis defines the weakest direction in which the station's adjusted position is known. In other words, it lies in the direction of maximum error in the station's coordinates. The v axis is orthogonal to u and defines the strongest direction in which the station's position is known, or the direction of minimum error. For any station, the value of t that orients the ellipse to provide these maximum and minimum values can be determined after the adjustment from the elements of the covariance matrix.

The exact probability level of the standard error ellipse is dependent on the number of degrees of freedom in the adjustment. This standard error ellipse can be modified in dimensions through the use of F statistical values to represent an error probability at any percentage selected. It will be shown later that the percent probability within the boundary of standard error ellipse for a simple closed traverse is only 35%. Surveyors often use the 95% probability since this affords a high level of confidence.

19.2 COMPUTATION OF ELLIPSE ORIENTATION AND SEMIAXES

As shown in Figure 19.4, the method for calculating the orientation angle t that yields maximum and minimum semiaxes involves a two-dimensional co-

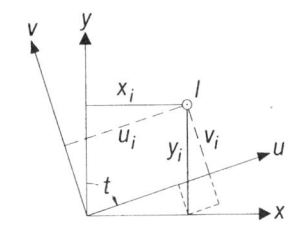

Figure 19.4 Two-dimensional rotation.

372 ERROR ELLIPSE

ordinate rotation. Notice that the t angle is defined as a clockwise angle from the y axis to the u axis. To propagate the errors in a point I from the xy system into an orthogonal uv system, the generalized law of the propagation of variances, discussed in Chapter 6, is used. The specific value for t that yields the maximum error along the u axis must be determined. The following steps accomplish this task.

Step 1: Any point I in the uv system can be represented with respect to its xy coordinates as

$$u_i = x_i \sin t + y_i \cos t$$
$$v_i = -x_i \cos t + y_i \sin t \qquad (19.1)$$

Rewriting Equations (19.1) in matrix form yields

$$\begin{bmatrix} u_i \\ v_i \end{bmatrix} = \begin{bmatrix} \sin t & \cos t \\ -\cos t & \sin t \end{bmatrix} \begin{bmatrix} x_i \\ y_i \end{bmatrix} \qquad (19.2)$$

or in simplified matrix notation,

$$Z = RX \qquad (19.3)$$

Step 2: Assume that for the adjustment problem in which I appears, there is a Q_{xx} matrix for the xy coordinate system. The problem is to develop, from the Q_{xx} matrix, a new covariance matrix Q_{zz} for the uv coordinate system. This can be done using the generalized law for the propagation of variances, given in Chapter 6 as

$$\Sigma_{zz} = S_0^2 R Q_{xx} R^T \qquad (19.4)$$

where

$$R = \begin{bmatrix} \sin t & \cos t \\ -\cos t & \sin t \end{bmatrix}$$

Since S_0^2 is a scalar, it can be dropped temporarily and recalled again after the derivation. Thus,

$$Q_{zz} = R Q_{xx} R^T = \begin{bmatrix} q_{uu} & q_{uv} \\ q_{uv} & q_{vv} \end{bmatrix} \qquad (19.5)$$

where

19.2 COMPUTATION OF ELLIPSE ORIENTATION AND SEMIAXES

$$Q_{xx} = \begin{bmatrix} q_{xx} & q_{xy} \\ q_{xy} & q_{yy} \end{bmatrix}$$

Step 3: Expanding Equation (19.5), the elements of the Q_{zz} matrix are

$$Q_{zz} = \begin{bmatrix} \begin{pmatrix} q_{xx}\sin^2 t + q_{xy}\cos t \sin t \\ + q_{xy}\sin t \cos t + q_{yy}\cos^2 t \end{pmatrix} & \begin{pmatrix} -q_{xx}\sin t \cos t - q_{xy}\cos^2 t \\ +q_{xy}\sin^2 t + q_{yy}\cos t \sin t \end{pmatrix} \\ \begin{pmatrix} -q_{xx}\cos t \sin t + q_{xy}\sin^2 t \\ -q_{xy}\cos^2 t + q_{yy}\sin t \cos t \end{pmatrix} & \begin{pmatrix} q_{xx}\cos^2 t - q_{xy}\sin t \cos t \\ -q_{xy}\cos t \sin t + q_{yy}\sin^2 t \end{pmatrix} \end{bmatrix}$$

$$= \begin{bmatrix} q_{uu} & q_{uv} \\ q_{uv} & q_{vv} \end{bmatrix} \quad (19.6)$$

Step 4: The q_{uu} element of Equation (19.6) can be rewritten as

$$q_{uu} = q_{xx}\sin^2 t + 2q_{xy}\cos t \sin t + q_{yy}\cos^2 t \quad (19.7)$$

The following trigonometric identities are useful in developing an equation for t:

$$\sin 2t = 2\sin t \cos t \quad (a)$$

$$\cos 2t = \cos^2 t - \sin^2 t \quad (b)$$

$$\cos^2 t + \sin^2 t = 1 \quad (c)$$

Substituting identity (*a*) into Equation (19.7) yields

$$q_{uu} = q_{xx}\sin^2 t + q_{yy}\cos^2 t + 2q_{xy}\frac{\sin 2t}{2} \quad (19.8)$$

Regrouping the first two terms and adding the necessary terms to Equation (19.8) to maintain equality yields

$$q_{uu} = \frac{q_{xx} + q_{yy}}{2}(\sin^2 t + \cos^2 t) + \frac{q_{xx}\sin^2 t}{2} + \frac{q_{yy}\cos^2 t}{2}$$

$$- \frac{q_{yy}\sin^2 t}{2} - \frac{q_{xx}\cos^2 t}{2} + q_{xy}\sin 2t \quad (19.9)$$

Substituting identity (*c*) and regrouping Equation (19.9) results in

$$q_{uu} = \frac{q_{xx} + q_{yy}}{2} + \frac{q_{yy}}{2}(\cos^2 t - \sin^2 t)$$

$$- \frac{q_{xx}}{2}(\cos^2 t - \sin^2 t) + q_{xy} \sin 2t \quad (19.10)$$

Finally, substituting identity (b) into Equation (19.10) yields

$$q_{uu} = \frac{q_{xx} + q_{yy}}{2} + \frac{q_{yy} - q_{xx}}{2} \cos 2t + q_{xy} \sin 2t \quad (19.11)$$

To find the value of t that maximizes q_{uu}, differentiate q_{uu} in Equation (19.8) with respect to t and set the results equal to zero. This results in

$$\frac{dq_{uu}}{dt} = -\frac{q_{yy} - q_{xx}}{2} 2 \sin 2t + 2q_{xy} \cos 2t = 0 \quad (19.12)$$

Reducing Equation (19.12) yields

$$(q_{yy} - q_{xx}) \sin 2t = 2q_{xy} \cos 2t \quad (19.13)$$

Finally, dividing Equation (19.13) by $\cos 2t$ yields

$$\frac{\sin 2t}{\cos 2t} = \tan 2t = \frac{2q_{xy}}{q_{yy} - q_{xx}} \quad (19.14a)$$

Equation (19.14a) is used to compute $2t$ and hence the desired angle t that yields the maximum value of q_{uu}. Note that the correct quadrant of $2t$ must be determined by noting the sign of the numerator and denominator in Equation (19.14a) before dividing by 2 to obtain t. Table 19.1 shows the proper quadrant for the different possible sign combinations of the

TABLE 19.1 Selection of the Proper Quadrant for $2t$[a]

Algebraic Sign of		
sin 2t	cos 2t	Quadrant
+	+	1
+	−	2
−	−	3
−	+	4

[a] When calculating for t, always remember to select the proper quadrant of $2t$ before dividing by 2.

19.2 COMPUTATION OF ELLIPSE ORIENTATION AND SEMIAXES

numerator and denominator.

Table 19.1 can be avoided by using the atan2 function available in most software packages. This function returns a value between $-180° \leq 2t \leq 180°$. If the value returned is negative, the correct value for $2t$ is obtained by adding 360°. Correct use of the atan2 function is

$$2t = \tan(q_{yy} - q_{xx}, 2q_{xy}) \tag{19.14b}$$

The use of Equation (19.14b) is demonstrated in the Mathcad worksheet for this chapter on the CD that accompanies this book.

Correlation between the latitude and departure of a station was discussed in Chapter 8. Similarly, the adjusted coordinates of a station are also correlated. Computing the value of t that yields the maximum and minimum values for the semiaxes is equivalent to rotating the covariance matrix until the off-diagonals are nonzero. Thus, the u and v coordinate values will be uncorrelated, which is equivalent to setting the q_{uv} element of Equation (19.6) equal to zero. Using the trigonometric identities noted previously, the element q_{uv} from Equation (19.6) can be written as

$$q_{uv} = \frac{q_{xx} - q_{yy}}{2} \sin 2t + q_{xy} \cos 2t \tag{19.15}$$

Setting q_{uv} equal to zero and solving for t gives us

$$\frac{\sin 2t}{\cos 2t} = \tan 2t = \frac{2q_{xy}}{q_{yy} - q_{xx}} \tag{19.16}$$

Note that this yields the same result as Equation (19.14), which verifies the removal of the correlation.

Step 5: In a fashion similar to that demonstrated in step 4, the q_{vv} element of Equation (19.6) can be rewritten as

$$q_{vv} = q_{xx} \cos^2 t - 2q_{xy} \cos t \sin t + q_{yy} \sin^2 t \tag{19.17}$$

In summary, the t angle, semimajor ellipse axis (q_{uu}), and semiminor axis (q_{vv}) are calculated using Equations (19.14), (19.7), and (19.17), respectively. These equations are repeated here, in order, for convenience. Note that these equations use only elements from the covariance matrix.

$$\tan 2t = \frac{2q_{xy}}{q_{yy} - q_{xx}} \tag{19.18}$$

$$q_{uu} = q_{xx} \sin^2 t + 2q_{xy} \cos t \sin t + q_{yy} \cos^2 t \tag{19.19}$$

376 ERROR ELLIPSE

$$q_{vv} = q_{xx} \cos^2 t - 2q_{xy} \cos t \sin t + q_{yy} \sin^2 t \qquad (19.20)$$

Equation (19.18) gives the t angle that the u axis makes with the y axis. Equation (19.19) yields the numerical value for q_{uu}, which when multiplied by the reference variance S_0^2 gives the variance along the u axis. The square root of the variance is the *semimajor axis* of the standard error ellipse. Equation (19.20) yields the numerical value for q_{vv}, which when multiplied by S_0^2 gives the variance along the v axis. The square root of this variance is the *semiminor axis* of the standard error ellipse. Thus, the semimajor and semiminor axes are

$$S_u = S_0\sqrt{q_{uu}} \quad \text{and} \quad S_v = S_0\sqrt{q_{vv}} \qquad (19.21)$$

19.3 EXAMPLE PROBLEM OF STANDARD ERROR ELLIPSE CALCULATIONS

In this section the error ellipse data for the trilateration example in Section 14.5 are calculated. From the computer listing given for the solution of that problem, the following values are recalled:

1. $S_0 = \pm 0.136$ ft
2. The unknown X and covariance Q_{xx} matrices were

$$X = \begin{bmatrix} dX_{\text{Wisconsin}} \\ dY_{\text{Wisconsin}} \\ dX_{\text{Campus}} \\ dY_{\text{Campus}} \end{bmatrix}$$

$$Q_{xx} = \begin{bmatrix} 1.198574 & -1.160249 & -0.099772 & -1.402250 \\ -1.160249 & 2.634937 & 0.193956 & 2.725964 \\ -0.099772 & 0.193956 & 0.583150 & 0.460480 \\ -1.402250 & 2.725964 & 0.460480 & 3.962823 \end{bmatrix}$$

19.3.1 Error Ellipse for Station Wisconsin

The tangent of $2t$ is

$$\tan 2t = \frac{2(-1.160249)}{2.634937 - 1.198574} = -1.6155$$

Note that the sign of the numerator is negative and the denominator is positive. Thus, from Table 19.1, angle $2t$ is in the fourth quadrant and $360°$ must be added to the computed angle. Hence,

19.3 EXAMPLE PROBLEM OF STANDARD ERROR ELLIPSE CALCULATIONS

$$2t = \tan^{-1}(-1.6155) = -58°14.5' + 360° = 301°45.5'$$

$$t = 150°53'$$

Substituting the appropriate values into Equation (19.21), S_u is

$$S_u = \pm 0.136$$
$$\times \sqrt{1.198574 \sin^2 t + 2(-1.160249) \cos t \sin t + 2.634937 \cos^2 t}$$
$$= \pm 0.25 \text{ ft}$$

Similarly substituting the appropriate values into Equation (19.21), S_v is

$$S_v = \pm 0.136$$
$$\times \sqrt{1.198574 \cos^2 t - 2(-1.160249) \cos t \sin t + 2.634937 \sin^2 t}$$
$$= \pm 0.10 \text{ ft}$$

Note that the standard deviations in the coordinates, as computed by Equation (13.24), are

$$S_x = S_0 \sqrt{q_{xx}} = \pm 0.136 \sqrt{1.198574} = \pm 0.15 \text{ ft}$$

$$S_y = S_0 \sqrt{q_{yy}} = \pm 0.136 \sqrt{2.634937} = \pm 0.22 \text{ ft}$$

19.3.2 Error Ellipse for Station Campus

Using the same procedures as in Section 19.3.1, the error ellipse data for station Campus are

$$2t = \tan^{-1} \frac{2 \times 0.460480}{3.962823 - 0.583150} = 15°14'$$

$$t = 7°37'$$

$$S_u = \pm 0.136$$
$$\times \sqrt{0.583150 \sin^2 t + 2(0.460480) \cos t \sin t + 3.962823 \cos^2 t}$$
$$= \pm 0.27 \text{ ft}$$

$$S_v = \pm 0.136$$
$$\times \sqrt{0.583150 \cos^2 t - 2(0.460480) \cos t \sin t + 3.962823 \sin^2 t}$$
$$= \pm 0.10 \text{ ft}$$

$$S_x = S_0\sqrt{q_{xx}} = \pm 0.136\sqrt{0.583150} = \pm 0.10 \text{ ft}$$

$$S_y = S_0\sqrt{q_{yy}} = \pm 0.136\sqrt{3.962823} = \pm 0.27 \text{ ft}$$

19.3.3 Drawing the Standard Error Ellipse

To draw the error ellipses for stations Wisconsin and Campus of Figure 19.5, the error rectangle is first constructed by laying out the values of S_x and S_y using a convenient scale along the x and y adjustment axes, respectively. For this example, an ellipse scale of 4800 times the map scale was selected. The t angle is laid off clockwise from the positive y axis to construct the u axis. The v axis is drawn 90° counterclockwise from u to form a right-handed coordinate system. The values of S_u and S_v are laid off along the U and V axes, respectively, to locate the semiaxis points. Finally, the ellipse is constructed so that it is tangent to the error rectangle and passes through its semiaxes points (refer to Figure 19.3).

19.4 ANOTHER EXAMPLE PROBLEM

In this section, the standard error ellipse for station u in the example of Section 16.4 is calculated. For the adjustment, $S_0 = \pm 1.82$ ft, and the X and Q_{xx} matrices are

$$X = \begin{bmatrix} dx_u \\ dy_u \end{bmatrix} \quad Q_{xx} = \begin{bmatrix} q_{xx} & q_{xy} \\ q_{xy} & q_{yy} \end{bmatrix} = \begin{bmatrix} 0.000532 & 0.000602 \\ 0.000602 & 0.000838 \end{bmatrix}$$

Error ellipse calculations are

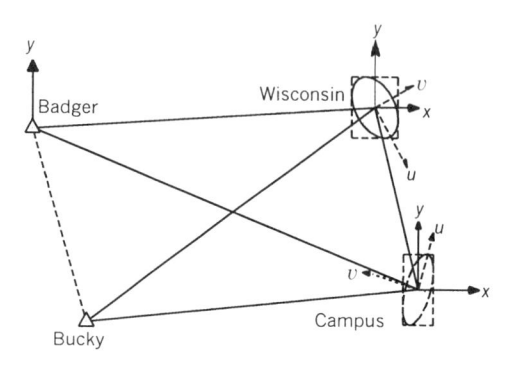

Figure 19.5 Graphical representation of error ellipses.

$$2t = \tan^{-1} \frac{2 \times 0.000602}{0.000838 - 0.000532} = 75°44'$$

$$t = 37°52'$$

$$S_u = \pm 1$$
$$\times \sqrt{0.000532 \sin^2 t + 2(0.000602) \cos t \sin t + 0.000838 \cos^2 t}$$
$$= \pm 0.036 \text{ ft}$$

$$S_v = \pm 1$$
$$\times \sqrt{0.000532 \cos^2 t - 2(0.000602) \cos t \sin t + 0.000838 \sin^2 t}$$
$$= \pm 0.008 \text{ ft}$$

$$S_x = \pm 1\sqrt{0.000532} = \pm 0.023 \text{ ft}$$

$$S_y = \pm 1\sqrt{0.000838} = \pm 0.029 \text{ ft}$$

Figure 19.6 shows the plotted standard error ellipse and its error rectangle.

19.5 ERROR ELLIPSE CONFIDENCE LEVEL

The calculations in Sections 19.3 and 19.4 produce *standard error ellipses.* These ellipses can be modified to produce error ellipses at any confidence level α by using an F statistic with two numerator degrees of freedom and the degrees of freedom for the adjustment in the denominator. Since the F statistic represents variance ratios for varying degrees of freedom, it can be

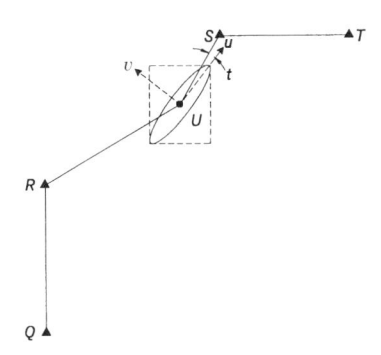

Figure 19.6 Graphical representation of error ellipse.

TABLE 19.2 $F_{\alpha,2,\text{degrees of freedom}}$ **Statistics for Selected Probability Levels**

Degrees of Freedom	Probability		
	90%	95%	99%
1	49.50	199.5	4999.50
2	9.00	19.00	99.00
3	5.46	9.55	30.82
4	4.32	6.94	18.00
5	3.78	5.79	13.27
10	2.92	4.10	7.56
15	2.70	3.68	6.36
20	2.59	3.49	5.85
30	2.49	3.32	5.39
60	2.39	3.15	4.98

expected that with increases in the number of degrees of freedom, there will be corresponding increases in precision. The $F_{\alpha,2,\text{degrees of freedom}}$ statistic modifier for various confidence levels is listed in Table 19.2. Notice that as the degrees of freedom increase, the F statistic modifiers decrease rapidly and begin to stabilize for larger degrees of freedom. The confidence level of an error ellipse can be increased to any level by using the multiplier

$$c = \sqrt{2F_{(\alpha,2,\text{degrees of freedom})}} \qquad (19.22)$$

Using the following equations, the percent probability for the semimajor and semiminor axes can be computed as

$$S_{u\%} = S_u c = S_u \sqrt{2F_{(\alpha,2,\text{degrees of freedom})}}$$

$$S_{v\%} = S_v c = S_v \sqrt{2F_{(\alpha,2,\text{degrees of freedom})}}$$

From the foregoing it should be apparent that as the number of degrees of freedom (redundancies) increases, precision increases, and the error ellipse sizes decrease. Using the techniques discussed in Chapter 4, the values listed in Table 19.2 are for the F distribution at 90% ($\alpha = 0.10$), 95% ($\alpha = 0.05$), and 99% ($\alpha = 0.01$) probability. These probabilities are most commonly used. Values from this table can be substituted into Equation (19.22) to determine the value of c for the probability selected and the given number of redundancies in the adjustment. This table is for convenience only and does not contain the values necessary for all situations that might arise.

Example 19.1 Calculate the 95% error ellipse for station Wisconsin of Section 19.3.1.

SOLUTION Using Equations (19.23) yields

$$S_{u95\%} = \pm 0.25\sqrt{2(199.50)} = \pm 4.99 \text{ ft}$$

$$S_{v95\%} = \pm 0.10\sqrt{2(199.50)} = \pm 2.00 \text{ ft}$$

$$S_{x95\%} = \sqrt{2(199.50)}S_x = \pm(19.97 \times 0.15) \pm 3.00 \text{ ft}$$

$$S_{y95\%} = \sqrt{2(199.50)}S_y = \pm(19.97 \times 0.22) \pm 4.39 \text{ ft}$$

The probability of the standard error ellipse can be found by setting the multiplier $2F_{\alpha,v_1,v_2}$ equal to 1 so that $F_{\alpha,v_1,v_2} = 0.5$. For a simple closed traverse with three degrees of freedom, this means that $F_{\alpha,2,3} = 0.5$. The value of α that satisfies this condition is 0.65 and was found by trial-and-error procedures using the program STATS. Thus, the percent probability of the standard error ellipse in a simple closed traverse is $(1 - 0.65) \times 100\%$, or 35%. The reader is encouraged to verify this result using program STATS. It is left as an exercise for the reader to show that the percent probability for the standard error ellipse ranges from 35% to only 39% for horizontal surveys that have fewer than 100 degrees of freedom.

19.6 ERROR ELLIPSE ADVANTAGES

Besides providing critical information regarding the precision of an adjusted station position, a major advantage of error ellipses is that they offer a method of making a visual comparison of the relative precisions between any two stations. By viewing the shapes, sizes, and orientations of error ellipses, various surveys can be compared rapidly and meaningfully.

19.6.1 Survey Network Design

The sizes, shapes, and orientations of error ellipses depend on (1) the control used to constrain the adjustment, (2) the observational precisions, and (3) the geometry of the survey. The last two of these three elements are variables that can be altered in the design of a survey to produce optimal results. In designing surveys that involve the traditional observations of distances and angles, estimated precisions can be computed for observations made with differing combinations of equipment and field procedures. Also, trial variations in station placement, which creates the network geometry, can be made. Then these varying combinations of observations and geometry can be processed through least squares adjustments and the resulting station error ellipses computed, plotted, and checked against the desired results. Once acceptable goals are achieved in this process, the observational equipment, field procedures, and network geometry that provide these results can be adopted. This overall process is called *network design*. It enables surveyors to select the equipment and field techniques, and to decide on the number and locations of stations that provide the highest precision at lowest cost. This can be done in the office using simulation software before bidding a contract or entering the field.

In designing networks to be surveyed using the traditional observations of distance, angle, and directions, it is important to understand the relationships of those observations to the resulting positional uncertainties of the stations. The following relationships apply:

1. Distance observations strengthen the positions of stations in directions collinear with the lines observed.
2. Angle and direction observations strengthen the positions of stations in directions perpendicular to the lines of sight.

A simple analysis made with reference to Figure 19.1 should clarify the two relationships above. Assume first that the length of line AB was measured precisely but its direction was not observed. Then the positional uncertainty of station B should be held within close tolerances by the distance observed, but it would only be held in the direction collinear with AB. The distance observation would do nothing to keep line AB from rotating, and in fact the position of B would be weak perpendicular to AB. On the other hand, if the direction of AB had been observed precisely but its length had not been measured, the positional strength of station B would be strongest in the direction perpendicular to AB. But an angle observation alone does nothing to fix distances between observed stations, and thus the position of station B would be weak along line AB. If both the length and direction AB were observed with equal precision, a positional uncertainty for station B that is more uniform in all directions would be expected. In a survey network that consists of many interconnected stations, analyzing the effects of observations is not quite as simple as was just demonstrated for the single line AB. Nevertheless, the two basic relationships stated above still apply.

Uniform positional strength in all directions for all stations is the desired goal in survey network design. This would be achieved if, following least squares adjustment, all error ellipses were circular in shape and of equal size. Although this goal is seldom completely possible, by diligently analyzing various combinations of geometric figures together with different combinations of observations and precisions, this goal can be approached. Sometimes, however, other overriding factors, such as station accessibility, terrain, and vegetation, preclude actual use of an optimal design.

The network design process discussed above is significantly aided by the use of aerial photos and/or topographic maps. These products enable the layout of trial station locations and permits analysis of the accessibility and intervisibility of these stations to be investigated. However, a field reconnaissance should be made before adopting the final design.

The global positioning system (GPS) has brought about dramatic changes in all areas of surveying, and network design is not an exception. Although GPS does require overhead visibility at each receiver station for tracking satellites, problems of intervisibility between ground stations is eliminated. Thus, networks having uniform geometry can normally be laid out. Because each station in the network is occupied in a GPS survey, the *XYZ* coordinates of the stations can be determined. This simplifies the problem of designing networks to attain error ellipses of uniform shapes and sizes. However, the geometric configuration of satellites is an important factor that affects station precisions. The *positional dilution of precision* (PDOP) can be a guide to the geometric strength of the observed satellites. In this case, the lower the PDOP, the stronger the satellite geometry. For more discussion on designing GPS surveys, readers are referred to books devoted to the subject of GPS surveying.

19.6.2 Example Network

Figure 19.7 shows error ellipses for two survey networks. Figure 19.7(*a*) illustrates the error ellipses from a trilateration survey with the nine stations,

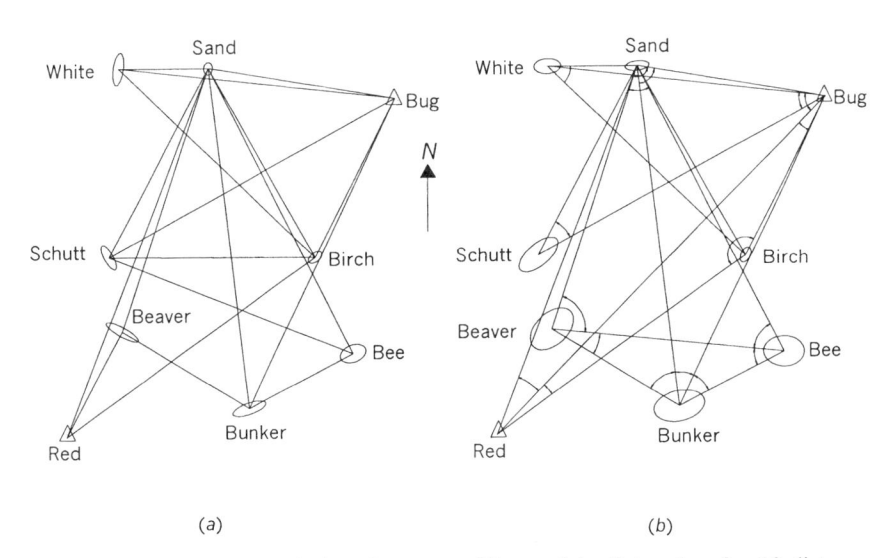

Figure 19.7 Network analysis using error ellipses: (*a*) trilateration for 19 distances; (*b*) triangulation for 19 angles.

two of which (Red and Bug) were control stations. The survey includes 19 distance observations and five degrees of freedom. Figure 19.7(b) shows the error ellipses of the same network that was observed using triangulation and a baseline from stations Red to Bug. This survey includes 19 observed angles and thus also has five degrees of freedom.

With respect to these two figures, and keeping in mind that the smaller the ellipse, the higher the precision, the following general observations can be made:

1. In both figures, stations Sand and Birch have the highest precisions. This, of course, is expected due to their proximity to control station Bug and because of the density of observations made to these stations, which included direct measurements from both control stations.
2. The large size of error ellipses at stations Beaver, Schutt, Bunker, and Bee of Figure 19.7(b) show that they have lower precision. This, too, is expected because there were fewer observations made to those stations. Also, neither Beaver nor Bee was connected directly by observation to either of the control stations.
3. Stations White and Schutt of Figure 19.7(a) have relatively high east–west precisions and relatively low north–south precisions. Examination of the network geometry reveals that this could be expected. Distance measurements to those two points from station Red, plus an observed distance between White and Schutt, would have greatly improved the north–south precision.
4. Stations Beaver and Bunker of Figure 19.7(a) have relatively low precisions east–west and relatively high precisions north–south. Again, this is expected when examining the network geometry.
5. The smaller error ellipses of Figure 19.7(a) suggest that the trilateration survey will yield superior precision to the triangulation survey of Figure 19.7(b). This is expected since the EDM had a stated uncertainty of ±(5 mm + 5 ppm). In a 5000-ft distance this yields an uncertainty of ±0.030 ft. To achieve the same precision, the comparable angle uncertainty would need to be

$$\theta'' = \frac{S}{R} \rho = \pm \frac{\pm 0.030}{5000} 206{,}264.8''/\text{rad} = \pm 1.2''$$

The proposed instrument and field procedures for the project that yielded the error ellipses of Figure 19.7(b) had an expected uncertainty of only ±6". Very probably, this ultimate design would include both observed distances and angles.

These examples serve to illustrate the value of computing station error ellipses in an a priori analysis. The observations were made easily and quickly

TABLE 19.3 Other Measures of Two-Dimensional Positional Uncertainties

Probability (%)	c	Name
35–39	1.00	Standard error ellipse
50.0	1.18	Circular error probable (CEP)
63.2	1.41	Distance RMS (DRMS)
86.5	2.00	Two-sigma ellipse
95.0	2.45	95% confidence level
98.2	2.83	2DRMS
98.9	3.00	Three-sigma ellipse

by comparison of the ellipses in the two figures. Similar information would have been difficult, if not impossible, to determine from standard deviations. By varying the survey it is possible ultimately to find a design that provides optimal results in terms of meeting a uniformly acceptable precision and survey economy.

19.7 OTHER MEASURES OF STATION UNCERTAINTY

Other measures of accuracies are sometimes called for in specifications. As discussed in Section 19.5, the standard error ellipse has a c-multiplier of 1.00 and a probability between 35 and 39%. Other common errors and probabilities are given in Table 19.3.

As demonstrated in the Mathcad worksheet on the CD that accompanies this book, the process of rotating the 2 × 2 block diagonal matrix for a station is the mathematical equivalent of orthogonalization. This process can be performed by computing eigenvalues and eigenvectors. For example, the eigenvalues of the 2 × 2 block diagonal matrix for station Wisconsin in Example 19.2 are 0.55222 and 3.28129, respectively. Thus, $S_{U\text{-Wis}}$ is $0.136\sqrt{3.28129} = \pm 0.25$ ft and $S_{V\text{-Wis}}$ is $0.136\sqrt{0.55222} = \pm 0.101$ ft.

TABLE 19.4 Measures of Three-Dimensional Positional Uncertainties

Probability (%)	c	Name
19.9	1.00	Standard ellipsoid
50.0	1.53	Spherical error probable (SEP)
61.0	1.73	Mean radical spherical error (MRSE)
73.8	2.00	Two-sigma ellipsoid
95.0	2.80	95% confidence level
97.1	3.00	Three-sigma ellipsoid

386 ERROR ELLIPSE

This property can be used to compute the error ellipsoids for three-dimensional coordinates from a GPS adjustment or the three-dimensional geodetic network adjustment discussed in Chapter 23. That is, the uncertainties along the three orthogonal axes of the error ellipsoid can be computed using eigenvalues of the 3×3 block diagonal matrix appropriate for each station. The common measures for ellipsoids are listed in Table 19.4.

PROBLEMS

Note: For problems requiring least squares adjustment, if a computer program is not distinctly specified for use in the problem, it is expected that the least squares algorithm will be solved using the program MATRIX, which is included on the CD supplied with the book.

19.1 Calculate the semiminor and semimajor axes of the standard error ellipse for the adjusted position of station U in Example 14.2. Plot the figure using a scale of 1:12,000 and the error ellipse using an appropriate scale.

19.2 Calculate the semiminor and semimajor axes of the 95% confidence error ellipse for Problem 19.1. Plot this ellipse superimposed over the ellipse of Problem 19.1.

19.3 Repeat Problem 19.1 for Example 15.1.

19.4 Repeat Problem 19.2 for Problem 19.3.

19.5 Repeat Problem 19.1 for the adjusted position of stations B and C of Example 15.3. Use a scale of 1:24,000 for the figure and plot the error ellipses using an appropriate multiplication factor.

Calculate the error ellipse data for the unknown stations in each problem.

19.6 Problem 15.2

19.7 Problem 15.5

19.8 Problem 15.6

19.9 Problem 15.9

19.10 Problem 15.12

19.11 Problem 16.2

19.12 Problem 16.3

19.13 Problem 16.7

19.14 Problem 16.9

Using a level of significance of 0.05, compute the 95% probable error ellipse for the stations in each problem.

19.15 Problem 15.2

19.16 Problem 16.3

19.17 Using the program STATS, determine the percent probability of the standard error ellipse for a horizontal survey with:
  (a) 3 degrees of freedom.
  (b) 9 degrees of freedom.
  (c) 20 degrees of freedom.
  (d) 50 degrees of freedom.
  (e) 100 degrees of freedom.

Programming Problems

19.18 Develop a computational program that takes the Q_{xx} matrix and S_0^2 from a horizontal adjustment and computes error ellipse data for the unknown stations.

19.19 Develop a computational program that does the same as described for Problems 19.15 and 19.16.

CHAPTER 20

CONSTRAINT EQUATIONS

20.1 INTRODUCTION

When doing an adjustment, it is sometimes necessary to fix an observation to a specific value. For instance, in Chapter 14 it was shown that the coordinates of a control station can be fixed by setting its dx and dy corrections to zero, and thus the corrections and their corresponding coefficients in the J matrix were removed from the solution. This is called a *constrained adjustment*. Another constrained adjustment occurs when the direction or length of a line is held to a specific value or when an elevation difference between two stations is fixed in differential leveling. In this chapter methods available for developing observational constraints are discussed. However, before discussing constraints, the procedure for including control station coordinates in an adjustment is described.

20.2 ADJUSTMENT OF CONTROL STATION COORDINATES

In examples in preceding chapters, when control station coordinates were excluded from the adjustments and hence their values were held fixed, constrained adjustments were being performed. That is, the observations were being forced to fit the control coordinates. However, control is not perfect and not all control is of equal reliability. This fact is evidenced by the fact that different orders of accuracy are used to classify control.

When more than minimal control is held fixed in an adjustment, the observations are forced to fit this control. For example, if the coordinates of two control stations are held fixed but their actual positions are not in agreement with the values given by their held coordinates, the observations will be adjusted to match the erroneous control. Simply stated, precise observations may be forced to fit less precise control. This was not a major problem in the days of transits and tapes, but it does happen with modern instrumentation. This topic is discussed in more detail in Chapter 21.

To clarify the problem further, suppose that a new survey is tied to two existing control stations set from two previous surveys. Assume that the precision of the existing control stations is only 1:10,000. A new survey uses equipment and field procedures designed to produce a survey of 1:20,000. Thus, it will have a higher accuracy than either of the control stations to which it must fit. If both existing control stations are fixed in the adjustment, the new observations must distort to fit the errors of the existing control stations. After the adjustment, their residuals will show a lower-order fit that matches the control. In this case it would be better to allow the control coordinates to adjust according to their assigned quality so that the observations are not distorted. However, it should be stated that the precision of the new coordinates relative to stations not in the adjustment can be only as good as the initial control.

The observation equations for control station coordinates are

$$x' = x + v_x \qquad (20.1)$$
$$y' = y + v_y$$

In Equation (20.1) x' and y' are the observed coordinate values of the control station, x and y the published coordinate values of the control station, and v_x and v_y the residuals for the respective published coordinate values.

To allow the control to adjust, Equations (20.1) must be included in the adjustment for each control station. To *fix* a control station in this scheme, high weights are assigned to the station's coordinates. Conversely, low weights will allow a control station's coordinates to adjust. In this manner, all control stations are allowed to adjust in accordance with their expected levels of accuracy. In Chapter 21 it is shown that when the control is included as observations, *poor* observations and control stations can be isolated in the adjustment by using weights.

Example 20.1 A trilateration survey was completed for the network shown in Figure 20.1 and the following observations collected:

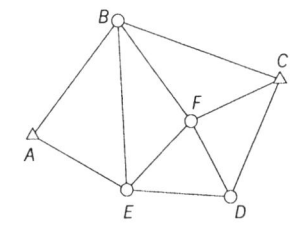

Figure 20.1 Trilateration network.

390 CONSTRAINT EQUATIONS

Control stations

Station	X (ft)	Y (ft)
A	10,000.00	10,000.00
C	12,487.08	10,528.65

Distance observations

From	To	Distance (ft)	σ (ft)	From	To	Distance (ft)	σ (ft)
A	B	1400.91	0.023	B	E	1644.29	0.023
A	E	1090.55	0.022	B	F	1217.54	0.022
B	C	1723.45	0.023	D	F	842.75	0.022
C	F	976.26	0.022	D	E	1044.99	0.022
C	D	1244.40	0.023	E	F	930.93	0.022

Perform a least squares adjustment of this survey, holding the control coordinates of stations A and C by appropriate weights (assume that these control stations have a precision of 1:10,000).

SOLUTION The J, X, and K matrices formed in this adjustment are

$$J = \begin{bmatrix}
\frac{\partial D_{AB}}{\partial x_A} & \frac{\partial D_{AB}}{\partial y_A} & \frac{\partial D_{AB}}{\partial x_B} & \frac{\partial D_{AB}}{\partial y_B} & 0 & 0 & 0 & 0 & 0 & 0 & 0 & 0 \\
\frac{\partial D_{AE}}{\partial x_A} & \frac{\partial D_{AE}}{\partial y_A} & 0 & 0 & 0 & 0 & 0 & 0 & \frac{\partial D_{AE}}{\partial x_E} & \frac{\partial D_{AE}}{\partial y_E} & 0 & 0 \\
0 & 0 & \frac{\partial D_{BE}}{\partial x_B} & \frac{\partial D_{BE}}{\partial y_B} & 0 & 0 & 0 & 0 & \frac{\partial D_{BE}}{\partial x_E} & \frac{\partial D_{BE}}{\partial y_E} & 0 & 0 \\
0 & 0 & \frac{\partial D_{BF}}{\partial x_B} & \frac{\partial D_{BF}}{\partial y_B} & 0 & 0 & 0 & 0 & 0 & 0 & \frac{\partial D_{BF}}{\partial x_F} & \frac{\partial D_{BF}}{\partial y_F} \\
0 & 0 & \frac{\partial D_{BC}}{\partial x_B} & \frac{\partial D_{BC}}{\partial y_B} & \frac{\partial D_{BC}}{\partial x_C} & \frac{\partial D_{BC}}{\partial y_C} & 0 & 0 & 0 & 0 & 0 & 0 \\
0 & 0 & 0 & 0 & \frac{\partial D_{CF}}{\partial x_C} & \frac{\partial D_{CF}}{\partial y_C} & 0 & 0 & 0 & 0 & \frac{\partial D_{CF}}{\partial x_F} & \frac{\partial D_{CF}}{\partial y_F} \\
0 & 0 & 0 & 0 & \frac{\partial D_{CD}}{\partial x_C} & \frac{\partial D_{CD}}{\partial y_C} & \frac{\partial D_{CD}}{\partial x_D} & \frac{\partial D_{CD}}{\partial y_D} & 0 & 0 & 0 & 0 \\
0 & 0 & 0 & 0 & 0 & 0 & \frac{\partial D_{DF}}{\partial x_D} & \frac{\partial D_{DF}}{\partial y_D} & 0 & 0 & \frac{\partial D_{DF}}{\partial x_F} & \frac{\partial D_{DF}}{\partial y_F} \\
0 & 0 & 0 & 0 & 0 & 0 & \frac{\partial D_{DE}}{\partial y_D} & \frac{\partial D_{DE}}{\partial y_D} & \frac{\partial D_{DE}}{\partial x_E} & \frac{\partial D_{DE}}{\partial y_E} & 0 & 0 \\
0 & 0 & 0 & 0 & 0 & 0 & 0 & 0 & \frac{\partial D_{EF}}{\partial x_E} & \frac{\partial D_{EF}}{\partial y_E} & \frac{\partial D_{EF}}{\partial x_F} & \frac{\partial D_{EF}}{\partial y_F} \\
1 & 0 & 0 & 0 & 0 & 0 & 0 & 0 & 0 & 0 & 0 & 0 \\
0 & 1 & 0 & 0 & 0 & 0 & 0 & 0 & 0 & 0 & 0 & 0 \\
0 & 0 & 0 & 0 & 1 & 0 & 0 & 0 & 0 & 0 & 0 & 0 \\
0 & 0 & 0 & 0 & 0 & 1 & 0 & 0 & 0 & 0 & 0 & 0 \\
\end{bmatrix}$$

20.2 ADJUSTMENT OF CONTROL STATION COORDINATES

$$X = \begin{bmatrix} dx_A \\ dy_A \\ dx_B \\ dy_B \\ dx_C \\ dy_C \\ dx_D \\ dy_D \\ dx_E \\ dy_E \\ dx_F \\ dy_F \end{bmatrix} \quad K = \begin{bmatrix} L_{AB} - AB_0 \\ L_{AE} - AE_0 \\ L_{BE} - BE_0 \\ L_{BF} - BF_0 \\ L_{BC} - BC_0 \\ L_{CF} - CF_0 \\ L_{CD} - CD_0 \\ L_{DF} - DF_0 \\ L_{DE} - DE_0 \\ L_{EF} - EF_0 \\ x_A - x_{A_0} \\ y_A - y_{A_0} \\ x_C - x_{C_0} \\ y_C - y_{C_0} \end{bmatrix}$$

Notice that the last four rows of the J matrix correspond to observation equations (20.1) for the coordinates of control stations A and C. Each coordinate has a row with *one* in the column corresponding to its correction. Obviously, by including the control station coordinates, four unknowns have been added to the adjustment: dx_A, dy_A, dx_C, and dy_C. However, four observations have also been added. Therefore, the number of redundancies is unaffected by adding the coordinate observation equations. That is, the adjustment has the same number of redundancies with or without the control equations.

It is possible to weight a control station according to the precision of its coordinates. Unfortunately, control stations are published with distance precisions rather than the covariance matrix elements that are required for weighting. However, estimates of the standard deviations of the coordinates can be computed from the published distance precisions. That is, if the distance precision between stations A and C is $1:10,000$ or better, their coordinates should have estimated errors that yield a distance precision of $1:10,000$ between the stations. To find the estimated errors in the coordinates that yield the appropriate distance precision between the stations, Equation (6.16) can be applied to the distance formula, resulting in

$$\sigma_{D_{ij}}^2 = \left(\frac{\partial D_{ij}}{\partial x_i}\right)^2 \sigma_{x_i}^2 + \left(\frac{\partial D_{ij}}{\partial y_i}\right)^2 \sigma_{y_i}^2 + \left(\frac{\partial D_{ij}}{\partial x_j}\right)^2 \sigma_{x_j}^2 + \left(\frac{\partial D_{ij}}{\partial y_j}\right)^2 \sigma_{y_j}^2 \quad (20.2)$$

In Equation (20.2), $\sigma_{D_{ij}}^2$ is the variance in distance D_{ij}, and $\sigma_{x_i}^2$, $\sigma_{y_i}^2$, $\sigma_{x_j}^2$, and $\sigma_{y_j}^2$ are the variances in the coordinates of the endpoints of the line. Assuming that the estimated errors for the coordinates of Equation (20.2) are equal and substituting in the appropriate partial derivatives yields

$$\sigma_{D_{ij}}^2 = 2\left(\frac{\partial D_{ij}}{\partial x}\right)^2 \sigma_x^2 + 2\left(\frac{\partial D_{ij}}{\partial y}\right)^2 \sigma_y^2 = 2\left(\frac{\Delta x}{IJ}\sigma_c\right)^2 + 2\left(\frac{\Delta x}{IJ}\sigma_c\right)^2 \quad (20.3)$$

392 CONSTRAINT EQUATIONS

In Equation (20.3), σ_c is the standard deviation in the x and y coordinates. (Note that the partial derivatives appearing in Equation (20.3) were described in Section 14.2.) Factoring $2\sigma_c^2$ from Equation (20.3) yields

$$\sigma_{D_{ij}}^2 = 2\sigma_c^2 \frac{\Delta x^2 + \Delta y^2}{IJ^2} = 2\sigma_c^2 \tag{20.4}$$

where

$$\sigma_c^2 = \sigma_x^2 = \sigma_y^2.$$

From the coordinates of A and C, distance AC is 2542.65 ft. To get a distance precision of 1:10,000, a maximum distance error of ± 0.25 ft exists. Assuming equal coordinate errors,

$$\pm 0.25 = \sigma_x \sqrt{2} = \sigma_y \sqrt{2}$$

Thus, $\sigma_x = \sigma_y = \pm 0.18$ ft. The standard deviations computed are used to weight the control in the adjustment. The weight matrix for this adjustment is

$$W = \begin{bmatrix}
\frac{1}{0.023^2} & 0 & 0 & 0 & 0 & 0 & 0 & 0 & 0 & 0 & 0 & 0 & 0 & 0 \\
0 & \frac{1}{0.022^2} & 0 & 0 & 0 & 0 & 0 & 0 & 0 & 0 & 0 & 0 & 0 & 0 \\
0 & 0 & \frac{1}{0.023^2} & 0 & 0 & 0 & 0 & 0 & 0 & 0 & 0 & 0 & 0 & 0 \\
0 & 0 & 0 & \frac{1}{0.022^2} & 0 & 0 & 0 & 0 & 0 & 0 & 0 & 0 & 0 & 0 \\
0 & 0 & 0 & 0 & \frac{1}{0.023^2} & 0 & 0 & 0 & 0 & 0 & 0 & 0 & 0 & 0 \\
0 & 0 & 0 & 0 & 0 & \frac{1}{0.022^2} & 0 & 0 & 0 & 0 & 0 & 0 & 0 & 0 \\
0 & 0 & 0 & 0 & 0 & 0 & \frac{1}{0.022^2} & 0 & 0 & 0 & 0 & 0 & 0 & 0 \\
0 & 0 & 0 & 0 & 0 & 0 & 0 & \frac{1}{0.022^2} & 0 & 0 & 0 & 0 & 0 & 0 \\
0 & 0 & 0 & 0 & 0 & 0 & 0 & 0 & \frac{1}{0.022^2} & 0 & 0 & 0 & 0 & 0 \\
0 & 0 & 0 & 0 & 0 & 0 & 0 & 0 & 0 & \frac{1}{0.022^2} & 0 & 0 & 0 & 0 \\
0 & 0 & 0 & 0 & 0 & 0 & 0 & 0 & 0 & 0 & \frac{1}{0.18^2} & 0 & 0 & 0 \\
0 & 0 & 0 & 0 & 0 & 0 & 0 & 0 & 0 & 0 & 0 & \frac{1}{0.18^2} & 0 & 0 \\
0 & 0 & 0 & 0 & 0 & 0 & 0 & 0 & 0 & 0 & 0 & 0 & \frac{1}{0.18^2} & 0 \\
0 & 0 & 0 & 0 & 0 & 0 & 0 & 0 & 0 & 0 & 0 & 0 & 0 & \frac{1}{0.18^2}
\end{bmatrix}$$

20.2 ADJUSTMENT OF CONTROL STATION COORDINATES

The adjustment results, obtained using the program ADJUST, are shown below.

```
         **** Adjusted Distance Observations ****
No. |    From  |   To  |   Distance   |  Residual  |
================================================================
  1|     A    |   B   |   1,400.910  |   -0.000   |
  2|     A    |   E   |   1,090.550  |   -0.000   |
  3|     B    |   E   |   1,644.288  |   -0.002   |
  4|     B    |   F   |   1,217.544  |    0.004   |
  5|     B    |   C   |   1,723.447  |   -0.003   |
  6|     C    |   F   |     976.263  |    0.003   |
  7|     C    |   D   |   1,244.397  |   -0.003   |
  8|     D    |   E   |   1,044.988  |   -0.002   |
  9|     E    |   F   |     930.933  |    0.003   |
 10|     D    |   F   |     842.753  |    0.003   |
=================

          ****** Adjusted Control Stations ******
No. | Sta. |   Northing  |   Easting   |  N Res  |  E Res  |
================================================================
  1|   C  |  10,528.650 |  12,487.080 |  0.000  |  0.002  |
  2|   A  |  10,000.000 |  10,000.000 | -0.000  | -0.002  |
================================================================

          Reference Standard Deviation = 0.25
                Degrees of Freedom = 2


          ******* Adjusted Unknowns ******
Station| Northing  | Easting  |σ North |σ East | t ang° |A axis |B axis |
================================================================
  A| 10,000.000|  9,999.998|  0.033|  0.045| 168.000|  0.046|  0.033|
  B| 11,103.933| 10,862.483|  0.039|  0.034|  65.522|  0.040|  0.033|
  C| 10,528.650| 12,487.082|  0.033|  0.045| 168.000|  0.046|  0.033|
  D|  9,387.462| 11,990.882|  0.040|  0.038|  49.910|  0.044|  0.033|
  E|  9,461.900| 10,948.549|  0.039|  0.034| 110.409|  0.039|  0.033|
  F| 10,131.563| 11,595.223|  0.033|  0.034|  17.967|  0.034|  0.033|
================================================================
```

Notice that the control stations were adjusted slightly, as evidenced by their residuals. Also note the error ellipse data computed for each control station.

20.3 HOLDING CONTROL STATION COORDINATES AND DIRECTIONS OF LINES FIXED IN A TRILATERATION ADJUSTMENT

As demonstrated in Example 14.1, the coordinates of a control station are easily fixed during an adjustment. This is accomplished by assigning values of zero to the coefficients of the dx and dy correction terms. This method removes their corrections from the equations. In that particular example, each observation equation had only two unknowns, since one end of each observed distance was a control station that was held fixed during the adjustment. This was a special case of a method known as *solution by elimination of constraints*.

This method can be shown in matrix notation as

$$A_1 X_1 + A_2 X_2 = L_1 + V \qquad (20.5)$$

$$C_1 X_1 + C_2 X_2 = L_2 \qquad (20.6)$$

In Equation (20.6), A_1, A_2, X_1, X_2, L_1, and L_2 are the A, X, and L matrices partitioned by the constraint equations, as shown in Figure 20.2; C_1 and C_2 are the partitions of the matrix C, consisting of the coefficients of the constraint equations; and V is the residual matrix. In this method, matrices A, C, and X are partitioned into two matrix equations that separate the constrained and unconstrained observations. Careful consideration should be given to the partition of C_1 since this matrix cannot be singular. If singularity exists, a new set of constraint equations that are mathematically independent must be determined. Also, since each constraint equation will remove one parameter from the adjustment, the number of constraints must not be so large that the remaining A_1 and X_1 have no independent equations or are themselves singular.

From Equation (20.6), solve for X_1 in terms of C_1, C_2, X_2, and L_2 as

$$X_1 = C_1^{-1}(L_2 - C_2 X_2) \qquad (20.7)$$

Substituting Equation (20.7) into Equation (20.5) yields

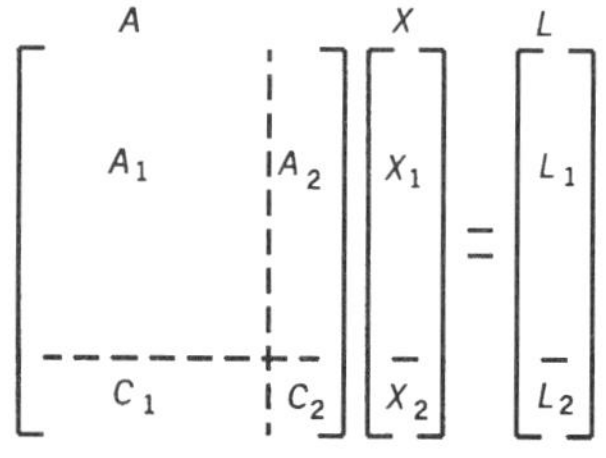

$$\begin{bmatrix} A_1 & | & A_2 \\ \hline C_1 & | & C_2 \end{bmatrix} \begin{bmatrix} X_1 \\ \hline X_2 \end{bmatrix} = \begin{bmatrix} L_1 \\ \hline L_2 \end{bmatrix}$$

Figure 20.2 A, X, and L matrices partitioned by constraint equations.

20.3 HOLDING CONTROL STATION COORDINATES AND DIRECTIONS OF LINES FIXED

$$A_1 [C_1^{-1}(L_2 - C_2 X_2)] + A_2 X_2 = L_1 + V \tag{20.8}$$

Rearranging Equation (20.8), regrouping, and dropping V for the time being gives

$$(-A_1 C_1^{-1} C_2 + A_2) X_2 = L_1 - A_1 C_1^{-1} L_2 \tag{20.9}$$

Letting $A' = -A_1 C_1^{-1} C_2 + A_2$, Equation (20.9) can be rewritten as

$$A' X_2 = L_1 - A_1 C_1^{-1} L_2 \tag{20.10}$$

Now Equation (20.10) can be solved for X_2, which, in turn, is substituted into Equation (20.7) to solve for X_1.

It can be seen that in the solution by elimination of constraint, the constraints equations are used to eliminate unknown parameters from the adjustment, thereby fixing certain geometric conditions during the adjustment. This method was used when the coordinates of the control stations were *removed* from the adjustments in previous chapters. In the following subsection, this method is used to hold the azimuth of a line during an adjustment.

20.3.1 Holding the Direction of a Line Fixed by Elimination of Constraints

Using this method, constraint equations are written and then functionally substituted into the observation equations to eliminate unknown parameters. To illustrate, consider that in Figure 20.3, the direction of a line *IJ* is to be held fixed during the adjustment. Thus, the position of *J* is constrained to move linearly along *IJ* during the adjustment. If *J* moves to *J'* after adjustment, the relationship between the direction of *IJ* and dx_j and dy_j is

$$dx_j = dy_j \tan \alpha \tag{20.11}$$

For example, suppose that the direction of line *AB* in Figure 20.4 is to be held fixed during a trilateration adjustment. Noting that station *A* is to be held

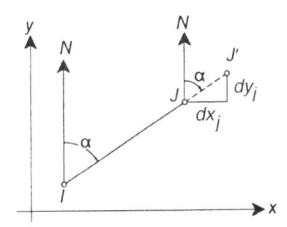

Figure 20.3 Holding direction *IJ* fixed.

396 CONSTRAINT EQUATIONS

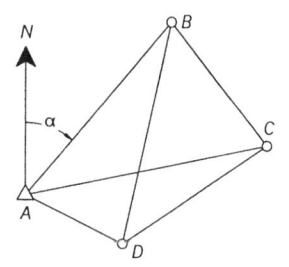

Figure 20.4 Holding direction AB fixed in a trilateration adjustment.

fixed and using prototype equation (14.9), the following observation equation results for observed distance *AB*:

$$k_{l_{ab}} + v_{l_b} = \frac{x_{b_0} - x_a}{AB_0} dx_b + \frac{y_{b_0} - y_a}{AB_0} dy_b \tag{20.12}$$

Now based on Equation (20.11), the following relationship is written for line *AB*:

$$dx_b = dy_b \tan \alpha \tag{20.13}$$

Substituting Equation (20.13) into Equation (20.12) yields

$$k_{l_{ab}} + v_{l_{ab}} = \frac{x_{b_0} - x_a}{AB_0} \tan \alpha \, dy_b + \frac{y_{b_0} - y_a}{AB_0} dy_b \tag{20.14}$$

Factoring dy_b in Equation (20.14), the constrained observation equation is

$$k_{l_{ab}} + v_{l_{ab}} = \frac{(x_{b_0} - x_a) \tan \alpha + (y_{b_0} - y_a)}{AB_0} dy_b \tag{20.15}$$

Using this same method, the coefficients of dy_b for lines *BC* and *BD* are also determined, resulting in the *J* matrix shown in Table 20.1.

For this example, the *K*, *X*, and *V* matrices are

$$K = \begin{bmatrix} AB - AB_0 \\ AC - AC_0 \\ AD - AD_0 \\ BC - BC_0 \\ BD - BD_0 \\ CD - CD_0 \end{bmatrix} \quad X = \begin{bmatrix} dy_b \\ dy_c \\ dx_c \\ dy_d \\ dx_d \end{bmatrix} \quad V = \begin{bmatrix} v_{ab} \\ v_{ac} \\ v_{ad} \\ v_{bb} \\ v_{bd} \\ v_{cd} \end{bmatrix}$$

TABLE 20.1 J Matrix of Figure 20.3

Distance	dy_b	dy_c	dx_c	dx_d	dy_d
AB	$[(x_b - x_a)\tan\alpha + (y_b - y_a)]/AB$	0	0	0	0
AC	0	$(y_c - y_a)/AC$	$(x_c - x_a)/AC$	0	0
AD	0	0	0	$(y_d - y_a)/AD$	$(x_d - x_a)/AD$
BC	$[(x_b - x_c)\tan\alpha + (y_b - y_c)]/BC$	$(y_c - y_b)/BC$	$(x_c - x_b)/BC$	0	0
BD	$[(x_b - x_d)\tan\alpha + (y_b - y_d)]/BD$	0	0	$(y_d - y_b)/BD$	$(x_d - x_b)/BD$
CD	0	$(y_c - y_d)/CD$	$(x_c - x_d)/CD$	$(y_d - y_c)/CD$	$(x_d - x_c)/CD$

Unknown

20.4 HELMERT'S METHOD

Another method of introducing constraints was presented by F. R. Helmert in 1872. In this procedure, the constraint equation(s) border the reduced normal equations as

$$\begin{bmatrix} A^TWA & \vdots & C^T \\ \cdots & \cdots & \cdots \\ C & \vdots & 0 \end{bmatrix} \begin{bmatrix} X_1 \\ \cdots \\ X_2 \end{bmatrix} = \begin{bmatrix} A^TWL_1 \\ \cdots \\ L_2 \end{bmatrix} \qquad (20.16)$$

To establish this matrix, the normal matrix and its matching constants matrix are formed, as has been done in Chapters 13 through 19. Following this, the observation equations for the constraints are formed. These observation equations are then included in the normal matrix as additional rows $[C]$ and columns $[C^T]$ in Equation (20.16) and their constants are added to the constants matrix as additional rows $[L_2]$ in Equation (20.16). The inverse of this bordered normal matrix is computed. The matrix solution of the Equation (20.16) is

$$\begin{bmatrix} X_1 \\ \cdots \\ X_2 \end{bmatrix} = \begin{bmatrix} A^TWA & \vdots & C^T \\ \cdots & \cdots & \cdots \\ C & \vdots & 0 \end{bmatrix}^{-1} \begin{bmatrix} A^TWL_1 \\ \cdots \\ L_2 \end{bmatrix} \qquad (20.17)$$

In Equation (20.17) X_2 is not used in the subsequent solution for the unknowns. This procedure is illustrated in the following examples.

Example 20.2 **Constrained Differential Leveling Adjustment** In Figure 20.5, differential elevations were observed for a network where the elevation difference between stations B and E is to be held at -17.60 ft. The elevation of A is 1300.62 ft, and the elevation differences observed for each line are shown below.

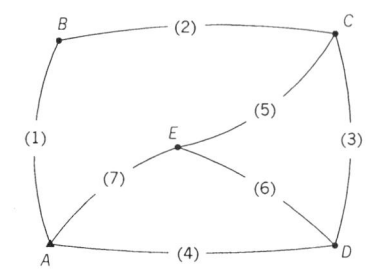

Figure 20.5 Differential leveling network.

20.4 HELMERT'S METHOD   399

Line	From	To	Elevation (ft)	S (ft)
1	A	B	25.15	0.07
2	B	C	−10.57	0.05
3	C	D	−1.76	0.03
4	D	A	−12.65	0.08
5	C	E	−7.06	0.03
6	E	D	5.37	0.05
7	E	A	−7.47	0.05

Perform a least squares adjustment of this level net constraining the required elevation difference.

SOLUTION   The A, X, and L matrices are

$$A = \begin{bmatrix} 1 & 0 & 0 & 0 \\ -1 & 1 & 0 & 0 \\ 0 & -1 & 1 & 0 \\ 0 & 0 & -1 & 0 \\ 0 & -1 & 0 & 1 \\ 0 & 0 & 1 & -1 \\ 0 & 0 & 0 & -1 \end{bmatrix} \quad X = \begin{bmatrix} B \\ C \\ D \\ E \end{bmatrix} \quad L = \begin{bmatrix} 1325.77 \\ -10.55 \\ -1.76 \\ -1313.27 \\ -7.06 \\ 5.37 \\ 1308.09 \end{bmatrix}$$

The weight matrix (W) is

$$W = \begin{bmatrix} 204.08 & 0 & 0 & 0 & 0 & 0 & 0 \\ 0 & 400 & 0 & 0 & 0 & 0 & 0 \\ 0 & 0 & 1111.11 & 0 & 0 & 0 & 0 \\ 0 & 0 & 0 & 156.25 & 0 & 0 & 0 \\ 0 & 0 & 0 & 0 & 1111.11 & 0 & 0 \\ 0 & 0 & 0 & 0 & 0 & 400 & 0 \\ 0 & 0 & 0 & 0 & 0 & 0 & 400 \end{bmatrix}$$

The reduced normal equations are

$$\begin{bmatrix} 604.08 & -400.00 & 0.00 & 0.00 \\ -400.00 & 2622.22 & -1111.11 & -1111.11 \\ 0.00 & -1111.11 & 1667.36 & -400.00 \\ 0.00 & -1111.11 & -400.00 & 1911.11 \end{bmatrix} \begin{bmatrix} B \\ C \\ D \\ E \end{bmatrix} = \begin{bmatrix} 274{,}793.30 \\ 5572.00 \\ 205{,}390.90 \\ 513{,}243.60 \end{bmatrix} \quad (a)$$

The reduced normal matrix is now bordered by the constraint equation

$$E - B = -17.60$$

which has a matrix form of

400 CONSTRAINT EQUATIONS

$$[-1 \ 0 \ 0 \ 1] \begin{bmatrix} B \\ C \\ D \\ E \end{bmatrix} = [-17.60] \quad (b)$$

The left side of Equation (b) is now included as an additional row and column to the reduced normal matrix in Equation (a). The lower-right corner diagonal element of the newly bordered normal matrix is assigned a value of 0. Similarly, the right-hand side of Equation (b) is added as an additional row in the right-hand side of Equation (a). Thus, the bordered-normal equations are

$$\begin{bmatrix} 604.80 & -400.00 & 0.00 & 0.00 & -1 \\ -400.00 & 2622.22 & -1111.11 & -1111.11 & 0 \\ 0.00 & -1111.11 & 1667.36 & -400.00 & 0 \\ 0.00 & -1111.11 & -400.00 & 1911.11 & 1 \\ -1 & 0 & 0 & 1 & 0 \end{bmatrix} \begin{bmatrix} B \\ C \\ D \\ E \\ X_2 \end{bmatrix} = \begin{bmatrix} 274{,}793.30 \\ 5572.00 \\ 205{,}390.90 \\ 513{,}243.60 \\ -17.60 \end{bmatrix} \quad (c)$$

Notice in Equation (c) that an additional unknown, X_2, is added at the bottom of the X to make the X matrix dimensionally consistent with the bordered-normal matrix of (c). Similarly, on the right-hand side of the equation, the k value of constraint equation (b) is added to the bottom of the matrix. Using Equation (20.17), the resulting solution is

$$X = \begin{bmatrix} 1325.686 \\ 1315.143 \\ 1313.390 \\ 1308.086 \\ -28.003 \end{bmatrix} \quad (d)$$

From the X matrix in Equation (d), elevation of station B is 1325.686 and that for station E is 1308.086. Thus, the elevation difference between stations B and E is exactly -17.60, which was required by the constraint condition.

Example 20.3 Constraining the Azimuth of a Line Helmert's method can also be used to constrain the direction of a line. In Figure 20.6 the bearing of line AB is to remain at its record value of N 0°04′ E. The data for this trilaterated network are

Control station

Station	X (m)	Y (m)
A	1000.000	1000.000

Initial approximations

Station	X (m)	Y (m)
B	1003.07	3640.00
C	2323.07	3638.46
D	2496.08	1061.74

20.4 HELMERT'S METHOD

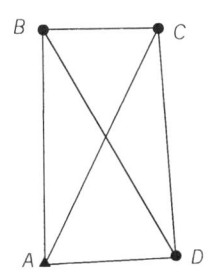

Figure 20.6 Network for Example 20.3.

Distance observations

From	To	Distance (m)	σ (m)	From	To	Distance (m)	σ (m)
A	C	2951.604	0.025	C	D	2582.534	0.024
A	B	2640.017	0.024	D	A	1497.360	0.021
B	C	1320.016	0.021	B	D	2979.325	0.025

Adjust the figure by the method of least squares holding the direction of the line AB using Helmert's method.

SOLUTION Using procedures discussed in Chapter 14, the reduced normal equations for the trilaterated system are

$$\begin{bmatrix} 2690.728 & -706.157 & -2284.890 & 0.000 & -405.837 & 706.157 \\ -706.157 & 2988.234 & 0.000 & 0.000 & 706.157 & -1228.714 \\ -2284.890 & 0.000 & 2624.565 & 529.707 & -8.737 & 124.814 \\ 0.000 & 0.000 & 529.707 & 3077.557 & 124.814 & -1783.060 \\ 405.837 & 706.157 & -8.737 & 124.814 & 2636.054 & -742.112 \\ 706.157 & -1228.714 & 124.814 & -1783.060 & -742.112 & 3015.328 \end{bmatrix}$$

$$\times \begin{bmatrix} dx_b \\ dy_b \\ dx_c \\ dy_c \\ dx_d \\ dy_d \end{bmatrix} = \begin{bmatrix} -6.615 \\ 6.944 \\ 21.601 \\ 17.831 \\ 7.229 \\ 11.304 \end{bmatrix} \qquad (e)$$

Following prototype equation (15.9), the linearized equation for the azimuth of line AB is

$$[78.13 \quad -0.09 \quad 0 \quad 0 \quad 0 \quad 0] \begin{bmatrix} dx_b \\ dy_b \\ dx_c \\ dy_c \\ dx_d \\ dy_d \end{bmatrix} = [0.139] \qquad (f)$$

402 CONSTRAINT EQUATIONS

The observation equation for the constrained direction [Equation (f)] is then added to the border of the matrix of reduced normal equations (e), which yields

$$\begin{bmatrix} 2690.728 & -706.157 & -2284.890 & 0.000 & -405.837 & 706.157 & 78.13 \\ -706.157 & 2988.234 & 0.000 & 0.000 & 706.157 & -1228.714 & -0.09 \\ -2284.890 & 0.000 & 2624.565 & 529.707 & -8.737 & 124.814 & 0.000 \\ 0.000 & 0.000 & 529.707 & 3077.557 & 124.814 & -1783.060 & 0.000 \\ 405.837 & 706.157 & -8.737 & 124.814 & 2636.054 & -742.112 & 0.000 \\ 706.157 & -1228.714 & 124.814 & -1783.060 & -742.112 & 3015.328 & 0.000 \\ 78.13 & -0.09 & 0.000 & 0.000 & 0.000 & 0.000 & 0.000 \end{bmatrix}$$

$$\times \begin{bmatrix} dx_b \\ dy_b \\ dx_c \\ dy_c \\ dx_d \\ dy_d \\ dx_2 \end{bmatrix} = \begin{bmatrix} -6.615 \\ 6.944 \\ 21.601 \\ 17.831 \\ 7.229 \\ 11.304 \\ 0.139 \end{bmatrix}$$

(g)

This is a nonlinear problem, and thus the solution must be iterated until convergence. The first two iterations yielded the X matrices listed as X_1 and X_2 below. The third iteration resulted in negligible corrections to the unknowns. The total of these corrections in shown below as X_T.

$$X_1 = \begin{bmatrix} 0.00179 \\ 0.00799 \\ 0.00636 \\ 0.01343 \\ 0.00460 \\ 0.01540 \\ -0.00337 \end{bmatrix} \quad X_2 = \begin{bmatrix} -0.00001 \\ -0.00553 \\ 0.00477 \\ -0.00508 \\ -0.00342 \\ -0.00719 \\ -0.00359 \end{bmatrix} \quad X_T = \begin{bmatrix} 0.00178 \\ 0.00247 \\ 0.01113 \\ 0.00835 \\ 0.00117 \\ 0.00821 \\ -0.00696 \end{bmatrix}$$

Adding the coordinate corrections of X_T to the initial approximations results in the final coordinates for stations B, C, and D of

B: (1003.072, 3640.003) C: (2323.081, 3638.468) D: (2496.081, 1061.748)

Checking the solution: Using Equation (15.1), check to see that the direction of line AB was held to the value of the constraint.

$$Az_{AB} = \tan^{-1} \frac{3.072}{2640.003} = 0°04'00'' \quad \text{(Check!)}$$

20.5 REDUNDANCIES IN A CONSTRAINED ADJUSTMENT

The number of redundancies in an adjustment increases by one for each parameter that is removed by a constraint equation. An expression for determining the number of redundancies is

$$r = m - n + c \tag{20.18}$$

where r is the number of redundancies (degrees of freedom) in the system, m the number of observations in the system, n the number of unknown parameters in the system, and c the number of mathematically independent constraints applied to the system. In Example 20.2 there were seven observations in a differential leveling network that had four stations with unknown elevations. One constraint was added to the system of equations that fixed the elevation difference between B and E as -17.60. In this way, the elevation of B and E became mathematically dependent. By applying Equation (20.18), it can be seen that the number of redundancies in the system is $r = 7 - 4 + 1 = 4$. Without the aforementioned constraint, this adjustment would have only $7 - 4 = 3$ redundancies. Thus, the constraint added one degree of freedom to the adjustment while making the elevations of B and E mathematically dependent.

Care must be used when adding constraints to an adjustment. It is possible to add as many mathematically independent constraint equations as there are unknown parameters. If that is done, all unknowns are constrained or fixed, and it is impossible to perform an adjustment. Furthermore, it is also possible to add constraints that are mathematically dependent equations. Under these circumstances, even if the system of equations has a solution, two mathematically dependent constraints would remove only one unknown parameter, and thus the redundancies in the system would increase by only one.

20.6 ENFORCING CONSTRAINTS THROUGH WEIGHTING

The methods described above for handling constraint equations can often be avoided simply by overweighting the observations to be constrained in a weighted least squares adjustment. This was done in Example 16.2 to fix the direction of a line. As a further demonstration of the procedure of enforcing constraints by overweighting, Example 20.3 will be adjusted by writing observation equations for azimuth AB and the control station coordinates X_A and Y_A. These observations will be fixed by assigning a 0.001″ standard deviation to the azimuth of line AB and standard deviations of 0.001 ft to the coordinates of station A.

CONSTRAINT EQUATIONS

The *J*, *K*, and *W* matrices for the first iteration of this problem are listed below. Note that the numbers have been rounded to three-decimal places for publication purposes only.

$$J = \begin{bmatrix}
-0.448 & -0.894 & 0.000 & 0.000 & 0.448 & 0.894 & 0.000 & 0.000 \\
-0.001 & -1.000 & 0.001 & 1.000 & 0.000 & 0.000 & 0.000 & 0.000 \\
0.000 & 0.000 & -1.000 & 0.001 & 1.000 & -0.001 & 0.000 & 0.000 \\
0.000 & 0.000 & 0.000 & 0.000 & -0.067 & 0.998 & 0.067 & -0.997 \\
-0.999 & -0.041 & 0.000 & 0.000 & 0.000 & 0.000 & 0.999 & 0.041 \\
0.000 & 0.000 & -0.050 & 0.865 & 0.000 & 0.000 & 0.050 & -0.865 \\
-78.130 & 0.091 & 78.130 & -0.091 & 0.000 & 0.000 & 0.000 & 0.000 \\
1.000 & 0.000 & 0.000 & 0.000 & 0.000 & 0.000 & 0.000 & 0.000 \\
0.000 & 1.000 & 0.000 & 0.000 & 0.000 & 0.000 & 0.000 & 0.000
\end{bmatrix}$$

$$K = \begin{bmatrix} -0.003 \\ 0.015 \\ 0.015 \\ 0.012 \\ 0.007 \\ -0.021 \\ 0.139 \\ 0.000 \\ 0.000 \end{bmatrix}$$

$$W = \begin{bmatrix}
\frac{1}{0.025^2} & 0 & 0 & 0 & 0 & 0 & 0 & 0 & 0 \\
0 & \frac{1}{0.024^2} & 0 & 0 & 0 & 0 & 0 & 0 & 0 \\
0 & 0 & \frac{1}{0.021^2} & 0 & 0 & 0 & 0 & 0 & 0 \\
0 & 0 & 0 & \frac{1}{0.024^2} & 0 & 0 & 0 & 0 & 0 \\
0 & 0 & 0 & 0 & \frac{1}{0.021^2} & 0 & 0 & 0 & 0 \\
0 & 0 & 0 & 0 & 0 & \frac{1}{0.025^2} & 0 & 0 & 0 \\
0 & 0 & 0 & 0 & 0 & 0 & \frac{\rho^2}{0.001^2} & 0 & 0 \\
0 & 0 & 0 & 0 & 0 & 0 & 0 & \frac{1}{0.001^2} & 0 \\
0 & 0 & 0 & 0 & 0 & 0 & 0 & 0 & \frac{1}{0.001^2}
\end{bmatrix}$$

The results of the adjustment (from the program ADJUST) are presented below.

20.6 ENFORCING CONSTRAINTS THROUGH WEIGHTING

```
*****************
Adjusted stations
*****************

                              Standard error ellipses computed
Station     X         Y         Sx      Sy      Su      Sv      t
===================================================================
   A     1,000.000 1,000.000  0.0010  0.0010  0.0010  0.0010  135.00°
   B     1,003.072 3,640.003  0.0010  0.0217  0.0217  0.0010    0.07°
   C     2,323.081 3,638.468  0.0205  0.0248  0.0263  0.0186  152.10°
   D     2,496.081 1,061.748  0.0204  0.0275  0.0281  0.0196   16.37°


******************************
Adjusted Distance Observations
******************************

Station     Station
Occupied    Sighted     Distance          V              S
================================================================
   A           C       2,951.620        0.0157         0.0215
   A           B       2,640.004       -0.0127         0.0217
   B           C       1,320.010       -0.0056         0.0205
   C           D       2,582.521       -0.0130         0.0215
   D           A       1,497.355       -0.0050         0.0206
   B           D       2,979.341        0.0159         0.0214


*****************************
Adjusted Azimuth Observations
*****************************

Station     Station
Occupied    Sighted      Azimuth           V             S"
================================================================
   A           B       0° 04' 00"         0.0"          0.0"


******************************************
              Adjustment Statistics
******************************************

                    Iterations      =         2
                    Redundancies    =         1
                    Reference Variance =    1.499
                    Reference So    =       ±1.2

            Passed X² test at 95.0% significance level!
                    X² lower value = 0.00
```

406 CONSTRAINT EQUATIONS

```
            X² upper value = 5.02
  A priori value of 1 used for reference variance
             in computations of statistics.
                     Convergence!
```

Notice in the adjustment above that the control station coordinates remained fixed and the residual of the azimuth of line *AB* is zero. Thus, the azimuth of line *AB* was held fixed without the inclusion of any constraint equation. It was simply constrained by overweighting the observation. Also note that the final adjusted coordinates of stations *B*, *C*, and *D* match the solution in Example 20.3.

PROBLEMS

Note: For problems requiring least squares adjustment, if a computer program is not distinctly specified for use in the problem, it is expected that the least squares algorithm will be solved using the program MATRIX, which is included on the CD supplied with the book.

20.1 Given the following lengths observed in a trilateration survey, adjust the survey by least squares using the elimination of constraints method to hold the coordinates of *A* at $x_a = 30,000.00$ and $y_a = 30,000.00$, and the azimuth of line *AB* to $30°00'00.00'' \pm 0.001''$ from north. Find the adjusted coordinates of *B*, *C*, and *D*.

Distance observations

Course	Distance (ft)	S (ft)	Course	Distance (ft)	S (ft)
AB	22,867.12	0.116	DA	29,593.60	0.149
BC	22,943.74	0.116	AC	30,728.64	0.155
CD	28,218.26	0.142	BD	41,470.07	0.208

Initial coordinates

Station	X (ft)	Y (ft)
B	41,433.56	49,803.51
C	60,054.84	36,399.65
D	51,386.93	9,545.64

20.2 Do Problem 20.1 using Helmert's method.

20.3 For the following traverse data, use Helmert's method and perform an adjustment holding the coordinates of station *A* fixed and azimuth of line *AB* fixed at $269°28'11''$. Assume that all linear units are feet.

PROBLEMS

Control station				Initial coordinates		
Station	X	Y		Station	X	Y
A	15,123.65	9803.10		B	14,423.26	9796.61
				C	12,620.56	9066.30

Distance observations				Angle observations		
Course	Distance	S		Stations	Angle	S (″)
AB	700.42	0.020		ABC	158°28′34″	4.5
BC	1945.01	0.022		BCA	5°39′06″	2.9
CA	2609.24	0.024		CAB	15°52′22″	4.1

20.4 Given the following differential leveling data, adjust it using Helmert's method. Hold the elevation of A to 100.00 ft and the elevation difference ΔElev_{AD} to −5.00 ft.

From	To	ΔElev (ft)	S	From	To	ΔElev (ft)	S
A	B	19.997	0.005	D	E	9.990	0.006
B	C	−19.984	0.008	E	A	−5.000	0.004
C	D	−4.998	0.003				

20.5 Using the method of weighting discussed in Section 20.6, adjust the data in:
(a) Problem 20.1.
(b) Problem 20.3.
(c) Problem 20.4.
(d) Compare the results with those of Problem 20.1, 20.3, or 20.4 as appropriate.

20.6 Do Problem 13.14 holding distance BC to 100.00 ft using:
(a) the elimination of constraints method.
(b) Helmert's method.
(c) Compare the results of the adjustments from the different methods.

20.7 Do Problem 14.15 holding the elevation difference between V and Z to 3.600 m using:
(a) the elimination of constraints method.
(b) Helmert's method.
(c) the method of overweighting technique.
(d) Compare the results of the adjustments from the various methods.

20.8 Do Problem 12.11 holding the difference in elevation between stations 2 and 8 to 66.00 ft. Use:

(a) the elimination of constraints method.

(b) Helmert's method.

(c) the method of overweighting technique.

20.9 Assuming that stations A and D are second-order class I horizontal control (1:50,000), do Problem 14.8 by including the control in the adjustment.

20.10 Do Problem 15.7 assuming that station A is first-order horizontal control (1:100,000) and B is second-order class I horizontal control (1:50,000) using the overweighting method.

20.11 Do Problem 15.9 assuming that the control stations are second-order class II horizontal control (1:20,000) using the overweighting method.

20.12 Do Problem 15.12 assuming that stations A and D are third-order class I control (1:10,000) using the overweighting method.

Practical Problems

20.13 Develop a computational program that computes the coefficients for the J matrix in a trilateration adjustment with a constrained azimuth. Use the program to solve Problem 20.8.

20.14 Develop a computational program that computes a constrained least squares adjustment of a trilateration network using Helmert's method. Use this program to solve Problem 20.11.

CHAPTER 21

BLUNDER DETECTION IN HORIZONTAL NETWORKS

21.1 INTRODUCTION

Up to this point, data sets are assumed to be free of blunders. However, when adjusting real observations, the data sets are seldom blunder free. Not all blunders are large, but no matter their sizes, it is desirable to remove them from the data set. In this chapter, methods used to detect blunders before and after an adjustment are discussed.

Many examples can be cited that illustrate mishaps that have resulted from undetected blunders in survey data. However, few could have been more costly and embarrassing than a blunder of about 1 mile that occurred in an early nineteenth-century survey of the border between the United States and Canada near the north end of Lake Champlain. Following the survey, construction of a U.S. military fort was begun. The project was abandoned two years later when the blunder was detected and a resurvey showed that the fort was actually located on Canadian soil. The abandoned facility was subsequently named Fort Blunder!

As discussed in previous chapters, observations are normally distributed. This means that occasionally, large random errors will occur. However, in accordance with theory, this should seldom happen. Thus, large errors in data sets are more likely to be blunders than random errors. Common blunders in data sets include number transposition, entry and recording errors, station misidentifications, and others. When blunders are present in a data set, a least squares adjustment may not be possible or will, at a minimum, produce poor or invalid results. To be safe, the results of an adjustment should never be accepted without an analysis of the post-adjustment statistics.

21.2 A PRIORI METHODS FOR DETECTING BLUNDERS IN OBSERVATIONS

In performing adjustments, it should always be assumed that there are possible observational blunders in the data. Thus, appropriate methods should be used to isolate and remove them. It is especially important to eliminate blunders when the adjustment is nonlinear because they can cause the solution to diverge. In this section, several methods are discussed that can be used to isolate blunders in a horizontal adjustment.

21.2.1 Use of the *K* Matrix

In horizontal surveys, the easiest method available for detecting blunders is to use the redundant observations. When initial approximations for station coordinates are computed using standard surveying methods, they should be *close* to their final adjusted values. Thus, the difference between observations computed from these initial approximations and their observed values (*K* matrix) are expected to be small in size. If an observational blunder is present, there are two possible situations that can occur with regard to the *K*-matrix values. If the observation containing a blunder is not used to compute initial coordinates, its corresponding *K*-matrix value will be relatively large. However, if an observation with a blunder is used in the computation of the initial station coordinates, the remaining redundant observations to that station should have relatively large values.

Figure 21.1 shows the two possible situations. In Figure 21.1(*a*), a distance blunder is present in line *BP* and is shown by the length *PP'*. However, this distance was not used in computing the coordinates of station *P*, and thus the *K*-matrix value for $BP' - BP_0$ will suggest the presence of a blunder by its relatively large size. In Figure 21.1(*b*), the distance blunder in *BP* was used to compute the initial coordinates of station *P'*. In this case, the redundant angle and distance observations connecting *P* with *A*, *C*, and *D* may show

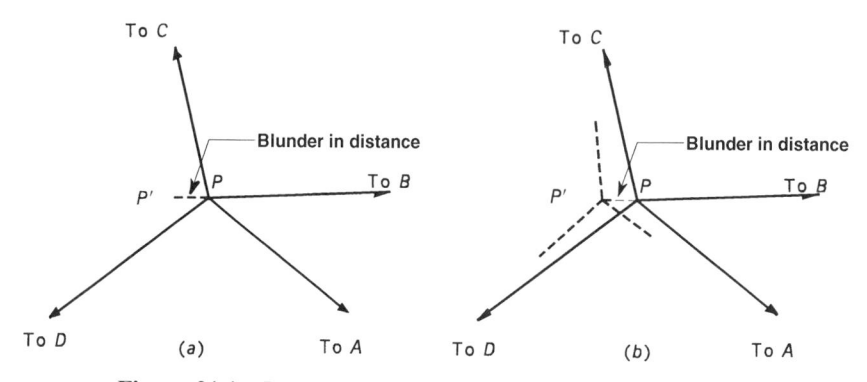

Figure 21.1 Presence of a distance blunder in computations.

21.2 A PRIORI METHODS FOR DETECTING BLUNDERS IN OBSERVATIONS

large discrepancies in the K-matrix. In the latter case, it is possible that some redundant observations may agree reasonably with their computed values since a shift in a station's position can occur along a sight line for an angle or along a radius for a distance. Still, most redundant observations will have large K-matrix values and thus raise suspicions that a blunder exists in one of the observations used to compute the coordinates of station P.

21.2.2 Traverse Closure Checks

As mentioned in Chapter 8, errors can be propagated throughout a traverse to determine the anticipated closure. Large complex networks can be broken into smaller link and loop traverses to check estimated closures against their actual values. When a loop fails to meet its estimated closure, the observations included in the computations should be checked for blunders.

Figure 21.2(a) and (b) show a graphical technique to isolate a traverse distance blunder and an angular blunder, respectively. In Figure 21.2(a), a blunder in distance CD is shown. Notice that the remaining courses, DE and EA, are translated by the blunder in the direction of course CD. Thus, the length of closure line ($A'A$) will be nearly equal to the length of the blunder in CD with a direction that is consistent with the azimuth of CD. Since other observations contain small random errors, the length and direction of the closure line, $A'A$, will not match the blunder exactly. However, when one blunder is present in a traverse, the misclosure and the blunder will be close in both length and direction.

In the traverse of Figure 21.2(b), the effect of an angular blunder at traverse station D is illustrated. As shown, the courses DE, EF, and FA' will be rotated about station D. Thus, the perpendicular bisector of the closure line AA' will point to station D. Again, due to small random errors in other observations, the perpendicular bisector may not intersect the blunder precisely, but it should be close enough to identify the angle with the blunder. Since the angle at the initial station is not used in traverse computations, it is possible to

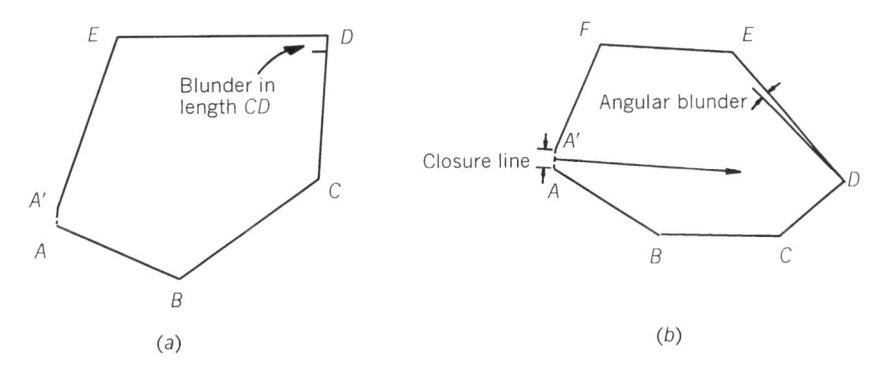

Figure 21.2 Effects of a single blunder on traverse closure.

isolate a single angular blunder by beginning traverse computations at the station with the suspected blunder. In this case, when the blunder is not used in the computations, estimated misclosure errors (see Chapter 8) will be met and the blunder can be isolated to the single unused angle. Thus, in Figure 21.2(*b*), if the traverse computations were started at station *D* and used an assumed azimuth for the course of *CD*, the traverse misclosure when returning to *D* would be within estimated tolerance since the angle at *D* is not used in the traverse computations.

21.3 A POSTERIORI BLUNDER DETECTION

When doing a least squares adjustment involving more than the minimum amount of control, both a *minimally* and *fully constrained* adjustment should be performed. In a *minimally constrained adjustment,* the data need to satisfy the appropriate geometric closures and are not influenced by control errors. After the adjustment, a χ^2 test[1] can be used to check the a priori value of the reference variance against its a posteriori estimate. However, this test is not a good indicator of the presence of a blunder since it is sensitive to poor relative weighting. Thus, the a posteriori residuals should also be checked for the presence of large discrepancies. If no large discrepancies are present, the observational weights should be altered and the adjustment rerun. Since this test is sensitive to weights, the procedures described in Chapters 7 through 10 should be used for building the stochastic model of the adjustment.

Besides the sizes of the residuals, the *signs of the residuals* may also indicate a problem in the data. From normal probability theory, residuals are expected to be small and randomly distributed. A small section of a larger network is shown in Figure 21.3. Notice that the distance residuals between stations *A* and *B* are all positive. This is not expected from normally distributed data. Thus, it is possible that either a blunder or a systematic error is present in some or all of the survey. If both *A* and *B* are control stations, part of the problem could stem from control coordinate discrepancies. This possibility can be isolated by doing a minimally constrained adjustment.

Although residual sizes can suggest observational errors, they do not necessarily identify the observations that contain blunders. This is due to the fact that least squares generally spreads a large observational error or blunder out radially from its source. However, this condition is not unique to least squares adjustments since any arbitrary adjustment method, including the compass rule for traverse adjustment, will also spread a single observational error throughout the entire observational set.

---

[1] Statistical testing was discussed in Chapter 4.

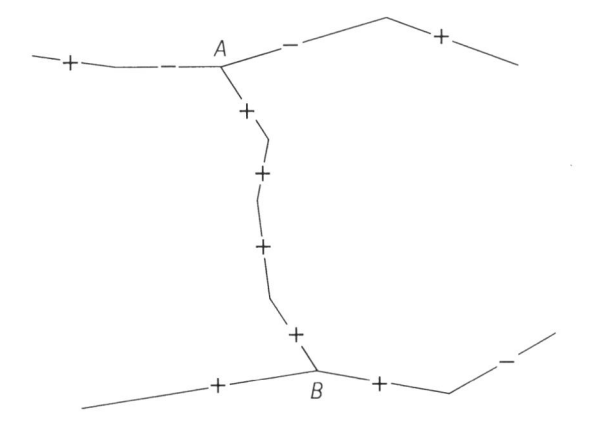

Figure 21.3 Distribution of residuals by sign.

Although an abnormally large residual may suggest the presence of a blunder in an observation, this is not always true. One reason for this could be poor relative weighting in the observations. For example, suppose that angle *GAH* in Figure 21.4 has a small blunder but has been given a relatively high weight. In this case the largest residual may well appear in a length between stations *G* and *H*, *B* and *H*, *C* and *F*, and most noticeably between *D* and *E*, due to their distances from station *A*. This is because the angular blunder will cause the network to spread or compress. When this happens, the signs of the distance residuals between *G* and *H*, *B* and *H*, *C* and *F*, and *D* and *E* may all be the same and thus indicate the problem. Again this situation can

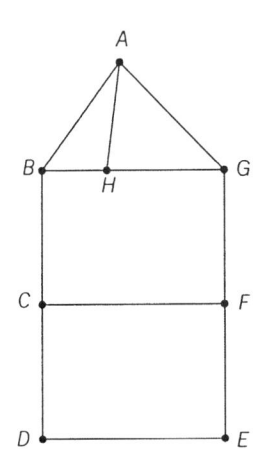

Figure 21.4 Survey network.

be minimized by using proper methods to determine observational variances so that they truly reflect the estimated errors in the observations.

21.4 DEVELOPMENT OF THE COVARIANCE MATRIX FOR THE RESIDUALS

In Chapter 5 it was shown how a sample data set could be tested at any confidence level to isolate observational residuals that were too large. The concept of statistical blunder detection in surveying was introduced in the mid-1960s and utilizes the cofactor matrix for the residuals. To develop this matrix, the adjustment of a linear problem can be expressed in matrix form as

$$L + V = AX + C \tag{21.1}$$

where C is a constants vector, A the coefficient matrix, X the estimated parameter matrix, L the observation matrix, and V the residual vector. Equation (21.1) can be rewritten in terms of V as

$$V = AX - T \tag{21.2}$$

where $T = L - C$, which has a covariance matrix of $W^{-1} = S^2 Q_{ll}$. The solution of Equation (21.2) results in the expression

$$X = (A^T W A)^{-1} A^T W T \tag{21.3}$$

Letting ε represent a vector of true errors for the observations, Equation (21.1) can be written as

$$L - \varepsilon = A\overline{X} + C \tag{21.4}$$

where $\overline{X}$ is the true value for the unknown parameter X and thus

$$T = L - C = A\overline{X} + \varepsilon \tag{21.5}$$

Substituting Equations (21.3) and (21.5) into Equation (21.2) yields

$$V = A(A^T W A)^{-1} A^T W (AX + \varepsilon) - (A\overline{X} + \varepsilon) \tag{21.6}$$

Expanding Equation (21.6) results in

$$V = A(A^T W A)^{-1} A^T W \varepsilon - \varepsilon + A(A^T W A)^{-1} A^T W A \overline{X} - A\overline{X} \tag{21.7}$$

Since $(A^T W A)^{-1} = A^{-1} W^{-1} A^{-T}$, Equation (21.7) can be simplified to

21.4 DEVELOPMENT OF THE COVARIANCE MATRIX FOR THE RESIDUALS

$$V = A(A^TWA)^{-1}A^TW\varepsilon - \varepsilon + A\overline{X} - A\overline{X} \qquad (21.8)$$

Factoring $W\varepsilon$ from Equation (21.8) yields

$$V = -[W^{-1} - A(A^TWA)^{-1}A^T]W\varepsilon \qquad (21.9)$$

Recognizing $(A^TWA)^{-1} = Q_{xx}$ and defining $Q_{vv} = W^{-1} - AQ_{xx}A^T$, Equation (21.9) can be rewritten as

$$V = -Q_{vv}W\varepsilon \qquad (21.10)$$

where $Q_{vv} = W^{-1} - AQ_{xx}A^T = W^{-1} - Q_{ll}$.

The Q_{vv} matrix is both singular and *idempotent*. Being singular, it has no inverse. When a matrix is idempotent, the following properties exist for the matrix: (a) The square of the matrix is equal to the original matrix (i.e., $Q_{vv} Q_{vv} = Q_{vv}$), (b) every diagonal element is between zero and 1, and (c) the sum of the diagonal elements, known as the *trace of the matrix*, equals the degrees of freedom in the adjustment. The latter property is expressed mathematically as

$$q_{11} + q_{22} + \cdots + q_{mm} = \text{degrees of freedom} \qquad (21.11)$$

(d) The sum of the square of the elements in any single row or column equals the diagonal element. That is,

$$q_{ii} = q_{i1}^2 + q_{i2}^2 + \cdots + q_{im}^2 = q_{1i}^2 + q_{2i}^2 + \cdots + q_{mi}^2 \qquad (21.12)$$

Now consider the case when all observations have zero errors except for a particular observation l_i that contains a blunder of size Δl_i. A vector of the true errors is expressed as

$$\Delta\varepsilon = \Delta l_i\, \varepsilon_i = \begin{bmatrix} 0 \\ 0 \\ \vdots \\ 0 \\ \Delta l_i \\ 0 \\ \vdots \\ 0 \end{bmatrix} = \Delta l_i \begin{bmatrix} 0 \\ 0 \\ \vdots \\ 0 \\ 1 \\ 0 \\ \vdots \\ 0 \end{bmatrix} \qquad (21.13)$$

If the original observations are uncorrelated, the specific correction for Δv_i can be expressed as

$$\Delta v_i = -q_{ii}\, w_{ii}\, \Delta l_i = -r_i\, \Delta l_i \qquad (21.14)$$

where q_{ii} is the ith diagonal of the Q_{vv} matrix, w_{ii} the ith diagonal term of the weight matrix, W, and $r_i = q_{ii} w_{ii}$ is the observational *redundancy number*.

When the system has a unique solution, r_i will equal zero, and if the observation is fully constrained, r_i would equal 1. The redundancy numbers provide insight into the geometric strength of the adjustment. An adjustment that in general has low redundancy numbers will have observations that lack sufficient checks to isolate blunders, and thus the chance for undetected blunders to exist in the observations is high. Conversely, a high overall redundancy number enables a high level of internal checking of the observations and thus there is a lower chance of accepting observations that contain blunders. The quotient of r/m, where r is the total number of redundant observations in the system and m is the number of observations, is called the *relative redundancy* of the adjustment.

21.5 DETECTION OF OUTLIERS IN OBSERVATIONS

Equation (21.10) defines the covariance matrix for the vector of residuals, v_i. From this the *standardized residual* is computed using the appropriate diagonal element of the Q_{vv} matrix as

$$\bar{v}_i = \frac{v_i}{\sqrt{q_{ii}}} \tag{21.15}$$

where $\bar{v}_i$ is the standardized residual, v_i the computed residual, and q_{ii} the diagonal element of the Q_{vv} matrix. Using the Q_{vv} matrix, the standard deviation in the residual is $S_0 \sqrt{q_{ii}}$. Thus, if the denominator of Equation (21.15) is multiplied by S_0, a t statistic is defined. If the residual is significantly different from zero, the observation used to derive the statistic is considered to be a blunder. The test statistic for this hypothesis test is

$$t_i = \frac{v_i}{S_0 \sqrt{q_{ii}}} = \frac{v_i}{S_v} = \frac{\bar{v}_i}{S_0} \tag{21.16}$$

Baarda (1968) computed rejection criteria for various significance levels (see Table 21.1) determining the α and β levels for Type I and Type II errors. The interpretation of these criteria is shown in Figure 21.5. When a blunder is present in the data set, the t distribution is shifted, and a statistical test for this shift may be performed. As with any other statistical test, two types of errors can occur. A Type I error occurs when data are rejected that do not contain blunders, and a Type II error occurs when a blunder is not detected in a data set where one is actually present. The rejection criteria are represented by the vertical line in Figure 21.5 and their corresponding significance

TABLE 21.1 Rejection Criteria with Corresponding Significance Levels

α	1 − α	β	1 − β	Rejection Criterion
0.05	0.95	0.80	0.20	2.8
0.001	0.999	0.80	0.20	4.1
0.001	0.999	0.999	0.001	6.6

levels are shown in Table 21.1. In practice, authors[2] have reported that 3.29 also works as a criterion for rejection of blunders.

Thus, the approach is to use a rejection level given by a t distribution with $r - 1$ degrees of freedom. The observation with the largest absolute value of t_i as given by Equation (21.17) is rejected when it is greater than the rejection level. That is, the observation is rejected when

$$\frac{|v_i|}{S_0\sqrt{q_{ii}}} > \text{rejection level} \qquad (21.17)$$

Since the existence of any blunder in the data set will affect the remaining observations and since Equation (21.18) depends on S_0, whose value was computed from data containing blunders, all observations that are detected as blunders should not be removed in a single pass. Instead, only the largest or largest independent group of observations should be deleted. Furthermore, since Equation (21.18) depends on S_0, it is possible to rewrite the equation so that it can be computed during the final iteration of a nonlinear adjustment. In this case the appropriate equation is

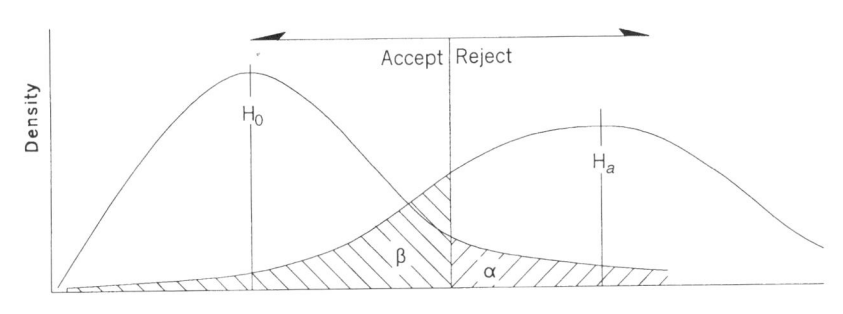

t density functions for the H_0 and H_a hypothesis

Figure 21.5 Effects of a blunder on the t distribution.

---
[2] References relating to the use of 3.29 as the rejection criterion are made in Amer (1979) and Harvey (1994).

$$\bar{v}_i = \frac{|v_i|}{\sqrt{q_{ii}}} > S_0 \times \text{rejection level} \qquad (21.18)$$

A summary of procedures for this manner of blunder detection is as follows:

Step 1: Locate all standardized residuals that meet the rejection criteria of Equation (21.17) or (21.18).
Step 2: Remove the largest detected blunder or unrelated blunder groups.
Step 3: Rerun the adjustment.
Step 4: Continue steps 1 through 3 until all detected blunders are removed.
Step 5: If more than one observation is removed in steps 1 through 4, reenter the observations in the adjustment *one at a time*. Check the observation after each adjustment to see if it is again detected as a blunder. If it is, remove it from the adjustment or have that observation reobserved.

Again it should be noted that this form of blunder detection is sensitive to improper relative weighting in observations. Thus, it is important to use weights that are reflective of the observational errors. Proper methods of computing estimated errors in observations, and weighting, were discussed in Chapters 7 through 10.

21.6 TECHNIQUES USED IN ADJUSTING CONTROL

As discussed in Chapter 20, some control is necessary in each adjustment. However, since control itself is not perfect, this raises the question of how control should be managed. If control stations that contain errors are heavily weighted, the adjustment will improperly associate the control errors with the observations. This effect can be removed by using only the minimum amount of control required to fix the project. Table 21.2 lists the type of survey versus the minimum amount of control. Thus, in a horizontal adjustment, if the coordinates of only one station and the direction of only one line are held

TABLE 21.2 Requirements for a Minimally Constrained Adjustment

Survey Type	Minimum Amount of Control
Differential leveling	1 benchmark
Horizontal survey	1 point with known xy coordinates
	1 course with known azimuth
GPS survey	1 point with known geodetic coordinates

fixed, the observations will not be constricted by the control. That is, the observations will need to satisfy the internal geometric constraints of the network only. If more than minimum control is used, these additional constraints will be factored into the adjustment.

Statistical blunder detection can help identify weak control or the presence of systematic errors in observations. Using a minimally constrained adjustment, the data set is screened for blunders. After becoming confident that the blunders are removed from the data set, a fully constrained adjustment is performed. Following the fully constrained adjustment, an F test is used to compare the ratio of the minimally and fully constrained reference variances. The ratio should be 1.[3] If the two reference variances are found to be statistically different, two possible causes might exist. The first is that there are errors in the control that must be isolated and removed. The second is that the observations contain systematic error. Since systematic errors are not compensating in nature, they will appear as blunders in the fully constrained adjustment. If systematic errors are suspected, they should be identified and removed from the original data set and the entire adjustment procedure redone. If no systematic errors are identified,[4] different combinations of control stations should be used in the constrained adjustments until the problem is isolated. By following this type of systematic approach, a single control station that has questionable coordinates can be isolated.

With this stated, it should be realized that the ideal amount of control in each survey type is greater than the minimum. In fact, for all three survey types, a minimum of three controls is always preferable. For example, in a differential leveling survey with only two benchmarks, it would be impossible to isolate the problem simply by removing one benchmark from the adjustment. However, if three benchmarks are used, a separate adjustment containing only two of the benchmarks can be run until the offending benchmark is isolated.

Extreme caution should always be used when dealing with control stations. Although it is possible that a control station was disturbed or that the original published coordinates contained errors, that is seldom the case. A prudent surveyor should check for physical evidence of disturbance and talk with other surveyors before deciding to discard control. If the station was set by a local, state, or federal agency, the surveyor should contact the proper

---

[3] The ratio of the reference variances from the minimally and fully constrained adjustments should be 1, since both reference variances should be statistically equal. That is, $\sigma^2_{\text{minimally constrained}} = \sigma^2_{\text{fully constrained}}$.
[4] When adjusting data that cover a large region (e.g., spherical excess, reduction to the ellipsoid) it is essential that geodetic corrections to the data be considered and applied where necessary. These corrections are systematic in nature and can cause errors when fitting to more than minimal control.

authorities and report any suspected problems. People in the agency familiar with the control may help explain any apparent problem. For example, it is possible that the control used in the survey was established by two previously nonconnecting surveys. In this case, the relative accuracy of the stations was never checked when they were established. Another problem with control common in surveys is the connection of two control points from different datums. As an example, suppose that a first-order control station and a high-accuracy reference network (HARN) station are used as control in a survey. These two stations come from different national adjustments and are thus in different datums. They will probably not agree with each other in an adjustment.

21.7 DATA SET WITH BLUNDERS

Example 21.1 The network shown in Figure 21.6 was established to provide control for mapping in the area of stations 1 through 6. It began from two National Geodetic Survey second-order class II (1:20,000 precision) control stations, 2000 and 2001. The data for the job were gathered by five field crews in a class environment. The procedures discussed in Chapter 7 were used to estimate the observational errors. The problem is to check for blunders in the data set using a rejection level of $3.29S_0$.

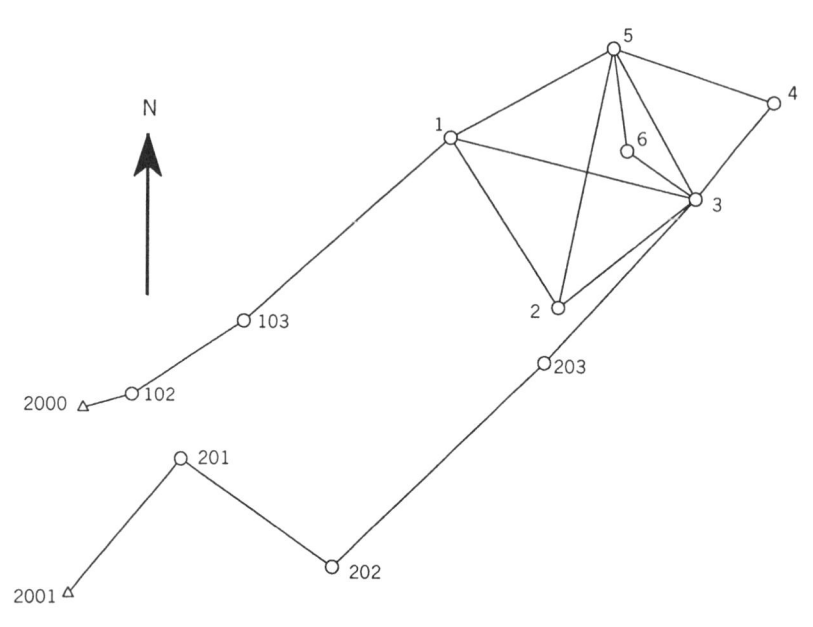

Figure 21.6 Data set with blunders.

21.7 DATA SET WITH BLUNDERS

Control stations

Station	Northing (ft)	Easting (ft)
2000	419,710.09	2,476,334.60
2001	419,266.82	2,476,297.98

Angle observations

Backsight	Occupied	Foresight	Angle	S (")
102	2000	2001	109°10'54.0"	25.5
2000	102	103	162°58'16.0"	28.9
102	103	1	172°01'43.0"	11.8
2000	2001	201	36°04'26.2"	7.4
2001	201	202	263°54'18.7"	9.7
201	202	203	101°49'55.0"	8.1
202	203	3	176°49'10.0"	8.4
203	3	2	8°59'56.0"	6.5
2	1	3	316°48'00.5"	6.3
3	5	4	324°17'44.0"	8.1
6	5	3	338°36'38.5"	10.7
1	5	3	268°49'32.5"	9.8
2	5	3	318°20'54.5"	7.0
2	3	1	51°07'11.0"	7.2
2	3	5	98°09'36.5"	10.3
2	3	6	71°42'51.5"	15.1
2	3	4	167°32'28.0"	14.5

Distance observations

From	To	Distance (ft)	S (ft)
2001	201	425.90	0.022
201	202	453.10	0.022
202	203	709.78	0.022
203	3	537.18	0.022
5	3	410.46	0.022
5	4	397.89	0.022
5	6	246.61	0.022
5	1	450.67	0.022
5	2	629.58	0.022
3	2	422.70	0.022
3	1	615.74	0.022
3	5	410.44	0.022
3	6	201.98	0.022
3	4	298.10	0.022
1	2	480.71	0.022
1	3	615.74	0.022
2000	102	125.24	0.022
102	103	327.37	0.022
103	1	665.79	0.022

422 BLUNDER DETECTION IN HORIZONTAL NETWORKS

Initial approximations were computed for the stations as follows:

Station	Northing (ft)	Easting (ft)
1	420,353.62	2,477,233.88
2	419,951.98	2,477,497.99
3	420,210.17	2,477,832.67
4	420,438.88	2,478,023.86
5	420,567.44	2,477,630.64
6	420,323.31	2,477,665.36
102	419,743.39	2,476,454.17
103	419,919.69	2,476,728.88
201	419,589.24	2,476,576.25
202	419,331.29	2,476,948.76
203	419,819.56	2,477,463.90

SOLUTION Do the a priori check of the computed observations versus their K-matrix values. In this check, only one angle is detected as having a difference great enough to suspect that it contains a blunder. This is angle 3–5–4, which was measured as 324°17'44.0" but was computed as 317°35'31.2". Since this difference should not create a problem with convergence during the adjustment, the angle remained in the data set and the adjustment was attempted. The results of the first trial adjustment are shown below. The software used the rejection criteria procedure based on Equation (21.18) for its blunder detection. A rejection level of $3.29S_0$ is used for comparison against the standardized residuals. The column headed Std. Res. represents the standardized residual of the observation as defined by Equation (21.15) and the column headed Red. Num. represents the redundancy number of the observation as defined by Equation (21.14).

```
     **** Adjusted Distance Observations ****
No. |From | To |Distance |Residual | Std. Res. |Red. Num. |
=================================================================
  1|   1  |  3 | 616.234 |   0.494 |    26.148 |   0.7458 |
  2|   1  |  2 | 480.943 |   0.233 |    12.926 |   0.6871 |
  3|   1  |  3 | 616.234 |   0.494 |    26.148 |   0.7458 |
  4|   3  |  4 | 267.044 | -31.056 | -1821.579 |   0.6169 |
  5|   3  |  6 | 203.746 |   1.766 |   107.428 |   0.5748 |
  6|   3  |  5 | 413.726 |   3.286 |   171.934 |   0.7719 |
  7|   3  |  2 | 422.765 |   0.065 |     3.500 |   0.7312 |
  8|   5  |  2 | 630.949 |   1.369 |    75.909 |   0.6791 |
  9|   5  |  1 | 449.398 |  -1.272 |   -79.651 |   0.5377 |
 10|   5  |  6 | 247.822 |   1.212 |    75.418 |   0.5488 |
 11|   5  |  4 | 407.125 |   9.235 |   631.032 |   0.4529 |
```

21.7 DATA SET WITH BLUNDERS

```
12|    5 |   3 | 413.726 |    3.266 |  170.888 |  0.7719 |
13|  102 | 103 | 327.250 |   -0.120 |  -17.338 |  0.1018 |
14|  103 |   1 | 665.702 |   -0.088 |  -12.395 |  0.1050 |
15|  201 | 202 | 453.362 |    0.262 |   91.903 |  0.0172 |
16|  202 | 203 | 709.856 |    0.076 |   10.737 |  0.1048 |
17|  203 |   3 | 537.241 |    0.061 |    8.775 |  0.1026 |
18| 2000 | 102 | 125.056 |   -0.184 |  -28.821 |  0.0868 |
19| 2001 | 201 | 425.949 |    0.049 |    7.074 |  0.1008 |
```

```
             **** Adjusted Angle Observations ****
No. |From | Occ |  To |    Angle     |Residual | Std. Res. |Red Num |
=====================================================================
  1|   2 |   1 |   3 | 316°49'55.1" |  114.6" |    28.041 | 0.4164 |
  2|   2 |   3 |   4 | 167°36'00.2" |  212.2" |    25.577 | 0.3260 |
  3|   2 |   3 |   6 |  71°43'01.5" |   10.0" |     1.054 | 0.3990 |
  4|   2 |   3 |   5 |  97°55'09.3" | -867.2" |  -101.159 | 0.6876 |
  5|   2 |   3 |   1 |  51°06'14.6" |  -56.4" |   -11.156 | 0.4985 |
  6| 203 |   3 |   2 |   8°59'36.3" |  -19.7" |   -13.003 | 0.0550 |
  7|   2 |   5 |   3 | 318°25'14.4" |  259.9" |    44.471 | 0.6949 |
  8|   1 |   5 |   3 | 268°58'49.8" |  557.3" |    78.590 | 0.5288 |
  9|   6 |   5 |   3 | 338°42'53.4" |  374.9" |    63.507 | 0.3058 |
 10|   3 |   5 |   4 | 322°02'24.7" |-8119.3" | -1781.060 | 0.3197 |
 11| 2000| 102 | 103 | 162°23'50.9" |-2065.1" |  -110.371 | 0.4194 |
 12|  102| 103 |   1 | 171°57'46.9" | -236.1" |  -112.246 | 0.0317 |
 13| 2001| 201 | 202 | 263°58'31.6" |  252.9" |   104.430 | 0.0619 |
 14|  201| 202 | 203 | 101°52'56.4" |  181.4" |    57.971 | 0.1493 |
 15|  202| 203 |   3 | 176°50'15.9" |   65.9" |    23.278 | 0.1138 |
 16|  102|2000 |2001 | 109°40'18.6" | 1764.6" |   106.331 | 0.4234 |
 17| 2000|2001 | 201 |  36°07'56.4" |  210.2" |   104.450 | 0.0731 |
```

```
             ****** Adjustment Statistics ******
       Adjustment's Reference Standard Deviation = 487.79
                  Rejection Level = 1604.82
```

The proper procedure for removing blunders is to remove the single observation that is greater in magnitude than the rejection level selected for the adjustment and is greater in magnitude than the value of any other standardized residual in the adjustment. This procedure prevents removing observations that are connected to blunders and thus are inherently affected by their presence. By comparing the values of the standardized residuals against the rejection level of the adjustment, it can be seen that both a single distance (3–4) and an angle (3–5–4) are possible blunders since their standardized residuals are greater than the rejection level chosen. However, upon inspection of Figure 21.6, it can be seen that a blunder in distance 3–4 will directly affect angle 3–5–4, and distance 3–4 has the standardized residual that is

424 BLUNDER DETECTION IN HORIZONTAL NETWORKS

greatest in magnitude. This explains the previous a priori rejection of this angle observation. That is, distance 3-4 directly affects the size of angle 3-5-4 in the adjustment. Thus, only distance 3-4 should be removed from the observations. After removing this distance from the observations, the adjustment was rerun with the results shown below.

```
              **** Adjusted Distance Observations ****
No. |From |  To |Distance |Residual |Std. Res. |Red. Num. |
====================================================================
   1|    1 |   3 | 615.693 |  -0.047 |   -2.495 |   0.7457 |
   2|    1 |   2 | 480.644 |  -0.066 |   -3.647 |   0.6868 |
   3|    1 |   3 | 615.693 |  -0.047 |   -2.495 |   0.7457 |
   4| 2001 | 201 | 425.902 |   0.002 |    0.265 |   0.1009 |
   5|    3 |   6 | 201.963 |  -0.017 |   -1.032 |   0.5765 |
   6|    3 |   5 | 410.439 |  -0.001 |   -0.032 |   0.7661 |
   7|    3 |   2 | 422.684 |  -0.016 |   -0.858 |   0.7314 |
   8|    5 |   2 | 629.557 |  -0.023 |   -1.280 |   0.6784 |
   9|    5 |   1 | 450.656 |  -0.014 |   -0.858 |   0.5389 |
  10|    5 |   6 | 246.590 |  -0.020 |   -1.241 |   0.5519 |
  11|    5 |   4 | 397.885 |  -0.005 |   -0.380 |   0.4313 |
  12|    5 |   3 | 410.439 |  -0.021 |   -1.082 |   0.7661 |
  13|  102 | 103 | 327.298 |  -0.072 |  -10.380 |   0.1018 |
  14|  103 |   1 | 665.751 |  -0.039 |   -5.506 |   0.1049 |
  15|  201 | 202 | 453.346 |   0.246 |   86.073 |   0.0172 |
  16|  202 | 203 | 709.807 |   0.027 |    3.857 |   0.1049 |
  17|  203 |   3 | 537.193 |   0.013 |    1.922 |   0.1027 |
  18| 2000 | 102 | 125.101 |  -0.139 |  -21.759 |   0.0868 |

              **** Adjusted Angle Observations ****
No. |From | Occ |  To |   Angle      | Residual | Std. Res.| Red Num|
====================================================================
   1|    2 |   1 |   3 |316°47'54.2" |    -6.3" |   -1.551 |  0.4160 |
   2|    2 |   3 |   4 |167°32'31.0" |     3.0" |    0.380 |  0.2988 |
   3|    2 |   3 |   6 | 71°42'46.0" |    -5.5" |   -0.576 |  0.3953 |
   4|    2 |   3 |   5 | 98°09'18.6" |   -17.9" |   -2.088 |  0.6839 |
   5|    2 |   3 |   1 | 51°07'04.1" |    -6.9" |   -1.360 |  0.4978 |
   6|  203 |   3 |   2 |  8°59'26.7" |   -29.3" |  -19.340 |  0.0550 |
   7|    2 |   5 |   3 |318°20'51.4" |    -3.1" |   -0.532 |  0.6933 |
   8|    1 |   5 |   3 |268°50'03.4" |    30.9" |    4.353 |  0.5282 |
   9|    6 |   5 |   3 |338°36'37.1" |    -1.4" |   -0.238 |  0.3049 |
  10|    3 |   5 |   4 |324°17'43.6" |    -0.4" |   -0.381 |  0.0160 |
  11| 2000 | 102 | 103 |162°24'10.2" | -2045.8" | -109.353 |  0.4193 |
  12|  102 | 103 |   1 |171°57'51.2" |  -231.8" | -110.360 |  0.0316 |
  13| 2001 | 201 | 202 |263°58'20.3" |   241.6" |   99.714 |  0.0619 |
```

```
14|  201 |  202 | 203 |101°52'34.7" |  159.7" |  51.023 | 0.1494 |
15|  202 |  203 |   3 |176°49'56.1" |   46.1" |  16.273 | 0.1138 |
16|  102 | 2000 |2001 |109°40'17.7" | 1763.7" | 106.280 | 0.4233 |
17| 2000 | 2001 | 201 | 36°07'46.9" |  200.7" |  99.688 | 0.0732 |
```

```
****** Adjustment Statistics ******
Adjustment's Reference Standard Deviation = 30.62
              Rejection Level = 100.73
```

After this adjustment, analysis of standardized residuals indicates that the angles most likely still to contain blunders are observations 11, 12, and 16. Of these, observation 12 displays the highest standardized residual. Looking at Figure 21.6, it is seen that this angle attaches the northern traverse leg to control station 2000. This is a crucial observation in the network if any hopes of redundancy in the orientation of the network are to be maintained. Since this is a flat angle (i.e., nearly 180°), it is possible that the backsight and foresight stations were reported incorrectly, which can be checked by reversing stations 102 and 1. However, without further field checking, it cannot be guaranteed that this occurred. A decision must ultimately be made about whether this angle should be reobserved. However, for now, this observation will be discarded and another trial adjustment made. In this stepwise blunder detection process, it is always wise to remove as few observations as possible. In no case should observations that are blunder-free be deleted. This can and does happen, however, in trial blunder detection adjustments. But through persistent and careful processing, ultimately only those observations that contain blunders can be identified and eliminated. The results of the adjustment after removing the angle 12 are shown below.

```
******************
Adjusted stations
******************
                                        Standard error ellipses computed
Station        X             Y       Sx      Sy      Su      Sv       t
===============================================================================
      1   2,477,233.72  420,353.59  0.071   0.069   0.092   0.036   133.47°
      2   2,477,497.89  419,951.98  0.050   0.083   0.090   0.037   156.01°
      3   2,477,832.55  420,210.21  0.062   0.107   0.119   0.034   152.80°
      4   2,477,991.64  420,400.58  0.077   0.121   0.138   0.039   149.71°
      5   2,477,630.43  420,567.45  0.088   0.093   0.123   0.036   136.74°
      6   2,477,665.22  420,323.32  0.071   0.096   0.114   0.036   145.44°
    102   2,476,455.89  419,741.38  0.024   0.018   0.024   0.017    80.86°
    103   2,476,735.05  419,912.42  0.051   0.070   0.081   0.031   147.25°
    201   2,476,576.23  419,589.23  0.020   0.022   0.024   0.017    37.73°
    202   2,476,948.74  419,331.29  0.029   0.041   0.042   0.029    14.24°
    203   2,477,463.84  419,819.58  0.040   0.077   0.081   0.032   160.84°
```

426 BLUNDER DETECTION IN HORIZONTAL NETWORKS

```
*******************************
Adjusted Distance Observations
*******************************
 Station   Station
Occupied   Sighted   Distance        V    Std.Res.   Red.#
=================================================
   2001      201      425.88      -0.023   -3.25    0.102
    201      202      453.09      -0.005   -3.25    0.006
    202      203      709.76      -0.023   -3.25    0.104
    203        3      537.16      -0.023   -3.25    0.103
      5        3      410.45      -0.011   -0.60    0.767
      5        4      397.89      -0.003   -0.19    0.436
      5        6      246.60      -0.014   -0.83    0.556
      5        1      450.68       0.013    0.80    0.542
      5        2      629.58       0.003    0.15    0.678
      3        2      422.70       0.003    0.16    0.736
      3        1      615.75       0.008    0.40    0.745
      3        5      410.45       0.009    0.44    0.767
      3        6      201.97      -0.013   -0.78    0.580
      1        2      480.71      -0.003   -0.19    0.688
      1        3      615.75       0.008    0.40    0.745
   2000      102      125.26       0.020    3.25    0.082
    102      103      327.39       0.023    3.25    0.101
    103        1      665.81       0.023    3.25    0.104


***************************
Adjusted Angle Observations
***************************
   Station    Station     Station
Backsighted  Occupied  Foresighted     Angle         V    Std.Res.   Red.#
=================================================================
      102      2000       2001    109°11'11.1"    17.06"    3.25    0.042
     2000       102        103    162°58'05.1"   -10.95"   -3.25    0.014
     2000      2001        201     36°04'23.8"    -2.45"   -3.25    0.010
     2001       201        202    263°54'15.7"    -2.97"   -3.25    0.009
      201       202        203    101°49'46.3"    -8.72"   -3.25    0.110
      202       203          3    176°49'01.0"    -8.98"   -3.25    0.109
      203         3          2      8°59'51.1"    -4.91"   -3.25    0.054
        2         1          3    316°48'02.8"     2.29"    0.57    0.410
        3         5          4    324°17'43.8"    -0.19"   -0.19    0.016
        6         5          3    338°36'37.0"    -1.51"   -0.26    0.302
        1         5          3    268°49'43.7"    11.20"    1.57    0.528
        2         5          3    318°20'51.1"    -3.44"   -0.59    0.691
        2         3          1     51°07'14.4"     3.45"    0.68    0.497
        2         3          5     98°09'22.0"   -14.55"   -1.71    0.680
        2         3          6     71°42'48.5"    -2.97"   -0.31    0.392
        2         3          4    167°32'29.5"     1.48"    0.19    0.294
```

21.7 DATA SET WITH BLUNDERS

```
*******************************************
             Adjustment Statistics
*******************************************
                   Iterations = 4
                 Redundancies = 12
             Reference Variance = 1.316
                 Reference So = ±1.1
     Possible blunder in observations with Std.Res. > 4
                   Convergence!
```

From analysis of the results, all observations containing blunders appear to have been removed. However, it should also be noted that several remaining distance and angle observations have very low redundancy numbers. This identifies them as unchecked observations, which is also an undesirable situation. Thus, good judgment dictates reobservation of the measurements deleted. This weakness can also be seen in the size of the standard error ellipses for the stations shown in Figure 21.7.

Note, especially, rotation of the error ellipses. That is, the uncertainty is primarily in a direction perpendicular to the line to stations 1 and 102. This condition is predictable since the angle 102–103–1 has been removed from the data set. Furthermore, the crew on the northern leg never observed an

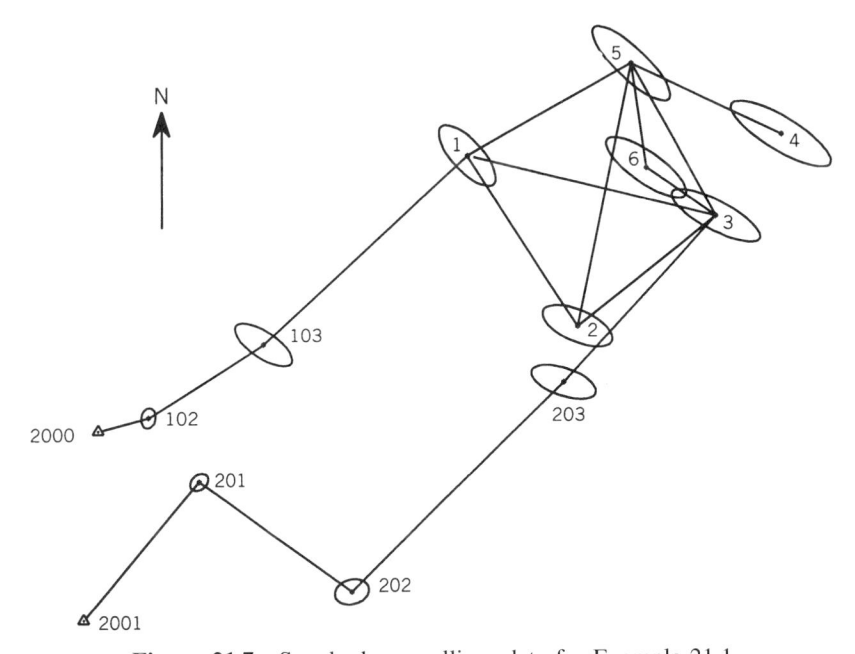

Figure 21.7 Standard error ellipse data for Example 21.1.

angle at station 1 that would tie into station 103, and thus the position of station 103 was found by the intersection of two distances that nearly form a straight line. This results in a larger error in the direction perpendicular to the lines at this station.

This example demonstrates the process used to statistically detect and remove observational blunders. Whether the observations should be remeasured depends on the intended use of the survey. Obviously, additional observations would strengthen the network and probably reduce the size of the error ellipses.

Observations between stations 102 and 201 also contribute to the overall strength in the network. However, because a building obstructs that line, these observations could not be obtained. This is a common problem in network design. That is, it is sometimes physically impossible to gather observations that would contribute to the total network strength. Thus, a compromise must be made between the *ideal* network and what is physically obtainable. Balancing these aspects requires careful planning before the observations are collected. Of course, line obstructions that occur due to terrain, vegetation, or buildings, can now be overcome by using GPS.

21.8 SOME FURTHER CONSIDERATIONS

Equation (21.14) shows the relationship between blunders and their effects on residuals as $\Delta v_i = -r_i \, \Delta l_i$. From this relationship note that the effect of the blunder, Δl_i, on the residual, Δv_i, is directly proportional to the redundancy number, r_i. Therefore:

1. If r_i is large (≈ 1), the blunder greatly affects the residual and should be easy to find.
2. If r_i is small (≈ 0), the blunder has little affect on the residual and will be difficult to find.
3. If $r_i = 0$, the blunder is undetectable and the parameters will be incorrect since the error has not been detected.

Since redundancy numbers can range from 0 to 1, it is possible to compute the minimum detectable error for a single blunder. For example, suppose that a value of 4.0 is used to isolate observational blunders. Then, if the reference variance of the adjustment is 6, all observations that have standardized residuals greater than 24.0 (4.0 × 6) are possible blunders. However, from Equation (21.14), it can be seen that for an observation with a redundancy number of 0.2 ($r_i = 0.2$) and a standardized residual of $\Delta v_i = 24.0$, the minimum detectable error is 24.0/0.2, or 120! Thus, a blunder, Δl_i, in this observation as large as five times the desired level can go undetected due to its low redundancy number. This situation can be extended to observations that have no observational checks; that is, r_i is 0. In this case, Equation (21.14) shows that it is impossible to detect any blunder, Δl_i, in the observation since $\Delta v_i / \Delta r_i$ is indeterminate.

With this taken into consideration, it has been shown that a marginally detectable blunder in an individual observation is

$$\Delta l_i = S \sqrt{\frac{\lambda_0}{q_{ii} w_{ii}^2}} \quad (21.19)$$

where λ_0 is the mean of the noncentral normal distribution shown in Figure 21.5, known as the *noncentrality parameter*. This parameter is the translation of the normal distribution that leads to rejection of the null hypothesis, whose values can be taken from nomograms developed by Baarda (1968). The sizes of the values obtained from Equation (21.19) provide a clear insight into weak areas of the network.

21.8.1 Internal Reliability

Internal reliability is found by examining how well observations check geometrically with each other. As mentioned previously, if a station is determined uniquely, q_{ii} will be zero in Equation (21.19), and the computed value of Δl_i is infinity. This indicates the lack of measurement self-checking. Since Equation (21.19) is independent of the actual observations, it provides a method of detecting weak areas in networks. To minimize the sizes of the undetected blunders in a network, the redundancy numbers of the individual observations should approach their maximum value of 1. Furthermore, for uniform network strength, the individual redundancy numbers, r_i, should be close to the global relative redundancy of r/m, where r is the number of redundant observations and m is the number of observations in the network. Weak areas in the network are located by finding regions where the redundancy numbers become small in comparison to relative redundancy.

21.8.2 External Reliability

An undetected blunder of size Δl_i has a direct effect on the adjusted parameters. External reliability is the effect of the undetected blunders on these parameters. As Δl_i (a blunder) increases, so will its effect on ΔX. The value of ΔX is given by

$$\Delta X = (A^T W A)^{-1} A^T W \Delta \varepsilon \quad (21.20)$$

Again, this equation is datum independent. From Equation (21.20) it can be seen that to minimize the value of ΔX_i, the size of redundancy numbers must be increased. Baarda suggested using average coordinate values in determining the effect of an undetected blunder with the following equation

$$\lambda = \Delta X^T Q_{xx} \Delta X \qquad (21.21)$$

where λ represents the noncentrality parameter.

The noncentrality parameter should remain as small as possible to minimize the effects of undetected blunders on the coordinates. Note that as the redundancy numbers on the observations become small, the effects of undetected blunders become large. Thus, the effect of the coordinates of a station from a undetected blunder is greater when the redundancy number is low. In fact, an observation with a high redundancy number is likely to be detected as a blunder.

A traverse sideshot can be used to explain this phenomenon. Since the angle and distance to the station are unchecked in a sideshot, the coordinates of the station will change according to the size of the observational blunders. The observations creating the sideshot will have redundancy numbers of zero since they are geometrically unchecked. This situation is neither good nor acceptable in a well-designed observational system. In network design, one should always check the redundancy numbers of the anticipated observations and strive to achieve uniformly high values for all observations. Redundancy numbers above 0.5 are generally sufficient to provide well-checked observations.

21.9 SURVEY DESIGN

In Chapters 8 and 19, the topic of observational system design was discussed. Redundancy numbers can now be added to this discussion. A well-designed network will provide sufficient geometric checks to allow observational blunders to be detected and removed. In Section 21.8.1 it was stated that if blunder removal is to occur uniformly throughout the system, the redundancy numbers should be close to the system's global relative redundancy. Furthermore, in Section 21.8.2 it was noted that redundancy numbers should be greater than 0.5. By combining these two additional concepts with the error ellipse sizes and shapes, and stochastic model planning, an overall methodology for designing observational systems can be obtained.

To begin the design process, the approximate positions for stations to be included in the survey must be determined. These locations can be determined from topographic maps, photo measurements, or previous survey data. The approximate locations of the control stations should be dictated by their desired locations, the surrounding terrain, vegetation, soils, sight-line obstructions, and so on. Field reconnaissance at this phase of the design process is generally worthwhile to verify sight lines and accessibility of stations. Moving a station only a small distance from the original design location may greatly enhance visibility to and from other stations but not change the geometry of the network significantly. By using topographic maps in this process, sight-line ground clearances can be checked by constructing profiles between stations.

When approximate station coordinates are determined, a stochastic model for the observational system can be designed following the procedures discussed in Chapter 7. In this design process, considerations should be given to the abilities of the field personnel, quality of the equipment, and observational procedures. After the design is completed, specifications for field crews can be written based on these design parameters. These specifications should include the type of instrument used, number of turnings for angle observations, accuracy of instrument leveling and centering, misclosure requirements, and many other items.

Once the stochastic model is designed, simulated observations are computed from the station coordinates, and a least squares adjustment of the observations is to be done. Since actual observations have not been made, their values are computed from the station coordinates. The adjustment will converge in a single iteration, with all residuals equaling zero. Thus, the reference variance must be assigned the a priori value of 1 to compute the error ellipse axes and coordinate standard deviations. Having completed the adjustment, the network can be checked for geometrically weak areas, unacceptable error ellipse sizes or shapes, and so on. This inspection may dictate the need for any or all of the following: (1) more observations, (2) different observational procedures, (3) different equipment, (4) more stations, (5) different network geometry, and so on. In any event, a clear picture of results obtainable from the observational system will be provided by the simulated adjustment and additional observations, or different network geometry can be used.

It should be noted that what is expected from the design may not actually occur, for numerous and varied reasons. Thus, systems are generally overdesigned. However, this tendency to overdesign should be tempered with the knowledge that it will raise the costs of the survey. Thus, a balance should be found between the design and costs. Experienced surveyors know what level of accuracy is necessary for each job and design observational systems to achieve the accuracy desired. It would be cost prohibitive and foolish always to design an observational system for maximum accuracy regardless of the survey's intended use. As an example, the final adjustment of the survey in Section 21.7 had sufficient accuracy to be used in a mapping project with a final scale of 1:1200 since the largest error ellipse semimajor axis (0.138 ft) would only plot as 0.0014 in. and is thus less than the width of a line on the final map product.

For convenience, the steps involved in network design are summarized below.

Step 1: Using a topographic map or aerial photos, lay out possible positions for stations.

Step 2: Use the topographic map together with air photos to check sight lines for possible obstructions and ground clearance.

Step 3: Do field reconnaissance, checking sight lines for obstructions not shown on the map or photos, and adjust positions of stations as necessary.

Step 4: Determine approximate coordinates for the stations from the map or photos.

Step 5: Compute values of observations using the coordinates from step 4.

Step 6: Using methods discussed in Chapter 6, compute the standard deviation of each observation based on available equipment and field measuring procedures.

Step 7: Perform a least squares adjustment, to compute observational redundancy numbers, standard deviations of station coordinates, and error ellipses at a specified percent probability.

Step 8: Inspect the solution for weak areas based on redundancy numbers and ellipse shapes. Add or remove observations as necessary, or reevaluate measurement procedures and equipment.

Step 9: Evaluate the costs of the survey, and determine if some other method of measurement (e.g., GPS) may be more cost-effective.

Step 10: Write specifications for field crews.

PROBLEMS

Note: For problems requiring least squares adjustment, if a computer program is not distinctly specified for use in the problem, it is expected that the least squares algorithm will be solved using the program MATRIX, which is included on the CD supplied with the book.

21.1 Discuss the effects of a distance blunder on a traverse closure and explain how it can be identified.

21.2 Discuss the effects of an angle blunder on a traverse closure, and explain how it can be identified.

21.3 Explain why a well-designed network has observational redundancy numbers above 0.5 and approximately equal.

21.4 Create a list of items that should be included in field specifications for a crew in a designed network.

21.5 Summarize the general procedures used in isolating observational blunders.

21.6 How are control problems isolated in an adjustment?

21.7 Discuss possible causes for control problems in an adjustment.

21.8 Why is it recommended that there be at least three control stations in a least squares adjustment?

21.9 Outline the procedures used in survey network design.

21.10 Using the procedures discussed in this chapter, analyze the data in Problem

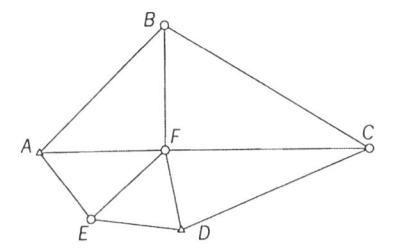

Figure P21.11

21.11 In Figure P21.11 the following data were gathered. Assuming that the control stations have a published precision of 1:20,000, apply the procedures discussed in this chapter to isolate and remove any apparent blunders in the data.

Control stations

Station	Easting	Northing
A	982.083	1000.204
D	2686.270	58.096

Approximate station coordinates

Station	Easting	Northing
B	2507.7	2500.6
C	4999.9	998.6
E	1597.6	200.0
F	2501.0	1009.6

Distance observations

From	To	Distance (m)	S (m)	From	To	Distance (m)	S (m)
A	B	2139.769	0.023	E	F	1231.086	0.021
A	F	1518.945	0.021	E	A	1009.552	0.020
B	C	2909.771	0.025	D	F	969.386	0.020
B	F	1491.007	0.021	D	E	1097.873	0.021
C	D	2497.459	0.023	C	F	2498.887	0.023

Angle observations

Backsight	Occupied	Foresight	Angle	S (")
F	A	E	52°47'12.3"	3.4
B	A	F	44°10'04.6"	2.8
F	B	A	45°13'12.5"	2.8
C	B	F	59°10'54.5"	2.7
F	C	B	27°30'00.0"	2.4
D	C	F	22°22'28.2"	2.5
F	D	C	78°53'43.8"	3.3
E	D	F	71°33'18.7"	3.8
A	E	F	85°42'04.9"	3.7
F	E	D	49°17'32.8"	3.4
A	F	B	90°36'40.5"	3.1
B	F	C	89°59'37.6"	2.8
C	F	D	78°43'47.8"	3.3
D	F	E	59°09'10.5"	3.6
E	F	A	41°30'43.3"	3.1

21.12 Apply to procedures discussed in this chapter to isolate any blunders in the following data.

Control stations

Station	X (ft)	Y (ft)
A	44,680.85	78,314.23
H	169,721.77	214,157.12

Approximate coordinates

Station	X (ft)	Y (ft)
B	21,112.93	151,309.33
C	49,263.50	213,175.12
D	107,855.97	145,744.68
E	113,747.95	78,968.90
F	186,743.04	78,314.24
G	206,710.31	138,216.04

Distance observations

Course	Distance (ft)	S (ft)	Course	Distance (ft)	S (ft)
AB	76,705.18	0.23	BC	67,969.04	0.20
CD	89,330.51	0.27	DE	67,035.17	0.20
EF	72,998.23	0.22	FG	63,141.95	0.19
GH	84,470.07	0.25	CH	120,461.89	0.36
AE	69,070.47	0.21	AD	92,401.26	0.28
DH	92,236.72	0.28	BD	86,921.33	0.26
DG	99,140.58	0.30	DF	103,778.98	0.31

Angle Observations

Backsight	Occupied	Foresight	Angle	S (″)
B	A	D	61°01′40″	2.1
D	B	A	68°26′08″	2.1
C	B	D	69°12′11″	2.1
B	C	D	65°27′20″	2.1
H	C	D	49°28′44″	2.1
D	H	C	47°24′34″	2.1
G	H	D	68°05′33″	2.1
D	G	H	59°40′32″	2.1
F	G	D	75°55′10″	2.1
D	F	G	67°54′48″	2.1
E	F	D	40°00′31″	2.1
A	E	D	85°30′03″	2.1
D	E	F	95°33′32″	2.3
D	A	E	46°19′22″	2.1
A	D	B	50°32′12″	2.1
B	D	C	45°20′27″	2.1
C	D	H	83°06′44″	2.1
H	D	G	52°13′51″	2.1
G	D	F	36°10′04″	2.1
F	D	E	44°26′04″	2.1
E	D	A	48°10′35″	2.1

21.13 Using the data set *Blunder.dat* on the CD that accompanies this book, isolate any blunders that are detectable with a rejection criteria of 3.29.

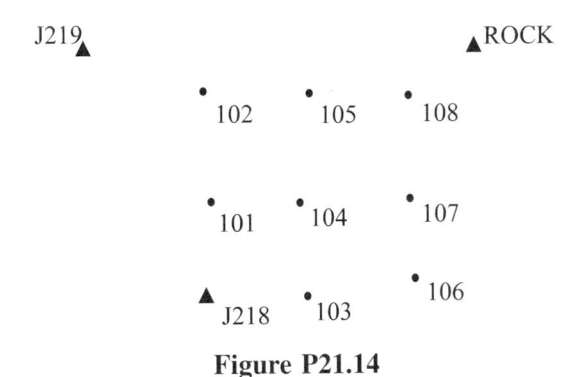

Figure P21.14

21.14 As shown in Figure P21.14, the following approximate station coordinates were determined from a map of an area where a second-order class I survey (1:50,000) is to be performed. All sight lines to neighboring stations have been checked and are available for conventional

436 BLUNDER DETECTION IN HORIZONTAL NETWORKS

observations of distances, angles, and azimuths. The control stations are visible to their nearest neighbors. Design a control network that will meet the specified accuracy at a 95% confidence level and have sufficient checks to ensure the reliability of the results. Stations J218, J219, and ROCK are first-order control stations (1:100,000).

Control stations

Station	X (m)	Y (m)
J218	283,224.223	116,202.946
J219	154,995.165	330,773.314
ROCK	521,287.251	330,276.310

Approximate station locations

Station	X (m)	Y(m)
101	280,278	194,109
102	276350	278,887
103	360,147	121,768
104	356,219	195,090
105	352,291	274,304
106	455,728	132,570
107	446,563	198,036
108	440,671	270,700

21.15 Using the criteria in Problem 21.14, design a carrier-phase GPS network to establish the control network. Assume that all stations have no obstructions above 15°.

Practical Problems

21.16 Design a 6 mi × 6 mi control network having a minimum of eight control stations using a topographic map of your local area. Design a traditional measurement network made up of angles, azimuths, and distances so that the largest ellipse axis at a 95% confidence level is less than 0.20 ft and so that all observations have redundancy numbers greater than 0.5. In the design, specify the shortest permissible sight distance, the largest permissible errors in pointing, reading, and instrument and target setup errors, the number of repetitions necessary for each angle measurement, and the necessary quality of angle and distance measuring instruments. Use realistic values for the instruments. Plot profiles of sight lines for each observation.

21.17 Design a 6 mile × 6 mile GPS control network to be established by differential GPS which has a minimum of eight control stations, using a topographic map of your local area to select station locations. Design the survey so that all baseline observations included in the network have redundancy numbers greater than 0.5. In the design, use a unit matrix for the covariance matrix of the baselines.

CHAPTER 22

GENERAL LEAST SQUARES METHOD AND ITS APPLICATION TO CURVE FITTING AND COORDINATE TRANSFORMATIONS

22.1 INTRODUCTION TO GENERAL LEAST SQUARES

When fitting points to a straight line, it must be recognized that both the x and y coordinates contain errors. Yet in the mathematical model presented in Section 11.11.1, illustrated in Figure 11.2, the residuals are applied only to the y coordinate. Because both coordinates contain errors, this mathematical model fails to account for the x coordinate being a measurement. In this chapter the *general least squares method* is presented and its use in performing adjustments where the observation equations involve more than a single measurement is demonstrated.

22.2 GENERAL LEAST SQUARES EQUATIONS FOR FITTING A STRAIGHT LINE

Consider the data illustrated in Figure 11.2. To account properly for both the x and y coordinates being measurements, the observation equation must contain residuals for both measurements. That is, Equation (11.40) must be rewritten as

$$F(x,y) = (y + v_y) - m(x + v_x) - b = 0 \qquad (22.1)$$

In Equation (22.1), x and y are a point's coordinate pair with residuals v_x and v_y, respectively, m is the slope of the line, and b is the y intercept. Equation (22.1) contains v_x, v_y, m, and b as unknowns and is nonlinear. Thus, its so-

438 GENERAL LEAST SQUARES METHOD

lution is obtained by using the methods outlined in Section 11.10. The resulting linearized form of Equation (22.1) is

$$\frac{\partial F}{\partial x} v_x + \frac{\partial F}{\partial y} v_y + \frac{\partial F}{\partial m} dm + \frac{\partial F}{\partial b} db = -(m_0 x + b_0 - y) \tag{22.2}$$

where the partial derivatives are

$$\frac{\partial F}{\partial x} = -m \qquad \frac{\partial F}{\partial y} = 1 \qquad \frac{\partial F}{\partial m} = -x \qquad \frac{\partial F}{\partial b} = -1 \tag{22.3}$$

For the four data points A, B, C, and D, substituting into Equations (22.3) into (22.2) the following four equations can be written:

$$\begin{aligned}
-m_0 v_{x_A} + v_{y_A} - x_A dm - db &= -(m_0 x_A + b_0 - y_A) \\
-m_0 v_{x_B} + v_{y_B} - x_B dm - db &= -(m_0 x_B + b_0 - y_B) \\
-m_0 v_{x_C} + v_{y_C} - x_C dm - db &= -(m_0 x_C + b_0 - y_C) \\
-m_0 v_{x_D} + v_{y_D} - x_D dm - db &= -(m_0 x_D + b_0 - y_D)
\end{aligned} \tag{22.4}$$

In matrix form, Equations (22.4) can be written as

$$BV + JX = K \tag{22.5}$$

where

$$B = \begin{bmatrix} -m_0 & 1 & 0 & 0 & 0 & 0 & 0 & 0 \\ 0 & 0 & -m_0 & 1 & 0 & 0 & 0 & 0 \\ 0 & 0 & 0 & 0 & -m_0 & 1 & 0 & 0 \\ 0 & 0 & 0 & 0 & 0 & 0 & -m_0 & 0 \end{bmatrix} \qquad J = \begin{bmatrix} -x_A & -1 \\ -x_B & -1 \\ -x_C & -1 \\ -x_D & -1 \end{bmatrix}$$

$$V = \begin{bmatrix} v_{x_A} \\ v_{y_A} \\ v_{x_B} \\ v_{y_B} \\ v_{x_C} \\ v_{y_C} \\ v_{x_D} \\ v_{y_D} \end{bmatrix} \qquad X = \begin{bmatrix} dm \\ db \end{bmatrix} \qquad K = \begin{bmatrix} -(m_0 x_A + b_0 - y_A) \\ -(m_0 x_B + b_0 - y_B) \\ -(m_0 x_C + b_0 - y_C) \\ -(m_0 x_D + b_0 - y_D) \end{bmatrix}$$

(22.6)

Now since both x and y are measured coordinates, they may each have individual estimated standard errors. Assuming that the coordinates are from independent observations, the four points will have eight measured coordinates and a cofactor matrix of

$$Q = \begin{bmatrix} \sigma_{x_A}^2 & 0 & 0 & 0 & 0 & 0 & 0 & 0 \\ 0 & \sigma_{y_A}^2 & 0 & 0 & 0 & 0 & 0 & 0 \\ 0 & 0 & \sigma_{x_B}^2 & 0 & 0 & 0 & 0 & 0 \\ 0 & 0 & 0 & \sigma_{y_B}^2 & 0 & 0 & 0 & 0 \\ 0 & 0 & 0 & 0 & \sigma_{x_C}^2 & 0 & 0 & 0 \\ 0 & 0 & 0 & 0 & 0 & \sigma_{y_C}^2 & 0 & 0 \\ 0 & 0 & 0 & 0 & 0 & 0 & \sigma_{x_D}^2 & 0 \\ 0 & 0 & 0 & 0 & 0 & 0 & 0 & \sigma_{y_D}^2 \end{bmatrix}$$

22.3 GENERAL LEAST SQUARES SOLUTION

In solving the general least squares problem, an *equivalent solution* is achieved. For this solution, the following *equivalent weight matrix* is created for the system:

$$W_e = (BQB^T)^{-1} \tag{22.7}$$

where B is as defined in Equation (22.6). Using the equivalent weight matrix in Equation (22.7), the *equivalent matrix system* is

$$J^T W_e J X = J^T W_e K \tag{22.8}$$

Equation (22.8) has the solution

$$X = (J^T W_e J)^{-1} J^T W_e K \tag{22.9}$$

Since this is a nonlinear equation system, the corrections in matrix X are applied to the initial approximations, and the method is repeated until the system converges. The *equivalent residuals* vector V_e is found following the usual procedure of

$$V_e = JX - K \tag{22.10}$$

Using Equation (22.10), the observational residuals are

$$V = QB^T W_e V_e \tag{22.11}$$

Also, since this is a nonlinear problem and the observations are being adjusted, the observations should also be updated according to their residuals. Thus, the updated observations for the second iteration are

$$l_i' = l_i + v_i \tag{22.12}$$

where l_i' are the second iterations observations, l_i the original observations, and v_i the observations' corresponding residuals. Generally in practice, the original observations are "close" to their final adjusted values, and thus Equation (22.12) is not actually used since the second iteration is only a check for convergence.

Finally, the reference variance for the adjustment can be computed using the equivalent residuals and weight matrix employing the equation

$$S_0^2 = \frac{V_e^T W_e V_e}{r} \tag{22.13}$$

where r is the number of redundancies in the system.

It should be noted that the same results can be obtained using the observational residuals and the following expression:

$$S_0^2 = \frac{V^T W V}{r} \tag{22.14}$$

Example 22.1 **Numerical Solution of the Straight-Line-Fit Problem.** Recall the least squares fit of points to a line in Section 11.11.1. In that example, the measured coordinate pairs were

A: (3.00, 4.50)   B: (4.25, 4.35)   C: (5.50, 5.50)   D: (8.00, 5.50)

and the solution for the slope of the line, m, and y intercept, b, were

$$m = 0.246$$
$$b = 3.663 \tag{a}$$

Additionally, the residuals were

$$V = AX - L = \begin{bmatrix} -0.10 \\ 0.46 \\ -0.48 \\ 0.13 \end{bmatrix}$$

Now this problem will be solved using the general least squares method. Assume that the following Q matrix is given:

$$Q = \begin{bmatrix} 0.020^2 & 0 & 0 & 0 & 0 & 0 & 0 & 0 \\ 0 & 0.015^2 & 0 & 0 & 0 & 0 & 0 & 0 \\ 0 & 0 & 0.023^2 & 0 & 0 & 0 & 0 & 0 \\ 0 & 0 & 0 & 0.036^2 & 0 & 0 & 0 & 0 \\ 0 & 0 & 0 & 0 & 0.033^2 & 0 & 0 & 0 \\ 0 & 0 & 0 & 0 & 0 & 0.028^2 & 0 & 0 \\ 0 & 0 & 0 & 0 & 0 & 0 & 0.016^2 & 0 \\ 0 & 0 & 0 & 0 & 0 & 0 & 0 & 0.019^2 \end{bmatrix}$$

22.3 GENERAL LEAST SQUARES SOLUTION

SOLUTION The step-by-step procedure for solving this problem using general least squares follows.

Step 1: Compute the initial approximations. Initial approximations for both m and b are found by using two points and solving the unique system. For this example the values from the solution of Section 11.11.1 given above will be used.

Step 2: Develop the appropriate matrices. In accordance with Equation (22.6), the B matrix is

$$B = \begin{bmatrix} -0.246 & 1 & 0 & 0 & 0 & 0 & 0 & 0 \\ 0 & 0 & -0.246 & 1 & 0 & 0 & 0 & 0 \\ 0 & 0 & 0 & 0 & -0.246 & 1 & 0 & 0 \\ 0 & 0 & 0 & 0 & 0 & 0 & -0.246 & 1 \end{bmatrix}$$

Using Equation (22.7), the equivalent weight matrix is

$$W_e = (BQB^T)^{-1} = \frac{1}{10{,}000}\begin{bmatrix} 2.5 & 0 & 0 & 0 \\ 0 & 13.3 & 0 & 0 \\ 0 & 0 & 8.5 & 0 \\ 0 & 0 & 0 & 3.8 \end{bmatrix}$$

$$= \begin{bmatrix} 4012.7 & 0 & 0 & 0 \\ 0 & 753.0 & 0 & 0 \\ 0 & 0 & 1176.6 & 0 \\ 0 & 0 & 0 & 2656.1 \end{bmatrix}$$

Step 3: Solve the system. The first iteration corrections are found using Equation (22.9) as

$$X = (J^T W_e J)^{-1} J^T W_e K = \begin{bmatrix} -0.0318 \\ 0.1907 \end{bmatrix}$$

where the J, K, and $J^T W_e J$ matrices for the first iteration were

$$J = \begin{bmatrix} -3.00 & -1 \\ -4.25 & -1 \\ -5.50 & -1 \\ -8.00 & -1 \end{bmatrix} \quad K = \begin{bmatrix} 0.246(3.00) + 3.663 - 4.50 \\ 0.246(4.25) + 3.663 - 4.25 \\ 0.246(5.50) + 3.663 - 5.50 \\ 0.246(8.00) + 3.663 - 5.50 \end{bmatrix} = \begin{bmatrix} -0.099 \\ 0.458 \\ -0.484 \\ 0.131 \end{bmatrix}$$

$$J^T W_e J = \begin{bmatrix} 255{,}297.91 & 42{,}958.45 \\ 42{,}958.45 & 8{,}598.40 \end{bmatrix} \quad J^T W_e K = \begin{bmatrix} 72.9737 \\ 273.5321 \end{bmatrix}$$

Step 4: Apply the corrections to m_0 and b_0 to update the initial approximations for the second iteration,

442 GENERAL LEAST SQUARES METHOD

$$m = 0.246 - 0.0318 = 0.2142$$

$$b = 3.663 + 0.1907 = 3.8537$$

Second iteration. During the second iteration, only the unknown parameters are updated, and thus only B, W_e, and K matrices differ from their first iteration counterparts. Their second iteration values are

$$B = \begin{bmatrix} -0.2142 & 1 & 0 & 0 & 0 & 0 & 0 & 0 \\ 0 & 0 & -0.2142 & 1 & 0 & 0 & 0 & 0 \\ 0 & 0 & 0 & 0 & -0.2142 & 1 & 0 & 0 \\ 0 & 0 & 0 & 0 & 0 & 0 & -0.2142 & 1 \end{bmatrix}$$

$$W_e = \begin{bmatrix} 4109.3 & 0 & 0 & 0 \\ 0 & 757.4 & 0 & 0 \\ 0 & 0 & 199.1 & 0 \\ 0 & 0 & 0 & 2682.8 \end{bmatrix} \quad K = \begin{bmatrix} -0.00370 \\ 0.51405 \\ -0.46820 \\ 0.06730 \end{bmatrix}$$

The corrections after this iteration are

$$X = \begin{bmatrix} 0.00002 \\ 0.00068 \end{bmatrix}$$

Thus, m and b are 0.214 and 3.854 to three decimal places. Using Equation (22.10), the equivalent residual vector is

$$V_e = \begin{bmatrix} -0.0030 \\ 0.5148 \\ -0.4674 \\ 0.0681 \end{bmatrix}$$

Using Equation (22.11), the observation residual vector is

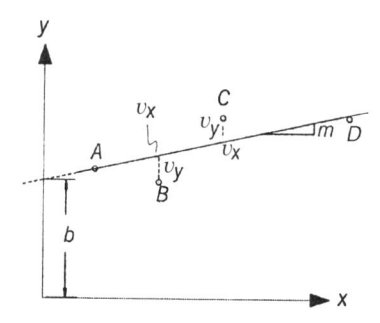

Figure 22.1 General least squares fit of points to a line.

22.4 TWO-DIMENSIONAL COORDINATE TRANSFORMATION BY GENERAL LEAST SQUARES

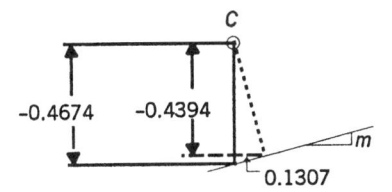

Figure 22.2 Residuals for point C.

$$V = QB^TW_eV_e = \begin{bmatrix} 0.0010 \\ -0.0027 \\ -0.0442 \\ 0.5053 \\ 0.1307 \\ -0.4394 \\ -0.0100 \\ 0.0660 \end{bmatrix}$$

A graphical interpretation of the residuals is shown in Figure 22.1. Notice how the equivalent residuals are aligned with the y axis, and that the observational residuals exist in the primary x- and y-axis directions. These residuals are shown more clearly in Figure 22.2, which is an enlarged view of the portion of Figure 22.1 that surrounds point C. The equivalent residual of C is -0.4674 from the line, and the observational residuals $v_{x_c} = 0.1307$ and $v_{y_c} = -0.4394$ are parallel to the x and y axes, respectively. This general solution is more appropriate for adjusting the actual coordinate measurements. Note that this solution is slightly different from that determined in Example 11.3.

Using Equation (22.13), the reference standard deviation is

$$S_0 = \sqrt{\frac{475.2}{4-2}} = \pm 15.4$$

22.4 TWO-DIMENSIONAL COORDINATE TRANSFORMATION BY GENERAL LEAST SQUARES

As presented in Chapter 18, two-dimensional coordinate transformations are commonly used to convert points from one two-dimensional coordinate system to another. Again, the general least squares method is a more appropriate method for these transformations since the coordinates in both systems are measurements that contain errors.

22.4.1 Two-Dimensional Conformal Coordinate Transformation

The two-dimensional conformal coordinate transformation, presented in Chapter 18, has four unknowns, consisting of a scale factor, rotation angle, and two translations. Equations (18.5) express this transformation, and they are repeated here for convenience. The transformation equations are

$$X = ax - by + c$$
$$Y = bx + ay + d \tag{22.15}$$

Equations (22.15) can be rearranged as

$$F: ax - by + c - X = 0$$
$$G: bx + ay + d - Y = 0 \tag{22.16}$$

Since the coordinates from both the xy and XY systems contain errors, Equations (22.16) are rewritten as

$$F(x, y, X, Y) = a(x + v_x) - b(y + v_y) + c - (X + v_X) = 0$$
$$G(x, y, X, Y) = b(x + v_x) + a(y + v_y) + d - (Y + v_Y) = 0 \tag{22.17}$$

These equations are nonlinear in terms of their observations and residuals. They are solved by linearizing the equations and iterating to a solution. The partial derivatives with respect to each unknown are

$$\frac{\partial f}{\partial x} = a \quad \frac{\partial f}{\partial y} = -b \quad \frac{\partial f}{\partial X} = -1 \quad \frac{\partial g}{\partial x} = b \quad \frac{\partial g}{\partial y} = a \quad \frac{\partial g}{\partial Y} = -1$$
$$\frac{\partial f}{\partial a} = x \quad \frac{\partial f}{\partial b} = -y \quad \frac{\partial f}{\partial c} = 1 \quad \frac{\partial g}{\partial a} = y \quad \frac{\partial g}{\partial b} = x \quad \frac{\partial g}{\partial d} = 1 \tag{22.18}$$

Using the partial derivatives in Equation (22.18), a matrix for each point can be built as

$$\begin{bmatrix} a_0 & -b_0 & 1 & 0 \\ b_0 & a_0 & 0 & 1 \end{bmatrix} \begin{bmatrix} v_x \\ v_y \\ v_X \\ v_Y \end{bmatrix} + \begin{bmatrix} x & -y & 1 & 0 \\ y & x & 0 & 1 \end{bmatrix} \begin{bmatrix} da \\ db \\ dc \\ dd \end{bmatrix}$$
$$= \begin{bmatrix} X - (a_0 x + b_0 y + T_x) \\ Y - (b_0 x + a_0 y + T_y) \end{bmatrix} \tag{22.19}$$

For a redundant system, the matrices are solved following the matrix procedures outlined in Section 22.3.

22.4 TWO-DIMENSIONAL COORDINATE TRANSFORMATION BY GENERAL LEAST SQUARES

Example 22.2 Four fiducial points are digitized from an aerial photo, and their measured (x,y) and control (X,Y) values are listed in Table 22.1. The standard deviations of these measurements are also listed. What are the most probable values for the transformation parameters and the resulting residuals?

SOLUTION In this problem, initial approximations for a, b, c, and d must first be computed. These values can be found using a standard least squares adjustment, or by solving the system with only two points. The former procedure was demonstrated in Example 18.1. Using standard least squares, initial approximations for the parameters are determined to be

$$a_0 = 25.386458$$
$$b_0 = -0.8158708$$
$$c_0 = -137.216$$
$$d_0 = -150.600$$

Now the B, Q, J, and K matrices described in Section 21.3 can be formed. They are listed below (note that the numbers are rounded to three decimal places for publication purposes only):

$$B = \begin{bmatrix} 25.386 & 0.816 & -1 & 0 & 0 & 0 & 0 & 0 & 0 & 0 & 0 & 0 & 0 & 0 & 0 & 0 \\ -0.816 & 25.386 & 0 & -1 & 0 & 0 & 0 & 0 & 0 & 0 & 0 & 0 & 0 & 0 & 0 & 0 \\ 0 & 0 & 0 & 0 & 25.386 & 0.816 & -1 & 0 & 0 & 0 & 0 & 0 & 0 & 0 & 0 & 0 \\ 0 & 0 & 0 & 0 & -0.816 & 25.386 & 0 & -1 & 0 & 0 & 0 & 0 & 0 & 0 & 0 & 0 \\ 0 & 0 & 0 & 0 & 0 & 0 & 0 & 0 & 25.386 & 0.816 & -1 & 0 & 0 & 0 & 0 & 0 \\ 0 & 0 & 0 & 0 & 0 & 0 & 0 & 0 & -0.816 & 25.386 & 0 & -1 & 0 & 0 & 0 & 0 \\ 0 & 0 & 0 & 0 & 0 & 0 & 0 & 0 & 0 & 0 & 0 & 0 & 25.386 & 0.816 & -1 & 0 \\ 0 & 0 & 0 & 0 & 0 & 0 & 0 & 0 & 0 & 0 & 0 & 0 & -0.816 & 25.386 & 0 & -1 \end{bmatrix}$$

$$J = \begin{bmatrix} 0.764 & -5.960 & 1 & 0 \\ 5.960 & 0.764 & 0 & 1 \\ 5.062 & -10.541 & 1 & 0 \\ 10.541 & 5.062 & 0 & 1 \\ 9.663 & -6.243 & 1 & 0 \\ 6.243 & 9.663 & 0 & 1 \\ 5.350 & -1.654 & 1 & 0 \\ 1.654 & 5.350 & 0 & 1 \end{bmatrix} \quad K = \begin{bmatrix} -0.03447270 \\ -0.08482509 \\ 0.11090026 \\ 0.13190015 \\ -0.18120912 \\ -0.00114251 \\ 0.04999940 \\ -0.02329275 \end{bmatrix}$$

TABLE 22.1 Data for a Two-Dimensional Conformal Coordinate Transformation

Point	$X \pm S_X$	$Y \pm S_Y$	$x \pm s_x$	$y \pm s_y$
1	-113.000 ± 0.002	0.003 ± 0.002	0.7637 ± 0.026	5.9603 ± 0.028
3	0.001 ± 0.002	112.993 ± 0.002	5.0620 ± 0.024	10.5407 ± 0.030
5	112.998 ± 0.002	0.003 ± 0.002	9.6627 ± 0.028	6.2430 ± 0.022
7	0.001 ± 0.002	-112.999 ± 0.002	5.3500 ± 0.024	1.6540 ± 0.026

GENERAL LEAST SQUARES METHOD

Also, the Q matrix is

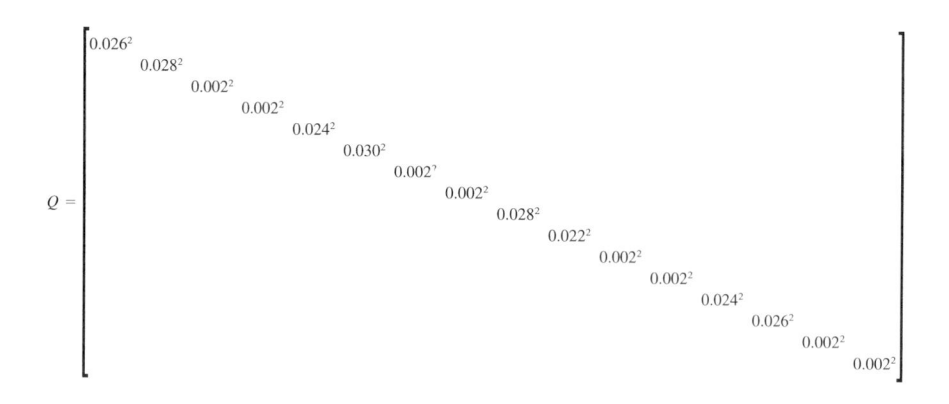

$$Q = \begin{bmatrix} 0.026^2 & & & & & & & & & & & & & & & \\ & 0.028^2 & & & & & & & & & & & & & & \\ & & 0.002^2 & & & & & & & & & & & & & \\ & & & 0.002^2 & & & & & & & & & & & & \\ & & & & 0.024^2 & & & & & & & & & & & \\ & & & & & 0.030^2 & & & & & & & & & & \\ & & & & & & 0.002^2 & & & & & & & & & \\ & & & & & & & 0.002^2 & & & & & & & & \\ & & & & & & & & 0.028^2 & & & & & & & \\ & & & & & & & & & 0.022^2 & & & & & & \\ & & & & & & & & & & 0.002^2 & & & & & \\ & & & & & & & & & & & 0.002^2 & & & & \\ & & & & & & & & & & & & 0.024^2 & & & \\ & & & & & & & & & & & & & 0.026^2 & & \\ & & & & & & & & & & & & & & 0.002^2 & \\ & & & & & & & & & & & & & & & 0.002^2 \end{bmatrix}$$

The solution for the first iteration is

$$X = \begin{bmatrix} -0.000124503 \\ -0.000026212 \\ -0.000325016 \\ -0.000029546 \end{bmatrix}$$

Adding these corrections to the initial approximations, yields

$$a = 25.38633347$$
$$b = -0.815897012$$
$$c = -137.2163$$
$$d = -150.6000$$

In the next iteration, only minor corrections occur, and thus the system has converged to a solution. The residuals and reference variance are computed as before. Although the solution has changed only slightly from the standard least squares method, it properly considers the fact that each observation equation contains four measurements. Of course, once the transformation parameters have been determined, any points that exist only in the xy system can be transformed into the XY system by substitution into Equations (22.15). This part of the problem is not demonstrated in this example. For the remainder of the chapter, only the adjustment model is developed.

22.4 TWO-DIMENSIONAL COORDINATE TRANSFORMATION BY GENERAL LEAST SQUARES

22.4.2 Two-Dimensional Affine Coordinate Transformation

As discussed in Section 17.5, the main difference between conformal and affine transformations is that the latter allows for different scales along the x and y axes and accounts for nonorthogonality in the axes. This results in six parameters. Equations (18.9) express the affine transformation, and they are repeated here for convenience.

$$X = ax + by + c$$
$$Y = dx + ey + f \qquad (22.20)$$

Equation (22.20) can be rewritten as

$$F(x,y,X,Y) = ax + by + c - X = 0$$
$$G(x,y,X,Y) = dx + ey + f - Y = 0 \qquad (22.21)$$

Again, Equation (22.21) consists of observations in both the x and y coordinates, and thus it is more appropriate to use the general least squares method. Therefore, these equations can be rewritten as

$$F(x,y,X,Y) = a(x + v_x) + b(y + v_y) + c - (X + v_X) = 0$$
$$G(x,y,X,Y) = d(x + v_x) + e(y + v_y) + f - (Y + v_Y) = 0 \qquad (22.22)$$

For each point, the linearized equations in matrix form are

$$\begin{bmatrix} a_0 & b_0 & -1 & 0 \\ d_0 & e_0 & 0 & -1 \end{bmatrix} \begin{bmatrix} v_x \\ v_y \\ v_X \\ v_Y \end{bmatrix} + \begin{bmatrix} x & y & 1 & 0 & 0 & 0 \\ 0 & 0 & 0 & x & y & 1 \end{bmatrix} \begin{bmatrix} da \\ db \\ dc \\ dd \\ de \\ df \end{bmatrix}$$

$$= \begin{bmatrix} X - (a_0 x + b_0 y + c_0) \\ Y - (d_0 x + e_0 y + f_0) \end{bmatrix} \qquad (22.23)$$

Two observation equations, like those of Equation (22.23), result for each control point. Since there are six unknown parameters, three control points are needed for a unique solution. With more than three, a redundant system exists, and the solution is obtained following the least squares procedures outlined in Section 22.3.

22.4.3 Two-Dimensional Projective Transformation

The two-dimensional projective coordinate transformation converts a projection of one plane coordinate system into another nonparallel plane system. This transformation was developed in Section 18.6. Equations (18.12) for this transformation are repeated here for convenience.

$$X = \frac{a_1 x + b_1 y + c_1}{a_3 x + b_3 y + 1}$$
$$Y = \frac{a_2 x + b_2 y + c_2}{a_3 x + b_3 y + 1}$$
(22.24)

Unlike the conformal and affine types, this transformation is nonlinear in its standard form. In their general form, the projective equations become

$$F(x,y,X,Y) = \frac{a_1(x + v_x) + b_1(y + v_y) + c_1}{a_3(x + v_x) + b_3(y + v_y) + 1} - (X + v_X) = 0$$
$$G(x,y,X,Y) = \frac{a_2(x + v_x) + b_2(y + v_y) + c_2}{a_3(x + v_x) + b_3(y + v_y) + 1} - (Y + v_Y) = 0$$
(22.25)

Again a linearized form for Equations (22.25) is needed. The partial derivatives for the unknown parameters were given in Section 18.6, and the remaining partial derivatives are given below.

$$\frac{\partial F}{\partial x} = \frac{a_1(b_3 y + 1) - a_3(b_1 y + c_1)}{(a_3 x + b_3 y + 1)^2} \qquad \frac{\partial F}{\partial y} = \frac{b_1(a_3 x + 1) - b_3(a_1 y + c_1)}{(a_3 x + b_3 y + 1)^2}$$

$$\frac{\partial G}{\partial x} = \frac{a_2(b_3 y + 1) - a_3(b_2 y + c_2)}{(a_3 + b_3 y + 1)^2} \qquad \frac{\partial G}{\partial y} = \frac{b_2(a_3 x + 1) - b_3(a_2 y + c_2)}{(a_3 x + b_3 y + 1)^2}$$

In matrix form, the linearized equations for each point are

$$\begin{bmatrix} \frac{\partial F}{\partial x} & \frac{\partial F}{\partial y} & -1 & 0 \\ \frac{\partial G}{\partial x} & \frac{\partial G}{\partial y} & 0 & -1 \end{bmatrix} \begin{bmatrix} v_x \\ v_y \\ v_X \\ v_Y \end{bmatrix} + \begin{bmatrix} \frac{\partial F}{\partial a_1} & \frac{\partial F}{\partial b_1} & \frac{\partial F}{\partial c_1} & 0 & 0 & 0 & \frac{\partial F}{\partial a_3} & \frac{\partial F}{\partial b_3} \\ 0 & 0 & 0 & \frac{\partial G}{\partial a_2} & \frac{\partial G}{\partial b_2} & \frac{\partial G}{\partial c_2} & \frac{\partial G}{\partial a_3} & \frac{\partial G}{\partial b_3} \end{bmatrix} \begin{bmatrix} da_1 \\ db_1 \\ dc_1 \\ da_2 \\ db_2 \\ dc_2 \\ da_3 \\ db_3 \end{bmatrix} = K$$

(22.26)

Equation (22.26) gives the observation equations for the two-dimensional projective transformation for one control point and the K matrix is defined in

Equation (18.13). Since there are eight unknown parameters, four control points are needed for a unique solution. More than four control points yields a redundant system that can be solved following the steps outlined in Section 22.3.

22.5 THREE-DIMENSIONAL CONFORMAL COORDINATE TRANSFORMATION BY GENERAL LEAST SQUARES

As explained in Section 18.7, this coordinate transformation converts points from one three-dimensional coordinate system to another. Equations (18.15) express this transformation, and the matrix form of those equations is

$$X = SRx + T \tag{22.27}$$

where the individual matrices are as defined in Section 18.7.

Equation (18.15) gives detailed expressions for the three-dimensional coordinate transformation. Note that these equations involve six observations, *xyz* and *XYZ*. In general least squares, these equations can be rewritten as

$$f(x,y,z,X,Y,Z) = S[r_{11}(x + V_x) + r_{21}(y + V_y) + r_{31}(z + V_z)] - (X + V_X) = 0$$
$$g(x,y,z,X,Y,Z) = S[r_{12}(x + V_x) + r_{22}(y + V_y) + r_{32}(z + V_z)] - (Y + V_Y) = 0$$
$$h(x,y,z,X,Y,Z) = S[r_{13}(x + V_x) + r_{23}(y + V_y) + r_{33}(z + V_z)] - (Z + V_Z) = 0$$
$$\tag{22.28}$$

Equations (22.28) are for a single point, and again, they are nonlinear. They can be expressed in linearized matrix form as

$$\begin{bmatrix} \frac{\partial f}{\partial x} & \frac{\partial f}{\partial y} & \frac{\partial f}{\partial z} & -1 & 0 & 0 \\ \frac{\partial g}{\partial x} & \frac{\partial g}{\partial y} & \frac{\partial g}{\partial z} & 0 & -1 & 0 \\ \frac{\partial h}{\partial x} & \frac{\partial h}{\partial y} & \frac{\partial h}{\partial z} & 0 & 0 & -1 \end{bmatrix} \begin{bmatrix} v_x \\ v_y \\ v_z \\ v_X \\ v_Y \\ v_Z \end{bmatrix}$$
$$+ \begin{bmatrix} \frac{\partial f}{\partial S} & 0 & \frac{\partial f}{\partial \theta_2} & \frac{\partial f}{\partial \theta_3} & 1 & 0 & 0 \\ \frac{\partial g}{\partial S} & \frac{\partial g}{\partial \theta_1} & \frac{\partial g}{\partial \theta_2} & \frac{\partial g}{\partial \theta_3} & 0 & 1 & 0 \\ \frac{\partial h}{\partial S} & \frac{\partial h}{\partial \theta_1} & \frac{\partial h}{\partial \theta_2} & \frac{\partial h}{\partial \theta_3} & 0 & 0 & 1 \end{bmatrix} \begin{bmatrix} dS \\ d\theta_1 \\ d\theta_2 \\ d\theta_3 \\ dT_X \\ dT_Y \\ dT_Z \end{bmatrix} = \begin{bmatrix} 0 \\ 0 \\ 0 \end{bmatrix} \tag{22.29}$$

450 GENERAL LEAST SQUARES METHOD

where

$$\frac{\partial f}{\partial x} = Sr_{11} \qquad \frac{\partial f}{\partial y} = Sr_{21} \qquad \frac{\partial f}{\partial z} = Sr_{31}$$

$$\frac{\partial g}{\partial x} = Sr_{12} \qquad \frac{\partial g}{\partial y} = Sr_{22} \qquad \frac{\partial g}{\partial z} = Sr_{32}$$

$$\frac{\partial h}{\partial x} = Sr_{13} \qquad \frac{\partial h}{\partial y} = Sr_{23} \qquad \frac{\partial h}{\partial z} = Sr_{33}$$

The remaining partial derivatives were given in Section 18.7. This system is solved using the methods discussed in Section 22.3.

Example 22.3 Estimated errors were added to the control coordinates in Example 18.4. The control data are repeated in Table 22.2 and standard deviations of the control coordinates, needed to form the Q matrix, are also listed. Following Table 22.2, output from the program ADJUST is listed. Note that the solution differs somewhat from the one obtained by standard least squares in Example 18.4.

SOLUTION

```
3D Coordinate Transformation of File:
using generalized least squares method
=========================================================

Measured Points
=========================================================
NAME        x          y          z         Vx        Vy       Vz
=========================================================
  1    1094.883    820.085   109.821   -0.001   -0.001    0.000
  2     503.891   1598.698   117.685   -0.002    0.000    0.000
  3    2349.343    207.658   151.387   -0.001    0.000    0.000
  4    1395.320   1348.853   215.261    0.001    0.000   -0.001
```

TABLE 22.2 Control Data for the Three-Dimensional Conformal Coordinate Transformation

Point	X	Y	Z	SX	SY	SZ
1	10,037.81	5262.09	772.04	0.05	0.06	0.05
2	10,956.68	5128.17	783.00	0.04	0.06	0.09
3	8,780.08	4840.29	782.62	0.02	0.04	0.02
4	10,185.80	4700.21	851.32	0.03	0.05	0.03

CONTROL POINTS
```
===============================================================
NAME         X         VX         Y         VY         Z         VZ
===============================================================
  1    10037.810   -0.063   5262.090   -0.026   772.040   -0.001
  2    10956.680   -0.019   5128.170    0.063   783.000   -0.027
  3     8780.080    0.000   4840.290    0.038   782.620   -0.001
  4    10185.800    0.032   4700.210   -0.085   851.320    0.007
```

Transformation Coefficients
```
===============================================================
Scale = 0.94996 +/- 0.00002
Omega =   2° 17' 00.0" +/- 0° 00' 26.7"
  Phi =  -0° 33' 05.6" +/- 0° 00' 06.1"
Kappa = 224° 32' 11.5" +/- 0° 00' 07.7"
   Tx = 10233.855 +/- 0.066
   Ty =  6549.964 +/- 0.055
   Tz =   720.867 +/- 0.219
```

Reference Standard Deviation: 1.293
Degrees of Freedom: 5
Iterations: 3

Transformed Coordinates
```
===============================================================
NAME         X         Sx         Y         Sy         Z         Sz
===============================================================
  1    10037.874   0.082   5262.116   0.063   772.041   0.275
  2    10956.701   0.087   5128.106   0.070   783.027   0.286
  3     8780.080   0.098   4840.251   0.087   782.622   0.314
  4    10185.767   0.096   4700.296   0.072   851.313   0.324
  5    10722.016   0.075   5691.210   0.062   766.067   0.246
  6    10043.245   0.074   5675.887   0.060   816.857   0.246
```

PROBLEMS

Note: For problems requiring least squares adjustment, if a computer program is not distinctly specified for use in the problem, it is expected that the least squares algorithm will be solved using the program MATRIX, which is included on the CD supplied with the book.

22.1 Solve Problem 11.9 using the general least squares method. Assign all coordinates a standard deviation of 1.

452 GENERAL LEAST SQUARES METHOD

22.2 Do Problem 11.11 using the general least squares method. Assign the computed coordinates the following standard deviations.

Station	S_X	S_Y		Station	S_X	S_Y
A	0.001	0.001		B	0.020	0.012
C	0.020	0.013		D	0.020	0.021
E	0.020	0.027		F	0.020	0.028
G	0.020	0.035		H	0.020	0.040
I	0.020	0.046				

22.3 Solve Problem 11.12 using the general least squares method. Assign all coordinates a standard deviation of 0.1.

22.4 Do Example 18.1 using the general least squares method. Assume estimated standard deviations for the coordinates of $S_X = S_Y = \pm 0.05$ and $S_x = S_y = \pm 0.005$.

22.5 Do Problem 18.1 using the general least squares method. Assume estimated standard deviations for all the coordinates of ± 0.05. Compare this solution with that of Problem 22.4.

22.6 Solve Problem 18.2 using the general least squares method. Assume estimated standard deviations of ± 0.003 mm for all the control coordinates.

22.7 Use the general least squares method to transform the points from the measured system to the control system using a two-dimensional:
(a) conformal transformation.
(b) affine transformation.
(c) projective transformation.

	Measured coordinates				Control coordinates			
Point	x (mm)	y (mm)	S_x (mm)	S_y (mm)	X (mm)	Y (mm)	S_X (mm)	S_Y (mm)
6	103.55	−103.670	0.003	0.004	103.951	−103.969	0.009	0.009
7	0.390	−112.660	0.005	0.005	0.001	−112.999	0.009	0.009
3	0.275	111.780	0.004	0.007	0.001	112.993	0.009	0.009
4	103.450	102.815	0.003	0.004	103.956	103.960	0.009	0.009
5	112.490	−0.395	0.003	0.003	112.998	0.003	0.009	0.009
32	18.565	−87.580						
22	−5.790	2.035						
12	6.840	95.540						
13	86.840	102.195						
23	93.770	2.360						
33	92.655	−90.765						

22.8 Given the following data, transform the points from the measured system to the control system using a three-dimensional conformal coordinate transformation. Assume that all coordinates have estimated standard deviations of ±0.05 m. Use the general least squares method.

	Measured coordinates			Control coordinates		
Point	x (m)	y (m)	z (m)	X (m)	Y (m)	Z (m)
1	607.54	501.63	469.09	390.13	499.74	469.32
2	598.98	632.36	82.81	371.46	630.95	81.14
3	643.65	421.28	83.50	425.43	419.18	82.38
4	628.58	440.51	82.27			
5	666.27	298.16	98.29			
6	632.59	710.62	103.01			

22.9 Solve Problem 18.14 using general least squares. Assume that all the control coordinates have estimated standard deviations of 0.005 m.

22.10 Do Problem 22.1 using the ADJUST program.

Solve each problem with the ADJUST program using the standard and general least squares adjustment options. Compare and explain any differences noted in the solutions.

22.11 Problem 22.6

22.12 Problem 22.7

22.13 Problem 22.8

Programming Problems

Develop a computational program that constructs the B, Q, J, and K matrices for each of the following transformations.

22.14 A two-dimensional conformal coordinate transformation.

22.15 A two-dimensional affine coordinate transformation.

22.16 A two-dimensional projective coordinate transformation.

22.17 A three-dimensional conformal coordinate transformation.

CHAPTER 23

THREE-DIMENSIONAL GEODETIC NETWORK ADJUSTMENT

23.1 INTRODUCTION

With the advent of total station instruments, survey data are being collected in three dimensions. Thus, it is advantageous to develop an adjustment model that works in three dimensions. Rigorous triangulation adjustment models date back to Bruns (1878). The main observational data consist of horizontal angles, vertical angles, azimuths, and slant distances. It is also possible to include differential leveling in the model. Since all data are collected on Earth's surface, the local geodetic coordinate system provides a natural system in which to perform the adjustment.

As shown in Figure 23.1, the local geodetic system is oriented such that the n axis points along the meridian of the ellipse (local geodetic north), the u axis is aligned along the normal of the ellipsoid, and the e axis creates a left-handed coordinate system. The local geodetic coordinate system can be related to the geocentric coordinate system (see Section 17.5) through a series of three-dimensional rotations discussed in Section 18.7. To align the X axis with the e axis, the Z axis is rotated by an amount of $\lambda - 180°$. Then the Z axis is aligned with the u axis by a rotation of $\phi - 90°$ about the once-rotated X axis. Changing the sign of e to convert to a right-handed coordinate system, the resulting expression is

$$\begin{bmatrix} \Delta n \\ -\Delta e \\ \Delta u \end{bmatrix} = R_2\,(\phi - 90°)R_3(\lambda - 180°) \begin{bmatrix} \Delta X \\ \Delta Y \\ \Delta Z \end{bmatrix} \quad (23.1)$$

23.1 INTRODUCTION

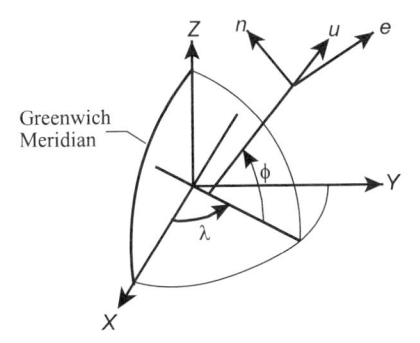

[Figure: diagram showing Z, n, u, e axes with Greenwich Meridian, X, Y axes, and angles φ, λ]

Figure 23.1 Relationship between the geocentric and local geodetic coordinate systems.

In Equation (23.1), changes in the local geodetic coordinate system and geocentric coordinate system are represented by $(\Delta n, \Delta e, \Delta u)$ and $(\Delta X, \Delta Y, \Delta Z)$, respectively. Multiplying the rotation matrices given in Section 18.7 and simplifying yields

$$\begin{bmatrix} \Delta n \\ -\Delta e \\ \Delta u \end{bmatrix} = \begin{bmatrix} -\sin\phi \cos\lambda & -\sin\phi \sin\lambda & \cos\phi \\ -\sin\lambda & \cos\lambda & 0 \\ \cos\phi \cos\lambda & \cos\phi \sin\lambda & \sin\phi \end{bmatrix} \begin{bmatrix} \Delta X \\ \Delta Y \\ \Delta Z \end{bmatrix} \quad (23.2)$$

The changes in the coordinates of the local geodetic system can be determined from the observation of azimuth Az, slant distance s, and vertical angle v. As shown in Figure 23.2, the changes in the local geodetic coordinate system can be computed as

$$\Delta n = s \cos v \cos Az$$
$$\Delta e = s \cos v \sin Az \quad (23.3)$$
$$\Delta u = s \sin v$$

From Figure 23.2, the following inverse relationships can also be developed:

$$s = \sqrt{\Delta n^2 + \Delta e^2 + \Delta u^2}$$
$$Az = \tan^{-1} \frac{\Delta e}{\Delta n} \quad (23.4)$$
$$v = \sin^{-1} \frac{\Delta u}{s}$$

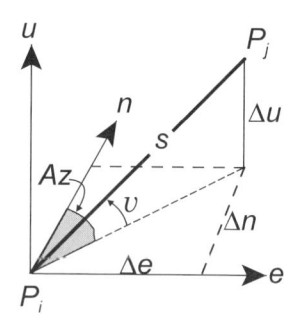

Figure 23.2 Reduction of observations in a local geodetic coordinate system.

By combining Equations (23.2) and (23.4), the reduced observations can be computed using changes in the geocentric coordinates. The resulting equations are

$$IJ = \sqrt{\Delta X^2 + \Delta Y^2 + \Delta Z^2} \tag{23.5}$$

$$Az_{ij} = \tan^{-1} \frac{-\Delta X \sin \lambda_i + \Delta Y \cos \lambda_i}{-\Delta X \sin \phi_i \cos \lambda_i - \Delta Y \sin \phi_i \sin \lambda_i - \Delta Z \cos \phi_i} \tag{23.6}$$

$$v_{ij} = \sin^{-1} \frac{\Delta X \cos \phi_i \cos \lambda_i + \Delta Y \cos \phi_i \sin \lambda_i + \Delta Z \sin \phi_i}{\sqrt{\Delta X^2 + \Delta Y^2 + \Delta Z^2}} \tag{23.7}$$

In Equations (23.5) to (23.7), ϕ_i and λ_i are the latitude and longitude of the observing station P_i, ΔX is $X_j - X_i$, ΔY is $Y_j - Y_i$, and ΔZ is $Z_j - Z_i$. For completeness, the equation for the zenith angle is

$$z_{ij} = \cos^{-1} \frac{\Delta X \cos \phi_i \cos \lambda_i + \Delta Y \cos \phi_i \sin \lambda_i + \Delta Z \sin \phi_i}{\sqrt{\Delta X^2 + \Delta Y^2 + \Delta Z^2}} \tag{23.8}$$

Furthermore, since an angle is the difference between two azimuths, Equation (23.6) can be applied to horizontal angles.

23.2 LINEARIZATION OF EQUATIONS

Equations (23.5) to (23.8) can be linearized with respect to the local geodetic coordinates. The development of these equations is covered in Vincenty (1989) and Leick (2004). The development of two coefficients are demonstrated and the final linearized prototype equations for slant distances, azimuths, horizontal angles, and vertical angles are listed in the following

23.2 LINEARIZATION OF EQUATIONS

subsections. It is important to note that the residuals exist implicitly in each equation in order to make the equations consistent when using observations containing random errors.

23.2.1 Slant Distance Observations

In the three-dimensional model, slant distance (also known as *slope distance*) does not need to be reduced to either the station ground marks or ellipsoid in the functional model. However, if they are not reduced, the values computed from Equations (23.5) to (23.8) are based on the instrument and reflector locations and not on the station's ground marks. The prototype equation for the slant distances observed is

$$a_1\, dn_i + a_2\, de_i + a_3\, du_i + a_4\, dn_j + a_5\, de_j + a_6\, du_j = s_{ij} - IJ_0 \quad (23.9)$$

In Equation (23.9), the coefficients a_1 to a_6 are defined in Table 23.1, IJ_0 is the slant distance computed using Equation (23.5), and s_{ij} is the slant distance observed.

To demonstrate the derivation of the coefficients, a_1 is derived by taking the partial derivative of the slant distance formula in Equation (23.4) with respect to n_i:

$$\frac{\partial s}{\partial n_i} = -\frac{\Delta n}{s} \quad (a)$$

By substituting Δn from Equation (23.3), the resulting equation for a_1 is

$$\frac{\partial s}{\partial n_i} = -\cos v_{ij} \cos Az_{ij} \quad (b)$$

Following similar procedures, the remaining coefficients for Equation (23.9) are derived.

23.2.2 Azimuth Observations

The prototype equation for the observed azimuths is

$$b_1\, dn_i + b_2\, de_i + b_3\, du_i + b_4\, dn_j + b_5\, de_j + b_6\, du_j = \alpha - Az_0 \quad (23.10)$$

In Equation (23.10), α is the observed azimuth and Az_0 is its computed value based on Equation (23.6) and approximate values for the station coordinates.

As an example, the coefficient for b_1 is computed by taking the partial derivative of the azimuth formula in Equation (23.4) with respect to n_i:

TABLE 23.1 Coefficients for Linearized Equations in Equations (23.9) to (23.11)

(a) $a_1 = -(\cos v_{ij} \cos Az_{ij})_0$ (b) $a_2 = -(\cos v_{ij} \sin Az_{ij})_0$ (c) $a_3 = -(\sin v_{ij})_0$

(d) $a_4 = -(\cos v_{ji} \cos Az_{ji})_0$ (e) $a_5 = -(\cos v_{ji} \sin Az_{ji})_0$ (f) $a_6 = -(\sin v_{ji})_0$

(g) $b_1 = \left(\dfrac{\sin Az_{ij}}{IJ \cos v_{ij}}\right)_0$ (h) $b_2 = -\left(\dfrac{\cos Az_{ij}}{IJ \cos v_{ij}}\right)_0$ (i) $b_3 = 0$

(j) $b_4 = -\left\{\dfrac{\sin Az_{ij}}{IJ \cos v_{ij}}\left[\cos(\phi_j - \phi_i) + \dfrac{\sin \phi_j \sin(\lambda_j - \lambda_i)}{\tan Az_{ij}}\right]\right\}_0$

(k) $b_5 = \left\{\dfrac{\cos Az_{ij}}{IJ \cos v_{ij}}[\cos(\lambda_j - \lambda_i) - \sin \phi_i \sin(\lambda_j - \lambda_i) \tan Az_{ij}]\right\}_0$

(l) $b_6 = \left\{\dfrac{\cos Az_{ij} \cos \phi_j}{IJ \cos v_{ij}}[\sin(\lambda_j - \lambda_i) + (\sin \phi_i \cos(\lambda_j - \lambda_i) - \cos \phi_i \tan \phi_j) \tan Az_{ij}]\right\}_0$

(m) $c_1 = \left(\dfrac{\sin v_{ij} \cos Az_{ij}}{IJ}\right)_0$ (n) $c_2 = \left(\dfrac{\sin v_{ij} \sin Az_{ij}}{IJ}\right)_0$ (o) $c_3 = -\left(\dfrac{\cos v_{ij}}{IJ}\right)_0$

(p) $c_4 = \left[\dfrac{-\cos \phi_i \sin \phi_j \cos(\lambda_j - \lambda_i) + \sin \phi_i \cos \phi_j + \sin v_{ij} \cos v_{ji} \cos Az_{ji}}{IJ \cos v_{ij}}\right]_0$

(q) $c_5 = \left[\dfrac{-\cos \phi_i \sin(\lambda_j - \lambda_i) + \sin v_{ij} \cos v_{ji} \sin Az_{ji}}{IJ \cos v_{ij}}\right]_0$

(r) $c_6 = \left[\dfrac{\sin v_{ij} \sin v_{ji} + \sin \phi_i \sin \phi_j + \cos \phi_i \cos \phi_j \cos(\lambda_j - \lambda_i)}{IJ \cos v_{ij}}\right]_0$

$$\dfrac{\partial Az_{ij}}{\partial n_i} = \dfrac{\Delta n^2}{\Delta n^2 + \Delta e^2} \dfrac{\Delta e}{\Delta n^2} = \dfrac{\Delta e}{\Delta n^2 + \Delta e^2} \qquad (c)$$

By substituting the appropriate formulas from Equation (23.3), the resulting equation for b_1 is

$$\dfrac{\partial Az_{ij}}{\partial n_i} = \dfrac{s \cos v_{ij} \sin Az_{ij}}{s^2 \cos^2 v_{ij} \cos^2 Az_{ij} + s^2 \cos^2 v_{ij} \sin^2 Az_{ij}}$$

$$= \dfrac{s \cos v_{ij} \sin Az_{ij}}{s^2 \cos^2 v_{ij} (\cos^2 Az_{ij} + \sin^2 Az_{ij})}$$

$$= \dfrac{\sin Az_{ij}}{s \cos v_{ij}} \qquad (d)$$

23.2 LINEARIZATION OF EQUATIONS

The remaining coefficients of Equation (23.10) can be derived similarly. For the remaining linearized observation equations in this section, the derivation of the coefficients for each equation follows procedures similar to those presented in Equations (a) through (d). The derivations are left to the reader.

23.2.3 Vertical Angle Observations

Vertical angles are measured in the vertical plane and have a value of zero at the horizon. As discussed in Section 23.7, vertical angles can be subject to substantial systematic errors caused by deflection of the vertical and refraction. Due to these errors, vertical angles should not be used in an adjustment on a regular basis. If these observations must be used in an adjustment, it is important either to correct the observations for the systematic errors or to add the unknown parameters to the mathematical model to correct for the systematic errors. As discussed in Section 23.7, adding correction parameters to a typical survey runs the risk of overparameterization. This occurs when there are more unknowns at a particular station than there are observations. In this case the system is unsolvable. Thus, it is assumed that the corrections will be made before the adjustment and the appropriate prototype equation is

$$c_1 \, dn_i + c_2 \, de_i + c_3 \, du_i + c_4 \, dn_j + c_5 \, de_j + c_6 \, du_j = v_{ij} - v_0 \quad (23.11)$$

The coefficients for Equation (23.11) are listed in Table 23.1. Their values are evaluated using the approximate coordinate values, v_{ij} is the observed vertical angle, and v_0 is the vertical angle computed using Equation (23.5) and approximate station coordinates.

23.2.4 Horizontal Angle Observations

As stated earlier, horizontal angles are the difference in two azimuths. That is, θ_{bif} is computed as $Az_{if} - Az_{ib}$, where b is the backsight station, i the instrument station, and f the foresight station. The prototype equation for a horizontal angle is

$$d_1 \, dn_b + d_2 \, de_b + d_3 \, du_b + d_4 \, dn_i + d_5 \, de_i + d_6 \, du_i + d_7 \, dn_f$$
$$+ \, d_8 \, de_f + d_9 \, du_f = \theta_{bif} - \theta_0 \quad (23.12)$$

The coefficients d_1 through d_9 in Equation (23.12) are listed in Table 23.2, θ_{bif} is the horizontal angle observed, and θ_0 is the value computed for the angle based on difference between the foresight and backsight azimuths computed using Equation (23.6).

TABLE 23.2 Coefficients for Linearized Equation (23.12)

(s) $d_1 = \left\{ \dfrac{\sin Az_{ib}}{IB \cos v_{ib}} \left[\cos(\phi_b - \phi_i) + \dfrac{\sin \phi_b \sin(\lambda_b - \lambda_i)}{\tan Az_{ib}} \right] \right\}_0$

(t) $d_2 = -\left\{ \dfrac{\cos Az_{ib}}{IB \cos v_{ib}} \left[\cos(\lambda_b - \lambda_i) + \sin \phi_i \sin(\lambda_b - \lambda_i) \tan Az_{ib} \right] \right\}_0$

(u) $d_3 = -\left\{ \dfrac{\cos Az_{ib} \cos \phi_b}{IB \cos v_{ib}} [\sin(\lambda_b - \lambda_i) \right.$
$\left. + (\sin \phi_i \cos(\lambda_b - \lambda_i) - \cos \phi_i \tan \phi_b) \tan Az_{ib}] \right\}_0$

(v) $d_4 = \left(\dfrac{\sin Az_{if}}{IF \cos v_{if}} - \dfrac{\sin Az_{ib}}{IB \cos v_{ib}} \right)_0$ (w) $d_5 = \left(\dfrac{\cos Az_{ib}}{IB \cos v_{ib}} - \dfrac{\cos Az_{if}}{IF \cos v_{if}} \right)_0$ (x) $d_6 = 0$

(y) $d_7 = -\left\{ \dfrac{\sin Az_{if}}{IF \cos v_{if}} \left[\cos(\phi_f - \phi_i) + \dfrac{\sin \phi_f \sin(\lambda_f - \lambda_i)}{\tan Az_{if}} \right] \right\}_0$

(z) $d_8 = \left\{ \dfrac{\cos Az_{if}}{IF \cos v_{if}} \left[\cos(\lambda_f - \lambda_i) + \sin \phi_i \sin(\lambda_f - \lambda_i) \tan Az_{if} \right] \right\}_0$

(aa) $d_9 = \left\{ \dfrac{\cos Az_{if} \cos \phi_f}{IF \cos v_{if}} [\sin(\lambda_f - \lambda_i) \right.$
$\left. + (\sin \phi_i \cos(\lambda_f - \lambda_i) - \cos \phi_i \tan \phi_f) \tan Az_{if}] \right\}_0$

23.2.5 Differential Leveling Observations

Orthometric height differences as derived from differential leveling also can be included in the three-dimensional geodetic network adjustment model. However, as discussed in Section 23.7, the inclusion of this observation type requires a correction for geoidal height differences between the stations and the application of orthometric corrections. Since the adjustment model is nonlinear, the observation equation for elevation differences as given in Equation (11.1) must also be linearized. The prototype equation for differential leveling between stations I and J is

$$1 \, du_j - 1 \, du_i = \Delta H_{ij} + \Delta N_{ij} - \Delta h_{ij} \qquad (23.13)$$

In Equation (23.13), ΔH_{ij} is the elevation difference observed between the stations, ΔN_{ij} the difference in geoidal height between the stations, and Δh_{ij} the change in geodetic height.

23.2.6 Horizontal Distance Observations

As shown in Figure 23.3, the elevation differences of the endpoint stations and converging radii cause the horizontal distances observed from opposite

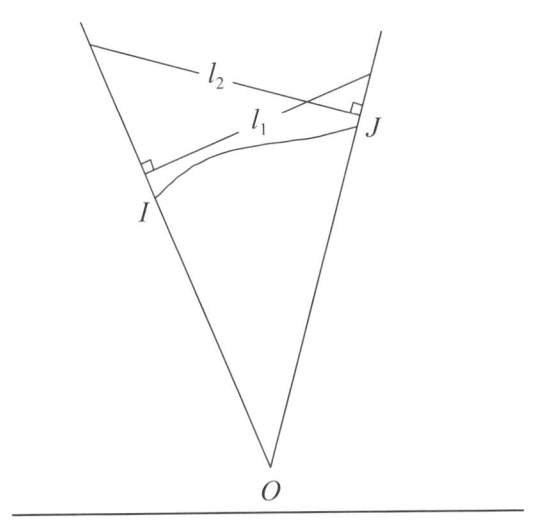

Figure 23.3 Comparison of horizontal distances from opposite ends of the line.

ends of the lines to start and terminate at different points. Thus, technically, l_1 and l_2 are not of the same length. In the local geodetic system, the observation equations for l_1 and l_2 are

$$l_1 = \sqrt{\Delta n_1^2 + \Delta e_1^2} \qquad (23.14)$$
$$l_2 = \sqrt{\Delta n_2^2 + \Delta e_2^2}$$

However, for the typically short distances observed in surveying practice, l_1 and l_2, Δn_1 and Δn_2, and Δe_1 and Δe_2 are approximately equal. Letting l equal l_1 and l_2, Δn equal Δn_1 and $-\Delta n_2$, and Δe equal Δe_1 and $-\Delta e_2$, the linearized observation equation in local geodetic system is

$$\left(\frac{\phi_j - \phi_i}{IJ} M_i\right)_0 dn_i + \left(\frac{\lambda_i - \lambda_j}{IJ} N_j\right) de_i - \left(\frac{\phi_j - \phi_i}{IJ} M_i\right)_0 dn_j$$
$$- \left(\frac{\lambda_i - \lambda_j}{IJ} N_j\right) de_j = l_{ij} - l_0 \qquad (23.15)$$

In Equation (23.15), the M_i is the radius in the meridian at the observing station I, N_i the radius in the normal, l_{ij} the horizontal distance observed, and l_0 the distance computed using approximate coordinates and Equation (23.14). The radii are computed as

$$M_i = \frac{a(1 - e^2)}{(1 - e^2 \sin^2 \phi_i)^{3/2}}$$

$$N_i = \frac{a}{\sqrt{1 - e^2 \sin^2 \phi_i}}$$

(23.16)

23.3 MINIMUM NUMBER OF CONSTRAINTS

A three-dimensional geodetic network adjustment requires both horizontal and vertical control. As discussed in Section 16.5, to fix the horizontal part of the adjustment requires one station fixed in position and one line of known direction. This can be accomplished by fixing the latitude and longitude of one station along with the azimuth of a line or the longitude of a second station. The vertical plane in the adjustment can be fixed with three benchmark stations. Since the adjustment is performed with geodetic heights, the orthometric height of the benchmark stations must be corrected using Equation (23.37).

As discussed in Chapter 20, control can be adjusted or fixed simply by setting the appropriate values in the stochastic model. In the case of vertical control, benchmarks are often given as orthometric heights. Since the geoid model is known only to about a centimeter, the standard deviations of geodetic heights for benchmark stations should not be set any better than this value.

23.4 EXAMPLE ADJUSTMENT

To illustrate a three-dimensional least squares adjustment, the simple network shown in Figure 23.4 will be used. The standard deviations for the coordinates are shown in Table 23.3 in the local geodetic system. This system was chosen

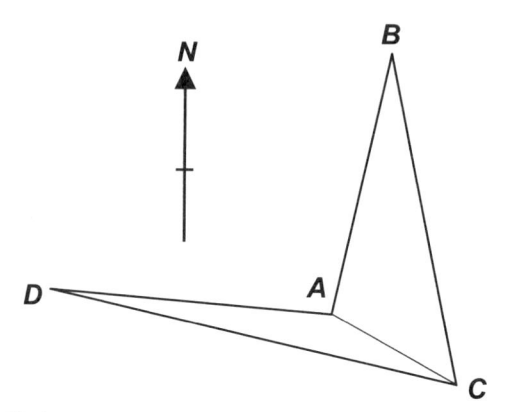

Figure 23.4 Example of a three-dimensional geodetic network.

TABLE 23.3 Data for the Three-Dimensional Geodetic Network

Geodetic positions

Station	φ	λ	H (m)	N (m)
A	41°18′26.04850″N	76°00′10.24860″W	372.221	−31.723
B	41°18′40.46660″N	76°00′05.50180″W	351.394	−31.713
C	41°18′22.04010″N	76°00′00.94390″W	362.865	−31.726
D	41°18′27.65860″N	76°00′31.38550″W	370.874	−31.722

Station	S_n (m)	S_e (m)	S_h (m)
A	±0.001	±0.001	±0.01
B	—	—	±0.01
C	—	—	±0.01

Geodetic azimuth

Course	Azimuth	S (″)
AB	13°56′26.9″	±0.001″

Slant distances

Course	Distance (m)	S (m)
AB	458.796	±0.005
AC	249.462	±0.005
CD	729.122	±0.006
DA	494.214	±0.005
BC	578.393	±0.005

Horizontal angles

Stations (bif)	Angles	S (″)
DAB	98°10′25″	±2.8
BAC	105°47′45″	±3.5
CAD	156°01′44″	±4.1
ABC	335°29′37″	±2.4
CBA	24°30′19″	±2.4
BCD	294°19′17″	±2.3

Zenith angles

Direction	Angle	S (″)
AC	92°09′01″	±2.5
CD	89°22′24″	±0.8
DA	89°50′44″	±1.2
AB	92°36′12″	±1.4
BC	88°52′01″	±1.1

Elevation differences

Stations	ΔElev (m)	S (m)
AC	−9.359	±0.005

because it is an intuitive system for assignment of realistic uncertainty values in the adjustment and since the adjustment will be performed in the local geodetic system. As shown in Table 23.3, the network is fixed in horizontal position by weighting of the northing and easting coordinates of station A and rotationally by overweighting the azimuth of line AB. The horizontal rotation of the network could have also been fixed by weighting the easting of station B.

The elevation datum was fixed in position and rotation by weighting the height components of stations A, B, and C. Since orthometric heights were given for the stations, the geoid separation was applied following Equation (23.37) to compute geodetic heights for the stations. The systematic errors discussed in Section 23.7 were removed previously from all angular observations. The results of the adjustment are shown in Figure 23.4.

Whereas the adjustment is performed in the local geodetic system, both geodetic and geocentric coordinates are required to compute the coefficients and computed observations. Since traditional observations are taken by an elevated instrument to some elevated target, the geodetic heights of each station must be increased by the setup heights when computing geodetic coordinates. This simple addition to the software removes the need for reducing observations to the station marks. In this example, the setup heights of the instruments and targets are assumed to be zero and thus do not need to be considered.

23.4.1 Addition of Slant Distances

Following prototype equation (23.9), each slant distance observation adds one row to the system of equations. As an example for the slant distance AC, the coefficients are computed as

$$-(\cos v_{AB} \cos Az_{AB})_0 \, dn_A - (\cos v_{AB} \sin Az_{AB})_0 \, de_A - (\sin v_{AB})_0 \, du_A$$
$$-(\cos v_{BA} \cos Az_{BA})_0 \, dn_B - (\cos v_{BA} \sin Az_{BA})_0 \, de_B - (\sin v_{BA})_0 \, du_B$$
$$= s_{AB} - AB_0 \qquad (23.17)$$

In Equation (23.17) the values for Az_{AB} and Az_{BA} are computed using Equation (23.6), v_{AB} and v_{BA} are computed using Equation (23.7), and AB using Equation (23.5). The numerical values, to five decimal places, for Equation (23.17) are

$$-0.96954 dn_A - 0.24067 de_A + 0.04541 du_A - 0.96954 dn_B$$
$$+ 0.24069 de_B - 0.04534 du_B = 0.0017 \qquad (23.18)$$

23.4 EXAMPLE ADJUSTMENT

For each slant distance, a similar observation equation is written. The reader should note in Equation (23.18) that unlike plane adjustments as presented in Chapter 14, the coefficients of the occupied and sighted stations vary slightly due to Earth curvature. Four more equations for distances AC, CD, DA, and BC are added to the system of equations.

23.4.2 Addition of Horizontal Angles

The angles observed in the network were corrected for the systematic errors caused by the height of targets and deflection of the vertical as given in Equations (23.30) and (23.32). Following prototype equation (23.12), an observation equation is written for each horizontal angle. As an example, the observation equation for angle DAB is

$$\left\{\frac{\sin Az_{AD}}{AD \cos v_{AD}}\left(\cos(\phi_D - \phi_A) + \frac{\sin \phi_D \sin(\lambda_D - \lambda_A)}{\tan Az_{AD}}\right)\right\}_0 dn_D$$

$$- \left\{\frac{\cos Az_{AD}}{AD \cos v_{AD}}[\cos(\lambda_D - \lambda_A) + \sin \phi_A \sin(\lambda_D - \lambda_A) \tan Az_{AD}]\right\}_0 de_D$$

$$- \left\{\frac{\sin Az_{AD} \cos \phi_D}{AD \cos v_{AD}}[\sin(\lambda_D - \lambda_A) + (\sin \phi_A \cos(\lambda_D - \lambda_A)\right.$$

$$\left. - \cos \phi_A \tan \phi_D) \tan Az_{AD}]\right\}_0 du_D$$

$$+ \left(\frac{\sin Az_{AB}}{AB \cos v_{AB}} - \frac{\sin Az_{AD}}{AD \cos v_{AD}}\right)_0 dn_A$$

$$+ \left(\frac{\cos Az_{AD}}{AD \cos v_{AD}} - \frac{\cos Az_{AB}}{AB \cos v_{AB}}\right)_0 de_A + 0du_A$$

$$- \left\{\frac{\sin Az_{AB}}{AB \cos v_{AB}}\left[\cos(\phi_B - \phi_A) + \frac{\sin \phi_B \sin(\lambda_B - \lambda_A)}{\tan Az_{AB}}\right]\right\}_0 dn_B$$

$$+ \left\{\frac{\cos Az_{AB}}{AB \cos v_{AB}}[\cos(\lambda_B - \lambda_A) + \sin \phi_A \sin(\lambda_B - \lambda_A) \tan Az_{AB}]\right\}_0 de_B$$

$$+ \left\{\frac{\sin Az_{AB} \cos \phi_B}{AB \cos v_{AB}}[\sin(\lambda_B - \lambda_A) + (\sin \phi_A \cos(\lambda_B - \lambda_A)\right.$$

$$\left. - \cos \phi_A \tan \phi_B) \tan Az_{AB}]\right\}_0 du_B = \theta_{DAB} - \theta_0 \qquad (23.19)$$

In Equation (23.19) the values in the braces are evaluated at their approximate coordinate values, θ_{DAB} is the observed angular value corrected for the systematic errors discussed in Section 23.7, and θ_0 is the value computed based on the difference in the computed values for the backsight (AD) and foresight (AB) azimuths. Substituting the appropriate values into Equation (23.19) and converting the radian values to units of s/m results in

$$-415.247 dn_D - 41.991 de_D - 0.00001 du_D + 523.669 dn_A - 394.828 de_A$$
$$+ 0 du_A - 108.432 dn_B + 436.790 de_B - 0.00003 du_B = 11.852'' \quad (23.20)$$

For each angle, a similar observation equation is written, resulting in a total of 10 equations to be added to the adjustment.

23.4.3 Addition of Zenith Angles

All the zenith angles in this example problem were corrected for deflection of the vertical, refraction, and target height as discussed in Section 23.7. Converting the zenith angle from A to C to an equivalent vertical angle and following Equation (23.11), the observation equation is

$$\left(\frac{\sin v_{AC} \cos Az_{AC}}{AC}\right)_0 dn_A + \left(\frac{\sin v_{AC} \sin Az_{AC}}{AC}\right)_0 de_A - \left(\frac{\cos v_{AC}}{AC}\right)_0 du_A$$

$$+ \left(\frac{-\cos \phi_A \sin \phi_C \cos(\lambda_C - \lambda_A) + \sin \phi_A \cos \phi_C}{AC \cos v_{AC}} + \sin v_{AC} \cos v_{CA} \cos Az_{CA}\right)_0 dn_C$$

$$+ \left(\frac{-\cos \phi_A \sin(\lambda_C - \lambda_A) + \sin v_{AC} \cos v_{CA} \sin Az_{CA}}{AC \cos v_{AC}}\right)_0 de_C$$

$$+ \left(\frac{\sin v_{AC} \sin v_{CA} + \sin \phi_A \sin \phi_C + \cos \phi_A \cos \phi_C \cos(\lambda_C - \lambda_A)}{AC \cos v_{AC}}\right)_0 du_C$$

$$= v_{AC} - v_0 \quad (23.21)$$

In Equation (23.21) the coefficients on the left side of the equations are evaluated using the approximate values of the coordinates in Equations (23.5) to (23.7). The numerical values for Equation (23.21) in units of s/m are

$$15.395 dn_A - 26.947 de_A - 826.229 du_A - 15.379 dn_C$$
$$+ 26.919 de_C + 826.230 du_C = 2.971''$$

For each zenith angle, a similar observation equation is written, resulting in a total of five equations to be added to the adjustment.

23.4.4 Addition of Observed Azimuths

As discussed in Section 23.7, observed azimuths must be corrected for deflection of the vertical and height of the target. In this example, the azimuth was already given as its geodetic value. Thus, the azimuth was used to fix the adjustment by weighting. The observation equation for the azimuth is

$$\left(\frac{\sin Az_{AB}}{AB \cos v_B}\right)_0 dn_A - \left(\frac{\cos Az_{AB}}{AB \cos v_{AB}}\right)_0 de_A + 0du_A$$

$$- \left\{\frac{\sin Az_{AB}}{AB \cos v_{AB}}[\cos(\phi_B - \phi_A) + \sin \phi_B \sin(\lambda_B - \lambda_A) \cos Az_B]\right\}_0 dn_B$$

$$+ \left\{\frac{\cos Az_{AB}}{AB \cos v_{AB}}[\cos(\lambda_B - \lambda_A) - \sin \phi_A \sin(\lambda_B - \lambda_A) \tan Az_{AB}]\right\}_0 de_B$$

$$+ \left\{\frac{\cos Az_{AB} \cos \phi_B}{AB \cos v_{AB}}[\sin(\lambda_B - \lambda_A) - (\sin \phi_A \cos(\lambda_B - \lambda_A)\right.$$

$$\left. - \cos \phi_A \tan \phi_B) \tan Az_{AB}]\right\}_0 du_B$$

$$= \alpha_{AB} - Az_{AB} \tag{23.22}$$

In Equation (23.22) the coefficients on the left side are evaluated from the approximate coordinate values using Equations (23.5) to (23.7). The numerical values of Equation (23.22), in units of s/m, are

$$108.42dn_A - 436.79de_A + 0du_A - 108.43dn_B$$
$$+ 436.79de_B - 0.00003du_B = -0.256''$$

23.4.5 Addition of Elevation Differences

Leveling is a process of determining heights above the geoid. The data for this example lists the orthometric and geoidal heights for each station. The difference in orthometric heights requires the application of Equation (23.38) to obtain the geodetic height differences between the stations. The observation equation for the elevation difference between stations A and C is

$$1du_C - 1du_A = \Delta H_{AC} + \Delta N_{AC} - \Delta h_{AC} \tag{23.23}$$

The resulting numerical values for Equation (23.23) are

$$1du_C - 1du_A = -9.359 + (-31.726 + 31.723)$$
$$- (331.139 - 340.498) = -0.003 \text{ m}$$

23.4.6 Adjustment of Control Stations

With the addition of the u coordinate, observation equations similar to those presented in Section 20.2 can be added for each control coordinate. As stated in Section 23.1, the local geodetic coordinate system has its origin at each instrument station. Thus it is not a true coordinate system with a single origin. Because of this, the approximate coordinates for each station must be stored in either the geocentric or geodetic coordinate system. In this discussion, it is assumed that the values are stored in the geodetic coordinate system. Again, nonlinear versions of the equations must be written to match the overall nonlinear nature of the adjustment. The observation equations are

$$1dn = N - n \qquad (23.24)$$

$$1de = E - e \qquad (23.25)$$

$$1du = U - u \qquad (23.26)$$

where (N,E,U) are the given control coordinate values and (n,e,u) are their adjusted values. In the first iteration, the control coordinate values and their adjusted values will be the same. In subsequent iterations, small variations between the control coordinate values and their adjusted counterparts will be observed. The observation equations for station A in the first iteration are

$$1dn = 0$$

$$1de = 0$$

$$1du = 0$$

Since the control coordinate values and approximate station coordinate values will be expressed in terms of geodetic coordinates, the initial K-matrix values given in Equations (23.24) to (23.26) will be expressed as changes in geodetic coordinates. These values must then be transformed into the local geodetic coordinate system. The relationships between changes in the two systems are

$$\begin{bmatrix} dn \\ de \\ du \end{bmatrix} = \begin{bmatrix} M+h & 0 & 0 \\ 0 & (N+h)\cos\phi & 0 \\ 0 & 0 & 1 \end{bmatrix} \begin{bmatrix} d\phi \\ d\lambda \\ dh \end{bmatrix} = R_{LG} \begin{bmatrix} d\phi \\ d\lambda \\ dh \end{bmatrix} \qquad (23.27)$$

In Equation (23.27), M represents the radius in the meridian and N the radius in the normal at a station with latitude ϕ as given by Equation (23.16), h the geodetic height, and R_{LG} the transformation matrix between changes in the geodetic and local geodetic coordinate systems. The K-matrix values expressed in terms of geodetic coordinate system can be transformed into K-matrix values of the local geodetic coordinate system using Equation (23.27).

23.4.7 Results of Adjustment

A partial list of the results of the adjustment from the ADJUST software package is given in Figure 23.5. The solution converged in two iterations. Note that the latitude and longitude of station A were held fixed, as was the azimuth of line AB.

23.4.8 Updating Geodetic Coordinates

At the completion of each iteration, the corrections of dn, de, and du will be determined for each station in the adjustment. However, geodetic coordinates are used to represent station positions. Thus, after each iteration the local geodetic coordinate system corrections of dn, de, and du must be transformed into changes in the geodetic system using the inverse relationship of Equation (23.27), or

$$d\phi_i = \frac{dn_i}{M_i + h_i} \qquad d\lambda_i = \frac{de_i}{(N_i + h_i)\cos\phi_i} \qquad dh_i = de_i \qquad (23.28)$$

In Equation (23.28), the corrections to the latitude, longitude, and geodetic height of station I are $d\phi_i$, $d\lambda_i$, and dh_i, respectively. All other terms are as defined previously. Similarly, the uncertainties for each station will be in the local geodetic system. Although these uncertainties can be used to represent northing, easting, and geodetic height errors at each station, Equation (6.13) must be used to transform these uncertainties into the geodetic system as

$$\Sigma_{\phi,\lambda,h} = R^{-1}\Sigma_{n,e,u}R^{-T} \qquad (23.29)$$

where

$$\Sigma_{\phi,\lambda,h} = \begin{bmatrix} \sigma_\phi^2 & \sigma_{\phi,\lambda} & \sigma_{\phi,h} \\ \sigma_{\phi,\lambda} & \sigma_\lambda^2 & \phi_{\lambda,h} \\ \sigma_{\phi,h} & \sigma_{\lambda,h} & \sigma_h^2 \end{bmatrix} \qquad R = \begin{bmatrix} M+h & 0 & 0 \\ 0 & (N+h)\cos\phi & 0 \\ 0 & 0 & 1 \end{bmatrix}$$

$$\Sigma_{n,e,u} = \begin{bmatrix} \sigma_n^2 & \sigma_{n,e} & \sigma_{n,u} \\ \sigma_{n,e} & \sigma_e^2 & \sigma_{e,u} \\ \sigma_{n,u} & \sigma_{e,u} & \sigma_u^2 \end{bmatrix}$$

Note that each station's uncertainties can be computed using the 3 × 3 block diagonal elements from the Q_{xx} matrix of the adjustment.

THREE-DIMENSIONAL GEODETIC NETWORK ADJUSTMENT

```
Adjusted Geodetic Coordinates
Station     Latitude        Longitude       Orth height      N        S-Lat(")    S-Lon(")    S-h
==================================================================================================
   A     41°18'26.04850"N  76°00'10.24860"W    372.2232    -31.72    0.000065    0.000087    0.0119
   B     41°18'40.46653"N  76°00'05.50185"W    351.3904    -31.71    0.000231    0.000113    0.0120
   C     41°18'22.04015"N  76°00'00.94395"W    362.8664    -31.73    0.000162    0.000270    0.0119
   D     41°18'27.65788"N  76°00'31.38547"W    370.8748    -31.72    0.000304    0.000353    0.0124

Station Statistics
Station      Sn              Se             t-ang°           Su           Sv           S-Elev#
==================================================================================================
   A       0.0020          0.0020          45.0000         0.0020       0.0020         0.0119
   B       0.0071          0.0026          13.9417         0.0073       0.0020         0.0120
   C       0.0050          0.0063          96.0029         0.0063       0.0050         0.0119
   D       0.0094          0.0082         175.3588         0.0094       0.0082         0.0124
#-Values do not include uncertainty in geoidal height.

Adjusted Slope Distances
Station                  Station
Occupied                 Sighted                        Distance                              V
==================================================================================================
   A                        B                           458.792                          -0.004
   A                        C                           249.468                           0.006
   C                        D                           729.116                          -0.006
   D                        A                           494.213                          -0.001
   B                        C                           578.399                           0.006

Adjusted Mark-to-Mark and Geodetic Distances
Station                  Station                      Mark-to-Mark                       Geodetic
Occupied                 Sighted                        Distance                         Distance
==================================================================================================
   A                        B                           458.792                          458.317
   A                        C                           249.468                          249.282
   C                        D                           729.116                          729.034
   D                        A                           494.213                          494.182
   B                        C                           578.399                          578.255

Adjusted Angle Observations
   Station               Station              Station
 Backsighted             Occupied           Foresighted              Angle               V(")
==================================================================================================
     D                      A                    B                98°10'22.13"          -2.87
     B                      A                    C               105°47'52.54"           7.54
     C                      A                    D               156°01'45.34"           1.34
     A                      B                    C               335°29'34.52"          -2.48
     C                      B                    A                24°30'25.48"           6.48
     B                      C                    D               294°19'09.93"          -7.07
     D                      C                    A                15°59'08.09"           7.09
     A                      C                    B                49°41'41.98"          -3.02
     C                      D                    A               352°00'53.43"           8.43
     A                      D                    C                 7°59'06.57"          -5.43

Adjusted Azimuth Observations
Station                  Station
Occupied                 Sighted                         Azimuth                              V
==================================================================================================
   A                        B                         13°56'26.9"                         0.00"
```

Figure 23.5 ADJUST listing of adjustment results for the example problem of Figure 23.4.

```
Adjusted Zenith Angle Observations
Station                  Station
Occupied                 Sighted              Zenith Angle                         V(")
===================================================================================
        A                   B                 92°36'12.2"                        -0.19"
        A                   C                 92°09'04.7"                        -3.69"
        C                   D                 89°22'25.1"                        -1.05"
        D                   A                 89°50'45.6"                        -1.61"
        B                   C                 88°52'01.3"                        -0.26"

Adjusted Elevation Difference Observations
Station                  Station             Elevation
Occupied                 Sighted              Diff.                                V
===================================================================================
        A                   C                  -9.363                          -0.0038
```

Iterations = 2
Redundancies = 15
Reference Variance = 4.075
Reference So = ±2.019

Figure 23.5 (*Continued*)

23.5 BUILDING AN ADJUSTMENT

Since this is a nonlinear adjustment, initial approximations are required for all station parameters. Very good initial values for horizontal coordinates can be determined using the procedures outlined in Chapter 16, with horizontal observations reduced to a map projection surface. Following the adjustment, grid coordinates can be converted to their geodetic equivalents. Similarly, orthometric heights of stations can be determined using the procedures discussed in Chapter 12. The resulting orthometric heights can be converted to geodetic heights using Equation (23.37). Since these are initial approximations, it is possible to use an average geoidal height N for the region when applying Equation (23.37). However, this procedure will result in more uncertainty in the resulting heights.

Before combining a large set of three-dimensional conventional observations, it is wise to perform adjustments using the horizontal and vertical functional models presented in Chapters 12 and 16 to isolate potential blunders. For large regional data sets, a smaller data subset should be adjusted to isolate blunders. After these smaller data sets are cleaned and adjusted, a combined adjustment can be attempted with better initial approximations for the unknowns and the knowledge that most if not all blunders have been removed from the data.

23.6 COMMENTS ON SYSTEMATIC ERRORS

Usually, small local surveys result in systematic error components that are small enough to be considered negligible. However as the size of the survey increases, variations in the direction of plumb lines must be taken into account. This error can be removed from angular observations prior to the adjustment with a priori knowledge of the deflection components η and ξ at each station. Modeled values for these components can be obtained using software such as DEFLEC99,[1] which is available from the National Geodetic Survey. The systematic error corrections to the zenith angles, azimuths, and horizontal angles are

$$z_{ij} = z'_{ij} + \xi_i \cos Az_{ij} + \eta_i \sin Az_{ij} \qquad (23.30)$$

$$Az_{ij} = Az_{obs} + \eta_i \tan \phi_i + (\xi_i \sin Az_{obs} - \eta_i \cos Az_{obs}) \cot z_{ij} \qquad (23.31)$$

$$\angle BIF = \angle BIF_{obs} + (\xi_i \sin Az_{if} - \eta_i \cos Az_{if}) \cot z_{if}$$

$$- (\xi_i \sin Az_{ib} - \eta_i \cos Az_{ib}) \cot z_{ib} \qquad (23.32)$$

where η_i and ξ_i are the deflection of the vertical components at the observation station I; z_{ij}' is the zenith angle observed from station I to J; z_{ij} the corrected zenith angle between the same stations; Az_{obs} the astronomical azimuth observed between station I and J; Az_{ij} the corrected azimuth; $\angle BIF_{obs}$ the angle observed where station B is the backsight station, I the instrument station, and F the foresight station; and $\angle BIF$ the corrected horizontal angle.

Additionally, target height differences must also be considered for both directions and angles. The azimuth of a line corrected for target height is

$$Az^c = \alpha + \frac{0.108'' \cos^2 \phi_i \sin 2\alpha \; h}{1000} \qquad (23.33)$$

In Equation (23.33), Az^c is the corrected azimuth, α the observed azimuth, ϕ_i the latitude of the observation station, and h the geodetic height of the target in meters. Since an angle is simply the difference between the foresight and backsight azimuths, the correction to an angle due to the height of the targets is

$$\theta^c = \theta_{bif} + 0.108'' \cos^2 \phi_i \left[\frac{h_f \sin 2\alpha_{if}}{1000} - \frac{\sin 2\alpha_{ib} \; h_b}{1000} \right] \qquad (23.34)$$

where θ^c is the corrected angle, θ_{bif} the observed angle, and α_{if} and α_{ib} the azimuths of the foresight and backsight lines, respectively.

Atmospheric refraction must be considered when including vertical or zenith angles in an adjustment. Since the correction for atmospheric refraction

---

[1] DEFLEC99 is available at http://www.ngs.noaa.gov/PC_PROD/pc_prod.shtml#DEFLEC99.

is so difficult to model, one can only hope to account for a portion of correction in an adjustment. The remainder of the correction must be modeled in the adjustment or be absorbed in the residuals of the observation. The correction of the first-order effect of refraction can be determined by observing simultaneously vertical or zenith angles at each end of the line *IJ*. From these observations, the first-order correction ΔZ_{ij} for reciprocal zenith angles is determined as

$$\Delta Z_{ij} \equiv \Delta Z_{ji} = 90° - [0.5(Z_{ij} + Z_{ji}) - \psi] \qquad (23.35)$$

In Equation (23.35), Z_{ij} and Z_{ji} are the reciprocal zenith angles observed simultaneously and ψ is the correction for the deflection of vertical, computed as

$$\psi = -(\xi_i \cos Az_{ij} + \eta_i \sin Az_{ij}) \qquad (23.36)$$

It is possible to create an adjustment model that includes correction terms for η, ξ, and atmospheric refraction. However, due to the limited number of observations in a typical survey network, there is a danger of overparameterization (Leick, 2004). Thus, it is recommended that these corrections be applied to the angles before carrying out the adjustment.

Unfortunately, to determine the deflection of the vertical components (η and ξ), the latitude and longitude of the station must be known before completion of the adjustment. Approximate values for latitude and longitude will suffice in most instances. However, in instances where good approximations for the geodetic coordinates are not known, it is possible to correct the original observations after an initial adjustment to ensure that the geodetic coordinates for each station are close to their final values. In this case, after correcting the observations, the adjustment could be run a second time.

A similar problem exists with differential leveling observations. Geodetic height h and orthometric height H differ by the geoidal height N, or

$$h = H + N \qquad (23.37)$$

Elevation differences as determined by differential leveling must account for the differences in geoidal height between the benchmark stations. This situation can be described mathematically as

$$h_j - h_i = (H_j + N_j) - (H_i + N_i)$$

$$\Delta h = \Delta H + \Delta N \qquad (23.38)$$

The correction to leveled height differences should be performed when these observations are included in the three-dimensional geodetic network adjustment. Values for geoidal height can be determined using the GEOID03 software available from the National Geodetic Survey.[2] Similar to deflection of the vertical parameters, it is possible to include geoidal height parameters in the mathematical model of the adjustment. Again, this approach runs a risk of overparameterization in the adjustment.

The orthometric correction O_C for a leveling line can be approximated using a formula derived by Bomford (1980) as

$$O_C = \frac{-H(\Delta\phi'')(0.0053 \sin 2\phi)}{\rho} \tag{23.39}$$

where H is the orthometric height at the instrument station in meters, $\Delta\phi''$ the change in latitude between the backsight and foresight stations in units of seconds, ϕ the latitude of the instrument station, and ρ the conversion from seconds to radians, which is approximately 206,264.8"/rad. Obviously, Equation (23.39) would be tedious to apply for a substantial north–south leveling circuit. However, for most typical surveys involving small regions, this correction can be very small and easily absorbed in the residuals of the adjustment. For instance, a 120-km north–south leveling line at approximately 42° latitude would result in an orthometric correction of about 2 cm.

PROBLEMS

Note: Unless otherwise specified, use the GRS 80 ellipsoidal parameters of a = 6.378.137.0 m and f = 1/298.257222101 for the following problems.

23.1 Using the data supplied in Figure 23.4, what value is computed for the slant distance AB?

23.2 Repeat Problem 23.1 for slant distance AC.

23.3 Using the data supplied in Figure 23.4, what value is computed for the azimuth AB?

23.4 Using the data supplied in Figure 23.4, what value is computed for the horizontal angle DAB?

23.5 Repeat Problem 23.4 for horizontal angle ACB.

23.6 Using the data supplied in Figure 23.4, what value is computed for v_{AB}?

[2]GEOID03 software is available at http://www.ngs.noaa.gov.

23.7 Repeat Problem 23.6 for the vertical angle v_{AC}.

23.8 Develop the observation equation for slant distance *AC* in Figure 23.4.

23.9 Repeat Problem 23.8 for slant distance *CD*.

23.10 Develop the observation equation for horizontal angle *BAC*.

23.11 Repeat Problem 23.10 for horizontal angle *ACB*.

23.12 Develop the observation equation for zenith angle *AB*.

23.13 The approximate values for the geodetic coordinates of the endpoints of line *AB* are A: (39°18′43.2000″ N, 106°46′40.8000″ W, 1191.161 m) and B: (39°18′38.5210″ N, 106°46′22.6822″ W, 1217.239 m). Using the WGS 84 ellipsoid, determine the computed:
(a) slant distance *AB*.
(b) azimuth *AB*.
(c) vertical angle *AB*.

23.14 Repeat Problem 23.13 with approximate values for the geodetic coordinates of A: (36°44′52.8000″ N, 119°46′15.6000″ W, 90.221 m) and B: (36°45′04.9159″ N, 119°46′17.9521″ W, 89.675 m).

23.15 Develop the observation equation for slant distance *AB* if the distance observed in Problem 23.13 is 458.237 m.

23.16 Develop the observation equation if the distance observed in Problem 23.14 is 378.011 m.

23.17 Develop the azimuth observation equation for Problem 23.13 if the azimuth observation is 108°23′19″. (Assume that all systematic errors are removed from the observation.)

23.18 Develop the azimuth observation equation for Problem 23.14 if the azimuth observation is 351°07′24″. (Assume that all systematic errors are removed from the observation.)

Use the following data and the GRS 80 ellipsoid in Problems 23.19 to 23.26.

Geodetic coordinates

Station	Latitude	Longitude	Height (m)	S_n (m)	S_e (m)	S_u (m)
A	40°25′28.7700″N	86°54′15.0464″W	92.314	0.001	0.001	0.01
B	40°25′46.3978″N	86°54′12.3908″W	80.004	—	—	0.01
C	40°25′43.3596″N	86°54′30.5172″W	87.268	—	—	0.01
D	40°25′33.6000″N	86°54′28.8000″W	95.585	—	—	—

Angle observations

B	I	F	Angle	S (″)	h_b (m)	h_i (m)	h_f (m)
D	A	C	26°18′07.6″	4.7	1.303	1.295	1.300
C	A	B	45°35′06.7″	4.1	1.402	1.405	1.398
B	A	D	288°06′42.4″	5.0	1.295	1.301	1.305
A	B	D	37°50′41.9″	4.1	1.305	1.398	1.299
D	B	C	33°13′03.8″	4.3	1.500	1.500	1.500
C	B	A	288°56′09.2″	4.6	1.206	1.210	1.208
B	C	A	63°20′53.4″	4.5	1.300	1.300	1.300
A	C	D	31°21′36.2″	5.1	1.425	1.423	1.420
D	C	B	265°17′26.4″	5.9	1.398	1.205	1.300
C	D	B	52°04′25.0″	5.3	1.500	1.500	1.500
B	D	A	70°15′56.7″	5.0	1.500	1.500	1.500
A	D	C	237°39′41.8″	6.6	1.500	1.400	1.500

Slant distance observations

From	To	Distance (m)	S (m)	hi (m)	hr (m)
A	B	547.433	0.008	1.400	1.500
B	C	437.451	0.008	1.497	1.595
C	D	303.879	0.008	1.500	1.500
D	A	356.813	0.008	1.302	1.296
A	C	579.263	0.008	1.300	1.500
B	D	552.833	0.008	1.500	1.400

Azimuth observations

From	To	Azimuth	S (″)	hi (m)	hr (m)
A	B	6°34′04.6″	0.001	1.500	1.500

Elevation differences

From	To	ΔElev (m)	S (m)
A	C	−5.053	0.012
B	D	13.585	0.012

23.19 Use a computational program to develop the coefficient matrix for the angular data in the table. List all coefficients in units of s/m.

23.20 Use a computational program to develop the coefficient matrix for the slant distance data in the table.

23.21 Use a computational program to develop the observation equation for the azimuth data in the table. List the coefficients in units of s/m.

23.22 Use a computational program to develop the observation equations for the elevation differences in the table.

23.23 Use ADJUST to determine the most probable values for the geodetic coordinates in the table using only the angle, azimuth, and slant distance observations.

23.24 Use ADJUST to determine the most probable values for the geodetic coordinates in the table using all the observations. Remember to account for the geoidal height at each station.

23.25 Repeat Problem 23.24 by removing azimuth *AB* from the adjustment and overweighting the easting of station *B*. The standard deviations of station *B* should be $S_n = \pm 1000$ m, $S_e = \pm 0.001$ m, and $S_u = \pm 0.01$ m.

23.26 What are the adjusted orthometric heights of the stations in the table?

23.27 If $\xi = 4.87''$ and $\eta = 0.01''$ are the deflection of the vertical components at station *A* in Section 23.4, what is the correction to the zenith angle going from station *A* to *B*?

23.28 Repeat Problem 23.27 for the zenith angle from station *A* to *C*.

23.29 If the zenith angle observed from *A* to *B* in Figure 23.6 is 91°04′54″ and the deflection of the vertical components are $\xi = 1.28''$ and $\eta = 3.58''$, what is the corrected zenith angle?

23.30 Using the data given in Problem 23.27, what is the deflection of the vertical correction for the azimuth *AB*?

23.31 Using the data given in Problem 23.29, what is the deflection of the vertical correction for the azimuth *AB*?

23.32 Using the data given in Problem 23.27, what is the deflection of the vertical correction for the angle *DAB*?

23.33 Using the data given in Problem 23.29, what is the deflection of the vertical correction for the angle *DAC*?

23.34 Do Problem 23.33 using angle *CAB*.

Programming Problems

23.35 Develop a computational program that corrects vertical angles for deflection of the vertical components.

23.36 Develop a computational program that computes the coefficients for the azimuths, angles, slant distances, and elevation differences. Check your program using data from Figure 23.6.

23.37 Modify the Mathcad worksheets provided for Chapter 23 on the CD that accompanies this book to incorporate target and instrument heights. Solve Problem 23.6.

CHAPTER 24

COMBINING GPS AND TERRESTRIAL OBSERVATIONS

24.1 INTRODUCTION

Ellipsoids define the mathematical shape of Earth or a portion thereof. Ellipsoids are commonly defined by the length of their semimajor axis, a, and the flattening factor, f. Commonly used ellipsoids are the Geodetic Reference System of 1980 (GRS 80) and the World Geodetic System of 1984 (WGS 84). Table 24.1 lists the length of the semimajor axis and flattening factor for these ellipsoids.

A network of points determined with respect to an ellipsoid is a *datum*. Datums define the geodetic coordinates of the points and thus the origin and orientation of the datum. Most datums are regional in nature since the network of points covers only a portion of the Earth. These are known as *local datums*. For example, the North American Datum of 1983 (NAD 83) is a local datum consisting of a network of points in Canada, the United States, Mexico, and some Caribbean islands. However, the International Terrestrial Reference Frame (ITRF) is an example of a datum defined by a multitude of points located on all of Earth's major land masses. This is known as a *global datum*. Similarly, the coordinates of the GPS satellites are determined by a global network of tracking stations on a unique datum. Stated incorrectly, GPS is said to be using the WGS 84 datum. This statement should be interpreted by the reader as the datum defined by the global network of tracking stations that use the WGS 84 ellipsoid.

Besides their reference ellipsoid, datums differ in origin (translation), scale, and rotations about the three cardinal axes. Many modern local datums and global datums have nearly aligned coordinate axes and differ by only a few meters in their origins. Thus, it is possible to use a three-dimensional coor-

TABLE 24.1 Defining Ellipsoidal Parameters

Ellipsoid	a (m)	$1/f$
GRS80	6,378,137.0	298.257222101
WGS84	6,378,137.0	298.257223563

dinate transformation to transform points from a local datum to a global datum and back again. When using GPS and absolute positioning techniques, the satellites serve as control points, and all points determined by this method are defined in either the ITRF or WGS 84 datum, depending on the source of ephemeris. However, this method of surveying is accurate only to the meter level. Thus, relative positioning techniques are generally used. However, if GPS receivers are placed on control points defined in a local datum such as NAD 83, the resultant points are defined in a hybrid of the WGS 84 or ITRF and the local datum. The differences in these systems are generally at the centimeter level. For lower-order surveys, this difference may not be of much importance. However, for higher-order control surveys, these differences must be taken into account.

To combine GPS and terrestrial observations, the control from two different datums must be reconciled. This can be done by transforming the local control coordinates into the GPS reference datum, or by transforming the GPS established points into the local datum.

As discussed in Chapter 18, a three-dimensional conformal coordinate transformation will take coordinates from one three-dimensional coordinate system into another. However, since the datums are nearly aligned, the rotational angles are generally very small. As shown in Section 24.2, the rotational process, as well as the entire transformation, can be simplified.

To perform the transformation, coordinates of common points in both systems must be placed into their respective geocentric coordinate systems (see Section 17.4). Following this conversion, a least squares adjustment can be performed to determine the transformation parameters between the two systems. Once the transformation parameters are determined, the coordinates of any remaining points can be transformed.

For example, assume that GPS is being used to stake out a highway alignment and that the highway alignment was designed using control points from the state plane coordinate system of 1983 (SPCS 83).[1] The real-time GPS datum is defined by a set of Department of Defense tracking stations using the WGS 84 ellipsoid. The geodetic control used in the local state plane coordinate system is based on a series of National Spatial Reference stations in the United States, Canada, and Mexico using the GRS 80 ellipsoid. The *NEH* coordinates of the highway design stations must be transformed from

---

[1] Map projection coordinates systems are discussed in Appendix F.

the SPCS 83 coordinate system and orthometric heights to geodetic coordinates. The geodetic coordinates are then transformed to geocentric coordinates[2] of $(X,Y,Z)_{LS}$. These points must be occupied using a GPS receiver and GPS coordinates $(X,Y,Z)_{GPS}$ derived from satellite observations. Once three common points are occupied, the transformation parameters can be determined using a *Helmert transformation*. However, the prudent surveyor will use four common points to increase the number of redundant measurements. Following this, all GPS coordinates are transferred into the local coordinate system, and stakeout of stations can be performed.

24.2 HELMERT TRANSFORMATION

Local datums such as NAD 83 are known as Earth-centered, Earth-fixed (ECEF) coordinate systems. This means that the Z axis is nearly aligned with the Conventional Terrestrial Pole, X axis with the Greenwich Meridian, and the origin is at the mass center of the Earth as derived by the datum points used in the definition. International datums such as the International Terrestrial Reference Frame use the same definitions for the axes, origin, and ellipsoid, but differ slightly due to the difference in the datum points used in its determination. Thus, the rotational parameters and translations between two ECEF coordinate systems are usually very small. The scale factor between two datums using the same units of measure should be nearly 1.

Since the sine of a very small angle is equal to the angle in radians, the cosine of a very small angle is nearly 1, and the product of two very small numbers is nearly zero, for two nearly aligned coordinate systems Equation (18.14) can be simplified as

$$R' = \begin{bmatrix} 1 & \theta_3 & -\theta_2 \\ -\theta_3 & 1 & \theta_1 \\ \theta_2 & -\theta_1 & 1 \end{bmatrix} = I + \begin{bmatrix} 0 & \Delta\theta_3 & -\Delta\theta_2 \\ -\Delta\theta_3 & 0 & \Delta\theta_1 \\ \Delta\theta_2 & -\Delta\theta_1 & 0 \end{bmatrix} = I + \Delta R$$

(24.1)

In Equation (24.1) $\Delta\theta_1$, $\Delta\theta_2$, and $\Delta\theta_3$ are in units of radians and have been separated from the full matrix by the addition of the unit matrix. The introduction of Δ indicates that these values are differentially small.

The transformation of coordinates from one local datum to another datum is performed as

$$X_{LD} = sR'X_{GD} + T$$

(24.2)

---

[2] Equations to convert geodetic coordinates to geocentric coordinates are given in Section 17.5. Appendix F has equations to convert geodetic coordinates to state plane coordinates.

24.2 HELMERT TRANSFORMATION

where s is the scale factor, X_{GD} the (x,y,z) coordinates from the global data (GD) to be transferred into the local datum (LD), and T the xyz translations needed to coincide the origins of the two datums. Similarly, the scale factor, s, and translation parameters, T, can be modified as

$$s = 1 + \Delta s \tag{24.3}$$

$$T = T_0 + \Delta T \tag{24.4}$$

In Equation (24.4), the approximate shift vector, T_0, can be computed as

$$T_0 = \begin{bmatrix} x \\ y \\ z \end{bmatrix}_{LD} - \begin{bmatrix} x \\ y \\ z \end{bmatrix}_{GD} \quad \text{and} \quad \Delta T = \begin{bmatrix} \Delta Tx \\ \Delta Ty \\ \Delta Tz \end{bmatrix} \tag{24.5}$$

Since Equation (24.2) is nonlinear, a single common point or an average of all the common points can be used in Equation (24.5) to obtain initial approximations. For a single station, i, the linearized model for the corrections is

$$X_{LD_i} - X_{GD_i} - T_0 = J_i\, dx \tag{24.6}$$

where

$$J_i = \begin{bmatrix} x_i & 0 & -z_i & y_i & 1 & 0 & 0 \\ y_i & z_i & 0 & -x_i & 0 & 1 & 0 \\ z_i & -y_i & x_i & 0 & 0 & 0 & 1 \end{bmatrix} \quad dx = \begin{bmatrix} \Delta s \\ \Delta \theta_1 \\ \Delta \theta_2 \\ \Delta \theta_3 \\ \Delta Tx \\ \Delta Ty \\ \Delta Tz \end{bmatrix} \tag{24.7}$$

Example 24.1 The geocentric coordinates in the North American Datum of 1983 (NAD 83) and, from a GPS adjustment using a precise ephemeris, from the International Terrestrial Reference Frame of 2000 (ITRF 00) datum are shown below. Determine the transformation parameters to transform the additional station into the local NAD 83 datum.

	NAD 83			ITRF 00		
Station	x (m)	y (m)	z (m)	x (m)	y (m)	z (m)
A	1,160,604.924	−4,655,917.607	4,188,338.994	1,160,374.046	−4,655,729.681	4,188,609.031
E	1,160,083.830	−4,655,634.217	4,188,722.346	1,159,852.942	−4,655,446.265	4,188,992.349
F	1,160,786.583	−4,655,564.279	4,188,637.038	1,160,555.699	−4,655,376.341	4,188,907.065
G	1,160,648.090	−4,656,191.149	4,188,040.857	1,160,417.220	−4,656,003.238	4,188,310.909
B				1,160,406.372	−4,655,402.864	4,188,929.400

482 COMBINING GPS AND TERRESTRIAL OBSERVATIONS

SOLUTION *First iteration:* Using station A, T_0 is

$$T_0 = \begin{bmatrix} 1{,}160{,}604.924 - 1{,}160{,}374.046 \\ -4{,}655{,}917.607 + 4{,}655{,}729.681 \\ 4{,}188{,}338.994 - 4{,}188{,}609.031 \end{bmatrix} = \begin{bmatrix} 230.877 \\ -187.926 \\ -270.037 \end{bmatrix}$$

The J and K matrices are

$$J = \begin{bmatrix} 1160374.046 & 0.000 & -4188609.031 & -4655729.681 & 1 & 0 & 0 \\ -4655729.681 & 4188609.031 & 0.000 & -1160374.046 & 0 & 1 & 0 \\ 4188609.031 & 4655729.681 & 1160374.0464 & 0.000 & 0 & 0 & 1 \\ 1159852.942 & 0.000 & -4188992.349 & -4655446.265 & 1 & 0 & 0 \\ -4655446.265 & 4188992.349 & 0.000 & -1159852.942 & 0 & 1 & 0 \\ 4188992.349 & 4655446.265 & 1159852.942 & 0.000 & 0 & 0 & 1 \\ 1160555.699 & 0.000 & -4188907.065 & -4655376.341 & 1 & 0 & 0 \\ -4655376.341 & 4188907.065 & 0.000 & -1160555.699 & 0 & 1 & 0 \\ 4188907.065 & 4655376.341 & 1160555.699 & 0.000 & 0 & 0 & 1 \\ 1160417.220 & 0.000 & -4188310.909 & -4656003.238 & 1 & 0 & 0 \\ -4656003.238 & 4188310.909 & 0.000 & -1160417.220 & 0 & 1 & 0 \\ 41883120.909 & 4656003.238 & 1160417.220 & 0.000 & 0 & 0 & 1 \end{bmatrix}$$

$$K = \begin{bmatrix} 0.000 \\ 0.000 \\ 0.000 \\ 0.011 \\ -0.026 \\ 0.034 \\ 0.006 \\ -0.012 \\ 0.010 \\ -0.008 \\ 0.015 \\ -0.015 \end{bmatrix}$$

Solving Equation (11.37) using the matrices above and adding X to the initial values results in

$$X = \begin{bmatrix} 0.0000000256 \\ -0.0000486262 \\ -0.0000388234 \\ -0.0000146664 \\ -230.9285 \\ 186.7768 \\ 271.3325 \end{bmatrix}$$

$s = 1 + 0.0000000256 = 1.0000000256$
$\theta_1 = -0°00'10.02987''$
$\theta_2 = -0°00'08.0079''$
$\theta_3 = -0°00'03.02517''$
$Tx = 230.877 - 230.92 = -0.052$
$Ty = -187.926 + 186.7768 = -1.149$
$Tz = -270.037 + 271.3325 = 1.295$

The next iteration resulted in negligible changes in the translations, and thus the transformation parameters are as listed. Using Equation (24.2), transformation parameters, and the GPS-derived ITRF 00 coordinates for station B, the NAD 83 coordinates for station B are

geocentric coordinates = (1,160,637.257 −4,655,590.805, 4,188,659.377)

geodetic coordinates = (41°18′40.46653″ N, 76°00′05.50185″ W,

319.677 m)

SPCS 83 coordinates = (746,397.796 m, 128,586.853 m)

The reader should note that the NAD 83 coordinates for station B are the same as those listed in Figure 23.5. However, slight differences in coordinates can be expected, due to random errors in the transformation.

Example 24.1 demonstrates the mathematical relationship between the two datums for this small set of points. This process should always be considered when combining traditional observations with GPS coordinates in higher-order surveys. That is, if GPS derived coordinates are to be entered into an adjustment, they should first be transformed into the local datum. In the United States, the National Geodetic Survey has developed *horizontal time-dependent positioning* (HTDP) software[3] that allows the users to transform coordinates between NAD 83, WGS 84, and ITRF 00. This software also takes into account plate tectonics by applying velocity vectors to stations when they are known.

The process of transforming points from global datums to local datums is important when performing RTK-GPS stakeout surveys. A GPS survey implicitly uses the points located in a global datum. Since engineering plans are generally developed in a local datum such as NAD 83, the GPS-derived coordinate values must be transformed into the local coordinate system. This process is known as *localization* (sometimes called *site calibration*) by manufacturers. From a design point of view, it is important to recognize that the best results will be obtained if points common in both datums surround the project area. After entering in the local datum coordinates, the GPS receiver should occupy each station. The software will then compute the transforma-

---

[3] A description of the HTDP software can be located on the NGS Web site at http://www.ngs.noaa.gov/TOOLS/Htdp/Htdp.shtml. A Mathcad worksheet that demonstrates this software is available on the CD that accompanies this book.

484 COMBINING GPS AND TERRESTRIAL OBSERVATIONS

tion parameters and use these to determine the coordinate values for the points to be surveyed.

24.3 ROTATIONS BETWEEN COORDINATE SYSTEMS

GPS uses the geocentric coordinate system and list its baseline vectors in this system. Often, it is preferable to obtain the coordinate changes in terms of geodetic coordinates. From geodesy, the relationship between changes in the geodetic coordinate system and the geocentric coordinate system is

$$dx = \begin{bmatrix} dX \\ dY \\ dZ \end{bmatrix}$$

$$= \begin{bmatrix} -(M+h)\sin\phi\sin\lambda & -(N+h)\cos\phi\sin\lambda & \cos\phi\cos\lambda \\ -(M+h)\sin\phi\sin\lambda & (N+h)\cos\phi\cos\lambda & \cos\phi\sin\lambda \\ (M+h)\cos\phi & 0 & \sin\phi \end{bmatrix} \begin{bmatrix} d\phi \\ d\lambda \\ dh \end{bmatrix}$$

$$= R_{XG} \begin{bmatrix} d\phi \\ d\lambda \\ dh \end{bmatrix} \tag{24.8}$$

where dx represents the changes in the geocentric coordinates, M is the radius in the meridian, and N the radius in the normal at latitude ϕ as given by Equation (23.18), h is the geodetic height of the point, and R_{XG} is the transformation matrix.

The transformation between changes in the geodetic coordinate system and the local geodetic coordinate system are given in Equation (23.29), which is repeated here for convenience:

$$\begin{bmatrix} dn \\ de \\ du \end{bmatrix} = \begin{bmatrix} M+h & 0 & 0 \\ 0 & (N+h)\cos\phi & 0 \\ 0 & 0 & 1 \end{bmatrix} \begin{bmatrix} d\phi \\ d\lambda \\ dh \end{bmatrix} = R_{LG} \begin{bmatrix} d\phi \\ d\lambda \\ dh \end{bmatrix} \tag{24.9}$$

where R_{LG} is the rotation matrix between the geodetic and local geodetic coordinate systems.

24.4 COMBINING GPS BASELINE VECTORS WITH TRADITIONAL OBSERVATIONS

As discussed in Chapter 23, the three-dimensional geodetic adjustment allows the adjustment of all traditional surveying observations. If the control is known in the local coordinate system, Equation (24.2) can be included in adjustment to account for datum differences.

24.4 COMBINING GPS BASELINE VECTORS WITH TRADITIONAL OBSERVATIONS

Baseline vectors are the geocentric coordinate differences between two points. Thus, the translation component of Equation (24.2) is removed, leaving only the scaling and rotational parameters, and the last three columns of the J_i matrix in Equations (24.6) and (24.7) are eliminated, leaving

$$J_i = \begin{bmatrix} x_i & 0 & -z_i & y_i \\ y_i & z_i & 0 & -x_i \\ z_i & -y_i & x_i & 0 \end{bmatrix} \quad dx = \begin{bmatrix} \Delta s \\ \Delta\theta_1 \\ \Delta\theta_2 \\ \Delta\theta_3 \end{bmatrix} \quad (24.10)$$

Equation (24.10) can be used in Equation (24.6). However, since the three-dimensional geodetic network adjustment is performed in the local geodetic coordinate system, the rotational elements of Equation (24.10) can be transformed about a single station in this system. Dropping the Δ symbol, the rotation about the single station in the local geodetic system becomes

$$R = R_3^T(\lambda_0)R_2^T(90 - \phi_0)R_3(\theta_3)R_2(\theta_2)R_1(\theta_1)R_2(90 - \phi_0)R_3(\lambda_0) \quad (24.11)$$

where θ_n is a rotation about the north axis of the local geodetic coordinate system; θ_e a rotation about the east axis of the local geodetic coordinate system; θ_u a rotation about the up axis of the local geodetic coordinate system; R_1, R_2, and R_3 are the rotation matrices defined in Section 18.7; and ϕ_0 and λ_0 are the geodetic coordinates of the rotational point. This point should be picked near the center of the project area.

Again, since in nearly aligned coordinate systems the rotations are small, the rotations above can be simplified to

$$R = \theta_n R_n + \theta_e R_e + \theta_u R_u + I \quad (24.12)$$

where

$$R_u = \begin{bmatrix} 0 & \sin\phi_0 & -\cos\phi_0 \sin\lambda_0 \\ -\sin\phi_0 & 0 & \cos\phi_0 \cos\lambda_0 \\ \cos\phi_0 \sin\lambda_0 & -\cos\phi_0 \cos\lambda_0 & 0 \end{bmatrix}$$

$$R_e = \begin{bmatrix} 0 & 0 & -\cos\lambda_0 \\ 0 & 0 & -\sin\lambda_0 \\ \cos\lambda_0 & \sin\lambda_0 & 0 \end{bmatrix}$$

$$R_n = \begin{bmatrix} 0 & -\cos\phi_0 & -\sin\phi_0 \sin\lambda_0 \\ \cos\phi_0 & 0 & \sin\phi_0 \cos\lambda_0 \\ \sin\phi_0 \sin\lambda_0 & -\sin\phi_0 \cos\lambda_0 & 0 \end{bmatrix}$$

I is a three-dimensional identity matrix; θ_n, θ_e, θ_u are in radian units; and ϕ_0 and λ_0 are the geodetic coordinates of the rotational point.

Finally, the transformation going from the observed GPS vector to its local geodetic equivalent between stations I and J is

$$\begin{bmatrix} \Delta X_{IJ} \\ \Delta Y_{IJ} \\ \Delta Z_{IJ} \end{bmatrix}_2 = (1 + s)R \begin{bmatrix} \Delta X_{IJ} \\ \Delta Y_{IJ} \\ \Delta Z_{IJ} \end{bmatrix}_1 \qquad (24.13)$$

where s represents the differential scale change between systems 1 and 2, R is defined in Equation (24.12), $[\cdot]_1$ represents the GPS baseline vector components in the local coordinate system, and $[\cdot]_2$ represents the GPS observed baseline vector components between stations I and J.

In Chapter 23 the three-dimensional geodetic network adjustment was developed in the local geodetic system. Thus, the addition of the GPS baseline vectors into this adjustment requires that the coefficient matrix be rotated into the same system. Recall from Chapter 17 that the coefficient matrix (A) for each GPS baseline vectors consisted of three rows containing -1, 0, and 1. This matrix must be rotated into the local geodetic system. Thus, for the baseline vector IJ, the new coefficient matrix values are derived as

For station I: $\qquad -(1 + s)R\ R_{XG}(\phi_i, \lambda_i)\ R_{LG}(\phi_i)^{-1}$ $\qquad (24.14)$

For station J: $\qquad (1 + s)R\ R_{XG}(\phi_j, \lambda_j)\ R_{LG}(\phi_j)^{-1}$

where s is the change in scale between the two systems, rotation matrix R is defined in Equation (24.12), rotation matrix R_{XG} is defined in Equation (24.8), rotation matrix R_{LG} is defined in Equation (24.9), and (ϕ_i, λ_i) and (ϕ_j, λ_j) are the geodetic coordinates from stations I and J, respectively.

When developing the matrices for a least squares adjustment, the unknown parameters for scale and rotation should be set to zero. After the first iteration these values will be modified and updated. At the end of the adjustment, these parameters can be checked for statistical significance as described in Section 18.8.

The elements of the coefficient matrix for the unknown rotation angles and scale are

$$J = \begin{bmatrix} X_J - X_I & r_{n_1} & r_{e_1} & r_{u_1} \\ Y_J - Y_I & r_{n_2} & r_{e_2} & r_{u_2} \\ Z_J - Z_I & r_{n_3} & r_{e_3} & r_{u_3} \end{bmatrix} \qquad (24.15)$$

where $X^T = [\Delta s\ \Delta\theta_n\ \Delta\theta_e\ \Delta\theta_u]$,

$$r_n = R_n \begin{bmatrix} X_J - X_I \\ Y_J - Y_I \\ Z_J - Z_I \end{bmatrix}, \qquad r_e = R_e \begin{bmatrix} X_J - X_I \\ Y_J - Y_I \\ Z_J - Z_I \end{bmatrix}, \qquad r_u = R_u \begin{bmatrix} X_J - X_I \\ Y_J - Y_I \\ Z_J - Z_I \end{bmatrix}$$

24.4 COMBINING GPS BASELINE VECTORS WITH TRADITIONAL OBSERVATIONS

Example 24.2 The local datum coordinates for station B are given in Example 24.1 and the GPS-derived and transformed approximate NAD 83 coordinates for station E are (41°18′43.9622″ N, 76°00′29.0384″ W, 292.354 m). Assuming initial values of zero for the scale and rotation parameters in Equation 24.13 and using point A as the rotational point, develop the coefficient (J) and constants (K) matrices for the transformation parameter and the following GPS-observed baseline.

Baseline	ΔX (m)	ΔY (m)	ΔZ (m)
EB	553.430	43.400	−62.949

SOLUTION From Example 24.1, the approximate NAD 83 geocentric and geodetic coordinates for stations E and B are

Station	ϕ	λ	h (m)
E	41°18′43.9622″	−76°00′29.0385″	292.354
B	41°18′40.4665″	−76°00′05.5019″	319.677
A	41°18′26.0485″	−76°00′10.2486″	—

By Equation (24.8), R_{XG} for station B is

$$R_{XG}^B = \begin{bmatrix} -1016184.4632 & 4655590.8060 & 0.1817 \\ 4076156.4570 & 1160637.2558 & -0.7288 \\ 4779908.9967 & 0.0000 & 0.6601 \end{bmatrix}$$

By Equation (24.8), R_{XG} for station E is

$$R_{XG}^E = \begin{bmatrix} -1015734.7290 & 4655634.2166 & 0.1816 \\ 4076334.1775 & 1160083.8286 & -0.7288 \\ 4779818.0865 & 0.0000 & 0.6602 \end{bmatrix}$$

By Equation (24.9), R_{LG} for station B is

$$R_{LG}^B = \begin{bmatrix} 6363584.8656 & 0.0000 & 0.0000 \\ 0.0000 & 4798083.4291 & 0.0000 \\ 0.0000 & 0.0000 & 1.0000 \end{bmatrix}$$

By Equation (24.9), R_{LG} for station E is

$$R_{LG}^E = \begin{bmatrix} 6363558.6197 & 0.0000 & 0.0000 \\ 0.0000 & 4797991.7099 & 0.0000 \\ 0.0000 & 0.0000 & 1.0000 \end{bmatrix}$$

Since the initial approximations for the three differential rotations and scale are initially 0, the R matrix in Equation (24.12) reduces to a 3×3 identity matrix for the first iteration. In subsequent iterations, this matrix will change.

Combining the data from Table (23.3) with the GPS baseline vectors of Example 24.1 yields a J matrix with 25 columns where the unknown parameters are $[dX_A \ dY_A \ dZ_A \ dX_B \ dY_B \ dZ_B \ dX_C \cdots dZ_D \ dX_E \ dY_E \ dZ_E \cdots s \ \theta_n \ \theta_e \ \theta_u]^T$. The J coefficient matrix for baseline EB is

$$J = \begin{bmatrix} 0 & 0 & 0 & a_{11} & a_{12} & a_{13} & 0 & \cdots & 0 & a_{14} & a_{15} & a_{16} & 0 & \cdots & 0 & b_{11} & b_{12} & b_{13} & b_{14} \\ 0 & 0 & 0 & a_{21} & a_{22} & a_{23} & 0 & \cdots & 0 & a_{24} & a_{25} & a_{26} & 0 & \cdots & 0 & b_{21} & b_{22} & b_{23} & b_{24} \\ 0 & 0 & 0 & a_{31} & a_{32} & a_{33} & 0 & \cdots & 0 & a_{34} & a_{35} & a_{36} & 0 & \cdots & 0 & b_{31} & b_{32} & b_{33} & b_{34} \end{bmatrix}$$

(24.16)

where, using Equation (24.14), the A-matrix coefficients are

$$A = \begin{bmatrix} -0.1597 & 0.9703 & 0.1817 & 0.1596 & -0.9703 & -0.1816 \\ 0.6405 & -0.2419 & -0.7288 & -0.6406 & 0.2418 & 0.7288 \\ 0.7511 & 0.0000 & 0.6601 & -0.7511 & 0.0000 & -0.6601 \end{bmatrix}$$

and, using Equation (24.15), the B-matrix coefficients for the transformation parameters are

$$B = \begin{bmatrix} 553.4271 & -72.9412 & 15.2308 & -17.2421 \\ 43.4106 & 405.6700 & -61.1002 & -376.7564 \\ -62.9699 & -361.3992 & 91.7378 & -411.2673 \end{bmatrix}$$

With the aid of Equations (17.2) to (17.4), the geocentric coordinates for stations E and B are

Station	X (m)	Y (m)	Z (m)
E	1,160,083.830	−4,655,634.217	4,188,722.346
B	1,160,637.257	−4,655,590.805	4,188,659.377

The values for the constants matrix (K) are

$$K = \begin{bmatrix} 553.430 - (1160637.257 - 1160083.830) \\ 43.400 - (-4655590.805 + 4655634.217) \\ -62.949 - (4188659.377 - 4188722.346) \end{bmatrix} = \begin{bmatrix} +0.003 \\ -0.012 \\ 0.020 \end{bmatrix}$$

Notice that the coefficient matrix for each GPS baseline vector component is no longer a matrix of -1, 0, and 1, but rather, contains noninteger values. Also note that the coefficients for the differential changes in scale and rotations are populated for each baseline vector component. The reader may wish to review the Mathcad worksheet on the CD that accompanies this book to explore the complete set of matrix operations.

24.5 OTHER CONSIDERATIONS

Using procedures similar to those shown in Chapters 23 and Example 24.2, a combined adjustment of the terrestrial and GPS baseline vectors can be performed. If GPS-derived station coordinates are to be fixed during the adjustment, they must first be transformed in the local datum to ensure consistency with any local control stations. However, if GPS-derived points are the only control in the adjustment, the entire adjustment can be performed using the global datum that was used to reduce the GPS observations.

In both chapters the adjustments are performed in the local geodetic coordinate system. This system was chosen since defining standard deviations for control stations in the (n,e,u) system is more intuitive to surveyors than either the geodetic or geocentric coordinate systems. References in the bibliography contain procedures for combining GPS and terrestrial observations using either the geocentric or geodetic systems.

PROBLEMS

24.1 Discuss what is meant by *local datum*.

24.2 Discuss what is meant by *global datum*.

24.3 How do datums differ?

24.4 Why is it important to perform a localization before staking out a highway alignment?

24.5 What is meant by *localization*?

24.6 Using the Helmert transformation parameters derived in Example 24.1, derive the NAD 83 geocentric coordinates (in meters) for a

490 COMBINING GPS AND TERRESTRIAL OBSERVATIONS

point having ITRF 00 coordinates of (1160652.008, −4655693.197, 4188423.986).

24.7 Repeat Problem 24.6 for a station having geocentric coordinates of (1160398.043, −4655803.184, 4188935.609).

24.8 Using the accompanying data, compute the Helmert transformation parameters to take the coordinate values from WGS 84 to NAD 83.

Station	NAD 83 x (m)	y (m)	z (m)	WGS 84 x (m)	y (m)	z (m)
100	1,160,097.952	−4,634,583.300	4,188,086.049	1,160,098.356	−4,634,583.248	4,188,086.233
101	1,160,285.844	−4,634,859.416	4,188,233.622	1,160,286.248	−4,634,859.364	4,188,233.806
102	1,159,986.652	−4,634,623.501	4,188,153.783	1,159,987.056	−4,634,623.449	4,188,153.967

24.9 Why is it important to have control points dispersed about the perimeter of a project area during a stakeout survey?

24.10 Using the appropriate information from Table 23.3, Example 24.2, and the accompanying baseline vector data, determine the nonzero elements of the coefficient (J) and constant (K) matrices.

Baseline	ΔX (m)	ΔY (m)	ΔZ (m)
ED	35.2573	−368.067	−347.063

24.11 Repeat Problem 24.10 for the following data. The approximate geodetic coordinates for station F are (41°18′39.7004″ N, 76°59′58.9973″ W, 312.731 m).

Baseline	ΔX (m)	ΔY (m)	ΔZ (m)
FB	−149.874	−25.079	22.222

24.12 Repeat Problem 24.10 for the following data. The approximate geodetic coordinates for station G are (41°18′12.8871″ N, 76°00′11.2922″ W, 350.935 m).

Baseline	ΔX (m)	ΔY (m)	ΔZ (m)
GD	−529.004	188.868	334.427

24.13 Using station *A* as the central point in a project and the baseline vectors given in Problem 24.10, what are the first iteration coefficients for the transformation parameters?

24.14 Repeat Problem 24.12 using the baseline from Problem 24.11.

24.15 Repeat Problem 24.12 using the baseline from Problem 24.12.

Programming Problem

24.16 Develop software that solves Problem 24.8.

CHAPTER 25

ANALYSIS OF ADJUSTMENTS

25.1 INTRODUCTION

An initial run of a least squares adjustment is not the end of the adjustment. Rather, it is the beginning of the data analysis process. Throughout this book, the mechanics of performing a least squares adjustment properly have been discussed. Additionally, statistical methods have been introduced to analyze the quality of observations. In this chapter we explore the procedures used in analyzing the results of an adjustment beginning with the analysis of residuals and reviewing data snooping as discussed in Chapter 21.

25.2 BASIC CONCEPTS, RESIDUALS, AND THE NORMAL DISTRIBUTION

The normal distribution and statistical testing were introduced in the beginning chapters using simple data sets. These basic concepts also apply in the analysis of data after an adjustment. When viewing Figure 25.1, the guiding principles used in analyzing observations from normally distributed data are:

1. Data tend to be clustered around a single value.
2. Errors tend to be equally distributed about this value.
3. Errors are equally distributed in sign.
4. Most errors tend to be small in magnitude and errors large in magnitude seldom occur.

25.2 BASIC CONCEPTS, RESIDUALS, AND THE NORMAL DISTRIBUTION

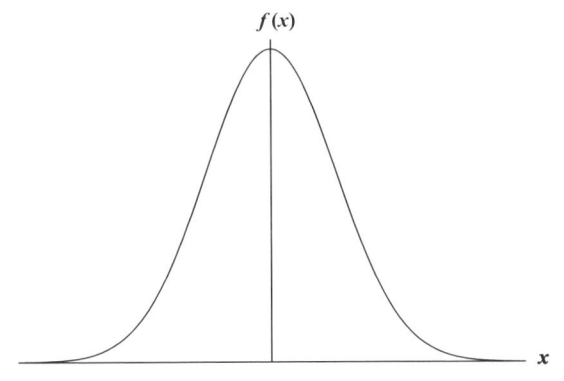

Figure 25.1 Normal distribution.

At the conclusion of a least squares adjustment, the residuals in the observations should be scanned and analyzed. However, terms such as *large in magnitude* must be defined statistically. This is accomplished by statistically comparing the observation's a priori standard deviation against its residual using a *t*-distribution multiplier. For example, if a distance observation has a residual of +0.25 ft when all other residuals in the adjustment are below 0.10 ft, this observation may be viewed as large. However, if the observation had an estimated standard deviation of only ±0.5 ft, its residual is actually well within its predicted range.

As an example, the results of the adjustment from Example 16.2 will be analyzed. A quick review of the residuals listed for the distance observations in Figure 25.2 show that the residual for distance *QR* is at least twice the size of the other distance residuals and might be viewed as large. However, the a priori standard deviation of this observation was ±0.026 ft from an adjustment with 13 redundant observations. The 95% confidence interval for this observational residual is computed as

$$v_{95} = \pm t_{0.025,13} S = \pm 2.16(0.026) = \pm 0.056 \text{ ft}$$

From this it can be shown that the residual of −0.038 ft for distance *QR* is well within the 95% confidence interval of ±0.056 ft.

When analyzing the angles, it can be seen *QTR* has a residual of 18.52″. This residual is more than three times larger than the next largest residual of 5.23″. Also note that this angle opposes distance *QR*. Therefore it may explain why distance *QR* had the largest distance residual. The a priori estimated standard deviation for angle *QTR* is ±4.0″. Using the same *t*-distribution critical value for a 95% confidence interval yields

$$v_{95} = 2.16(4.0'') = \pm 8.6''$$

```
*******************************
Adjusted Distance Observations
*******************************
Station           Station
Occupied          Sighted        Distance              V              S
===============================================================================
   Q                 R           1,639.978          -0.0384         0.0159
   R                 S           1,320.019           0.0176         0.0154
   S                 T           1,579.138           0.0155         0.0158
   T                 Q           1,664.528           0.0039         0.0169
   Q                 S           2,105.953          -0.0087         0.0156
   R                 T           2,266.045           0.0104         0.0163

***************************
Adjusted Angle Observations
***************************
Station           Station        Station
Backsighted       Occupied       Foresighted      Angle             V         S
===============================================================================
   R                 Q              S          38°48'52.8"       2.08"     1.75
   S                 Q              T          47°46'13.1"       0.70"     1.95
   T                 Q              R         273°24'54.1"      -2.37"     2.40
   Q                 R              S         269°57'34.0"       0.64"     2.26
   R                 S              T         257°32'57.3"       0.50"     2.50
   S                 T              Q         279°04'34.5"       3.32"     2.33
   S                 R              T          42°52'52.3"       1.34"     1.82
   S                 R              Q          90°02'26.0"      -0.74"     2.26
   Q                 S              R          51°08'41.3"      -3.73"     1.98
   T                 S              Q          51°18'21.4"       5.23"     2.04
   Q                 T              R          46°15'20.5"      18.52"     1.82
   R                 T              S          34°40'05.0"      -0.74"     1.72
```

Figure 25.2 Adjusted distances and angles from Example 16.2.

The actual residual for angle *QTR* is well outside the range of ±8.6″. In this case, the residual for angle *QTR* definitely fits the definition of *large* and it is a candidate for reobservation or removal from the adjustment.

Using the procedures outlined in Chapter 21, data snooping confirms that angle *QTR* is a detectable blunder. This angular observation can be removed from the data set and the adjustment rerun. The results of this readjustment are shown in Figure 25.3. Note that the residual for distance *QR* was reduced by about half of its previous value. This demonstrates the interrelationship between the distance *QR* and angle *QTR*. Also note that all the residuals are within a single standard deviation of their a priori estimated values. In fact, all the distance residuals are less than a third of their a priori estimated errors, and all the angle residuals are less than half their a priori estimates.

There are 18 observations in this adjustment. Thus, about half of the residuals (nine) should be positive and the others negative. In this adjustment

25.2 BASIC CONCEPTS, RESIDUALS, AND THE NORMAL DISTRIBUTION 495

```
Adjusted stations
                          Error ellipse confidence level at 0.950
Station      X         Y       Sx      Sy      Su      Sv       t
=================================================================
   R     1,003.06  2,640.01  0.000   0.006   0.017   0.000    0.11°
   S     2,323.06  2,638.47  0.005   0.007   0.019   0.014  156.28°
   T     2,661.74  1,096.09  0.006   0.007   0.021   0.015   26.18°

Adjusted Distance Observations
Station    Station
Occupied   Sighted    Distance       V        S      Std.Res.  Red.#
=====================================================================
   Q         R       1,640.008    -0.0081   0.0060   -0.409   0.576
   R         S       1,320.006     0.0054   0.0055    0.295   0.579
   S         T       1,579.133     0.0099   0.0056    0.510   0.597
   T         Q       1,664.514    -0.0097   0.0060   -0.495   0.569
   Q         S       2,105.966     0.0039   0.0056    0.162   0.702
   R         T       2,266.034    -0.0014   0.0058   -0.057   0.700

Adjusted Angle Observations
Station      Station     Station
Backsighted Occupied  Foresighted  Angle           V       S    Std.Res. Red.#
================================================================================
    R          Q           S      38°48'50.2"  -0.45"  0.64   -0.13   0.795
    S          Q           T      47°46'11.7"  -0.73"  0.69   -0.21   0.757
    T          Q           R     273°24'58.1"   1.58"  0.89    0.44   0.672
    Q          R           S     269°57'34.7"   1.31"  0.80    0.32   0.767
    R          S           T     257°32'56.9"   0.11"  0.88    0.03   0.716
    S          T           Q     279°04'30.3"  -0.91"  0.87   -0.24   0.700
    S          R           T      42°52'52.6"   1.58"  0.64    0.41   0.821
    S          R           Q      90°02'25.3"  -1.41"  0.80   -0.36   0.746
    Q          S           R      51°08'44.5"  -0.53"  0.73   -0.14   0.767
    T          S           Q      51°18'18.6"   2.43"  0.74    0.71   0.722
    R          T           S      34°40'04.3"  -1.37"  0.61   -0.38   0.814

Adjusted Azimuth Observations
Station    Station
Occupied   Sighted    Azimuth       V        S"     Std.Res.  Red.#
=====================================================================
   Q         R       0°06'24.5"   -0.00"    0.00"     0.0     0.000

Adjustment Statistics
=====================
        Iterations = 2
      Redundancies = 12
  Reference Variance = 0.1243
         Reference So = ±0.35

         Failed to pass X² test at 95.0% significance level!
                    X² lower value = 4.40
                    X² upper value = 23.34
```

Figure 25.3 Readjustment of the data in Example 16.2 after removing angle *QTR*.

there are exactly nine positive residuals and nine negative residuals. This follows the second guideline for normally distributed data. In this example, the residuals do not show any skewness, as discussed in Section 2.4. However, it should be remembered that since we are working with a sample of data, some variation from theory is acceptable.

Example 25.1 A GPS baseline vector determined by the rapid-static method (5 mm + 1 ppm) has a length of about 10.5 km. The instrument setups were estimated to be within ±0.003 mm of the true station location. Its combined XYZ residual is 0.021 m. At a 95% level of confidence, should this residual be considered too large if the number or redundancies in the adjustment is 20?

SOLUTION The a priori estimated error in this baseline using the rapid-static method is

$$\sigma = \sqrt{3^2 + 3^2 + 5^2 + \left(\frac{10{,}500{,}000}{1{,}000{,}000}\right)^2} = \pm 12.4 \text{ mm} \qquad t_{0.025,20} = 2.09$$

The acceptable 95% range is $R = 2.09(12.4) = \pm 25.9$ mm. The residual was 21 mm. This is inside of the 95% confidence interval and thus is an acceptable size for a combined residual vector with this baseline and surveying method.

25.3 GOODNESS-OF-FIT TEST

The results of the adjustment in Figure 25.3 appear to follow the normal distribution and the size of the residuals indicates that the data appear to be consistent. However, the adjustment failed to pass the goodness-of-fit (χ^2) test. This demonstrates a weakness in the goodness-of-fit test. That is, passing or failing the test is not a good indicator of the quality of the data or the presence of blunders. Nor does the test reveal the exact problem in data when the test fails. The χ^2 test should be viewed as a warning flag for an adjustment that requires further analysis, not as an indicator of bad data.

As discussed in Chapters 5 and 16, the χ^2 test compares the a prior reference variance, set equal to 1, against the reference variance computed from the data. An analysis of Equation (12.15) shows that the test will fail if the residuals are too large or too small compared to the weights of the observations. For example, an observation with a small residual should have a high weight and a large residual should have a low weight. If the residuals tend to be smaller than their a priori standard deviations, the resulting reference variance will probably be less than 1. This is an example of an incorrect stochastic model. If the residuals tend to be larger than their a priori standard deviations, the reference variance will be greater than 1 and the χ^2 test will fail in the upper bounds of the distribution.

The original data set of Example 16.2 failed the goodness-of-fit test. In the original adjustment, it failed because the computed reference variance was greater than 1. This was probably caused by the large residual for angle *QTR*. After angle *QTR* was removed from the data, the χ^2 test failed because the reference variance was statistically less than 1. Since the residuals appear to be much smaller than their a priori standard deviations, an incorrect stochastic model is to blame. In this case, the distance residuals were about three times less than the estimated standard deviations and the angles were one-half of their estimated standard deviations. Thus, to pass the χ^2 test with this adjustment, the stochastic model needs to be modified.

In developing the stochastic model for this problem, it was originally believed that the field crew had use a total station with a DIN 18723 accuracy of $\pm 4''$ and an EDM accuracy of 5 mm + 5 ppm. It was later discovered that the crew used an instrument having a DIN 18723 accuracy of $\pm 2''$ and an EDM accuracy of 2 mm + 2 ppm. Estimated standard deviations using this new information resulted in the adjustment shown in Figure 25.4. This adjustment passed the goodness-of-fit test. However, this did not change the adjustment results significantly. That is, the adjusted coordinate values and adjusted observations are nearly the same in the adjustments shown in Figures 25.3 and 25.4. What did change in Figure 25.4 were the statistical results. Since the statistical results of the adjustment are of little value to the surveyor once analyzed, the adjustment shown in Figure 25.3 is sufficient for most applications.

Practitioners often view the goodness-of-fit test as an indicator of a possible problem. If the goodness-of-fit test failed because the computed reference variance was too small, the test result is often ignored, since correcting it does not change the adjusted coordinates significantly. In this case, some practitioners will say that the adjustment failed the goodness-of-fit test on the "good" side. That is, the reference variance was too small because the residuals are too small for their given weights. However, if the reference variance computed is greater than 1, the results should be analyzed for possible blunders in the observations. After the identifiable blunders have been eliminated from the data set, the stochastic model might still exhibit a problem. In this case, if the computed reference variance is greater than 1, the estimated standard deviations for the observations should be increased. In the preceding example the computed reference variance is less than 1. To correct this, the estimated standard deviations for the observations were decreased.

This adjustment brings about another important discussion point. Had all the a priori standard deviations been reduced by a single scale factor, the results of the adjustment would be the same.[1] That is, if the a priori standard

---

[1] Readers are encouraged to rerun the data in Example 16.2 with the a priori standard deviations reduced by a factor of $\frac{1}{2}$ to compare the adjustment results.

498 ANALYSIS OF ADJUSTMENTS

```
Adjusted stations
                               Error ellipse confidence level at 0.950
Station       X         Y        Sx      Sy      Su      Sv        t
==============================================================================
   R      1,003.06   2,640.00  0.000   0.009   0.024   0.000     0.11°
   S      2,323.06   2,638.47  0.008   0.009   0.028   0.021   151.27°
   T      2,661.74   1,096.09  0.008   0.011   0.031   0.022    26.14°

Adjusted Distance Observations
Station     Station
Occupied    Sighted    Distance       V         S      Std.Res.   Red.#
==============================================================================
   Q           R      1,640.008   -0.0083   0.0086    -0.830     0.580
   R           S      1,320.006    0.0053   0.0081     0.527     0.603
   S           T      1,579.133    0.0100   0.0081     0.971     0.615
   T           Q      1,664.515   -0.0089   0.0087    -0.895     0.568
   Q           S      2,105.966    0.0039   0.0079     0.341     0.673
   R           T      2,266.034   -0.0011   0.0082    -0.097     0.658

Adjusted Angle Observations
Station       Station     Station
Backsighted  Occupied   Foresighted    Angle          V      S   Std.Res. Red.#
==============================================================================
   R            Q           S     38°48'50.2"   -0.46" 0.95   -0.24    0.808
   S            Q           T     47°46'11.7"   -0.73" 1.02   -0.38    0.776
   T            Q           R    273°24'58.1"    1.59" 1.29    0.87    0.667
   Q            R           S    269°57'34.7"    1.27" 1.14    0.64    0.751
   R            S           T    257°32'56.8"    0.03" 1.27    0.02    0.696
   S            T           Q    279°04'30.4"   -0.79" 1.25   -0.43    0.684
   S            R           T     42°52'52.6"    1.55" 0.94    0.77    0.820
   S            R           Q     90°02'25.3"   -1.37" 1.14   -0.69    0.751
   Q            S           R     51°08'44.4"   -0.57" 1.06   -0.29    0.772
   T            S           Q     51°18'18.7"    2.53" 1.09    1.34    0.751
   R            T           S     34°40'04.3"   -1.42" 0.90   -0.72    0.827

Adjusted Azimuth Observations
Station     Station
Occupied    Sighted    Azimuth         V        S"     Std.Res.   Red.#
==============================================================================
   Q           R      0°06'24.5"    -0.00"    0.00"      0.0      0.000

Adjustment Statistics
*********************
         Iterations = 2
       Redundancies = 12
   Reference Variance = 0.4572
       Reference So = ±0.68

              Passed X² test at 95.0% significance level!
                      X² lower value = 4.40
                      X² upper value = 23.34
```

Figure 25.4 Readjusted data from Example 16.2 with a different stochastic model.

deviations for all the observations had been cut in half, this would have resulted in increasing all the weights by a factor of 4. Since weights are relative, performing a scalar reduction in the standard deviations does not change the adjustment.

This example demonstrates the importance of selecting appropriate a priori standard deviations for the observations. Estimated standard deviations for observations cannot be selected from the recesses of one's mind. Doing this not only affects how the errors in the observations are distributed but also how the results of the adjustment are analyzed. For traditional surveys it is always best to select a stochastic model that reflects the estimated accuracies in the observations using procedures discussed in Chapters 7 and 9. Good estimates of setup errors will typically result in a sound stochastic model.

However, the network adjustment of GPS baseline vectors derives its stochastic model from the least squares reduction of each baseline. As discussed in Chapter 17, the stochastic model is part of the printout from the baseline reduction. Assuming proper field procedures, if the GPS network adjustment fails to pass the χ^2 test, there is little one can do to modify the model. In fact, if field procedures are consistent, scaling the entire stochastic model might be the only option. Although this procedure may result in a "passed" χ^2 test, it will not change the coordinates in the solution since weights are relative. In fact, it will only change the a posteriori statistics. Thus, scaling the entire stochastic model by a single factor is important only if the statistics are important. Since this is seldom true, there is little value in scaling the stochastic model.

Better adjustments could be obtained if the commercial GPS software employed the multipoint solution technique. This solution takes the individual pseudoranges from the satellites and computes baseline vectors and the unknown coordinates of the network stations simultaneously. Current software offerings compute the baseline vectors from the pseudoranges and then coordinates of the network stations from these computed vectors. This is a two-step process. The multipoint solution is similar to the three-dimensional geodetic network and the photogrammetric block-bundle adjustment in that all unknowns are computed in one adjustment. However, it is important to remove any large blunders before attempting a multipoint solution. Thus, the two-step solution process must be maintained to ferret out problems in the data. Since the main advantage of the multipoint solution lies in the development of the stochastic model, only a few examples of this solution technique exist.

25.4 COMPARISON OF RESIDUAL PLOTS

Massive quantities of data can be collected and reduced during the reduction of GPS carrier-phase observations. Software manufacturers typically plot pseudorange residuals against a time line. Examples of this are shown in

Figure 25.5, where plot (a) is for satellite vehicle 24 (SV 24) and (b) is for satellite vehicle 28 (SV 28). Notice that the residuals in Figure 25.5(b) have a slight downward slope at the beginning of the session. In fact, except for the first 3 minutes, the residuals are fairly uniform even though most are negative in sign. During the first 3 minutes of the session, satellite 28 was just clearing the set mask angle and suffered some loss of lock problems. This skewed the data.

The first 3 minutes of data from this satellite could be removed and the baseline reprocessed to correct this problem. However, this might not be practical given the shortness of the overall observation session. Since much field time is lost to travel, setup, and teardown, it is always wise to collect more data than are needed to resolve the position of the receiver. This is similar to observing distances and angles more than once.

Note that the residuals in Figure 25.5(a) are mostly positive. This could have several causes, including loss-of-lock problems. However, these types

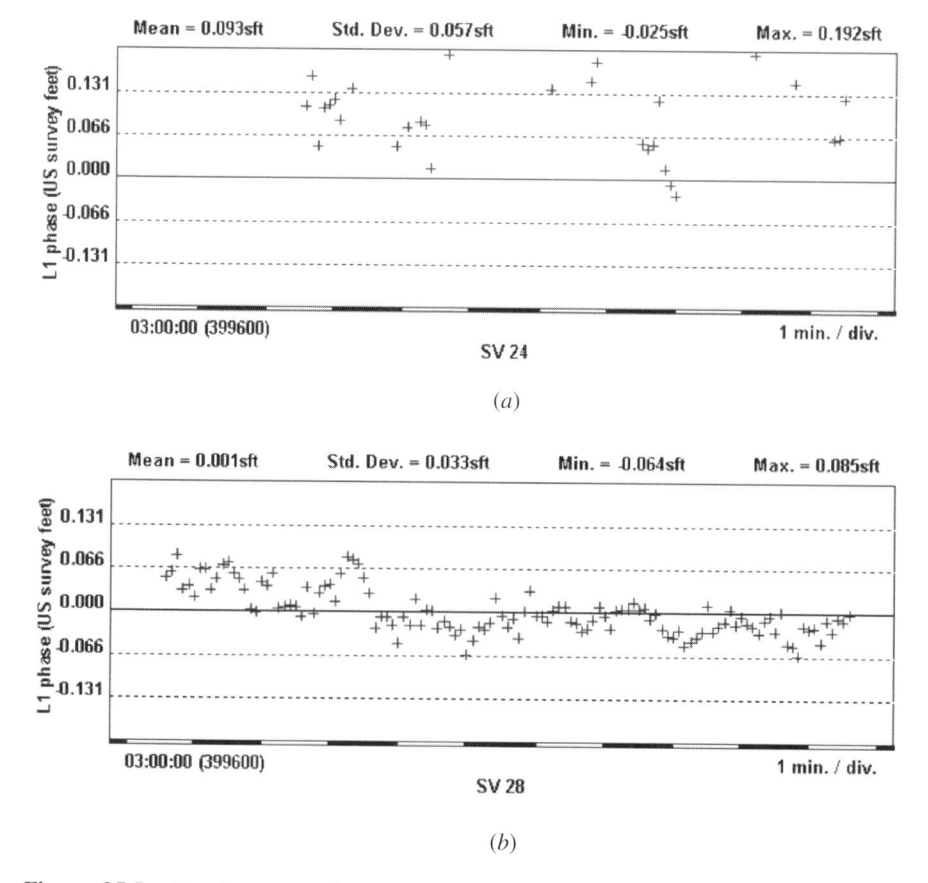

Figure 25.5 Pseudorange residual plots from satellites 24 (a) and 28 (b) versus time.

of plots are typically seen when the broadcast ephemeris is used in the reduction. The broadcast ephemeris is a near-future prediction of the location of the satellite. For various reasons, satellites often stray from their predicted paths. When this happens, the coordinates for the satellite's position at the time of broadcast are in error. This problem results in plots like Figure 25.5(a), where most of the residuals typically have one sign indicating a skewed data set. This problem can be corrected by downloading and processing with a precise ephemeris. In general, the best solutions can always be obtained with a precise ephemeris. Since the ultrarapid ephemeris is available within a few hours of data collection, it is wise to wait for this ephemeris before processing the data.

Figure 25.5 also depicts a problem with real-time kinematic GPS (RTK-GPS) surveys. That is, by its very nature, the broadcast ephemeris must be used in processing the data. Thus, the option of downloading a precise ephemeris is not available. RTK-GPS has its place in stakeout and mapping surveys. However, it should never be used to establish anything other than very low-order control. The RTK-GPS survey is the equivalent of a radial total station survey, leaving only minimal checks in the quality of the data. Since all GPS surveys can be seriously affected by solar activity, the RTK-GPS procedure leaves the surveyor open to poor results without any indication at the receiver.

25.5 USE OF STATISTICAL BLUNDER DETECTION

When more than one blunder exists in a given data set, the statistical blunder detection methods presented in Chapter 21 help isolate the offending observations. These methods along with graphical techniques, were presented thoroughly in that chapter.

The problem of having multiple blunders usually exists in large control surveys. When adjusting large data sets, it is wise to break the data into smaller subsets that can be checked for blunders before attempting the larger adjustment. By breaking a larger adjustment into smaller subsets of data, data problems can often be isolated, corrected, or removed from the observations.

GPS provides a similar example. When performing a large GPS adjustment, it is often wise initially to download and process GPS data by session. Since a GPS campaign typically involves several observational sessions, trying to combine all sessions into one large adjustment generally takes more time than individually processing the data by session. Processing GPS data by session simplifies the process of isolating poor and trivial baselines. After processing each session in a GPS survey satisfactorily, a combined network adjustment can be attempted and analyzed for blunders. It may be important at this point to reanalyze and possibly reprocess baselines to obtain the best solution. Again, statistical blunder detection will help analyze possible blunders in baselines. If the observation sessions are long enough, it is possible to eliminate data from the session to remove possible problems caused by

PROBLEMS

25.1 Following an adjustment, how is the term *large* defined in relation to residuals?

25.2 What are the basic concepts of a normal distribution, and how are these concepts used to analyze the results of a least squares adjustment?

25.3 An adjustment has 20 observations. Only nine of these residuals are negative in sign. Should this cause concern during the analysis?

25.4 Discuss why the goodness-of-fit test is not always a reliable indicator of a blunder in data.

25.5 What does it mean to "fail on the good side" of the χ^2 test?

25.6 Discuss the importance of a proper stochastic model.

25.7 Discuss why the scaling of the stochastic model does not affect the adjusted parameters.

25.8 What is the possible reason for the residual plot in Figure 25.5a?

25.9 Discuss the importance of collecting more than the minimum amount of data in a GPS survey.

25.10 Why should the RTK-GPS method not be used to establish high-order control stations?

25.11 Why should GPS baseline vectors be processed by session rather than by job?

25.12 A GPS baseline vector determined by the rapid-static method (5 mm + 1 ppm) has a length of about 5.5 km. The instrument setups were estimated to be within ± 0.003 m of the true station location. Its combined *XYZ* residual is 2.5 cm. At a 95% confidence level, should this residual be considered too large if the number redundancies in the adjustment is 15?

In Problems 25.13 through 25.20, analyze the results of the least squares adjustment, indicating if there are any areas of concern in the data or stochastic model. When possible, isolate and remove questionable data and stochastic models. After the changes, rerun the adjustment and compare the results. (*Note:* Not all of the problems have questionable data.)

25.13 Problem 12.1
25.14 Problem 12.13
25.15 Problem 13.15
25.16 Problem 15.9
25.17 Problem 16.7
25.18 Problem 16.9
25.19 Problem 16.11
25.20 Problem 17.8

CHAPTER 26

COMPUTER OPTIMIZATION

26.1 INTRODUCTION

Large amounts of computer time and storage requirements can be used when performing least squares adjustments. This is due to the fact that as the problems become more complex, the matrices become larger, and the storage requirements and time consumed in doing numerical operations both grow rapidly. As an example, in analyzing the storage requirements of a 25-station horizontal least squares adjustment that has 50 distance and 50 angle observations, the coefficient matrix would have dimensions of 100 rows and 50 columns. If this adjustment were done in double precision,[1] it would require 40,000 bytes of storage for the coefficient matrix alone. The weight matrix would require an additional 80,000 bytes of storage. Also, at least two additional intermediate matrices[2] must be formed in computing the solution. From this example it is easy to see that large quantities of computer time and computer memory can be required in least squares adjustments. Thus, when writing least squares software, it is desirable to take advantage of some storage and computing optimization techniques. In this chapter some of these techniques are described.

26.2 STORAGE OPTIMIZATION

Many matrices used in a surveying adjustment are large but sparse. Using the example above, a single row of the coefficient matrix for a distance obser-

---

[1] The storage requirements of a double-precision number is 8 bytes.
[2] The intermediate matrices developed are A^T and A^TW or J^T and J^TW, depending on whether the adjustment is linear or nonlinear.

vation would require a 50-element row for its four nonzero elements. In fact, the entire coefficient matrix is very sparsely populated by nonzero values. Similarly, the normal matrix is always symmetric, and thus nearly half its storage is used by duplicate entries. It is relatively easy to take advantage of these conditions to reduce the storage requirements.

For the normal matrix, only the upper or lower triangular portion of the matrix need be saved. In this storage scheme, the two-dimensional matrix is saved as a vector. An example of a 4 × 4 normal matrix is shown in Figure 26.1. Its upper and lower triangular portions are shown separated and their elements numbered for reference. The vector on the right of Figure 26.1 shows the storage scheme. This scheme eliminates the need to save the duplicate entries of the normal matrix but requires some form of *relational mapping* between the original matrix indices of row i and column j and the vector's index of row i. For the upper triangular portion of the matrix, it can be shown that the vector index for any (i,j) element is computed as

$$\text{Index}(i,j) = \tfrac{1}{2}[j \times (j-1)] + i \qquad (26.1)$$

Using the function given in Equation (26.1), the vector index for element (2,3) of the normal matrix would be computed as

$$\text{Index}(2,3) = \tfrac{1}{2}[3 \times (3-1)] + 2 = 5$$

An equivalent mapping formula for the lower triangular portion of the matrix is given as

$$\text{Index}(i,j) = \tfrac{1}{2}[i \times (i-1)] + j \qquad (26.2)$$

Equation (26.1) or (26.2) can be used to compute the storage location for each element in the upper or lower part of the normal matrix, respectively. Of course, it requires additional computational time to map the location of each element every time it is used. One method of minimizing computing

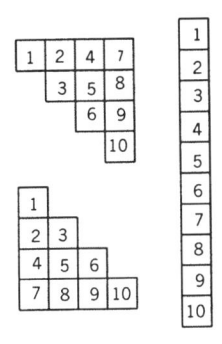

Figure 26.1 Normal matrix.

time is to utilize a *mapping table*. The mapping table is an integer vector that identifies the storage location for only the initial element of a row or column. To reduce total computational time, the indices are stored as one number less than their actual value. The example matrix stored in Figure 26.1 would have a relational mapping table of

$$VI^T = [0\ 1\ 3\ 6\ 10]$$

Although the storage location of the first element is 1, it can be seen above that it is identified as 0. Similarly, every other mapping index is one number less than its actual location in the vector, as shown in Figure 26.1. The mapping vector is found using the pseudocode shown in Table 26.1.

By using this type of mapping table, the storage location for the first element of any upper triangular column is computed as

$$\text{Index}(i,j) = VI(j) + i \qquad (26.3)$$

Using the mapping table, the storage location for the (2,3) element is

$$\text{Index}(2,3) = VI(3) + 2 = 3 + 2 = 5$$

The equivalent formula for the lower triangular portion is

$$\text{Index}(i,j) = VI(i) + j \qquad (26.4)$$

Obviously, this method of indexing an element's position requires additional storage in a mapping table, but the additional storage is offset by the decreased computational time that is created by the operations equation (26.1) or (26.2). Since the mapping table replaces the more time-consuming multiplication and division operations of Equations (26.1) and (26.2), it is prudent to use this indexing method whenever memory is available. Although every compiler and machine are different, a set of times to compare indexing methods are shown in Table 26.2. The best times were achieved by minimizing both the number of operations and calls to the mapping function. Thus, in method C the mapping table was created to be a global variable in the program, and the table was accessed directly as needed. For comparison, the time to access the matrix elements directly using the loop shown in method D was 1.204 seconds. Thus, method C uses only slightly more time than method D

TABLE 26.1 Creation of a Mapping Table

$VI(1) = 0$
For i going from 2 to the number of unknowns: $VI(i) = VI(i-1) + i - 1$

TABLE 26.2 Comparison of Indexing Methods

Method	Time (sec)	Extra Storage
A	4.932	0
B	3.937	2 bytes per element
C	1.638	2 bytes per element
D	1.204	Full matrix

Function A: Index = j(j − 1)/2 + i
Function B: Index = VI(j) + i;

Method A: Using function A
  i from 1 to 1000
    j from i to 1000
      k = Index(i,j)
Method B: Using function B
  i from 1 to 1000
    j from i to 1000
      k = Index(i,j)
Method C: Direct access to mapping table
  i from 1 to 1000
    j from i to 1000
      k = VI(j) + I
Method D: Direct matrix element access.
  i from 1 to 1000
    j from 1 to 1000
      k = A[i,j]

but helps avoid the additional memory that is required by the full matrix required for method *D*. Meanwhile, method *A* is slower than method *C* by a factor of 3. Since the matrix elements must be accessed repeatedly during a least squares solution, this difference can be a significant addition to the total solution time.

The additional overhead cost of having the mapping table is minimized by declaring it as an integer array so that each element requires only 2 bytes of storage. In the earlier example of 25 stations, this results in 100 bytes of additional storage for the mapping table. Speed versus storage is a decision that every programmer will face continually.

26.3 DIRECT FORMATION OF THE NORMAL EQUATIONS

The basic weighted observation equation form is $WAX = WL + WV$. The least squares solution for this equation is $X = (A^TWA)^{-1}(A^TWL) = N^{-1}C$, where A^TWA is the normal equations matrix N, and A^TWL is the constants matrix C. If the weight matrix is diagonal, formation of the A, W, L, and A^TW matrices is unnecessary when building the normal equations, and the storage requirements for them can be eliminated. This is accomplished by forming the normal matrix directly from the observations. The tabular method in Section 11.8 showed the feasibility of this method. Notice in Table 11.2 that the contribution of each observation to the normal matrix is computed individually and, subsequently, added. This shows that there is no need to form the coefficient, constants, weight, or any intermediate matrices when deriving the normal matrix. Conceptually, this is developed as follows:

Step 1: Zero the normal and constants matrix.
Step 2: Zero a single row of the coefficient matrix.
Step 3: Based on the values of a single row of the coefficient matrix, add the proper value to the appropriate normal and constant elements.
Step 4: Repeat steps 3 and 4 for all observations.

This procedure works in all situations that involve a diagonal weight matrix. Procedural modifications can be developed for weight matrices with limited correlation between the observations. Computer algorithms in BASIC, FORTRAN, C, and Pascal for this method are shown in Table 26.3.

26.4 CHOLESKY DECOMPOSITION

Having formed only a triangular portion of the normal matrix, a Cholesky decomposition can be used to greatly reduce the time needed to find the solution. This procedure takes advantage of the fact that the normal matrix is always a positive definite[3] matrix. Due to this property, it can be expressed as the product of a lower triangular matrix and its transpose; that is,

$$N = LU = LL^T = U^T U$$

where L is a lower triangular matrix of the form

$$L = \begin{bmatrix} l_{11} & 0 & 0 & \cdots & 0 \\ l_{21} & l_{22} & 0 & \cdots & 0 \\ l_{31} & l_{32} & l_{33} & & 0 \\ \vdots & \vdots & \vdots & \ddots & 0 \\ l_{n1} & l_{n2} & l_{n3} & \cdots & l_{nn} \end{bmatrix}$$

When the normal matrix is stored as a lower triangular matrix, it can be factored using the following procedure. For $i = 1, 2, \ldots$, number of unknowns, compute

$$l_{ii} = (l_{ii} - \sum_{k=1}^{i-1} l_{ik}^2)^{1/2} \tag{26.5}$$

For $j = i + 1, i + 2, \ldots$, number of unknowns, compute

$$l_{ji} = \frac{l_{ji} - \sum_{k=1}^{i-1} l_{ij} l_{jk}}{l_{ii}}$$

The procedures above are shown in code form in Table 26.4.

---
[3] A positive definite square matrix, A, has the property that $X^T A X > 0$ for all nonzero vectors, X.

TABLE 26.3 Algorithms for Building the Normal Equations Directly from Their Observations

BASIC Language:
For i = 1 to unknown
  ix = VI(i): ixi = ix + i
  N(ixi) = N(ixi) + A(i)∧2 * W(i)
  C(i) = C(i) + A(i)*W(i)*L(i)
  For j = i+1 to unknown
    N(ix+j) = N(ix+j) + A(i)*W(i)*A(j)
  Next j
Next I

C Language:
for (i=1; i<=unknown; i++) {
  ix = VI[i]; ixi = ix + i;
  n[ixi] = n[ixi] + a[i]*a[i] * w[i];
  c[i] := c[i] + a[i]*w[i]*l[i];
  for (j=i+1; j<=unknown; j++)
    n[ix+j] = n[ix+j] + a[i]*w[i]*a[j];
}// for i

FORTRAN Language:
Do 100 i = 1, unknown
  ix = VI(i)
  ixi = ix + i
  N(ixi) = N(ixi) + A(i)**2 * W(i)
  C(i) = C(i) + A(i)*W(i)*L(i)
  Do 100 j = i+1, unknown
    N(ix+j) = N(ix+j) + A(i)*W(i)*A(j)
100 Continue

Pascal Language:
For i := 1 to unknown do begin
  ix := VI[i]; ixi := ix + i;
  N[ixi] := N[ixi] + Sqr(A[i]) * W[i];
  C[i] := C[i] + A[i]*W[i]*L[i];
  For j := i+1 to unknown do
    N[ix+j] := N[ix+j] + A[i]*W[i]*A[j]
End; {for i}

TABLE 26.4 Computer Algorithms for Computing Cholesky Factors of a Normal Matrix

BASIC Language:
```
FOR i = 1 TO unknown
  ix = VI(i): ixi = ix + i: s = 0#
  FOR k = 1 TO i - 1
    s = s + N(ix + k) ^ 2: NEXT k
  N(ixi) = SQR(N(ixi) - s)
  FOR j = i + 1 TO unknown
    s = 0#: jx = VI(j)
    FOR k = 1 TO i - 1
      s = s + N(jx + k) * N(ix + k): NEXT k
    N(jx + i) = (N(jx + i) - s)/N(ixi)
  NEXT j
NEXT i
```

C Language:
```
for (i=1; i<=unknown; i++) {
  ix = VI[i]; ixi = ix+i; s = 0.0;
  for (k=1; k<i; k++)
    s = s + n[ix+k]*n[ix+k];
  n[ixi] = sqrt(n[ixi] - s);
  for (j=i+1; j<=unknown; j++) {
    s = 0; jx = VI[j];
    for (k=1; k<i; k++)
      s = s + n[jx+k] * n[ix+k];
    n[jx+i] = (n[jx+i] - s) / n[ixi];
  } // for j
} // for I
```

FORTRAN Language:
```
      Do 30 i = 1,Unknown
        ix = VI(i)
        i1 = i - 1
        S = 0.0
        Do 10 k=1, i1
10        s = s + N(ix+k)**2
        N(ix+i) = Sqrt(N(ix+i)-s)
        Do 30 j = i+1, Unknown
          s = 0.0
          Do 20 k = 1, i1
20          s = s + N(VI[j]+k) * N(ix+k)
          N(VI[j]+i) = (N(VI[j]+i) - s)/N(ix+i)
30    Continue
```

Pascal Language:
```
For i := 1 to unknown do Begin
  ix := VI[i]; ixi := ix+i; S := 0;
  For k := 1 to Pred(i) do
    S := S + Sqr(N[ix+k]);
  N[ixi] := Sqrt(N[ixi] - S);
  For j := Succ(i) to unknown do Begin
    S := 0; jx := VI[j];
    For k := 1 to Pred(i) do
      S := S + N[jx+k]*N[ix+k];
    N[jx+i] := (N[jx+i] - S) / N[ixi]
  End; {for j}
End; {for i}
```

26.5 FORWARD AND BACK SOLUTIONS

Being able to factor the normal matrix into triangular matrices has the advantage that the matrix solution can be obtained without the use of an inverse. The equivalent triangular matrices representing the normal equations are

$$(A^TWA)X = NX = LUX = LL^TX$$

$$= \begin{bmatrix} l_{11} & 0 & 0 & \cdots & 0 \\ l_{21} & l_{22} & 0 & \cdots & 0 \\ l_{31} & l_{32} & l_{33} & & 0 \\ \vdots & \vdots & \vdots & \ddots & 0 \\ l_{n1} & l_{n2} & l_{n3} & \cdots & l_{nn} \end{bmatrix} \begin{bmatrix} l_{11} & l_{21} & l_{31} & \cdots & l_{n1} \\ 0 & l_{22} & l_{32} & \cdots & l_{n2} \\ 0 & 0 & l_{33} & \cdots & l_{n3} \\ \vdots & \vdots & & \ddots & \vdots \\ 0 & 0 & 0 & 0 & l_{nn} \end{bmatrix} \begin{bmatrix} x_1 \\ x_2 \\ x_3 \\ \vdots \\ x_n \end{bmatrix}$$

$$= \begin{bmatrix} c_1 \\ c_2 \\ c_3 \\ \vdots \\ c_n \end{bmatrix} = C \qquad (26.6)$$

Equation (26.6) can be rewritten as $LY = C$, where $L^TX = Y$. From this, the solution for Y can be found by taking advantage of the triangular form of L. This is known as a *forward substitution*. Steps involved in forward substitution are as follows:

Step 1: Solve for y_1 as $y_1 = c_1/l_{11}$.
Step 2: Substitute this value into row 2, and compute y_2 as $y_2 = (c_2 - l_{21}y_1)/l_{22}$.
Step 3: Repeat this procedure until all values for y are found using the algorithm $y_i = (c_i - \sum_{k=1}^{i-1} l_{ik}y_k)/l_{ii}$.

Having determined Y, the solution for the matrix system $L^TX = Y$ is computed in a manner similar to that above. However, this time the solution starts at the lower right corner and proceeds up the matrix L^T. Called a *backward substitution*, this is done with the following steps:

Step 1: Compute x_n as $x_n = y_n/l_{nn}$.
Step 2: Solve for x_{n-1} as $x_{n-1} = (y_{n-1} - l_{mn}x_n)/l_{(n-1)(n-1)}$.
Step 3: Repeat this procedure until all unknowns are computed using the algorithm $x_k = (y_k - \sum_{j=k+1}^{n} l_{kj}y_j)/l_{kk}$.

In this process, once the original values in the constant matrix are accessed and changed, they are not needed again and the original C matrix can be overwritten with Y and X matrices so that the entire process requires no additional storage. In fact, this method of solution also requires fewer operations than solving $X = (A^T WA)^{-1} A^T WL$ directly. Table 26.5 lists the computer codes for these algorithms.

26.6 USING THE CHOLESKY FACTOR TO FIND THE INVERSE OF THE NORMAL MATRIX

If necessary, the original normal matrix inverse can be found with the Cholesky factor. To derive this matrix, the inverse of the Cholesky factor is computed. The normal matrix inverse is the product of this inverse times its

TABLE 26.5 Computer Algorithms for Forward and Backward Substitutions

BASIC Language:
```
Rem Forward Substitution
For i = 1 to Unknown
   ix = VI(i)
   C(i) = C(i) / N(ix+i)
   For j = i+1 to Unknown
      C(j) = C(j) - N(ix+j) * C(i)
   Next j
Next i
Rem Backward Substitution
For i = Unknown to 1 Step -1
   ix = VI(i)
   For j = Unknown To i+1 Step -1
      C(i) = N(ix+j) * C(j)
   Next j
   C(i) = C(i) / N(ix+i)
Next i
```

C Language:
```
//Forward Substitution
for (i=1; i<=unknown; i++){
   ix = VI[i];
   c[i] = c[i]/n[ix+i];
   for (j=i+1; j<=unknown; j++){
      c[j] = c[j] - n[ix+j]*c[i];
   } //for j
} //for i
//backward substitution
for (i=unknown; i>=1; i--){
   ix = VI[i];
   for (j=unknown; j>=i+1; j--){
      c[i] = c[i] - n[ix+j] * c[j];
   } //for j
   c[i] = c[i]/n[ix+i];
} // for i
```

FORTRAN Language:
```
C Forward Substitution
   Do 100 i = 1,Unknown
      ix = VI(i)
      C(i) = C(i) / N(ix+i)
      Do 100 j = i+1, Unknown
         C(j) = C(j) - N(ix+j) *
C(i)
100 Continue

C Backward Substitution
   Do 110 i = Unknown, 1,-1
      ix = VI(i)
      Do 120 j = Unknown, i+1, -1
120      C(i) = C(i) - N(ix+j) * C(j)
      C(i) = C(i) / N(ix+i)
110 Continue
```

Pascal Language:
```
{Forward Substitution}
For i := 1 to Unknown Do Begin
   ix := VI[i];
   C[i] := C[i] / N[ix+i];
   For j := i+1 to Unknown Do
      C[j] := C[j] - N[ix+j] * C[i];
End; {for i}

{Backward Substitution}
For i := Unknown DownTo 1 do Begin
   ix := VI[i];
   For j := Unknown DownTo i+1 Do
      C[i] := C[i] - N[ix+j] * C[j];
   C[i] := C[i] / N[ix+i];
End; {for i}
```

transpose. The Cholesky factor inverse is determined using the algorithm in Table 26.6.

Code for this algorithm is shown in Table 26.7.

26.7 SPARENESS AND OPTIMIZATION OF THE NORMAL MATRIX

In the least squares adjustment of most surveying and photogrammetry problems, it is known that certain locations in the normal matrix will contain zeros. The network shown in Figure 26.2 can be used to demonstrate this fact. In that figure, assume that distances and angles were observed for every line and arc, respectively. The following observations are made about the network's connectivity. Station 1 is connected to stations 2 and 8 by distance observations and is also connected to stations 2, 3, and 8 by angles. Notice that its connection to station 3 is due to angles turned at both 2 and 8. Since these angles directly connect stations 1 and 2, it follows that if the coordinates of 3 change, so will the coordinates of 1, and thus with respect to station 1, the normal matrix can be expected to have nonzero elements corresponding to stations 1, 2, 3, and 8. Conversely, the positions corresponding to stations 4, 5, 6, and 7 will have zeros.

Using this analysis with station 3, because it is connected to stations 1, 2, 4, 7, and 8 by angles, zero elements can be expected in the normal matrix corresponding to stations 5 and 6. Similar analyses can be made for each station. The resulting symbolic normal matrix representation is shown in Figure 26.3.

A process known as *reordering the unknowns* can minimize both storage and computational time. Examine the matrix shown in Figure 26.4, which results from placing the unknowns of Figure 26.3 in the order 6, 5, 4, 7, 3, 1, 2, and 8. This new matrix has its nonzero elements immediately adjacent to the diagonal elements, and thus the known zero elements are grouped together and appear in the leftmost columns of their respective matrix rows. By modifying the mapping table, storing the known zero elements of the original matrix in Figure 26.3 can be avoided, as is shown in the column matrix of Figure 26.4.

This storage scheme requires a mapping table that provides the storage location for the first nonzero element of each row. Since the diagonal elements are always nonzero, they can be stored in a separate column matrix without

TABLE 26.6 Pseudocode for an Algorithm Computing the Inverse of a Cholesky Decomposed Matrix

```
for i going from the number of unknowns down to 1
    for k going from number of unknowns down to i + 1
        for j going from i+1 to k, sum the product N(i, j) * N(j, k)
        N(i, k) = -S / N(i, i)
    N(i, i) = 1/N(i, i)
```

TABLE 26.7 Computer Algorithms to Find the Inverse of a Cholesky Factored Matrix

BASIC Language:
{Inverse}
For i = Unknown To 1 Step −1
  ix = VI(i): ixi = ix + i
  For k = Unknown To i+1 Step −1
    S = 0!
    For j = i+1 To k
      S = S + N(ix+j) * N(VI(j)+k)
    Next j
    N(ix+k) = −S / N(ixi)
  Next k
  N(ixi) = 1.0 / N(ixi)
Next i

{Inverse * Transpose of Inverse}
For j = 1 to Unknown
  ixj = VI(j)
  For k = j to Unknown
    S = 0!
    For i = k to Unknown
      S = S + N(VI(k)+i) * N(ixj+i)
    Next i
    N(ixj+k) = S
  Next k
Next j

C Language:
{Inverse}
for (i=unknown; i>=1; i−−) {
  ix = VI[i]; ixi = ix + i;
  for(k=unknown; k>=i+1; k−−) {
    s = 0.0;
    for (j=i+1; j<=k; j++)
      s = s + n[ix+j] * n[VI[j]+k];
    n[ix+k] = −s / n[ixi];
  } // for k
  n[ixi] = 1.0 / n[ixi];
} // for i

{inverse * transpose of inverse}
for (j=1; j<=unknown; j++) {
  ixj = VI[j];
  for (k=j; k<=unknown; k++) {
    s = 0.0;
    For (i=k; i<=unknown; i++)
      s = s + n[VI[k]+i] * n[ixj+i];
    n[ixj+k] = s
  } // for k
} //for j

Fortran Language:
C Inverse
  Do 10 i = Unknown, 1,−1
    ix = VI(i)
    ixi = ix + i
    Do 11 k = Unknown, i+1,−1
      S = 0.0
      Do 12 j = i+1,k
12   S = S + N(ix+j) * N(VI(j)+k)
      N(ix+k) = −S / N(ixi)
11   Continue
    N(ixi) = 1.0 / N(ixi)
10 Continue

C Inverse * Transpose of Inverse}
  Do 20 j = 1, Unknown
    ixj = VI(j)
    Do 21 k = j, Unknown
      S = 0.0
      Do 22 i = k,Unknown
22   S = S + N(VI(k)+i) * N(ixj+i)
      N(ixj+k) = S
21   Continue
20 Continue

Pascal Language:
{Inverse}
For i := Unknown DownTo 1 do Begin
  ix := VI[i]; ixi := ix + i;
  For k := Unknown DownTo i+1 Do Begin
    S := 0.0;
    For j := i+1 To k Do
      S := S + N[ix+j] * N[VI[j]+k];
    N[ix+k] := −S / N[ixi];
  End; {For k}
  N[ixi] := 1.0 / N[ixi];
End; {For i}

{Inverse * Transpose of Inverse}
For j := 1 to Unknown Do Begin
  ixj := VI[j];
  For k := j to Unknown Do Begin
    S := 0.0;
    For i := k to Unknown Do
      S := S + N[VI[k]+i] * N[ixj+i];
    N[ixj+k] := S
  End; {For k}
End; {For j}

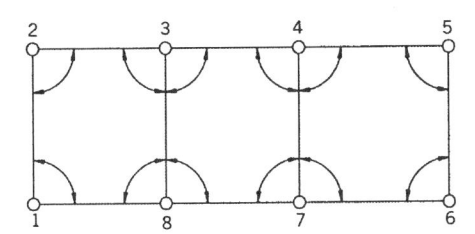

26.7 SPARENESS AND OPTIMIZATION OF THE NORMAL MATRIX 515

Figure 26.2 Series of connected traverses.

loss of efficiency. Ignoring the fact that each station has two unknowns, the mapping of the first element for each row is 1, 1, 2, 4, 7, 9, 10, 12, and 17. This mapping scheme enables finding the starting position for each row and allows for the determination of the off-diagonal length of the row. That is, row 1 starts at position 1 of the column matrix, but has a length of zero (1 − 1), indicating that there are no off-diagonal elements in this row. Row 2 starts at position 1 of the column matrix and has a length of (2 − 1) = 1. Row 5 starts at 7 and has a length of (9 − 7) = 2. By optimizing for matrix sparseness and symmetry, only 24 (16 + 8) elements of the original 64-element matrix need be stored.

This storage savings, when exploited, can also result in a savings of computational time. To understand this, first examine how the Cholesky factorization procedure processes the normal matrix. For a lower triangular matrix factorization process, Figure 26.5 shows the manner in which elements are accessed. Notice that when a particular column is modified, the elements to the left of this column are used. Also notice that no rows above the corresponding diagonal element are used.

To see how the reorganized sparse matrix can be exploited in the factorization process, the processing steps must be understood. In Figure 26.4 the

	1	2	3	4	5	6	7	8
1	X	X	X	0	0	0	0	X
2	X	X	X	0	0	0	0	X
3	X	X	X	X	0	0	X	X
4	0	0	X	X	X	X	X	X
5	0	0	0	X	X	X	X	0
6	0	0	0	X	X	X	X	0
7	0	0	X	X	X	X	X	X
8	X	X	X	X	0	0	X	X

Figure 26.3 Normal matrix.

516 COMPUTER OPTIMIZATION

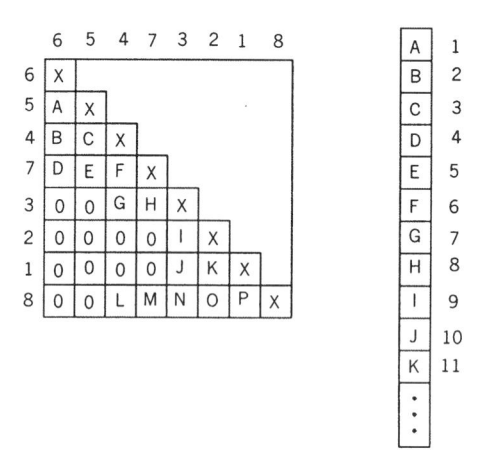

	6	5	4	7	3	2	1	8
6	X							
5	A	X						
4	B	C	X					
7	D	E	F	X				
3	0	0	G	H	X			
2	0	0	0	0	I	X		
1	0	0	0	0	J	K	X	
8	0	0	L	M	N	O	P	X

A	1
B	2
C	3
D	4
E	5
F	6
G	7
H	8
I	9
J	10
K	11
⋮	⋮

Figure 26.4 Reordered matrix.

first two elements of row 5 are known zeros and each summation loop in the factorization process should start with the third column. If the zeros in the off-diagonals are rearranged such that they occur in the leftmost columns of the rows, it is possible to avoid that portion of the operations for the rows. Notice how the rows near the lower portion of the matrix are optimized to minimize the computational effort. A savings of eight multiplications is obtained for computations of column 5. In a large system, the cost savings of this optimization technique can be enormous. These techniques are discussed completely by George and Lui (1981). A comparison of the operations performed when computing column 5 of Figure 26.4 in a Cholesky decomposition of both a full and an optimized solution routine are shown in Table 26.8. Note that computations for rows 6 and 7 do not exist since these rows had known zeros in the columns before column 5 after the reordering process. If only the more time-consuming multiplication and division operations are counted, the optimized solution requires five operations compared to 19 in the nonoptimized (full) solution.

Several methods have been developed to optimize the ordering of the unknowns. Two of the better known reordering schemes are the *reverse Cuthill–Mckee* and the *banker's algorithm*. Both of these algorithms reorder stations based on their connectivity. The example of Figure 26.2 will be used to

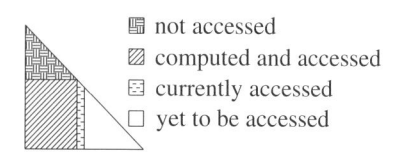

▦ not accessed
▨ computed and accessed
⊠ currently accessed
☐ yet to be accessed

Figure 26.5 Computation of a Cholesky factor.

26.7 SPARENESS AND OPTIMIZATION OF THE NORMAL MATRIX

TABLE 26.8 Comparison of the Number of Operations in Computing One Column

Full Operations	Optimized Operations
1. $s = n_{51}^2 + n_{52}^2 + \cdots + n_{54}^2$	$s = n_{53}^2 + n_{54}^2$
2. $n_{55} = $ square root of $(n_{55} - s)$	$n_{55} = $ square root of $(n_{55} - s)$
3. $j = 6$	$j = 6$
(a) $s = n_{61}*n_{51} + n_{62}*n_{52} + \cdots + n_{64}*n_{54}$	no operations necessary
(b) $n_{65} = (n_{65} - s)/n_{55}$	$j = 7$
$j = 7$	no operations necessary
(a) $s = n_{71}*n_{51} + n_{72}*n_{52} + \cdots + n_{74}*n_{54}$	$j = 8$
(b) $n_{75} = (n_{75} - s)/n_{55}$	$s = n_{83}*n_{53} + n_{84}*n_{54}$
$j = 8$	
(a) $s = n_{81}*n_{51} + n_{82}*n_{52} + \cdots + n_{84}*n_{54}$	
(b) $n_{85} = (n_{85} - s)/n_{55}$	

demonstrate the banker's algorithm. In this example, a connectivity matrix is first developed. This is simply a list of stations that are connected to a given station either by an angle or by distance observations. For instance, station 1 is connected to stations 2 and 8 by distances, and additionally to station 3 by angles. The complete connectivity matrix of Figure 26.2 is given in the first two columns of Table 26.9. In addition to the connectivity, the algorithms also need to know the number of connected points for each station. This is known as the *station's degree*.

To start the reordering, select the station with the lowest degree. In this example that could be station 1, 2, 5, or 6. For demonstration purposes, begin with station 6. This station is now removed from each station's connectivity matrix, and thus stations 4, 5, and 7 have their degrees reduced by one. This is shown in column 6. Now select the next station with the smallest degree.

TABLE 26.9 Connectivity Matrix

Station	Connectivity	Degree	6	5	4	7	3	1	2
1	2,3,8	3	3	3	3	3	2	—	—
2	1,3,8	3	3	3	3	3	2	1	—
3	1,2,4,7,8	5	5	5	4	3	—	—	—
4	3,5,6,7,8	5	4	3	—	—	—	—	—
5	4,6,7	3	2	—	—	—	—	—	—
6	4,5,7	3	—	—	—	—	—	—	—
7	3,4,5,6,8	5	4	3	2	—	—	—	—
8	1,2,3,4,7	5	5	5	4	3	2	1	0

Since station 5 had its degree reduced to 2 when station 6 was selected, it now has the smallest degree. This process is continued; that is, the next station with the smallest degree is selected and is removed from the connectivity of the remaining stations. If two or more stations have the smallest degree, the most recently changed station with the smallest degree is chosen. In this example, this happened after choosing station 5. Station 4 was selected over stations 1 and 2 simply because it just had its degree reduced. Although station 7 qualified equally to station 4, station 4 was higher on the list and was selected for that reason. This happened again after selecting station 7. Station 3 was the next choice since it was the first station reduced to a degree of 3. Note that the final selection of station 8 was omitted from the list.

A similar optimized reordering can be found by starting with station 1, 2, or 5. This is left as an exercise. For a computer algorithm of this reordering, the reader should refer to the NOAA Technical Memorandum NGS 4 by Richard Snay (1976); the computer algorithm for the *reverse Cuthill–McKee* is given by George and Lui (1981).

PROBLEMS

26.1 Explain why computer optimization techniques are necessary for doing large least squares adjustments.

26.2 When using double precision, how much computer memory is required in forming the J, W, K, J^T, J^TW, J^TWK, and J^TWJ matrices for a 20-station adjustment having 50 total observations?

26.3 If the normal matrix of Problem 26.2 were stored in lower triangular form and computed directly from the observations, how much storage would be required?

26.4 Discuss how computational savings are created when solving the normal equations by Cholesky factorization and substitution.

26.5 Write a comparison of an operations table similar to that in Table 26.8 for column 3 of the matrix shown in Figure 26.4.

26.6 Repeat Problem 26.5 for column 6.

26.7 Using the station reordering algorithm presented in Section 26.7, develop a connectivity matrix and reorder the stations when starting with station 1. Draw the normal matrix of the newly reordered station, which is similar to the sketch shown in Figure 26.4.

26.8 Repeat Problem 26.7, starting the reordering with station 2.

26.9 Repeat Problem 26.7, starting the reordering with station 5.

Programming Problems

26.10 Develop a least squares program similar to those developed in Chapter 16 that builds the normal equations directly from the observations and uses a Cholesky factorization procedure to find the solution.

26.11 Develop a Mathcad worksheet that will solve Problem 26.10.

APPENDIX A

INTRODUCTION TO MATRICES

A.1 INTRODUCTION

Matrix algebra provides at least two important advantages: (1) it enables reducing complicated systems of equations to simple expressions that can be visualized and manipulated more easily, and (2) it provides a systematic, mathematical method for solving problems that is well adapted to computers. Problems are frequently encountered in surveying, geodesy, and photogrammetry that require the solution of large systems of equations. This book deals specifically with the analysis and adjustment of redundant measurements that must satisfy certain geometric conditions. This frequently results in large equation systems which, when solved according to the least squares method, yield most probable estimates for adjusted observations and unknown parameters. As will be demonstrated, matrix methods are particularly well suited for least squares computations, and in this book they are used for analyzing and solving these equation systems.

A.2 DEFINITION OF A MATRIX

A *matrix* is a set of numbers or symbols arranged in a square or rectangular array of m rows and n columns. The arrangement is such that certain defined mathematical operations can be performed in a systematic and efficient manner. As an example of a matrix representation, consider the following system of three linear equations involving three unknowns:

$$a_{11}x_1 + a_{12}x_2 + a_{13}x_3 = c_1$$
$$a_{21}x_1 + a_{22}x_2 + a_{23}x_3 = c_2 \quad \text{(A.1)}$$
$$a_{31}x_1 + a_{32}x_2 + a_{33}x_3 = c_3$$

In Equation (A.1), the a's are coefficients of the unknowns x's, and the c's are the constant terms. The system above can be represented in summation notation as

$$\sum_{i=1}^{3} a_{1i}x_i = c_1$$

$$\sum_{i=1}^{3} a_{2i}x_i = c_2$$

$$\sum_{i=1}^{3} a_{3i}x_i = c_3$$

It can also be represented in matrix form as

$$\begin{bmatrix} a_{11} & a_{12} & a_{13} \\ a_{21} & a_{22} & a_{23} \\ a_{31} & a_{32} & a_{33} \end{bmatrix} \begin{bmatrix} x_1 \\ x_2 \\ x_3 \end{bmatrix} = \begin{bmatrix} c_1 \\ c_2 \\ c_3 \end{bmatrix} \quad \text{(A.2)}$$

In turn, Equation (A.2) can be represented in compact matrix notation as

$$AX = C \quad \text{(A.3)}$$

In Equation (A.3), capital letters (A, X, and C) are used to denote a matrix or an array of numbers or symbols. From simplified equation (A.3), it is immediately obvious that matrix methods provide a compact shorthand notation convenient for handling a large system of equations.

A.3 SIZE OR DIMENSIONS OF A MATRIX

The size or dimension of a matrix is specified by its number of rows m and its number of columns n. Thus, matrix $_2D^3$ below is a 2×3 matrix; that is, there are two rows and three columns (i.e., $m = 2$, $n = 3$). Notice that the subscript indicates the number of rows and the superscript indicates the number of columns in the notation $_2D^3$.

$$_2D^3 = \begin{bmatrix} d_{11} & d_{12} & d_{13} \\ d_{21} & d_{22} & d_{23} \end{bmatrix} = \begin{bmatrix} 3 & 2 & 6 \\ 5 & 4 & 1 \end{bmatrix}$$

Also, the matrix

$$_3E^2 = \begin{bmatrix} e_{11} & e_{12} \\ e_{21} & e_{22} \\ e_{31} & e_{32} \end{bmatrix} = \begin{bmatrix} 7 & 1 \\ 4 & 3 \\ 2 & 8 \end{bmatrix}$$

is a 3 × 2 matrix.

Note that the position of an element in a matrix is defined by a double subscript and that a lowercase letter is used to designate any particular element within a matrix. Thus, $d_{23} = 1$ is in row 2 and column 3 of the D matrix above. In general, the subscript ij indicates an element's position in a matrix, where i represents the row and j the column.

A.4 TYPES OF MATRICES

Several different types of matrices exist as described below. Various symbols can be used to designate them, as illustrated.

1. *Column matrix.* The number of rows can be any positive integer, but the number of columns is 1.

$$A = \begin{bmatrix} 3 \\ -2 \\ 5 \end{bmatrix}$$

2. *Row matrix.* The number of columns can be any positive integer, but the number of rows is 1.

$$A = \begin{bmatrix} 6 & -4 & 2 \end{bmatrix}$$

3. *Rectangular matrix.* The number of rows and columns are m and n, respectively, where m and n are any positive integers.

$$A = \begin{bmatrix} 3 & 2 & 6 \\ 5 & 4 & 1 \end{bmatrix}$$

4. *Square matrix.* The number of rows equals the number of columns.

$$A = \begin{bmatrix} 4 & 2 & -5 \\ -7 & 3 & 4 \\ 6 & -1 & 9 \end{bmatrix}$$

A square matrix, for which the determinant is zero, is termed *singular*. If the determinant is nonzero, it is termed *nonsingular*. (The determinant of a matrix is discussed in Section B.3.)

5. *Symmetric matrix*. The matrix is mirrored about the main diagonal going from top left to bottom right (i.e., element a_{ij} = element a_{ji}). A symmetric matrix is always a square matrix.

$$A = \begin{bmatrix} 2 & -4 & 6 \\ -4 & 7 & 3 \\ 6 & 3 & 5 \end{bmatrix}$$

6. *Diagonal matrix*. Only the elements on the main diagonal are not zero. The diagonal matrix is always a square matrix.

$$A = \begin{bmatrix} 7 & 0 & 0 \\ 0 & -3 & 0 \\ 0 & 0 & 6 \end{bmatrix}$$

7. *Unit matrix*. This is a diagonal matrix with 1's along the main diagonal. It is also called an *identity matrix* and is usually identified by the symbol I.

$$I = \begin{bmatrix} 1 & 0 & 0 \\ 0 & 1 & 0 \\ 0 & 0 & 1 \end{bmatrix}$$

8. *Transpose of a matrix*. This is obtained by interchanging rows and columns (i.e., element a_{ij}^T = element a_{ji}). Thus, the dimensions of A^T are the reverse of the dimensions of A. If

$$A = \begin{bmatrix} 2 & 4 & 7 \\ 5 & 3 & 1 \end{bmatrix}$$

the transpose of A, denoted A^T, is

$$A^T = \begin{bmatrix} 2 & 5 \\ 4 & 3 \\ 7 & 1 \end{bmatrix}$$

A.5 MATRIX EQUALITY

Two matrices are said to be *equal* only when they are equal element by element. Thus, the two matrices must be the same size or have the same dimensions.

$$A = \begin{bmatrix} 1 & 7 & 6 \\ 4 & 3 & 2 \end{bmatrix} = B = \begin{bmatrix} 1 & 7 & 6 \\ 4 & 3 & 2 \end{bmatrix}$$

A.6 ADDITION OR SUBTRACTION OF MATRICES

Matrices can be added or subtracted, but to do so, they must have the same dimensions. If two matrices have equal dimensions, they are said to be *conformable* for addition or subtraction. In adding or subtracting matrices, elements from each unique row/column position of the two matrices are added or subtracted systematically and the sum or difference is placed in the same unique row/column location of the resulting matrix. The following example illustrates this procedure.

$$_2A^3 + {_2B^3} = \begin{bmatrix} 7 & 3 & -1 \\ 2 & -5 & 6 \end{bmatrix} + \begin{bmatrix} 1 & 5 & 6 \\ -4 & -2 & 3 \end{bmatrix} = \begin{bmatrix} 8 & 8 & 5 \\ -2 & -7 & 9 \end{bmatrix} = {_2C^3}$$

Assuming that two matrices are conformable for addition or subtraction, the following are true:

(a) $A + B = B + A$ (commutative law)
(b) $A + (B + C) = (A + B) + C$ (associative law)

A.7 SCALAR MULTIPLICATION OF A MATRIX

Matrices can be multiplied by a scalar (i.e., a *constant*). Let k be any scalar quantity; then

$$kA = Ak$$

The following are examples.

$$4 \times \begin{bmatrix} 3 & -1 \\ 2 & 6 \\ 4 & 7 \\ 5 & 3 \end{bmatrix} = \begin{bmatrix} 3 & -1 \\ 2 & 6 \\ 4 & 7 \\ 5 & 3 \end{bmatrix} \times 4 = \begin{bmatrix} 12 & -4 \\ 8 & 24 \\ 16 & 28 \\ 20 & 12 \end{bmatrix}$$

As illustrated above, each element of the matrix A is multiplied by the scalar k to obtain the elements of C. Note that $4 \times A = A + A + A + A = A \times 4$.

A.8 MATRIX MULTIPLICATION

If matrix A is to be *postmultiplied* by matrix B (i.e., the product of A × B determined), the number of columns in matrix A must equal the number of rows in matrix B. This is a basic requirement for matrix multiplication. When this condition is satisfied, A and B are said to be *conformable* for multiplication. The product C will have the same number of rows as A and the same number of columns as B. Thus, the following multiplications are possible:

$$_4A^2 \times {_2B^3} = {_4C^3}$$

$$_1A^3 \times {_3B^1} = {_1C^1}$$

$$_3A^3 \times {_3B^1} = {_3C^1}$$

These multiplications are not possible:

$$_2B^3 \times {_4A^2}$$

$$_6A^2 \times {_6B^3}$$

To demonstrate the process of matrix multiplication, consider the following example:

$$_2A^3 \times {_3B^2} = \begin{bmatrix} a_{11} & a_{12} & a_{13} \\ a_{21} & a_{22} & a_{12} \end{bmatrix} \begin{bmatrix} b_{11} & b_{12} \\ b_{21} & b_{22} \\ b_{31} & b_{32} \end{bmatrix} = \begin{bmatrix} c_{11} & c_{12} \\ c_{21} & c_{22} \end{bmatrix} = {_2C^2}$$

The elements c_{ij} of matrix C are the total sums obtained by multiplying each element in row i of matrix A successively by the elements in column j of matrix B and then summing these products. Thus, for the example above,

$$c_{11} = a_{11} b_{11} + a_{12} b_{21} + a_{13} b_{31} = \sum_{i=1}^{3} a_{1i} b_{i1}$$

$$c_{12} = a_{11} b_{12} + a_{12} b_{22} + a_{13} b_{32} = \sum_{i=1}^{3} a_{1i} b_{i2}$$

$$c_{21} = a_{21} b_{11} + a_{22} b_{21} + a_{23} b_{31} = \sum_{i=1}^{3} a_{2i} b_{i1}$$

$$c_{22} = a_{21} b_{12} + a_{22} b_{22} + a_{23} b_{32} = \sum_{i=1}^{3} a_{2i} b_{i2}$$

The process above is seen more easily with a numerical example.

$$\begin{bmatrix} 1 & 2 & 3 \\ 4 & 2 & 7 \end{bmatrix} \begin{bmatrix} 4 & 8 \\ 6 & 2 \\ 5 & 3 \end{bmatrix} = \begin{bmatrix} c_{11} & c_{12} \\ c_{21} & c_{22} \end{bmatrix} = \begin{bmatrix} 31 & 21 \\ 63 & 57 \end{bmatrix}$$

$$c_{11} = 1 \times 4 + 2 \times 6 + 3 \times 5 = 31$$

$$c_{12} = 1 \times 8 + 2 \times 2 + 3 \times 3 = 21$$

$$c_{21} = 4 \times 4 + 2 \times 6 + 7 \times 5 = 63$$

$$c_{22} = 4 \times 8 + 2 \times 2 + 7 \times 3 = 57$$

The student should now verify the matrix representation of Equations (A.1) and (A.2). Notice that the product of a unit matrix, I, and a conformable matrix, A (one with the same number of rows as I), equals the original matrix A. Thus,

$$_2I^2 \times {_2A_2} = \begin{bmatrix} 1 & 0 \\ 0 & 1 \end{bmatrix} \begin{bmatrix} 5 & 6 \\ 7 & 8 \end{bmatrix} = \begin{bmatrix} 5 & 6 \\ 7 & 8 \end{bmatrix} = {_2A^2}$$

Assuming that matrices A, B, and C are conformable for multiplication and in the order indicated, the following are true:

(c) $A(B+C) = AB + AC$ (first distributive law)
(d) $(A+B)C = AC + BC$ (second distributive law)
(e) $A(BC) = (AB)C$ (associative law)
(f) $(AB)^T = B^T A^T$

The following cautions are also stated:

(g) AB is not generally equal to BA, and BA may not even be conformable.
(h) If $AB = 0$, neither A nor B necessarily $= 0$.
(i) If $AB = AC$, B does not necessarily $= C$.

$$\text{Let } A = \begin{bmatrix} 2 & 1 \\ 1 & 2 \end{bmatrix} \quad B = \begin{bmatrix} -1 & -1 \\ 2 & 2 \end{bmatrix} \quad C = \begin{bmatrix} 1 & 1 \\ 1 & 1 \end{bmatrix}$$

Example of (c):

$$A(B+C) = \begin{bmatrix} 2 & 1 \\ 1 & 2 \end{bmatrix} \times \begin{bmatrix} 3 & 3 \\ 6 & 6 \end{bmatrix} = AB + AC$$

$$= \begin{bmatrix} 0 & 0 \\ 3 & 3 \end{bmatrix} + \begin{bmatrix} 3 & 3 \\ 3 & 3 \end{bmatrix} = \begin{bmatrix} 3 & 3 \\ 6 & 6 \end{bmatrix}$$

Example of (d)[1]:

$$(A+B)C = \begin{bmatrix} 1 & 0 \\ 3 & 4 \end{bmatrix} \begin{bmatrix} 1 & 1 \\ 1 & 1 \end{bmatrix} = \begin{bmatrix} 1 & 1 \\ 7 & 7 \end{bmatrix} = AC + BC$$

$$= \begin{bmatrix} 3 & 3 \\ 3 & 3 \end{bmatrix} + \begin{bmatrix} -2 & -2 \\ 4 & 4 \end{bmatrix} = \begin{bmatrix} 1 & 1 \\ 7 & 7 \end{bmatrix}$$

Example of (e):

$$A(BC) = \begin{bmatrix} 2 & 1 \\ 1 & 2 \end{bmatrix} \begin{bmatrix} -2 & -2 \\ 4 & 4 \end{bmatrix} = \begin{bmatrix} 0 & 0 \\ 6 & 6 \end{bmatrix} = (AB)C$$

$$= \begin{bmatrix} 0 & 0 \\ 3 & 3 \end{bmatrix} \begin{bmatrix} 1 & 1 \\ 1 & 1 \end{bmatrix} = \begin{bmatrix} 0 & 0 \\ 6 & 6 \end{bmatrix}$$

Example of (f):

$$\text{Let } A = \begin{bmatrix} 2 & 6 & 4 \\ 1 & 2 & 7 \end{bmatrix} \text{ and } B = \begin{bmatrix} 3 & 2 \\ 9 & 0 \\ 1 & 3 \end{bmatrix} \text{ then } AB = \begin{bmatrix} 64 & 16 \\ 28 & 23 \end{bmatrix}$$

so

$$(AB)^{\mathrm{T}} = \begin{bmatrix} 64 & 28 \\ 16 & 23 \end{bmatrix} \text{ and } B^{\mathrm{T}} A^{\mathrm{T}} = \begin{bmatrix} 3 & 9 & 1 \\ 2 & 0 & 3 \end{bmatrix} \begin{bmatrix} 2 & 1 \\ 6 & 2 \\ 4 & 7 \end{bmatrix} = \begin{bmatrix} 64 & 28 \\ 16 & 23 \end{bmatrix}$$

Example of (g):

$$AB = \begin{bmatrix} 0 & 0 \\ 3 & 3 \end{bmatrix} \neq \begin{bmatrix} -3 & -3 \\ 6 & 6 \end{bmatrix} = BA$$

Example of (h):

$$\text{Let } A = \begin{bmatrix} 2 & 2 \\ 1 & 1 \end{bmatrix} \text{ and } B = \begin{bmatrix} -1 & -2 \\ 1 & 2 \end{bmatrix}$$

Then $AB = 0$, but neither A nor B equal 0.

---

[1] The multiplication symbol, ×, is generally not written in matrix equations. This convention of not using × is followed in this book.

Example of (i):

$$\text{Let } A = \begin{bmatrix} 2 & 2 \\ 1 & 1 \end{bmatrix}, \quad B = \begin{bmatrix} 1 & 2 \\ 1 & 3 \end{bmatrix}, \quad \text{and} \quad C = \begin{bmatrix} 4 & 10 \\ -2 & -5 \end{bmatrix}$$

Then $AB = AC$, but $B \neq C$, where

$$AB = \begin{bmatrix} 4 & 10 \\ 2 & 5 \end{bmatrix} = AC$$

A.9 COMPUTER ALGORITHMS FOR MATRIX OPERATIONS

It should be apparent that addition, subtraction, and multiplication of large matrices involves many arithmetic operations. These are very tedious when done by hand but can be done quickly by a computer. In this section, general mathematical expressions are developed for performing these operations using a computer. These general mathematical expressions, when programmed for computer solution, are called *algorithms*.

A.9.1 Addition or Subtraction of Two Matrices

Note that the two matrices A and B in Figure A.1 are *conformable* for addition. Find the sum of the two matrices and place the results in C.

Step 1: Add the first element (a_{11}) of the A matrix to the first element (b_{11}) of B, placing the result in the first element (c_{11}) of C. Repeat this process for each of the successive columns along the first row of A and B, or from $j = 1$ to $j = n$.

Step 2: Iterate step 1 for each successive row of the matrices, or for rows increasing from $i = 1$ to $i = m$. Table A.1 shows this entire operation of adding two matrices in the four computer languages BASIC, C, FORTRAN, and Pascal.

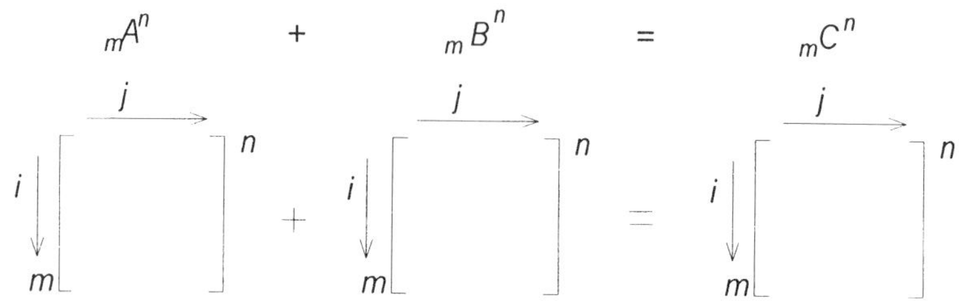

Figure A.1 Addition of matrices.

TABLE A.1 Addition Algorithm in BASIC, C, FORTRAN, and Pascal

BASIC Language:	FORTRAN Language:
1000 For i = 1 to M	Do 100 I = 1,M
1010 For j = 1 to N	Do 100 J = 1,N
1020 C(i,j) = A(i,j) + B(i,j)	C(i,j) = A(i,j) + B(i,j)
1030 Next j	100 Continue
1040 Next i	
C Language:	**Pascal Language:**
for (i=0; i<m; i++)	For i := 1 to M do
for (j=0; j<n; j++)	For j := 1 to N do
C[i][j] = A[i][j] + B[i][j];	C[i,j] := A[i,j] + B[i,j];

A.9.2 Matrix Multiplication

Consider the two matrices A and B in Figure A.2, which are conformable for multiplication. Find the product AB and place the results in C.

Step 1: Sum the products (A row $i = 1$) × (B column $k = 1$), with the elements of A and B being increased successively from $j = 1$ to $j = p$. Place the result in c_{ik}, or c_{11}, for this first step. Mathematically, this step is represented as

$$c_{ik} = \sum_{j=1}^{p} a_{ij} b_{jk} \quad \text{with} \quad i = 1 \text{ and } k = 1$$

Step 2: Increase k successively by 1, repeat step 1, and place the results in c_{ik}. Continue increasing k and repeating step 1 until $k = p$.

Step 3: Increase i from 1 to 2, and repeat steps 1 and 2 in entirety. Upon completion with $i = 2$, increment i by 1 and repeat steps 1 and 2 again in entirety. Continue this process through $i = m$. This completes the matrix multiplication.

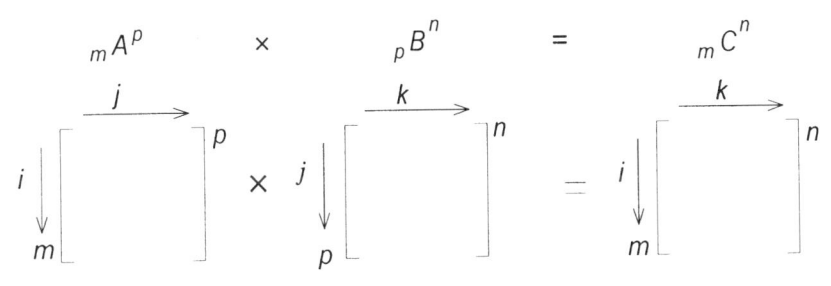

Figure A.2 Multiplication of matrices.

TABLE A.2 Multiplication Algorithm in BASIC, C, FORTRAN, and Pascal

BASIC Language:
```
For i = 1 to M
  For k = 1 to N
    C(i,k) = 0.0
    For j = 1 to P
      C(i,k) = C(i,k) + A(i,j)*B(j,k)
    Next j : Next k: Next I
```

C Language:
```
for (i=0; i<m; i++)
  for (k=0; k<n; k++) {
    C[i][k] = 0;
    for (j=0; j<p; j++)
      C[i][k] = C[i][k] + A[i][j]*B[j][k];
  } //for k
```

FORTRAN Language:
```
      Do 100 i = 1,M
        Do 100 k = 1,N
          C(i,k) = 0.0
          Do 100 j = 1,P
            C(i,k)=C(i,k)+A(i,j)*B(j,k)
100   Continue
```

Pascal Language:
```
For i := 1 to M do
  For k := 1 to N do Begin
    C[i,k] := 0.0;
    For j := 1 to P do
      C[i,k] := C[i,k] + A[i,j]*B[j,k]
  End; {for k}
```

This operation is shown Table A.2 in the four languages BASIC, C, FORTRAN, and Pascal. As a final note on multiplication, it is essential that the order in which the matrices are multiplied be specified.

A.10 USE OF THE MATRIX SOFTWARE

A software package called MATRIX is included on the CD provided with the book. It includes all the matrix operations that will be necessary to study the subject of adjustment computations and solve the after-chapter homework problems herein that require matrix manipulation. Instructions for use of the software are given in Appendix F.

PROBLEMS

A.1 Suppose that the following system of linear equations are to be represented in matrix form as $AX = B$. Formulate the three matrices.

$$2x_1 + x_2 + 5x_3 + x_4 = 5$$

$$x_1 + x_2 - 3x_3 + 4x_4 = -1$$

$$3x_1 + 6x_2 - 2x_3 + x_4 = 8$$

$$2x_1 + 2x_2 + 2x_3 - 3x_4 = 2$$

A.2 For Problem A.1, find the product of A^TA.

A.3 Do the following matrix operations.

(a) $\begin{bmatrix} 1 & 2 & -1 & 0 \\ 4 & 0 & 2 & 1 \\ 2 & -5 & 1 & 2 \end{bmatrix} + \begin{bmatrix} 3 & -4 & 1 & 2 \\ 1 & 5 & 0 & -3 \\ 2 & -2 & 3 & -1 \end{bmatrix}$

(b) $\begin{bmatrix} 1 & 2 & -1 & 0 \\ 4 & 0 & 2 & 1 \\ 2 & -5 & 1 & 2 \end{bmatrix} - \begin{bmatrix} 3 & -4 & 1 & 2 \\ 1 & 5 & 0 & 3 \\ 2 & -2 & 3 & -1 \end{bmatrix}$

(c) $3 \begin{bmatrix} 1 & 2 & -1 & 0 \\ 4 & 0 & 2 & 1 \\ 2 & -5 & 1 & 2 \end{bmatrix}$

(d) $\begin{bmatrix} 1 & 2 & 1 \\ 4 & 0 & 2 \end{bmatrix} \begin{bmatrix} 3 & -4 \\ 1 & 5 \\ -2 & 2 \end{bmatrix}$

A.4 Solve Problem A.2 using the MATRIX software.

A.5 Solve Problem A.3 using the MATRIX software.

A.6 Let

$$A = \begin{bmatrix} 2 & 0 \\ 3 & 1 \end{bmatrix} \quad B = \begin{bmatrix} 4 & -1 \\ 0 & 2 \end{bmatrix} \quad C = \begin{bmatrix} 1 & 0 \\ 0 & 1 \end{bmatrix} \quad D = \begin{bmatrix} 0 & 0 \\ 0 & 0 \end{bmatrix}$$

(a) Find AB.
(b) Find B^2.
(c) Find CB.
(d) Find BA.
(e) Find B^3.
(f) Find DB.
(g) Find C^3.

A.7 Do Problem A.6 using the MATRIX software.

A.8 Multiply the following, if possible. If not possible, give reasons why the multiplication cannot be done.

(a) $\begin{bmatrix} 2 & 1 \\ 4 & 0 \end{bmatrix} \begin{bmatrix} 1 & 2 \end{bmatrix}$

(b) $\begin{bmatrix} 3 & 1 & 3 \end{bmatrix} \begin{bmatrix} 4 \\ 0 \\ 9 \end{bmatrix}$

(c) $\begin{bmatrix} 2 & 3 & 4 & 4 \\ 1 & 0 & -1 & 6 \\ 0 & 1 & 2 & 9 \end{bmatrix} \begin{bmatrix} 0 & 2 \\ 3 & 1 \\ 1 & 0 \\ 0 & -1 \end{bmatrix}$

(d) $\begin{bmatrix} 2 \\ 0 \end{bmatrix} \begin{bmatrix} 3 & -1 \end{bmatrix}$

A.9 Expand the following summations into their equivalent algebraic expressions.

(a) $\sum_{k=1}^{5} k$

(b) $\sum_{k=3}^{7} (k - 2)$

(c) $\sum_{k=1}^{4} a_k$

(d) $\sum_{k=1}^{3} a_{2k} a_{k3}$

(e) $\sum_{i=1}^{3} \sum_{j=1}^{4} (a_{ij} + b_{ij})$

(f) $\sum_{i=1}^{2} \sum_{j=1}^{2} \sum_{k=1}^{3} a_{ik} b_{kj}$

A.10 Write the algebraic expressions for the system of equations represented by the following matrix notation.

$$\begin{bmatrix} 2 & 0 \\ 1 & 3 \\ 4 & 2 \end{bmatrix} \begin{bmatrix} x_1 \\ x_2 \end{bmatrix} = \begin{bmatrix} 2 \\ 1 \\ 3 \end{bmatrix}$$

A.11 Show that, in general, $(AB)^T = B^T A^T$.

Programming Problems

A.12 Write a program that will read and write the elements of a matrix, compute the transpose of the matrix, and write the solution. (*Hint:* Place the reading, writing, and transposition codes in separate subroutines/procedures/functions to be called from the main program.)

A.13 Select one of the matrix addition codes from Table A.1 and enter it and any necessary supporting code to solve Problem A.3(a). (*Hint:* Place the code in Table A.1 in a separate subroutine/procedure/function to be called from the main program.)

A.14 Select one of the matrix multiplication programs from Table A.2, and enter this code and any necessary supporting code to solve Problem A.3(d). (*Hint:* Place the code in Table A.2 in a separate subroutine/procedure/function to be called from the main program.)

A.15 Write a program that will read and write a file of matrices, and do the matrix operations of transposition, addition, subtraction, and multiplication. Demonstrate the program by doing Problems A.2 and A.3 (a), (b), and (d). (*Hint:* Place the reading, writing, transposition, addition, subtraction, and multiplication codes in separate subroutines/procedures/functions to be called from the main program.)

A.16 Write a Mathcad worksheet that solves Problems A.3 and A.6.

APPENDIX B

SOLUTION OF EQUATIONS BY MATRIX METHODS

B.1 INTRODUCTION

As stated in Appendix A, an advantage offered by matrix algebra is its adaptability to computer use. Using matrix algebra, large systems of simultaneous linear equations can be programmed for general computer solution using only a few systematic steps. The simplicities of programming matrix additions and multiplications, for example, were presented in Section A.9. To solve a system of equations using matrix methods, it is first necessary to define and compute the inverse matrix.

B.2 INVERSE MATRIX

If a square matrix is nonsingular (its determinant is not zero), it possesses an inverse matrix. When a system of simultaneous linear equations consisting of n equations and involving n unknowns is expressed as $AX = B$, the *coefficient matrix* (A) is a square matrix of dimensions $n \times n$. Consider the system of linear equations

$$AX = B \qquad (B.1)$$

The inverse of matrix A, symbolized as A^{-1}, is defined as

$$A^{-1}A = I \qquad (B.2)$$

where I is the identity matrix. Premultiplying both sides of matrix Equation (B.1) by A^{-1} gives

$$A^{-1}AX = A^{-1}B$$

Reducing yields

$$IX = A^{-1}B$$
$$X = A^{-1}B \tag{B.3}$$

Thus, the inverse is used to find the matrix of unknowns, X. The following points should be emphasized regarding matrix inversions:

1. Square matrices have inverses, with the exception noted below.
2. When the determinant of a matrix is zero, the matrix is said to be singular and its inverse cannot be found.
3. The inversion of even a small matrix is a tedious and time-consuming operation when done by hand. However, when done by a computer, the inverse can be found quickly and easily.

B.3 INVERSE OF A 2 × 2 MATRIX

Several general methods are available for finding a matrix inverse. Two are considered herein. Before proceeding with general cases, however, consider the specific case of finding the inverse for a 2 × 2 matrix using simple elementary matrix operations. Let any 2 × 2 matrix be symbolized as A. Also, let

$$A = \begin{bmatrix} a & b \\ c & d \end{bmatrix} \quad \text{and} \quad A^{-1} = \begin{bmatrix} w & x \\ y & z \end{bmatrix}$$

By applying Equation (B.2) and recalling the definition of an identity matrix I as given in Section A.4, it is possible to calculate w, x, y, and z in terms of a, b, c, and d of A^{-1}. Substituting in the appropriate values gives

$$\begin{bmatrix} a & b \\ c & d \end{bmatrix} \begin{bmatrix} w & x \\ y & z \end{bmatrix} = \begin{bmatrix} 1 & 0 \\ 0 & 1 \end{bmatrix}$$

By matrix multiplication

$$aw + by = 1 \tag{a}$$
$$ax + bz = 0 \tag{b}$$
$$cw + dy = 0 \tag{c}$$
$$cx + dz = 1 \tag{d}$$

The determinant of A is symbolized as $\|A\|$ and is equal to $ad - bc$.

SOLUTION OF EQUATIONS BY MATRIX METHODS

$$\left\{\begin{array}{l} \text{From } (a)\ y = \dfrac{1-aw}{b};\ \text{from } (b)\ y = -\dfrac{cw}{d} \\ \text{then } \dfrac{1-aw}{b} = -\dfrac{cw}{d};\ \text{reducing: } w = \dfrac{d}{da-bc} = \dfrac{d}{\|A\|} \end{array}\right\}$$

$$\left\{\begin{array}{l} \text{From } (b)\ z = -\dfrac{ax}{b};\ \text{from } (d)\ z = \dfrac{1-cx}{d} \\ \text{then } -\dfrac{ax}{b} = \dfrac{1-cx}{d};\ \text{reducing: } x = \dfrac{b}{-da+bc} = \dfrac{-b}{\|A\|} \end{array}\right\}$$

$$\left\{\begin{array}{l} \text{From } (a)\ w = \dfrac{1-by}{a};\ \text{from } (c)\ w = -\dfrac{dy}{c} \\ \text{then } -\dfrac{1-by}{a} = -\dfrac{dy}{c};\ \text{reducing: } y = \dfrac{c}{-ad+cb} = \dfrac{-c}{\|A\|} \end{array}\right\}$$

$$\left\{\begin{array}{l} \text{From } (b)\ x = -\dfrac{bz}{a};\ \text{from } (d)\ x = \dfrac{1-dz}{c} \\ \text{then } -\dfrac{bz}{a} = \dfrac{1-dz}{c};\ \text{reducing: } z = \dfrac{a}{ad-bc} = \dfrac{a}{\|A\|} \end{array}\right\}$$

Thus, for any 2 × 2 matrix composed of the elements $\begin{bmatrix} a & b \\ c & d \end{bmatrix}$, its inverse is simply

$$\begin{bmatrix} \dfrac{d}{\|A\|} & \dfrac{-b}{\|A\|} \\ \dfrac{-b}{\|A\|} & \dfrac{a}{\|A\|} \end{bmatrix} = \dfrac{1}{\|A\|}\begin{bmatrix} d & -b \\ -b & a \end{bmatrix}$$

Example B.1 If $A = \begin{bmatrix} 2 & 3 \\ 4 & 1 \end{bmatrix}$, find A^{-1}.

SOLUTION

$$A^{-1} = -\dfrac{1}{10}\begin{bmatrix} 1 & -3 \\ -4 & 2 \end{bmatrix}$$

A check on the inverse can be obtained by testing it against its definition, or $A^{-1}A = I$. Thus,

$$-\frac{1}{10}\begin{bmatrix} 1 & -3 \\ -4 & 2 \end{bmatrix}\begin{bmatrix} 2 & 3 \\ 4 & 1 \end{bmatrix} = \begin{bmatrix} 1 & 0 \\ 0 & 1 \end{bmatrix} = I$$

B.4 INVERSES BY ADJOINTS

The inverse of A can be found using the method of adjoints with the following equation:

$$A^{-1} = \frac{\text{adjoint of } A}{\text{determinant of } A} = \frac{\text{adjoint of } A}{\|A\|} \tag{B.4}$$

The *adjoint of A* is obtained by first replacing each matrix element by its signed minor or *cofactor,* and then transposing the resulting matrix. The cofactor of element a_{ij} equals $(-1)^{i+j}$ times the numerical value of the determinant for the remaining elements after row i and column j have been removed from the matrix. This procedure is illustrated in Figure B.1, where the cofactor of a_{12} is

$$(-1)^{1+2}(a_{21}a_{33} - a_{31}a_{23}) = a_{31}a_{23} - a_{21}a_{33}$$

Using this procedure, the inverse of the following A matrix is found:

$$A = \begin{bmatrix} a_{11} & a_{12} & a_{13} \\ a_{21} & a_{22} & a_{23} \\ a_{31} & a_{32} & a_{33} \end{bmatrix} = \begin{bmatrix} 4 & 3 & 2 \\ 3 & 4 & 1 \\ 2 & 3 & 4 \end{bmatrix}$$

For this A matrix, the cofactors are calculated as follows

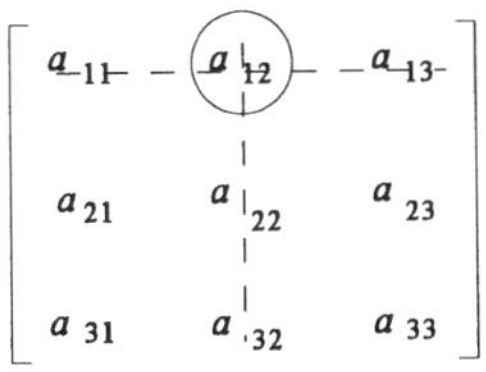

$$\begin{bmatrix} a_{11} & a_{12} & a_{13} \\ a_{21} & a_{22} & a_{23} \\ a_{31} & a_{32} & a_{33} \end{bmatrix}$$

Figure B.1 Cofactor of the a_{12} element.

cofactor of $a_{11} = (-1)^2 (4 \times 4 - 1 \times 3) = 13$

cofactor of $a_{21} = (-1)^3 (3 \times 4 - 2 \times 3) = -6$

cofactor of $a_{31} = (-1)^4 (3 \times 1 - 2 \times 4) = -5$

cofactor of $a_{12} = (-1)^3 (3 \times 4 - 1 \times 2) = -10$

Following the procedure above, the matrix of cofactors is

$$\text{matrix of cofactors} = \begin{bmatrix} 13 & -10 & 1 \\ -6 & 12 & -6 \\ -5 & 2 & 7 \end{bmatrix}$$

Transposing this cofactor matrix produces the following adjoint of A:

$$\text{adjoint of } A = \begin{bmatrix} 13 & -6 & -5 \\ -10 & 12 & 2 \\ 1 & -6 & 7 \end{bmatrix}$$

The *determinant* of A is the sum of the products of the elements in the first row of the original matrix times their respective cofactors. Since the cofactors were obtained in the previous step, this simplifies to

$$\|A\| = 4(13) + 3(-10) + 2(1) = 24$$

The inverse of A is now calculated as

$$A^{-1} = \frac{1}{24} \begin{bmatrix} 13 & -6 & -5 \\ -10 & 12 & 2 \\ 1 & -6 & 7 \end{bmatrix} = \begin{bmatrix} 13/24 & -1/4 & -5/24 \\ -5/12 & 1/2 & 1/12 \\ 1/24 & -1/4 & 7/24 \end{bmatrix}$$

Again, a check on the arithmetical work is obtained by using the definition of an inverse:

$$AA^{-1} = \begin{bmatrix} 13/24 & -1/4 & -5/24 \\ -5/12 & 1/2 & 1/12 \\ 1/24 & -1/4 & 7/24 \end{bmatrix} \begin{bmatrix} 4 & 3 & 2 \\ 3 & 4 & 1 \\ 2 & 3 & 4 \end{bmatrix} = \begin{bmatrix} 1 & 0 & 0 \\ 0 & 1 & 0 \\ 0 & 0 & 1 \end{bmatrix} = I$$

B.5 INVERSES BY ROW TRANSFORMATIONS

1. The multiplication of every element in any row by a nonzero scalar.
2. The addition (or subtraction) of the elements in any row to the elements of any other row.
3. Combinations of 1 and 2.

B.5 INVERSES BY ROW TRANSFORMATIONS

If elementary row transformations are performed successively on A such that A is transformed into I, and if throughout the procedure the same row transformations are also done to the same rows of the identity matrix I, the I matrix will be transformed into A^{-1}. This procedure is illustrated using the same matrix as that used to demonstrate the method of adjoints.

Initially, the original matrix and the identity matrix are listed side by side:

$$\begin{matrix} A & & I \\ \begin{bmatrix} 4 & 3 & 2 \\ 3 & 4 & 1 \\ 2 & 3 & 4 \end{bmatrix} & | & \begin{bmatrix} 1 & 0 & 0 \\ 0 & 1 & 0 \\ 0 & 0 & 1 \end{bmatrix} \end{matrix}$$

With the following three row transformations performed on A and I, they are transformed into matrices A_1 and I_1, respectively:

1. Multiply row 1 of matrices A and I by $1/a_{11}$, or 1/4. Place the results in row 1 of A_1 and I_1, respectively. This converts a_{11} of matrix A_1 to 1, as shown below.

$$\begin{bmatrix} 1 & 3/4 & 1/2 & | & 1/4 & 0 & 0 \\ 3 & 4 & 1 & | & 0 & 1 & 0 \\ 2 & 3 & 4 & | & 0 & 0 & 1 \end{bmatrix}$$

2. Multiply row 1 of matrices A_1 and I_1 by a_{21}, or 3. Subtract the resulting row from row 2 of matrices A and I and place the difference in row 2 of A_1 and I_1, respectively. This converts a_{21} of A_1 to zero.
3. Multiply row 1 of matrices A_1 and I_1 by a_{31}, or 2. Subtract the resulting row from row 3 of matrices A and I and place the difference in row 3 of A_1 and I_1, respectively. This changes a_{31} of A_1 to zero.

After doing these operations, the transformed matrices A_1 and I_1 are

$$\begin{matrix} A_1 & & I_1 \\ \begin{bmatrix} 1 & 3/4 & 1/2 \\ 0 & 7/4 & -1/2 \\ 0 & 3/2 & 3 \end{bmatrix} & | & \begin{bmatrix} 1/4 & 0 & 0 \\ -3/4 & 1 & 0 \\ -1/2 & 0 & 1 \end{bmatrix} \end{matrix}$$

Notice that the first column of A_1 has been made equivalent to the first column of a 3 × 3 identity matrix as a result of these three row transformations. For matrices having more than three rows, this same general procedure would be followed for each row to convert the first element in each row of A_1 to zero, with the exception to first row of A.

Next, the following three elementary row transformations are done on matrices A_1 and I_1 to transform them into matrices A_2 and I_2:

540 SOLUTION OF EQUATIONS BY MATRIX METHODS

1. Multiply row 2 of A_1 and I_1 by $1/a_{22}$, or 4/7, and place the results in row 2 of A_2 and I_2. This converts a_{22} to 1, as shown below.

$$\begin{bmatrix} 1 & 3/4 & 1/2 & | & 1/4 & 0 & 0 \\ 0 & 1 & -2/7 & | & -3/7 & 4/7 & 0 \\ 0 & 3/2 & 3 & | & -1/2 & 0 & 1 \end{bmatrix}$$

2. Multiply row 2 of A_2 and I_2 by a_{12}, or 3/4. Subtract the resulting row from row 1 of A_1 and I_1 and place the difference in row 1 of A_2 and I_2, respectively.
3. Multiply row 2 of A_2 and I_2 by a_{32}, or 3/2. Subtract the resulting row from row 3 of A_1 and I_1 and place the difference in row 3 of A_2 and I_2, respectively.

After doing these operations, the transformed matrices A_2 and I_2 are

$$\begin{array}{cc} A_2 & I_2 \end{array}$$
$$\begin{bmatrix} 1 & 0 & 5/7 & | & 4/7 & -3/7 & 0 \\ 0 & 1 & -2/7 & | & -3/7 & 4/7 & 0 \\ 0 & 0 & 24/7 & | & 1/7 & -6/7 & 1 \end{bmatrix}$$

Notice that after this second series of steps is completed, the second column of A_2 conforms to column two of a 3 × 3 identity matrix. Again, for matrices having more than three rows, this same general procedure would be followed for each row, to convert the second element in each row (except the second row) of A_2 to zero.

Finally, the following three row transformations are applied to matrices A_2 and I_2 to transform them into matrices A_3 and I_3. These three steps are:

1. Multiply row 3 of A_2 and I_2 by $1/a_{33}$, or 7/24 and place the results in row 3 of A_3 and I_3, respectively. This converts a_{33} to 1, as shown below.

$$\begin{bmatrix} 1 & 0 & 5/7 & | & 4/7 & -3/7 & 0 \\ 0 & 1 & -2/7 & | & -3/7 & 4/7 & 0 \\ 0 & 0 & 1 & | & 1/24 & -1/4 & 7/24 \end{bmatrix}$$

2. Multiply row 3 of A_2 and I_2 by a_{13}, or 5/7. Subtract the results from row 1 of A_2 and I_2 and place the difference in row 1 of A_3 and I_3, respectively.
3. Multiply row 3 of A_2 and I_2 by a_{23}, or $-2/7$. Subtract the results from row 2 of A_2 and I_2 and place the difference in row 2 of A_3 and I_3.

Following these operations, the transformed matrices A_3 and I_3 are

TABLE B.1 Inverse Algorithm in BASIC, C, FORTRAN, and PASCAL

BASIC Language:
REM INVERT A MATRIX
FOR k = 1 TO n
  FOR j = 1 TO n
    IF j<>k THEN A(k,j) = A(k,j)/A(k,k)
  NEXT j
  A(k,k) = 1/A(k,k)
  FOR i = 1 TO n
    IF i<>k THEN
      FOR j=1 TO n
        IF j<>k THEN A(i,j) = A(i,j) − A(i,k)*A(k,j)
      NEXT j
      A(i,k) = −A(i,k)*A(k,k)
    END IF
NEXT i: NEXT k

C Language:
```
for (k=0; k<n; k++) {
  for (j=0; j<n; j++)
    if (j!=k) A[k][j] = A[k][j]/A[k][k];
  A[k][k] = 1.0/A[k][k];
  for (i=0; i<n; i++)
    if (i!=k) {
      for (j=0; j<n; j++)
        if (j!=k) A[i][j] = A[i][j] − A[i][k]*A[k][j];
      A[i][k] = −A[i][k]*A[k][k];
    } //if i<>k
} //for k
```

FORTRAN Language:
```
      Do 560 k = 1,N
        Do 520 j = 1,N
          If (j.NE.k) Then
            A(k,j) = A(k,j)/A(k,k)
520     Continue
        A(K,K) = 1.0/A(K,K)
        Do 560 i = 1,N
          If (i.EQ.k) Then GOTO 560
          Do 550 j = 1,N
            If (j.NE.k) Then
              A(i,j) = A(i,j) − A(i,k)*A(k,j)
550       Continue
          A(i,k) = −A(i,k) * A(k,k)
560   Continue
```

Pascal Language:
For k := 1 to N do Begin
  For j := 1 to N do
    If (j<>k) then A[k,j] := A[k,j]/A[k,k];
  A[k,k] := 1.0/A[k,k];
  For i := 1 to N do
    If (i<>k) then Begin
      For j := 1 to N do
        If (j<>k) then A[i,j] := A[i,j] − A[i,k]*A[k,j];
      A[i,k] := −A[i,k]*A[k,k];
    End; {If i<>k}
End; {for k}

$$\begin{array}{cc} A_3 & I_3 = A^{-1} \\ \begin{bmatrix} 1 & 0 & 0 \\ 0 & 1 & 0 \\ 0 & 0 & 1 \end{bmatrix} & \begin{bmatrix} 12/24 & -1/4 & -5/24 \\ -5/12 & 1/2 & 1/12 \\ 1/24 & -1/4 & 7/24 \end{bmatrix} \end{array}$$

Notice that through of these nine elementary row transformations, the original A matrix is transformed into the identity matrix and the original identity matrix is transformed into A^{-1}. Also note that A^{-1} obtained by this method agrees exactly with the inverse obtained by the method of adjoints. This is because any nonsingular matrix has a unique inverse.

It should be obvious that the quantity of work involved in inverting matrices increases greatly with the matrix size, since the number of necessary row transformations is equal to the square of the number of rows or columns. Because of this, it is not considered practical to invert large matrices by hand. This work is done more conveniently with a computer. Since the procedure of elementary row transformations is systematic, it is easily programmed.

Table B.1 shows algorithms, written in the BASIC, C, FORTRAN, and Pascal programming languages, for calculating the inverse of any $n \times n$ nonsingular matrix A. Students should review the code in their preferred language to gain familiarity with computer procedures.

B.6 EXAMPLE PROBLEM

Example B.2 Suppose that an EDM instrument is placed at point A in Figure B.2 and a reflector is placed successively at B, C, and D. The observed values AB, AC, and AD are shown in the figure. Calculate the unknowns X_1, X_2, and X_3 by matrix methods. The values observed are

$$AB = 125.27$$
$$AC = 259.60$$
$$AD = 395.85$$

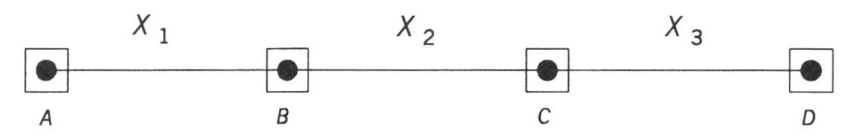

Figure B.2 Observation of a line.

SOLUTION Formulate the basic equations:

$$1X_1 + 0X_2 + 0X_3 = 125.27$$
$$1X_1 + 1X_2 + 0X_3 = 259.60$$
$$1X_1 + 1X_2 + 1X_3 = 395.85$$

Represented in matrix notation, these equations are $AX = L$. In this matrix equation, the individual matrices are

$$A = \begin{bmatrix} 1 & 0 & 0 \\ 1 & 1 & 0 \\ 1 & 1 & 1 \end{bmatrix} \quad X = \begin{bmatrix} x_1 \\ x_2 \\ x_3 \end{bmatrix} \quad L = \begin{bmatrix} 125.27 \\ 259.60 \\ 395.85 \end{bmatrix}$$

The solution in matrix notation is $X = A^{-1}L$. Using elementary row transformation, the inverse of A is

$$\begin{bmatrix} 1 & 0 & 0 & | & 1 & 0 & 0 \\ 1 & 1 & 0 & | & 0 & 1 & 0 \\ 1 & 1 & 1 & | & 0 & 0 & 1 \end{bmatrix} \rightarrow \begin{bmatrix} 1 & 0 & 0 & | & 1 & 0 & 0 \\ 0 & 1 & 0 & | & -1 & 1 & 0 \\ 0 & 0 & 1 & | & 0 & -1 & 1 \end{bmatrix}$$

Solving $X = A^{-1}L$, the unknowns are

$$X = A^{-1}L = \begin{bmatrix} 1 & 0 & 0 \\ -1 & 1 & 0 \\ 0 & -1 & 1 \end{bmatrix} \begin{bmatrix} 125.27 \\ 259.60 \\ 395.85 \end{bmatrix} = \begin{bmatrix} 125.27 \\ 134.33 \\ 136.25 \end{bmatrix}$$

$$X_1 = 125.27$$
$$X_2 = 134.33$$
$$X_3 = 136.25$$

PROBLEMS

B.1 Describe when a 2 × 2 matrix has no inverse.

B.2 Find the inverse of A using the method of adjoints.

$$A = \begin{bmatrix} 3 & -1 & -1 \\ -1 & 3 & -1 \\ -1 & -1 & 3 \end{bmatrix}$$

544 SOLUTION OF EQUATIONS BY MATRIX METHODS

B.3 Find the inverse of A in Problem B.2 using elementary row transformations.

B.4 Solve the following system of linear equations using matrix methods.

$$x + 5y = -8$$
$$-x - 2y = -1$$

B.5 Solve the following system of linear equations using matrix methods.

$$x + y - z = -8$$
$$3x - y + z = -4$$
$$-x + 2y + 2z = 21$$

B.6 Compute the inverses of the following matrices.

$$A = \begin{bmatrix} 8 & 5 \\ 3 & 12 \end{bmatrix} \quad B = \begin{bmatrix} 16 & 2 \\ -8 & 3 \end{bmatrix}$$

B.7 Compute the inverses of the following matrices.

$$A = \begin{bmatrix} 3 & -1 & 0 \\ -1 & 3 & -1 \\ 0 & -1 & 3 \end{bmatrix} \quad B = \begin{bmatrix} 4 & 3 & 7 \\ -1 & 0 & 4 \\ 2 & 8 & 10 \end{bmatrix}$$

B.8 Compute the inverses of the following matrices.

$$A = \begin{bmatrix} 13 & -6 & 0 \\ -6 & 18 & -6 \\ 0 & -6 & 16 \end{bmatrix} \quad B = \begin{bmatrix} 1 & 2 & 6 \\ 2 & -3 & 4 \\ 0 & 6 & -12 \end{bmatrix}$$

B.9 Solve the following matrix system.

$$\begin{bmatrix} 13 & -6 & 0 \\ -6 & 18 & -6 \\ 0 & -6 & 16 \end{bmatrix} \begin{bmatrix} A \\ B \\ C \end{bmatrix} = \begin{bmatrix} 740.02 \\ 612.72 \\ 1072.22 \end{bmatrix}$$

Use the MATRIX software to do each problem.

B.10 Problem B.4

B.11 Problem B.5

B.12 Problem B.6

B.13   Problem B.7

B.14   Problem B.8

B.15   Problem B.9

Programming Problems

B.16   Select one of the coded matrix inverse routines from Table B.1, enter the code into a computer, and use it to solve Problem B.7. (*Hint:* Place the code in Table B.1 in a separate subroutine/function/procedure to be called from the main program.)

B.17   Add a block of code to the inverse routine in the language of your choice that will inform the user when a matrix is singular.

B.18   Write a program that reads and writes a file with a nonsingular matrix; finds its inverse, and write the results. Use this program to solve Problem B.7. (*Hint:* Place the reading, writing, and inversing code in separate subroutines/functions/procedures to be called from the main program. Provide a way to identify each matrix in the output file.)

B.19   Write a program that reads and writes a file containing a system of equations written in matrix form, solves the system using matrix operations, and writes the solution. Use this program to solve Problem B.9. (*Hint:* Place the reading, writing, and inversing code in separate subroutines/functions/procedures to be called from the main program. Provide a way to identify each matrix in the output file.)

APPENDIX C

NONLINEAR EQUATIONS AND TAYLOR'S THEOREM

C.1 INTRODUCTION

In adjustment computations it is frequently necessary to deal with nonlinear equations. For example, some observation equations relate observed quantities to unknown parameters through the transcendental functions of sine, cosine, or tangent; others relate them through terms raised to second- and higher-order powers. The task of solving a system of nonlinear equations is formidable. To facilitate the solution, a first-order Taylor series approximation may be used to create a set of linear equations. The equations can then be solved by matrix methods discussed in Appendix B.

C.2 TAYLOR SERIES LINEARIZATION OF NONLINEAR EQUATIONS

Suppose that the following equation relates a observed value L to its unknown parameters x and y through nonlinear coefficients as

$$L = f(x,y) \tag{C.1}$$

By Taylor's theorem, the equation is represented as

$$\begin{aligned} L &= f(x,y) \\ &= f(x_0, y_0) + \frac{(\partial L/\partial x)_0}{1!} dx + \frac{(\partial^2 L/\partial x^2)_0}{2!} dx^2 + \cdots + \frac{(\partial^n L/\partial x^n)_0}{n!} dx^n \\ &\quad + \frac{(\partial L/\partial y)_0}{1!} dy + \frac{(\partial^2 L/\partial y^2)_0}{2!} dy^2 + \cdots + \frac{(\partial^n L/\partial y^n)_0}{n!} dy^n + R \end{aligned} \tag{C.2}$$

In Equation (C.2), x_0 and y_0 are approximations of x and y; $f(x_0,y_0)$ is the nonlinear function evaluated at these approximations; R is the remainder, and dx and dy are corrections to the initial approximations, such that

$$x = x_0 + dx \tag{C.3}$$
$$y = y_0 + dy$$

A more exact Taylor series approximation is obtained by increasing the value of n in Equation (C.2). However, as the order of each successive term increases, its significance in the overall expression decreases. If all terms containing derivatives higher than the first are dropped, the following linear expression is obtained:

$$L = f(x,y) = f(x_0,y_0) + \left(\frac{\partial L}{\partial x}\right)_0 dx + \left(\frac{\partial L}{\partial y}\right)_0 dy \tag{C.4}$$

Once the initial approximations are selected, the only unknowns in Equation (C.4) are the corrections dx and dy. Of course, by dropping the higher-order terms from the Taylor series, Equation (C.4) becomes only a good approximation of the original equation. However, in the solution, an iterative procedure can be followed that yields accurate answers. This iterative procedure uses the following steps:

Step 1: Determine initial approximations for the unknowns. They may be obtained by guessing or from observations. It should be understood that the closer the initial approximations are to the final solution, the faster it will be obtained. For some problems, initial approximations can be obtained from graphical solutions or computed from available data or observations. For others, the determination of initial approximations can involve considerable computational effort.

Step 2: Substitute the initial approximations into Equation (C.4) and solve for the corrections dx and dy.

Step 3: Calculate revised values of x and y using Equations (C.3).

Step 4: Using these newly revised values for x and y, repeat steps 2 and 3.

Step 5: Continue the procedure until the corrections dx and dy are small enough to bring x and y within tolerable accuracy. When this occurs, the solution is said to have *converged*.

C.3 NUMERICAL EXAMPLE

To clarify this procedure further, a numerical example will be solved.

548   NONLINEAR EQUATIONS AND TAYLOR'S THEOREM

Example C.1   Linearize the following pair of nonlinear equations containing the two unknowns x and y.

$$F(x,y) = x + y - 2y^2 = -4$$
$$G(x,y) = x^2 + y^2 = 8$$

SOLUTION   Determine the partial derivative for each equation with respect to each unknown.

$$\frac{\partial F}{\partial x} = 1 \qquad \frac{\partial F}{\partial y} = 1 - 4y$$

$$\frac{\partial G}{\partial x} = 2x \qquad \frac{\partial F}{\partial y} = 2y$$

Compute initial approximations for each unknown. An estimate of $x = 1$ and $y = 1$ is used for the approximations initially.

First iteration: Write the linearized equations in the form of Equation (C.4).

$$F(x,y) = 1 + 1 - 2(1)^2 + dx + [1 - 4(1)] \, dy = -4$$
$$G(x,y) = (1)^2 + (1)^2 + 2(1) \, dx + 2(1) \, dy = 8$$

From the two equations above, solve for the unknowns dx and dy according to Equation (C.3):

$$dx = 1.25 \quad \text{and} \quad dy = 1.75$$

Using this solution, determine updated values for x and y:

$$x = x_0 + dx = 1.00 + 1.25 = 2.25$$
$$y = y_0 + dy = 1.00 + 1.75 = 2.75$$

Second iteration: Continue the procedure demonstrated for the first iteration.

$$F = 2.25 + 2.75 - 2(2.75)^2 + dx + [1 - 4(2.75)] \, dy = -4$$
$$G = (2.25)^2 + (2.75)^2 + 2(2.25) \, dx + 2(2.75) \, dy = 8$$

From the two equations above, $dx = -0.25$ and $dy = -0.64$, from which

$$x = x_0 + dx = 2.25 - 0.25 = 2.00$$
$$y = y_0 + dy = 2.75 - 0.64 = 2.11$$

Third iteration:

$$F = 2 + 2.11 - 2(2.11)^2 + dx + [1 - 4(2.11)]\, dy = -4$$
$$G = (2)^2 + (2.11)^2 + 2(2)\, dx + 2(2.11)\, dy = 8$$

From the two equations above, $dx = 0.00$ and $dy = -0.11$, from which

$$x = x_0 + dx = 2.00 + 0.00 = 2.00$$
$$y = y_0 + dy = 2.11 - 0.11 = 2.00$$

Fourth iteration:

$$F = 2 + 2 - 2(2)^2 + dx + [1 - 4(2)]\, dy = -4$$
$$G = (2)^2 + (2)^2 + 2(2)dx + 2(2)\, dy = 8$$

Using the two equations above, the corrections to x and y are zero to the nearest hundredth. Thus, the solution has converged and the values of $x = 2.00$ and $y = 2.00$ are the desired unknowns. Note that the initial values for the initial approximations were relatively poor and four iterations were required to find the final solution. However, had better estimates been made (say, $x_0 = 2.1$ and $y_0 = 1.9$), the solution would have converged in one or two iterations and saved computational effort. Fortunately, there are accepted computational procedures to determine close approximations in many surveying problems.

C.4 USING MATRICES TO SOLVE NONLINEAR EQUATIONS

The example of Section C.3 could be solved using matrix methods. However, as in the algebraic approach, the equations must be linearized using Taylor's series. To facilitate linearization using Taylor's theorem, the *Jacobian matrix* (a matrix consisting of the partial derivatives taken with respect to the unknown variables) is formed. This is the coefficient matrix of the linearized equations. The Jacobian matrix for the example of Section C.3 is

$$J = \begin{bmatrix} \dfrac{\partial F}{\partial x} & \dfrac{\partial F}{\partial y} \\ \dfrac{\partial G}{\partial x} & \dfrac{\partial G}{\partial y} \end{bmatrix}$$

In the Jacobian matrix above, the first column contains the partial derivative for each equation with respect to x, and the second column contains the partial derivative of each equation with respect to y. The linearized form of the equations may then be expressed in matrix notation as

$$JX = K \tag{C.5}$$

In Equation (C.5), J is the Jacobian matrix, X the matrix of unknown corrections dx and dy, and K the matrix of constants. Specifically, for the example of Section C.3, these matrices are

$$J = \begin{bmatrix} 1 & 1 - 4y_0 \\ 2x_0 & 2y_0 \end{bmatrix} \quad X = \begin{bmatrix} dx \\ dy \end{bmatrix} \quad K = \begin{bmatrix} -4 - F(x_0, y_0) \\ 8 - G(x_0, y_0) \end{bmatrix}$$

where $F(x_0, y_0)$ and $G(x_0, y_0)$ are the equations F and G solved at the initial approximations of x_0 and y_0.

Beginning with a set of initial approximations x_0 and y_0, the J and K matrices of Equation (C.5) are formed. X is computed using the matrix methods presented in Appendix B. Having updated the unknowns according to Equations (C.3), the J and K matrices are formed again and the solution for X is computed. This procedure is iterated until convergence is achieved.

C.5 SIMPLE MATRIX EXAMPLE

Example C.2 Find the solution of the nonlinear system of equations shown below using matrix methods.

$$F(x, y) = x^2 + 3xy - 4y^2 = 6$$
$$G(x, y) = x + xy - y^2 = 3$$

SOLUTION The partial derivatives of functions F and G with respect to the unknowns, x and y, are

$$\frac{\partial F}{\partial x} = 2x + 3y \qquad \frac{\partial F}{\partial y} = 3x - 8y$$

$$\frac{\partial G}{\partial x} = 1 + y \qquad \frac{\partial G}{\partial y} = x - 2y$$

Thus, the Jacobian matrix is

$$J = \begin{bmatrix} \dfrac{\partial F}{\partial x} & \dfrac{\partial F}{\partial y} \\ \dfrac{\partial G}{\partial x} & \dfrac{\partial G}{\partial y} \end{bmatrix} = \begin{bmatrix} 2x_0 + 3y_0 & 3x_0 - 8y_0 \\ 1 + y_0 & x_0 - 2y_0 \end{bmatrix}$$

and the system of equations to solve is

$$\begin{bmatrix} 2x_0 + 3y_0 & 3x_0 - 8y_0 \\ 1 + y_0 & x_0 - 2y_0 \end{bmatrix} \begin{bmatrix} dx \\ dy \end{bmatrix} = \begin{bmatrix} 6 - F(x_0, y_0) \\ 3 - G(x_0, y_0) \end{bmatrix}$$

First iteration: Using $x_0 = 3$ and $y_0 = 0$ yields

$$\begin{bmatrix} 6 & 9 \\ 1 & 3 \end{bmatrix} \begin{bmatrix} dx \\ dy \end{bmatrix} = \begin{bmatrix} 6 - 9 \\ 3 - 3 \end{bmatrix} = \begin{bmatrix} -3 \\ 0 \end{bmatrix}$$

The determinant for the Jacobian matrix above is $3(6) - 9(1) = 9$, and thus the matrix solution is

$$\begin{bmatrix} dx \\ dy \end{bmatrix} = \dfrac{1}{9} \begin{bmatrix} 3 & 9 \\ -1 & 6 \end{bmatrix} \begin{bmatrix} -3 \\ 0 \end{bmatrix} = \begin{bmatrix} -1.0 \\ 0.3 \end{bmatrix}$$

Applying Equation (C.3), initial approximations for a second iteration are

$$\begin{bmatrix} x \\ y \end{bmatrix} = \begin{bmatrix} x_0 \\ y_0 \end{bmatrix} + \begin{bmatrix} dx \\ dy \end{bmatrix} = \begin{bmatrix} 3 \\ 0 \end{bmatrix} + \begin{bmatrix} -1 \\ 0.3 \end{bmatrix} = \begin{bmatrix} 2.0 \\ 0.3 \end{bmatrix}$$

Second iteration:

$$\begin{bmatrix} 4.9 & 3.6 \\ 1.3 & 1.4 \end{bmatrix} \begin{bmatrix} dx \\ dy \end{bmatrix} = \begin{bmatrix} 6 - 5.44 \\ 3 - 2.51 \end{bmatrix} = \begin{bmatrix} 0.56 \\ 0.49 \end{bmatrix}$$

From this, dx and dy are found to be 0.45 and 0.77, respectively. This makes the approximations of x and y for the third iteration 1.55 and 1.07, respectively. The procedures are followed until the final solution for x and y is found to be 2.00 and 1.00, respectively. Again, fewer iterations would have been required if the initial approximations had been closer to the final values.

C.6 PRACTICAL EXAMPLE

Example C.3 Assume that the x and y coordinates of three points on a circle have been observed. Their coordinates are (9.4, 5.6), (7.6, 7.2), and (3.8, 4.8), respectively. The equation for a circle with center (h,k) and radius r is

$(x - h)^2 + (y - k)^2 = r^2$. Determine the coordinates of the center of the circle and its radius.

SOLUTION The equation of a circle is rewritten as $C(h,k,r) = (x - h)^2 + (y - k)^2 - r^2 = 0$. The partial derivatives with respect to the unknowns h, k, and r are

$$\frac{\partial C}{\partial h} = -2(x - h) \qquad \frac{\partial C}{\partial k} = -2(y - k) \qquad \frac{\partial C}{\partial r} = -2r$$

For each point observed, one equation is written, resulting in a system of three equations and three unknowns. The general linearized form of these equations expressed using matrices is

$$\begin{bmatrix} \frac{\partial C_1}{\partial h} & \frac{\partial C_1}{\partial k} & \frac{\partial C_1}{\partial r} \\ \frac{\partial C_2}{\partial h} & \frac{\partial C_2}{\partial k} & \frac{\partial C_2}{\partial r} \\ \frac{\partial C_3}{\partial h} & \frac{\partial C_2}{\partial k} & \frac{\partial C_3}{\partial r} \end{bmatrix} \begin{bmatrix} dh \\ dk \\ dr \end{bmatrix} = \begin{bmatrix} 0 - [(x_1 - h_0)^2 + (y_1 - k_0)^2 - r_0^2] \\ 0 - [(x_2 - h_0)^2 + (y_2 - k_0)^2 - r_0^2] \\ 0 - [(x_3 - h_0)^2 + (y_3 - k_0)^2 - r_0^2] \end{bmatrix} \quad \text{(C.6)}$$

After taking partial derivatives, Equation (C.6) becomes

$$\begin{bmatrix} -2(x_1 - h_0) & -2(y_1 - k_0) & -2r_0 \\ -2(x_2 - h_0) & -2(y_2 - k_0) & -2r_0 \\ -2(x_3 - h_0) & -2(y_3 - k_0) & -2r_0 \end{bmatrix} \begin{bmatrix} dh \\ dk \\ dr \end{bmatrix}$$
$$= \begin{bmatrix} 0 - [(x_1 - h_0)^2 + (y_1 - k_0)^2 - r_0^2] \\ 0 - [(x_2 - h_0)^2 + (y_2 - k_0)^2 - r_0^2] \\ 0 - [(x_3 - h_0)^2 + (y_3 - k_0)^2 - r_0^2] \end{bmatrix} \quad \text{(C.7)}$$

Equations (C.7) can be simplified by multiplying each side by $-1/2$. The resulting equations are

$$\begin{bmatrix} (x_1 - h_0) & (y_1 - k_0) & r_0 \\ (x_2 - h_0) & (y_2 - k_0) & r_0 \\ (x_3 - h_0) & (y_3 - k_0) & r_0 \end{bmatrix} \begin{bmatrix} dh \\ dk \\ dr \end{bmatrix} = \begin{bmatrix} 0.5[(x_1 - h_0)^2 + (y_1 - k_0)^2 - r_0^2] \\ 0.5[(x_2 - h_0)^2 + (y_2 - k_0)^2 - r_0^2] \\ 0.5[(x_3 - h_0)^2 + (y_3 - k_0)^2 - r_0^2] \end{bmatrix}$$
(C.8)

Assuming approximate initial values for h, k, and r as 7, 4.5, and 3, respectively, Equations (C.8) are

C.6 PRACTICAL EXAMPLE

$$\begin{bmatrix} 9.4-7 & 5.6-4.5 & 3 \\ 7.6-7 & 7.2-4.5 & 3 \\ 3.8-7 & 4.8-4.5 & 3 \end{bmatrix} \begin{bmatrix} dh \\ dk \\ dr \end{bmatrix} = \begin{bmatrix} 0.5[(9.4-7)^2 + (5.6-4.5)^2 - 3^2] \\ 0.5[(7.6-7)^2 + (7.2-4.5)^2 - 3^2] \\ 0.5[(3.8-7)^2 + (4.8-4.5)^2 - 3^2] \end{bmatrix}$$

(C.9)

Simplifying Equations (C.9) yields

$$\begin{bmatrix} 2.4 & 1.1 & 3 \\ 0.6 & 2.7 & 3 \\ -3.2 & 0.3 & 3 \end{bmatrix} \begin{bmatrix} dh \\ dk \\ dr \end{bmatrix} = \begin{bmatrix} -1.015 \\ -0.675 \\ 0.665 \end{bmatrix}$$

Solving this system gives results of

$$\begin{bmatrix} dh \\ dk \\ dr \end{bmatrix} = \begin{bmatrix} -0.28462 \\ -0.10769 \\ -0.07115 \end{bmatrix}$$

After applying these changes to the initial approximations, updated values for h, k, and r of 6.7154, 4.3923, and 2.9288, respectively, are obtained. The second iteration results in corrections of 0, 0, and 0.014945. Since the correction for r is still comparatively large, the iteration process must be continued. After the third iteration, suitable convergence was achieved. The final values for h, k, and r are 6.72, 4.39, and 2.94, respectively, which are within 0.00001 of a perfect solution.

Sometimes, more than one method is available for solving a problem. For example, in Example C.3 an alternative linear form of the equation of a circle could have been used. That equation is $x^2 + y^2 + 2dx + 2ey + f = 0$, where the center of the circle is at $(-d, -e)$ and the circle's radius is $\sqrt{d^2 + e^2 - f}$. Note that the equation is linear in terms of its unknowns (d, e, f) and thus iterations are not necessary in solving for the unknowns. Writing a rearranged form of this linear equation for each of three measured sets of (x, y) coordinates yields

$$2dx_1 + 2ey_1 + f = -(x_1^2 + y_1^2)$$
$$2dx_2 + 2ey_2 + f = -(x_2^2 + y_2^2) \qquad (C.10)$$
$$2dx_3 + 2ey_3 + f = -(x_3^2 + y_3^2)$$

Equations (C.10) can in turn be represented in matrix notation as

$$\begin{bmatrix} 2x_1 & 2y_1 & 1 \\ 2x_2 & 2y_2 & 1 \\ 2x_3 & 2y_3 & 1 \end{bmatrix} \begin{bmatrix} d \\ e \\ f \end{bmatrix} = \begin{bmatrix} -(x_1^2 + y_1^2) \\ -(x_2^2 + y_2^2) \\ -(x_3^2 + y_3^2) \end{bmatrix}$$

Solving this matrix system, the center of the circle is again found to be (6.72, 4.39), and its radius is determined to be 2.94.

In this appendix, the Taylor series has been applied to solve for the unknowns in nonlinear equations. Many equations in surveying, geodesy, and photogrammetry are nonlinear. In surveying, examples include the distance and angle formulas, which are nonlinear in terms of station coordinates. The Taylor series is used to linearize these equations and find least squares solutions. Thus, when performing least squares adjustments of plane observations, the techniques presented in this appendix must be used in the solutions.

PROBLEMS

C.1 Solve for the unknowns x and y in the following nonlinear equations using Taylor's theorem. (Use $x_0 = 5$ and $y_0 = 5$ for initial approximations.)

$$x^2 y - 3x^2 = 75$$

$$x^2 - y = 19$$

C.2 Solve for the unknown values of x, y, and z in the following three nonlinear equations using the Taylor series. (Use $x_0 = y_0 = z_0 = 2$ for initial approximations.)

$$x^2 - y^2 + 2xy + z = 4$$

$$-x + y + z = 4$$

$$-2x^2 - y + z^3 = 23$$

C.3 Use the MATRIX software to solve Problem C.1.

C.4 Use the MATRIX software to solve Problem C.2.

C.5 Find the center and radius of a circle using the equation $(x - h)^2 + (y - k)^2 = r^2$ given the coordinates of points A, B, and C on the circle. Follow the procedures discussed in Section C.5. Use initial approximations of $h_0 = 5$, $k_0 = 4$, and $r_0 = 2$ for the first iteration.

A: (7.2, 5.2)   B: (4.0, 6.4)   C: (4.0, 2.4)

C.6 Repeat Problem C.5 using the linear equation: $x^2 + y^2 + 2dx + 2ey + f = 0$.

C.7 Repeat Problem C.5 using the points A: $(0.50, -0.70)$, B: $(1.00, 0.00)$, C: $(0.70, 0.70)$. Use initial approximations of $h_0 = 0$, $k_0 = 0$, and $r_0 = 1$.

C.8 Repeat Problem C.7 using the linear equation $x^2 + y^2 + 2dx + 2ey + f = 0$.

C.9 Use the ADJUST software to solve Problem C.5.

C.10 Use the ADJUST software to solve Problem C.7.

C.11 The distance formula between two stations i and j is $D_{ij} = \sqrt{(x_j - x_i)^2 + (y_j - y_i)^2}$. Write the linearized form of in terms of the variables x_i, y_i, x_j, and y_j.

C.12 The azimuth formula between two stations i and j is $\alpha_{ij} = \tan^{-1}[(x_j - x_i)/(y_j - y_i)]$. Write the linearized form of this equation in terms of the variables x_i, y_i, x_j, and y_j.

C.13 The formula for an angle $\angle jik$ is $\alpha_{ik} - \alpha_{ij}$, where α is defined in Problem C.12. Write the linearized form of this equation in terms of the variables x_i, y_i, x_j, y_j, x_k, and y_k.

Programming Problems

C.14 Create a programmed package that solves Problem C.6.

C.15 Create a programmed package that solves Problem C.5.

APPENDIX D

NORMAL ERROR DISTRIBUTION CURVE AND OTHER STATISTICAL TABLES

D.1 DEVELOPMENT OF THE NORMAL DISTRIBUTION CURVE EQUATION

In Section 2.4 the histogram and frequency polygon were presented as methods for graphical portrayal of random error distributions. If a large number of these distributions were examined for sets, measurements in surveying, geodesy, and photogrammetry, it would be found that they conform to normal (or Gaussian) distributions. The general laws governing normal distributions are stated as follows:

1. Positive and negative errors occur with equal probability and equal frequency.
2. Small errors are more common than large errors.
3. Large errors seldom occur, and there is a limit to the size of the greatest random error that will occur in any set of observations.

A curve that conforms to these laws, plotted with the size of the error on the abscissa and the probability of occurrence on the ordinate, is shown in Figure 3.3. This curve is repeated in Figure D.1 and is called the *normal distribution curve,* the *normal curve of error,* or simply the *probability curve.* A smooth curve of this same shape would be obtained if for a very large group of measurements, a histogram were plotted with an infinitesimally small class interval. In this section, the equation for this curve is developed.

Assume that the normal distribution curve is continuous and that the probability of an error occurring between x and $x + dx$ is given by the function

D.1 DEVELOPMENT OF THE NORMAL DISTRIBUTION CURVE EQUATION

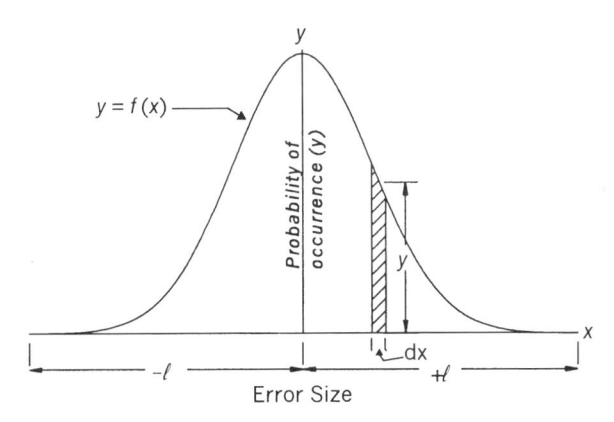

Figure D.1 Normal distribution curve.

$y = f(x)$. Further assume that this is the equation for the probability curve. The form of $f(x)$ will now be determined. Since as explained in Chapter 3, probabilities are equivalent to areas under the probability curve, the probabilities of errors occurring within the ranges $(x_1$ and $x_1 + dx_1)$, $(x_2$ and $x_2 + dx_2)$, and so on, are $f(x_1)\,dx_1$, $f(x_2)\,dx_2, \ldots, f(x_n)\,dx_n$. The total area under the probability curve represents the total probability or simply 1. Then for a finite number of possible errors,

$$f(x_1)\,dx_1 + f(x_2)\,dx_2 + \cdots + f(x_n)\,dx_n = 1 \tag{D.1}$$

If the total range of errors $x_1, x_2, \ldots, x_n$ is between plus and minus 1, considering an infinite number of errors that makes the curve continuous, the area under the curve can be set equal to

$$1 = \int_{-1}^{1} f(x)\,dx$$

But because the area under the curve from $+1$ to $+\infty$ and from -1 to $-\infty$ is essentially zero, the integration limits are extended to $\pm\infty$, as

$$1 = \int_{-\infty}^{\infty} f(x)\,dx = \int_{-\infty}^{\infty} y\,dx \tag{D.2}$$

Now suppose that the quantity M has been measured and that it is equal to some function of n unknown parameters $z_1, z_2, \ldots, z_n$ such that $M = f(z_1, z_2, \ldots, z_n)$. Also let $x_1, x_2, \ldots, x_m$ be the errors of m observations $M_1, M_2, \ldots, M_m$, and let $f(x_1)\,dx_1$, $f(x_2)\,dx_2, \ldots, f(x_m)\,dx_m$ be the probabilities of errors falling within the ranges $(x_1$ and $dx_1)$, $(x_2$ and $dx_2)$, and so on. By

558 NORMAL ERROR DISTRIBUTION CURVE AND OTHER STATISTICAL TABLES

Equation (3.1), the probability P of the simultaneous occurrence of all of these errors is equal to the product of the individual probabilities; thus,

$$P = [f(x_1)\,dx_1]\,[f(x_2)\,dx_2] \cdots [f(x_m)\,dx_m]$$

Then, by logs,

$$\log P = \log f(x_1) + \log f(x_2) + \cdots + \log f(x_m)$$
$$+ \log dx_1 + \log dx_2 + \cdots + \log dx_m \quad (D.3)$$

The most probable values of the errors will occur when P is maximized or when the log of P is maximized. To maximize a function, it is differentiated with respect to each unknown parameter z, and the results set equal to zero. After logarithmic differentiation of Equation (D.3), the following n equations result (note that the dx's are constants independent of the z's, and therefore their differentials with respect to the z's are zero):

$$\frac{1}{P}\frac{\partial P}{\partial z_1} = \frac{1}{f(x_1)}\frac{df(x_1)}{dx_1}\frac{dx_1}{dz_1} + \frac{1}{f(x_2)}\frac{df(x_2)}{dx_2}\frac{dx_2}{dz_1} + \cdots + \frac{1}{f(x_m)}\frac{df(x_m)}{dx_m}\frac{dx_m}{dz_1} = 0$$

$$\frac{1}{P}\frac{\partial P}{\partial z_2} = \frac{1}{f(x_1)}\frac{df(x_1)}{dx_1}\frac{dx_1}{dz_2} + \frac{1}{f(x_2)}\frac{df(x_2)}{dx_2}\frac{dx_2}{dz_2} + \cdots + \frac{1}{f(x_m)}\frac{df(x_m)}{dx_m}\frac{dx_m}{dz_2} = 0$$

$$\vdots$$

$$\frac{1}{P}\frac{\partial P}{\partial z_n} = \frac{1}{f(x_1)}\frac{df(x_1)}{dx_1}\frac{dx_1}{dz_n} + \frac{1}{f(x_2)}\frac{df(x_2)}{dx_2}\frac{dx_2}{dz_n} + \cdots + \frac{1}{f(x_m)}\frac{df(x_m)}{dx_m}\frac{dx_m}{dz_n} = 0$$

$$(D.4)$$

Now let

$$f'(x) = \frac{df(x)}{dx} \quad (D.5)$$

Substituting Equation (D.5) into (D.4) gives

$$\frac{f'(x_1)\,dx_1}{f(x_1)\,dz_1} + \frac{f'(x_2)\,dx_2}{f(x_2)\,dz_1} + \cdots + \frac{f'(x_m)\,dx_m}{f(x_m)\,dz_1} = 0$$

$$\frac{f'(x_1)\,dx_1}{f(x_1)\,dz_2} + \frac{f'(x_2)\,dx_2}{f(x_2)\,dz_2} + \cdots + \frac{f'(x_m)\,dx_m}{f(x_m)\,dz_2} = 0 \quad (D.6)$$

$$\vdots$$

$$\frac{f'(x_1)\,dx_1}{f(x_1)\,dz_n} + \frac{f'(x_2)\,dx_2}{f(x_2)\,dz_n} + \cdots + \frac{f'(x_m)\,dx_m}{f(x_m)\,dz_n} = 0$$

D.1 DEVELOPMENT OF THE NORMAL DISTRIBUTION CURVE EQUATION

Thus far, $f(x)$ and $f'(x)$ are general, regardless of the number of unknown parameters. Now consider the special case where there is only one unknown z and $M_1, M_2, \ldots, M_m$ are m observed values of z. If z^* is the true value of the quantity, the errors associated with the observations are

$$x_1 = z^* - M_1, \quad x_2 = z^* - M_2, \quad \ldots, \quad x_m = z^* - M_m \qquad (D.7)$$

Differentiating Equation (D.7) gives

$$1 = \frac{dx_1}{dz} = \frac{dx_2}{dz} = \cdots = \frac{dx_m}{dz} \qquad (D.8)$$

Then for this special case, substituting Equations (D.7) and (D.8) into Equations (D.6), they reduce to a single equation:

$$\frac{f'(z^* - M_1)}{f(z^* - M_1)} + \frac{f'(z^* - M_2)}{f(z^* - M_2)} + \cdots + \frac{f'(z^* - M_m)}{f(z^* - M_m)} = 0 \qquad (D.9)$$

Equation (D.9) for this special case in consideration is also general for any value of m and for any observed values $M_1, M_2, \ldots, M_m$. Thus, let the values of M be

$$M_2 = M_3 = \cdots = M_m = M_1 - mN$$

where N is chosen for convenience as $N = (M_1 - M_2)/m$.

The arithmetic mean is the most probable value for this case of a single quantity having been observed several times; therefore, z^*, the most probable value in this case, is

$$z^* = \frac{M_1 + M_2 + \cdots + M_n}{m}$$

$$= \frac{M_1 + (m - 1)(M_1 - mN)}{m}$$

$$= M_1 - mN + N$$

$$= M_1 - N(m - 1)$$

$$z^* - M_1 = -N(m - 1) = N(1 - m) \qquad (D.10)$$

Recall that $N = (M_1 - M_2)/m$, from which $M_1 = mN + M_2$. Substituting into Equation (D.10) gives

$$z^* - (mN + M_2) = N(1 - m)$$

$$z^* - M_2 = N$$

Similarly, since $N = (M_1 - M_3)/m = (M_1 - M_4)/m$, and so on,

$$z^* - M_3 = N$$
$$z^* - M_4 = N$$
$$\vdots$$

Substituting these expressions into Equation (D.9) yields

$$\frac{f'[n(1-m)]}{f[N(1-M)]} + \frac{(m-1)f'(N)}{f(N)} = 0 \qquad (D.11)$$

Rearranging yields

$$\frac{f[N(1-m)]}{Nf[N(1-m)](1-m)} = \frac{f(N)}{f(N)N} = \text{constant}$$

because N in this case is a constant. Thus,

$$\frac{f'(x)}{xf(x)} = \text{constant} = K \qquad (D.12)$$

Substituting Equation (D.5) into Equation (D.12) yields

$$f'(x) = xf(x)K = \frac{df(x)}{dx}$$

from which $df(x)/dx = xf(x)K$. Integrating gives

$$\log_e f(x) = \tfrac{1}{2} Kx^2 + C_1$$
$$f(x) = e^{C_1} e^{Kx^2/2}$$

But letting

$$e^{C_1} = C$$

then

$$f(x) = Ce^{(Kx^2)/2} \qquad (D.13)$$

In Equation (D.13), since $f(x)$ decreases as x increases, the exponent must be negative. Arbitrarily letting

$$h = \sqrt{\frac{K}{2}} \qquad (D.14)$$

D.1 DEVELOPMENT OF THE NORMAL DISTRIBUTION CURVE EQUATION

and incorporating the negative into Equation (D.13), there results

$$f(x) = Ce^{-h^2x^2} \tag{D.15}$$

To find the value of the constant C, substitute Equation (D.15) into Equation (D.2):

$$\int_{-\infty}^{\infty} Ce^{-h^2x^2}\, dx = 1$$

Also, arbitrarily set $t = hx$; then $dt = h\, dx$ and $dx = dt/h$, from which, after changing variables, we obtain

$$\frac{C}{h} = \int_{-\infty}^{\infty} e^{-t^2}dt = 1$$

The value of the definite integral is $\sqrt{\pi}$, from which[1]

$$\frac{C}{h\sqrt{\pi}} = 1$$

$$C = \frac{h}{\sqrt{\pi}} \tag{D.16}$$

Substituting Equation (D.16) into Equation (D.15) gives

$$f(x) = \frac{h}{\sqrt{\pi}} e^{-h^2x^2} \tag{D.17}$$

Note that from Equation (D.14) that $h = \sqrt{K/2}$. For the normal distribution, $K = 1/\sigma^2$. Substituting this into Equation (D.17) gives

$$f(x) = \frac{1}{\sqrt{2\sigma^2\pi}} e^{-(1/2\sigma^2)x^2} = \frac{1}{\sigma\sqrt{2\pi}} e^{-x^2/2\sigma^2} \tag{D.18}$$

where the terms are as defined for Equation (3.2).

This is the general equation for the probability curve, having been derived in this instance from the consideration of a special case. In Table D.1, values for areas under the standard normal distribution function from negative infinity to t are tabulated.

---

[1] The technique of integrating this nonelementary function is beyond the scope of this book but can be found in advanced references.

TABLE D.1 Percentage Points for the Standard Normal Distribution Function

$$N_z(t) = \int_{-\infty}^{\infty} \frac{1}{\sqrt{2\pi}} e^{-x^2/2}\, dx$$

t	0	1	2	3	4	5	6	7	8	9
−3.2	0.00069	0.00066	0.00064	0.00062	0.00060	0.00058	0.00056	0.00054	0.00052	0.00050
−3.1	0.00097	0.00094	0.00090	0.00087	0.00084	0.00082	0.00079	0.00076	0.00074	0.00071
−3.0	0.00135	0.00131	0.00126	0.00122	0.00118	0.00114	0.00111	0.00107	0.00104	0.00100
−2.9	0.00187	0.00181	0.00175	0.00169	0.00164	0.00159	0.00154	0.00149	0.00144	0.00139
−2.8	0.00256	0.00248	0.00240	0.00233	0.00226	0.00219	0.00212	0.00205	0.00199	0.00193
−2.7	0.00347	0.00336	0.00326	0.00317	0.00307	0.00298	0.00289	0.00280	0.00272	0.00264
−2.6	0.00466	0.00453	0.00440	0.00427	0.00415	0.00402	0.00391	0.00379	0.00368	0.00357
−2.5	0.00621	0.00604	0.00587	0.00570	0.00554	0.00539	0.00523	0.00508	0.00494	0.00480
−2.4	0.00820	0.00798	0.00776	0.00755	0.00734	0.00714	0.00695	0.00676	0.00657	0.00639
−2.3	0.01072	0.01044	0.01017	0.00990	0.00964	0.00939	0.00914	0.00889	0.00866	0.00842
−2.2	0.01390	0.01355	0.01321	0.01287	0.01255	0.01222	0.01191	0.01160	0.01130	0.01101
−2.1	0.01786	0.01743	0.01700	0.01659	0.01618	0.01578	0.01539	0.01500	0.01463	0.01426
−2.0	0.02275	0.02222	0.02169	0.02118	0.02068	0.02018	0.01970	0.01923	0.01876	0.01831
−1.9	0.02872	0.02807	0.02743	0.02680	0.02619	0.02559	0.02500	0.02442	0.02385	0.02330
−1.8	0.03593	0.03515	0.03438	0.03362	0.03288	0.03216	0.03144	0.03074	0.03005	0.02938
−1.7	0.04457	0.04363	0.04272	0.04182	0.04093	0.04006	0.03920	0.03836	0.03754	0.03673
−1.6	0.05480	0.05370	0.05262	0.05155	0.05050	0.04947	0.04846	0.04746	0.04648	0.04551
−1.5	0.06681	0.06552	0.06426	0.06301	0.06178	0.06057	0.05938	0.05821	0.05705	0.05592
−1.4	0.08076	0.07927	0.07780	0.07636	0.07493	0.07353	0.07215	0.07078	0.06944	0.06811
−1.3	0.09680	0.09510	0.09342	0.09176	0.09012	0.08851	0.08691	0.08534	0.08379	0.08226
−1.2	0.11507	0.11314	0.11123	0.10935	0.10749	0.10565	0.10383	0.10204	0.10027	0.09853
−1.1	0.13567	0.13350	0.13136	0.12924	0.12714	0.12507	0.12302	0.12100	0.11900	0.11702
−1.0	0.15866	0.15625	0.15386	0.15151	0.14917	0.14686	0.14457	0.14231	0.14007	0.13786
−0.9	0.18406	0.18141	0.17879	0.17619	0.17361	0.17106	0.16853	0.16602	0.16354	0.16109
−0.8	0.21186	0.20897	0.20611	0.20327	0.20045	0.19766	0.19489	0.19215	0.18943	0.18673
−0.7	0.24196	0.23885	0.23576	0.23270	0.22965	0.22663	0.22363	0.22065	0.21770	0.21476
−0.6	0.27425	0.27093	0.26763	0.26435	0.26109	0.25785	0.25463	0.25143	0.24825	0.24510
−0.5	0.30854	0.30503	0.30153	0.29806	0.29460	0.29116	0.28774	0.28434	0.28096	0.27760
−0.4	0.34458	0.34090	0.33724	0.33360	0.32997	0.32636	0.32276	0.31918	0.31561	0.31207
−0.3	0.38209	0.37828	0.37448	0.37070	0.36693	0.36317	0.35942	0.35569	0.35197	0.34827
−0.2	0.42074	0.41683	0.41294	0.40905	0.40517	0.40129	0.39743	0.39358	0.38974	0.38591
−0.1	0.46017	0.45620	0.45224	0.44828	0.44433	0.44038	0.43644	0.43251	0.42858	0.42465
−0.0	0.50000	0.49601	0.49202	0.48803	0.48405	0.48006	0.47608	0.47210	0.46812	0.46414

TABLE D.1 (*Continued*)

$$N_z(t) = \int_{-\infty}^{\infty} \frac{1}{\sqrt{2\pi}} e^{-x^2/2} \, dx$$

t	0	1	2	3	4	5	6	7	8	9
0.0	0.50000	0.50399	0.50798	0.51197	0.51595	0.51994	0.52392	0.52790	0.53188	0.53586
0.1	0.53983	0.54380	0.54776	0.55172	0.55567	0.55962	0.56356	0.56749	0.57142	0.57535
0.2	0.57926	0.58317	0.58706	0.59095	0.59483	0.59871	0.60257	0.60642	0.61026	0.61409
0.3	0.61791	0.62172	0.62552	0.62930	0.63307	0.63683	0.64058	0.64431	0.64803	0.65173
0.4	0.65542	0.65910	0.66276	0.66640	0.67003	0.67364	0.67724	0.68082	0.68439	0.68793
0.5	0.69146	0.69497	0.69847	0.70194	0.70540	0.70884	0.71226	0.71566	0.71904	0.72240
0.6	0.72575	0.72907	0.73237	0.73565	0.73891	0.74215	0.74537	0.74857	0.75175	0.75490
0.7	0.75804	0.76115	0.76424	0.76730	0.77035	0.77337	0.77637	0.77935	0.78230	0.78524
0.8	0.78814	0.79103	0.79389	0.79673	0.79955	0.80234	0.80511	0.80785	0.81057	0.81327
0.9	0.81594	0.81859	0.82121	0.82381	0.82639	0.82894	0.83147	0.83398	0.83646	0.83891
1.0	0.84134	0.84375	0.84614	0.84849	0.85083	0.85314	0.85543	0.85769	0.85993	0.86214
1.1	0.86433	0.86650	0.86864	0.87076	0.87286	0.87493	0.87698	0.87900	0.88100	0.88298
1.2	0.88493	0.88686	0.88877	0.89065	0.89251	0.89435	0.89617	0.89796	0.89973	0.90147
1.3	0.90320	0.90490	0.90658	0.90824	0.90988	0.91149	0.91309	0.91466	0.91621	0.91774
1.4	0.91924	0.92073	0.92220	0.92364	0.92507	0.92647	0.92785	0.92922	0.93056	0.93189
1.5	0.93319	0.93448	0.93574	0.93699	0.93822	0.93943	0.94062	0.94179	0.94295	0.94408
1.6	0.94520	0.94630	0.94738	0.94845	0.94950	0.95053	0.95154	0.95254	0.95352	0.95449
1.7	0.95543	0.95637	0.95728	0.95818	0.95907	0.95994	0.96080	0.96164	0.96246	0.96327
1.8	0.96407	0.96485	0.96562	0.96638	0.96712	0.96784	0.96856	0.96926	0.96995	0.97062
1.9	0.97128	0.97193	0.97257	0.97320	0.97381	0.97441	0.97500	0.97558	0.97615	0.97670
2.0	0.97725	0.97778	0.97831	0.97882	0.97932	0.97982	0.98030	0.98077	0.98124	0.98169
2.1	0.98214	0.98257	0.98300	0.98341	0.98382	0.98422	0.98461	0.98500	0.98537	0.98574
2.2	0.98610	0.98645	0.98679	0.98713	0.98745	0.98778	0.98809	0.98840	0.98870	0.98899
2.3	0.98928	0.98956	0.98983	0.99010	0.99036	0.99061	0.99086	0.99111	0.99134	0.99158
2.4	0.99180	0.99202	0.99224	0.99245	0.99266	0.99286	0.99305	0.99324	0.99343	0.99361
2.5	0.99379	0.99396	0.99413	0.99430	0.99446	0.99461	0.99477	0.99492	0.99506	0.99520
2.6	0.99534	0.99547	0.99560	0.99573	0.99585	0.99598	0.99609	0.99621	0.99632	0.99643
2.7	0.99653	0.99664	0.99674	0.99683	0.99693	0.99702	0.99711	0.99720	0.99728	0.99736
2.8	0.99744	0.99752	0.99760	0.99767	0.99774	0.99781	0.99788	0.99795	0.99801	0.99807
2.9	0.99813	0.99819	0.99825	0.99831	0.99836	0.99841	0.99846	0.99851	0.99856	0.99861
3.0	0.99865	0.99869	0.99874	0.99878	0.99882	0.99886	0.99889	0.99893	0.99896	0.99900
3.1	0.99903	0.99906	0.99910	0.99913	0.99916	0.99918	0.99921	0.99924	0.99926	0.99929
3.2	0.99931	0.99934	0.99936	0.99938	0.99940	0.99942	0.99944	0.99946	0.99948	0.99950

D.2 OTHER STATISTICAL TABLES

Three often-used statistical tables whose use and interpretation are discussed in detail in Chapters 4 and 5 are given below. The equations used to generate each table are also presented.

D.2.1 χ^2 Distribution

Chi square is a density function for the distribution of sample variances computed from sets with degrees of freedom selected for a population. The use of this distribution to construct confidence intervals for the population variance and to perform hypothesis testing involving the population variance are discussed in detail in Chapter 4. The χ^2 distribution is illustrated in Figure D.2.

The χ^2 distribution critical values given in Table D.2 were generated using the following function. (Critical χ^2 values for both tails of the distribution were derived with a program using numerical integration routines similar to those used in STATS.)

$$\alpha = \int_0^{\chi^2} \frac{1}{2^{v/2}\Gamma(v/2)} u^{(v-2)/2} e^{-u/2} \, du$$

where v is the degrees of freedom and Γ is the gamma function, which is defined as

$$\Gamma(v) = \int_0^\infty u^{v-1} e^{-u} \, du$$

It is computed as $\Gamma(v) = (v-1)! = (v-1)(v-2)(v-3) \cdots (3)(2)(1)$.

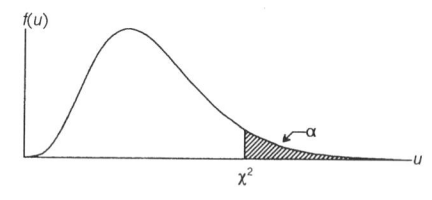

Figure D.2 χ^2 distribution.

TABLE D.2 Critical Values for the χ^2 Distribution

$\alpha \rightarrow$ $\nu \downarrow$	0.999	0.995	0.990	0.975	0.950	0.900	0.500	0.100	0.050	0.025	0.010	0.005	0.001
1	0.000002	0.000039	0.000157	0.000982	0.004	0.016	0.455	2.705	3.841	5.023	6.634	7.877	10.81
2	0.002	0.01	0.02	0.05	0.10	0.21	1.39	4.61	5.99	7.38	9.21	10.60	13.81
3	0.02	0.07	0.12	0.22	0.35	0.58	2.37	6.25	7.82	9.35	11.34	12.84	16.26
4	0.09	0.21	0.30	0.48	0.71	1.06	3.36	7.78	9.49	11.14	13.28	14.86	18.47
5	0.21	0.41	0.55	0.83	1.15	1.61	4.35	9.24	11.07	12.83	15.09	16.75	20.51
6	0.38	0.68	0.87	1.24	1.64	2.20	5.35	10.64	12.59	14.45	16.81	18.55	22.46
7	0.60	0.99	1.24	1.69	2.17	2.83	6.35	12.02	14.07	16.01	18.48	20.28	24.32
8	0.86	1.34	1.65	2.18	2.73	3.49	7.34	13.36	15.51	17.53	20.09	21.96	26.12
9	1.15	1.74	2.09	2.70	3.33	4.17	8.34	14.68	16.92	19.02	21.67	23.59	27.88
10	1.48	2.16	2.56	3.25	3.94	4.87	9.34	15.99	18.31	20.48	23.21	25.19	29.59
11	1.83	2.60	3.05	3.82	4.58	5.58	10.34	17.28	19.68	21.92	24.72	26.76	31.26
12	2.21	3.07	3.57	4.40	5.23	6.30	11.34	18.55	21.03	23.34	26.22	28.30	32.91
13	2.62	3.57	4.11	5.01	5.89	7.04	12.34	19.81	22.36	24.74	27.69	29.82	34.53
14	3.04	4.08	4.66	5.63	6.57	7.79	13.34	21.06	23.68	26.12	29.14	31.32	36.12
15	3.48	4.60	5.23	6.26	7.26	8.55	14.34	22.31	25.00	27.49	30.58	32.80	37.70
16	3.94	5.14	5.81	6.91	7.96	9.31	15.34	23.54	26.30	28.85	32.00	34.27	39.25
17	4.42	5.70	6.41	7.56	8.67	10.09	16.34	24.77	27.59	30.19	33.41	35.72	40.79
18	4.91	6.27	7.02	8.23	9.39	10.86	17.34	25.99	28.87	31.53	34.81	37.16	42.31
19	5.41	6.84	7.63	8.91	10.12	11.65	18.34	27.20	30.14	32.85	36.19	38.58	43.82
20	5.92	7.43	8.26	9.59	10.85	12.44	19.34	28.41	31.41	34.17	37.57	40.00	45.31
21	6.45	8.03	8.90	10.28	11.59	13.24	20.34	29.62	32.67	35.48	38.93	41.40	46.80
22	6.98	8.64	9.54	10.98	12.34	14.04	21.34	30.81	33.92	36.78	40.29	42.80	48.27
23	7.53	9.26	10.20	11.69	13.09	14.85	22.34	32.01	35.17	38.08	41.64	44.18	49.73
24	8.09	9.89	10.86	12.40	13.85	15.66	23.34	33.20	36.42	39.36	42.98	45.56	51.18
25	8.65	10.52	11.52	13.12	14.61	16.47	24.34	34.38	37.65	40.65	44.31	46.93	52.62
26	9.22	11.16	12.20	13.84	15.38	17.29	25.34	35.56	38.89	41.92	45.64	48.29	54.05
27	9.80	11.81	12.88	14.57	16.15	18.11	26.34	36.74	40.11	43.19	46.96	49.64	55.48
28	10.39	12.46	13.56	15.31	16.93	18.94	27.34	37.92	41.34	44.46	48.28	50.99	56.89
29	10.99	13.12	14.26	16.05	17.71	19.77	28.34	39.09	42.56	45.72	49.59	52.34	58.30
30	11.59	13.79	14.95	16.79	18.49	20.60	29.34	40.26	43.77	46.98	50.89	53.67	59.70
35	14.69	17.19	18.51	20.57	22.47	24.80	34.34	46.06	49.80	53.20	57.34	60.27	66.62
40	17.92	20.71	22.16	24.43	26.51	29.05	39.34	51.81	55.76	59.34	63.69	66.77	73.40
50	24.67	27.99	29.71	32.36	34.76	37.69	49.33	63.17	67.50	71.42	76.15	79.49	86.66
60	31.74	35.53	37.48	40.48	43.19	46.46	59.33	74.40	79.08	83.30	88.38	91.95	99.61
120	77.76	83.85	86.92	91.57	95.70	100.62	119.33	140.23	146.57	152.21	158.95	163.65	173.6

D.2.2 t Distribution

The t distribution (also known as *Student's distribution*), shown in Figure D.3, is used to derive confidence intervals for the population mean when the sample set is small. It is also used in hypothesis testing to check the validity of a sample mean against a population mean. The uses for this distribution are discussed in greater detail in Chapter 4.

The t distribution tables were generated using the following function. (Critical t values for the upper tail of the distribution were derived with a program using numerical integration routines similar to those available in STATS.)

$$\alpha = \int \frac{\Gamma(\nu + 1)/2}{\sqrt{\nu\pi}\,\Gamma(\nu/2)} \left(1 + \frac{x^2}{\nu}\right)^{-(\nu+1)/2} dx$$

where Γ is the gamma function as defined in Section D.2.1 and ν is the degrees of freedom in the function. In Table D.3, critical values of t are listed that are required to achieve the percentage points listed in the top row. The distribution is symmetrical, and thus

$$F(-t) = 1 - F(t)$$

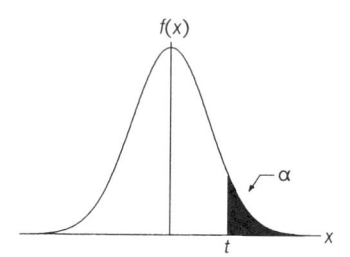

Figure D.3 t distribution.

TABLE D.3 Critical Values for the t Distribution

$\alpha \rightarrow$ $\nu \downarrow$	0.400	0.350	0.300	0.250	0.200	0.150	0.100	0.050	0.025	0.010	0.005	0.001	0.0005
1	0.325	0.510	0.727	1.000	1.376	1.963	3.078	6.314	12.705	31.816	63.639	318.20	636.18
2	0.289	0.445	0.617	0.816	1.061	1.386	1.886	2.920	4.303	6.964	9.925	22.401	31.579
3	0.277	0.424	0.584	0.765	0.978	1.250	1.638	2.353	3.183	4.541	5.842	10.216	12.954
4	0.271	0.414	0.569	0.741	0.941	1.190	1.533	2.132	2.776	3.748	4.604	6.897	8.610
5	0.267	0.408	0.559	0.727	0.920	1.156	1.476	2.015	2.571	3.365	4.032	5.895	6.880
6	0.265	0.404	0.553	0.718	0.906	1.134	1.440	1.943	2.447	3.143	3.708	5.208	5.961
7	0.263	0.402	0.549	0.711	0.896	1.119	1.415	1.895	2.365	2.998	3.500	4.785	5.408
8	0.262	0.399	0.546	0.706	0.889	1.108	1.397	1.860	2.306	2.896	3.356	4.510	5.041
9	0.261	0.398	0.543	0.703	0.883	1.100	1.383	1.833	2.262	2.821	3.250	4.304	4.781
10	0.260	0.397	0.542	0.700	0.879	1.093	1.372	1.812	2.228	2.764	3.169	4.149	4.605
11	0.260	0.396	0.540	0.697	0.876	1.088	1.363	1.796	2.201	2.718	3.106	4.029	4.452
12	0.259	0.395	0.539	0.695	0.873	1.083	1.356	1.782	2.179	2.681	3.055	3.933	4.329
13	0.259	0.394	0.538	0.694	0.870	1.079	1.350	1.771	2.160	2.650	3.012	3.854	4.230
14	0.258	0.393	0.537	0.692	0.868	1.076	1.345	1.761	2.145	2.624	2.977	3.789	4.148
15	0.258	0.393	0.536	0.691	0.866	1.074	1.341	1.753	2.131	2.602	2.947	3.734	4.079
16	0.258	0.392	0.535	0.690	0.865	1.071	1.337	1.746	2.120	2.583	2.921	3.688	4.021
17	0.257	0.392	0.534	0.689	0.863	1.069	1.333	1.740	2.110	2.567	2.898	3.647	3.970
18	0.257	0.392	0.534	0.688	0.862	1.067	1.330	1.734	2.101	2.552	2.878	3.611	3.926
19	0.257	0.391	0.533	0.688	0.861	1.066	1.328	1.729	2.093	2.539	2.861	3.580	3.887
20	0.257	0.391	0.533	0.687	0.860	1.064	1.325	1.725	2.086	2.528	2.845	3.553	3.853
21	0.257	0.391	0.532	0.686	0.859	1.063	1.323	1.721	2.080	2.518	2.831	3.528	3.822
22	0.256	0.390	0.532	0.686	0.858	1.061	1.321	1.717	2.074	2.508	2.819	3.506	3.795
23	0.256	0.390	0.532	0.685	0.858	1.060	1.319	1.714	2.069	2.500	2.807	3.486	3.770
24	0.256	0.390	0.531	0.685	0.857	1.059	1.318	1.711	2.064	2.492	2.797	3.467	3.748
25	0.256	0.390	0.531	0.684	0.856	1.058	1.316	1.708	2.060	2.485	2.787	3.451	3.727
26	0.256	0.390	0.531	0.684	0.856	1.058	1.315	1.706	2.056	2.479	2.779	3.435	3.708
27	0.256	0.389	0.531	0.684	0.855	1.057	1.314	1.703	2.052	2.473	2.771	3.421	3.691
28	0.256	0.389	0.530	0.683	0.855	1.056	1.313	1.701	2.048	2.467	2.763	3.409	3.675
29	0.256	0.389	0.530	0.683	0.854	1.055	1.311	1.699	2.045	2.462	2.756	3.397	3.661
30	0.256	0.389	0.530	0.683	0.854	1.055	1.310	1.697	2.042	2.457	2.750	3.385	3.647
35	0.255	0.388	0.529	0.682	0.852	1.052	1.306	1.690	2.030	2.438	2.724	3.340	3.592
40	0.255	0.388	0.529	0.681	0.851	1.050	1.303	1.684	2.021	2.423	2.704	3.307	3.552
60	0.254	0.387	0.527	0.679	0.848	1.045	1.296	1.671	2.000	2.390	2.660	3.232	3.461
120	0.254	0.386	0.526	0.677	0.845	1.041	1.289	1.658	1.980	2.358	2.617	3.160	3.374
∞	0.253	0.385	0.525	0.675	0.842	1.037	1.282	1.645	1.960	2.326	2.576	3.291	3.300

D.2.3 F Distribution

This F distribution (also known as the *Fisher distribution*), shown in Figure D.4, is used to derive confidence intervals for the ratio of two population variances. It is also used in hypothesis testing for this ratio. Uses for this distribution are discussed in Chapter 4.

Critical F values for the upper tail of the distribution were derived with a program using numerical integration routines similar to those used in STATS. The tables were generated using the following function:

$$\alpha = \int_F^\infty \frac{\Gamma(\nu_1 + \nu_2)/2}{\Gamma(\nu_1/2)\Gamma(\nu_2/2)} \left(\frac{\nu_1}{\nu_2}\right)^{\nu_1/2} \frac{x^{(\nu_1-2)/2}}{1 + (\nu_1/\nu_2)x^{(\nu_1+\nu_2)/2}} dx$$

where Γ is the gamma function as defined in Section D.2.1, ν_1 the numerator degrees of freedom, and ν_2 the denominator degrees of freedom.

For critical values in the lower tail of the distribution, the following relationship can be used in conjunction with the tabular values given in Table D.4.

$$F_{\alpha,\nu_1,\nu_2} = \frac{1}{F_{1-\alpha,\nu_2,\nu_1}}$$

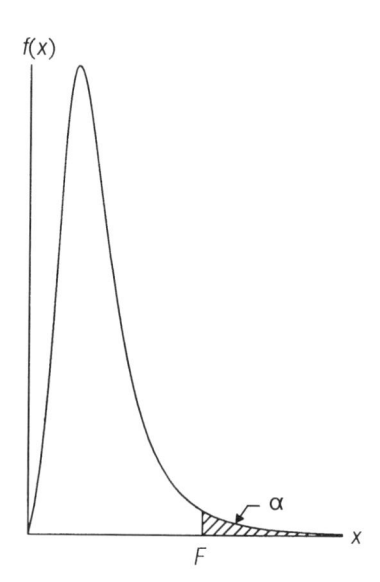

Figure D.4 F distribution.

TABLE D.4 Critical Values for the F Distribution

$\alpha = 0.20$

$v_2 \downarrow$ $v_1 \rightarrow$	1	2	3	4	5	6	7	8	9	10	12	15	20	24	30	40	60	120
1	9.47	12.00	13.06	13.64	14.01	14.26	14.44	14.58	14.68	14.77	14.90	15.04	15.17	15.24	15.31	15.37	15.44	15.51
2	3.56	4.00	4.16	4.24	4.28	4.32	4.34	4.36	4.37	4.38	4.40	4.42	4.43	4.44	4.45	4.46	4.46	4.47
3	2.68	2.89	2.94	2.96	2.97	2.97	2.97	2.98	2.98	2.98	2.98	2.98	2.98	2.98	2.98	2.98	2.98	2.98
4	2.35	2.47	2.48	2.48	2.48	2.47	2.47	2.47	2.46	2.46	2.46	2.45	2.44	2.44	2.44	2.44	2.43	2.43
5	2.18	2.26	2.25	2.24	2.23	2.22	2.21	2.20	2.20	2.19	2.18	2.18	2.17	2.16	2.16	2.15	2.15	2.14
6	2.07	2.13	2.11	2.09	2.08	2.06	2.05	2.04	2.03	2.03	2.02	2.01	2.00	1.99	1.98	1.98	1.97	1.96
7	2.00	2.04	2.02	1.99	1.97	1.96	1.94	1.93	1.93	1.92	1.91	1.89	1.88	1.87	1.86	1.86	1.85	1.84
8	1.95	1.98	1.95	1.92	1.90	1.88	1.87	1.86	1.85	1.84	1.83	1.81	1.80	1.79	1.78	1.77	1.76	1.75
9	1.91	1.93	1.90	1.87	1.85	1.83	1.81	1.80	1.79	1.78	1.76	1.75	1.73	1.72	1.71	1.70	1.69	1.68
10	1.88	1.90	1.86	1.83	1.80	1.78	1.77	1.75	1.74	1.73	1.72	1.70	1.68	1.67	1.66	1.65	1.64	1.63
11	1.86	1.87	1.83	1.80	1.77	1.75	1.73	1.72	1.70	1.69	1.68	1.66	1.64	1.63	1.62	1.61	1.60	1.59
12	1.84	1.85	1.80	1.77	1.74	1.72	1.70	1.69	1.67	1.66	1.65	1.63	1.61	1.60	1.59	1.58	1.56	1.55
13	1.82	1.83	1.78	1.75	1.72	1.69	1.68	1.66	1.65	1.64	1.62	1.60	1.58	1.57	1.56	1.55	1.53	1.52
14	1.81	1.81	1.76	1.73	1.70	1.67	1.65	1.64	1.63	1.62	1.60	1.58	1.56	1.55	1.53	1.52	1.51	1.49
15	1.80	1.80	1.75	1.71	1.68	1.66	1.64	1.62	1.61	1.60	1.58	1.56	1.54	1.53	1.51	1.50	1.49	1.47
16	1.79	1.78	1.74	1.70	1.67	1.64	1.62	1.61	1.59	1.58	1.56	1.54	1.52	1.51	1.49	1.48	1.47	1.45
17	1.78	1.77	1.72	1.68	1.65	1.63	1.61	1.59	1.58	1.57	1.55	1.53	1.50	1.49	1.48	1.46	1.45	1.43
18	1.77	1.76	1.71	1.67	1.64	1.62	1.60	1.58	1.56	1.55	1.53	1.51	1.49	1.48	1.46	1.45	1.43	1.42
19	1.76	1.75	1.70	1.66	1.63	1.61	1.58	1.57	1.55	1.54	1.52	1.50	1.48	1.46	1.45	1.44	1.42	1.40
20	1.76	1.75	1.70	1.65	1.62	1.60	1.58	1.56	1.54	1.53	1.51	1.49	1.47	1.45	1.44	1.42	1.41	1.39
21	1.75	1.74	1.69	1.65	1.61	1.59	1.57	1.55	1.53	1.52	1.50	1.48	1.46	1.44	1.43	1.41	1.40	1.38
22	1.75	1.73	1.68	1.64	1.61	1.58	1.56	1.54	1.53	1.51	1.49	1.47	1.45	1.43	1.42	1.40	1.39	1.37
23	1.74	1.73	1.68	1.63	1.60	1.57	1.55	1.53	1.52	1.51	1.49	1.46	1.44	1.42	1.41	1.40	1.38	1.36
24	1.74	1.72	1.67	1.63	1.59	1.57	1.55	1.53	1.51	1.50	1.48	1.46	1.43	1.42	1.40	1.39	1.37	1.35
25	1.73	1.72	1.66	1.62	1.59	1.56	1.54	1.52	1.51	1.49	1.47	1.45	1.42	1.41	1.39	1.38	1.36	1.34
26	1.73	1.71	1.66	1.62	1.58	1.56	1.53	1.52	1.50	1.49	1.47	1.44	1.42	1.40	1.39	1.37	1.35	1.33
27	1.73	1.71	1.66	1.61	1.58	1.55	1.53	1.51	1.49	1.48	1.46	1.44	1.41	1.40	1.38	1.36	1.35	1.33
28	1.72	1.71	1.65	1.61	1.57	1.55	1.52	1.51	1.49	1.48	1.46	1.43	1.41	1.39	1.37	1.36	1.34	1.32
29	1.72	1.70	1.65	1.60	1.57	1.54	1.52	1.50	1.49	1.47	1.45	1.43	1.40	1.39	1.37	1.35	1.33	1.31
30	1.72	1.70	1.64	1.60	1.57	1.54	1.52	1.50	1.48	1.47	1.45	1.42	1.39	1.38	1.36	1.35	1.33	1.31
50	1.69	1.66	1.60	1.56	1.52	1.49	1.47	1.45	1.43	1.42	1.39	1.37	1.34	1.32	1.30	1.28	1.26	1.24
60	1.68	1.65	1.59	1.55	1.51	1.48	1.46	1.44	1.42	1.41	1.38	1.35	1.32	1.31	1.29	1.27	1.24	1.22
80	1.67	1.64	1.58	1.53	1.50	1.47	1.44	1.42	1.41	1.39	1.37	1.34	1.31	1.29	1.27	1.25	1.22	1.19
120	1.66	1.63	1.57	1.52	1.48	1.45	1.43	1.41	1.39	1.37	1.35	1.32	1.29	1.27	1.25	1.23	1.20	1.17

(*continues*)

TABLE D.4 (*Continued*)
$\alpha = 0.10$

$\nu_1 \rightarrow$ $\nu_2 \downarrow$	1	2	3	4	5	6	7	8	9	10	12	15	20	24	30	40	60	120
1	39.85	49.49	53.59	55.83	57.23	58.20	58.90	59.43	59.85	60.19	60.70	61.21	61.73	61.99	62.26	62.52	62.79	63.05
2	8.53	9.00	9.16	9.24	9.29	9.33	9.35	9.37	9.38	9.39	9.41	9.42	9.44	9.45	9.46	9.47	9.47	9.48
3	5.54	5.46	5.39	5.34	5.31	5.28	5.27	5.25	5.24	5.23	5.22	5.20	5.18	5.18	5.17	5.16	5.15	5.14
4	4.54	4.32	4.19	4.11	4.05	4.01	3.98	3.95	3.94	3.92	3.90	3.87	3.84	3.83	3.82	3.80	3.79	3.78
5	4.06	3.78	3.62	3.52	3.45	3.40	3.37	3.34	3.32	3.30	3.27	3.24	3.21	3.19	3.17	3.16	3.14	3.12
6	3.78	3.46	3.29	3.18	3.11	3.05	3.01	2.98	2.96	2.94	2.90	2.87	2.84	2.82	2.80	2.78	2.76	2.74
7	3.59	3.26	3.07	2.96	2.88	2.83	2.78	2.75	2.72	2.70	2.67	2.63	2.59	2.58	2.56	2.54	2.51	2.49
8	3.46	3.11	2.92	2.81	2.73	2.67	2.62	2.59	2.56	2.54	2.50	2.46	2.42	2.40	2.38	2.36	2.34	2.32
9	3.36	3.01	2.81	2.69	2.61	2.55	2.51	2.47	2.44	2.42	2.38	2.34	2.30	2.28	2.25	2.23	2.21	2.18
10	3.28	2.92	2.73	2.61	2.52	2.46	2.41	2.38	2.35	2.32	2.28	2.24	2.20	2.18	2.16	2.13	2.11	2.08
11	3.23	2.86	2.66	2.54	2.45	2.39	2.34	2.30	2.27	2.25	2.21	2.17	2.12	2.10	2.08	2.05	2.03	2.00
12	3.18	2.81	2.61	2.48	2.39	2.33	2.28	2.24	2.21	2.19	2.15	2.10	2.06	2.04	2.01	1.99	1.96	1.93
13	3.14	2.76	2.56	2.43	2.35	2.28	2.23	2.20	2.16	2.14	2.10	2.05	2.01	1.98	1.96	1.93	1.90	1.88
14	3.10	2.73	2.52	2.39	2.31	2.24	2.19	2.15	2.12	2.10	2.05	2.01	1.96	1.94	1.91	1.89	1.86	1.83
15	3.07	2.70	2.49	2.36	2.27	2.21	2.16	2.12	2.09	2.06	2.02	1.97	1.92	1.90	1.87	1.85	1.82	1.79
16	3.05	2.67	2.46	2.33	2.24	2.18	2.13	2.09	2.06	2.03	1.99	1.94	1.89	1.87	1.84	1.81	1.78	1.75
17	3.03	2.64	2.44	2.31	2.22	2.15	2.10	2.06	2.03	2.00	1.96	1.91	1.86	1.84	1.81	1.78	1.75	1.72
18	3.01	2.62	2.42	2.29	2.20	2.13	2.08	2.04	2.00	1.98	1.93	1.89	1.84	1.81	1.78	1.75	1.72	1.69
19	2.99	2.61	2.40	2.27	2.18	2.11	2.06	2.02	1.98	1.96	1.91	1.86	1.81	1.79	1.76	1.73	1.70	1.67
20	2.97	2.59	2.38	2.25	2.16	2.09	2.04	2.00	1.96	1.94	1.89	1.84	1.79	1.77	1.74	1.71	1.68	1.64
21	2.96	2.57	2.36	2.23	2.14	2.08	2.02	1.98	1.95	1.92	1.87	1.83	1.78	1.75	1.72	1.69	1.66	1.62
22	2.95	2.56	2.35	2.22	2.13	2.06	2.01	1.97	1.93	1.90	1.86	1.81	1.76	1.73	1.70	1.67	1.64	1.60
23	2.94	2.55	2.34	2.21	2.11	2.05	1.99	1.95	1.92	1.89	1.84	1.80	1.74	1.72	1.69	1.66	1.62	1.59
24	2.93	2.54	2.33	2.19	2.10	2.04	1.98	1.94	1.91	1.88	1.83	1.78	1.73	1.70	1.67	1.64	1.61	1.57
25	2.92	2.53	2.32	2.18	2.09	2.02	1.97	1.93	1.89	1.87	1.82	1.77	1.72	1.69	1.66	1.63	1.59	1.56
26	2.91	2.52	2.31	2.17	2.08	2.01	1.96	1.92	1.88	1.86	1.81	1.76	1.71	1.68	1.65	1.61	1.58	1.54
27	2.90	2.51	2.30	2.17	2.07	2.00	1.95	1.91	1.87	1.85	1.80	1.75	1.70	1.67	1.64	1.60	1.57	1.53
28	2.89	2.50	2.29	2.16	2.06	2.00	1.94	1.90	1.87	1.84	1.79	1.74	1.69	1.66	1.63	1.59	1.56	1.52
29	2.89	2.50	2.28	2.15	2.06	1.99	1.93	1.89	1.86	1.83	1.78	1.73	1.68	1.65	1.62	1.58	1.55	1.51
30	2.88	2.49	2.28	2.14	2.05	1.98	1.93	1.88	1.85	1.82	1.77	1.72	1.67	1.64	1.61	1.57	1.54	1.50
50	2.81	2.41	2.20	2.06	1.97	1.90	1.84	1.80	1.76	1.73	1.68	1.63	1.57	1.54	1.50	1.46	1.42	1.38
60	2.79	2.39	2.18	2.04	1.95	1.87	1.82	1.77	1.74	1.71	1.66	1.60	1.54	1.51	1.48	1.44	1.40	1.35
80	2.77	2.37	2.15	2.02	1.92	1.85	1.79	1.75	1.71	1.68	1.63	1.57	1.51	1.48	1.44	1.40	1.36	1.31
120	2.75	2.35	2.13	1.99	1.90	1.82	1.77	1.72	1.68	1.65	1.60	1.55	1.48	1.45	1.41	1.37	1.32	1.26

$\alpha = 0.05$

$v_2 \downarrow$ \ $v_1 \rightarrow$	1	2	3	4	5	6	7	8	9	10	12	15	20	24	30	40	60	120
1	161.4	199.5	215.7	224.6	230.2	234.0	236.8	238.9	240.5	241.9	243.9	245.9	248	249	250	251	252	253.2
2	18.51	19.00	19.16	19.25	19.30	19.33	19.35	19.37	19.38	19.40	19.41	19.43	19.45	19.45	19.46	19.47	19.48	19.49
3	10.13	9.55	9.28	9.12	9.01	8.94	8.89	8.85	8.81	8.79	8.74	8.70	8.66	8.64	8.62	8.59	8.57	8.55
4	7.71	6.94	6.59	6.39	6.26	6.16	6.09	6.04	6.00	5.96	5.91	5.86	5.80	5.77	5.75	5.72	5.69	5.66
5	6.61	5.79	5.41	5.19	5.05	4.95	4.88	4.82	4.77	4.74	4.68	4.62	4.56	4.53	4.50	4.46	4.43	4.40
6	5.99	5.14	4.76	4.53	4.39	4.28	4.21	4.15	4.10	4.06	4.00	3.94	3.87	3.84	3.81	3.77	3.74	3.70
7	5.59	4.74	4.35	4.12	3.97	3.87	3.79	3.73	3.68	3.64	3.57	3.51	3.44	3.41	3.38	3.34	3.30	3.27
8	5.32	4.46	4.07	3.84	3.69	3.58	3.50	3.44	3.39	3.35	3.28	3.22	3.15	3.12	3.08	3.04	3.01	2.97
9	5.12	4.26	3.86	3.63	3.48	3.37	3.29	3.23	3.18	3.14	3.07	3.01	2.94	2.90	2.86	2.83	2.79	2.75
10	4.96	4.10	3.71	3.48	3.33	3.22	3.14	3.07	3.02	2.98	2.91	2.85	2.77	2.74	2.70	2.66	2.62	2.58
11	4.84	3.98	3.59	3.36	3.20	3.09	3.01	2.95	2.90	2.85	2.79	2.72	2.65	2.61	2.57	2.53	2.49	2.45
12	4.75	3.89	3.49	3.26	3.11	3.00	2.91	2.85	2.80	2.75	2.69	2.62	2.54	2.51	2.47	2.43	2.38	2.34
13	4.67	3.81	3.41	3.18	3.03	2.92	2.83	2.77	2.71	2.67	2.60	2.53	2.46	2.42	2.38	2.34	2.30	2.25
14	4.60	3.74	3.34	3.11	2.96	2.85	2.76	2.70	2.65	2.60	2.53	2.46	2.39	2.35	2.31	2.27	2.22	2.18
15	4.54	3.68	3.29	3.06	2.90	2.79	2.71	2.64	2.59	2.54	2.48	2.40	2.33	2.29	2.25	2.20	2.16	2.11
16	4.49	3.63	3.24	3.01	2.85	2.74	2.66	2.59	2.54	2.49	2.42	2.35	2.28	2.24	2.19	2.15	2.11	2.06
17	4.45	3.59	3.20	2.96	2.81	2.70	2.61	2.55	2.49	2.45	2.38	2.31	2.23	2.19	2.15	2.10	2.06	2.01
18	4.41	3.55	3.16	2.93	2.77	2.66	2.58	2.51	2.46	2.41	2.34	2.27	2.19	2.15	2.11	2.06	2.02	1.97
19	4.38	3.52	3.13	2.90	2.74	2.63	2.54	2.48	2.42	2.38	2.31	2.23	2.16	2.11	2.07	2.03	1.98	1.93
20	4.35	3.49	3.10	2.87	2.71	2.60	2.51	2.45	2.39	2.35	2.28	2.20	2.12	2.08	2.04	1.99	1.95	1.90
21	4.32	3.47	3.07	2.84	2.68	2.57	2.49	2.42	2.37	2.32	2.25	2.18	2.10	2.05	2.01	1.96	1.92	1.87
22	4.30	3.44	3.05	2.82	2.66	2.55	2.46	2.40	2.34	2.30	2.23	2.15	2.07	2.03	1.98	1.94	1.89	1.84
23	4.28	3.42	3.03	2.80	2.64	2.53	2.44	2.37	2.32	2.27	2.20	2.13	2.05	2.01	1.96	1.91	1.86	1.81
24	4.26	3.40	3.01	2.78	2.62	2.51	2.42	2.36	2.30	2.25	2.18	2.11	2.03	1.98	1.94	1.89	1.84	1.79
25	4.24	3.39	2.99	2.76	2.60	2.49	2.40	2.34	2.28	2.24	2.16	2.09	2.01	1.96	1.92	1.87	1.82	1.77
26	4.22	3.37	2.98	2.74	2.59	2.47	2.39	2.32	2.27	2.22	2.15	2.07	1.99	1.95	1.90	1.85	1.80	1.75
27	4.21	3.35	2.96	2.73	2.57	2.46	2.37	2.31	2.25	2.20	2.13	2.06	1.97	1.93	1.88	1.84	1.79	1.73
28	4.20	3.34	2.95	2.71	2.56	2.45	2.36	2.29	2.24	2.19	2.12	2.04	1.96	1.91	1.87	1.82	1.77	1.71
29	4.18	3.33	2.93	2.70	2.55	2.43	2.35	2.28	2.22	2.18	2.10	2.03	1.94	1.90	1.85	1.81	1.75	1.70
30	4.17	3.32	2.92	2.69	2.53	2.42	2.33	2.27	2.21	2.16	2.09	2.01	1.93	1.89	1.84	1.79	1.74	1.68
50	4.03	3.18	2.79	2.56	2.40	2.29	2.20	2.13	2.07	2.03	1.95	1.87	1.78	1.74	1.69	1.63	1.58	1.51
60	4.00	3.15	2.76	2.53	2.37	2.25	2.17	2.10	2.04	1.99	1.92	1.84	1.75	1.70	1.65	1.59	1.53	1.47
80	3.96	3.11	2.72	2.49	2.33	2.21	2.13	2.06	2.00	1.95	1.88	1.79	1.70	1.65	1.60	1.54	1.48	1.41
120	3.92	3.07	2.68	2.45	2.29	2.18	2.09	2.02	1.96	1.91	1.83	1.75	1.66	1.61	1.55	1.50	1.43	1.35

(continues)

TABLE D.4 (*Continued*)
$\alpha = 0.025$

$v_2 \downarrow$ \ $v_1 \rightarrow$	1	2	3	4	5	6	7	8	9	10	12	15	20	24	30	40	60	120
1	647.8	799.5	864.2	899.6	921.8	937.1	948.2	956.7	963.3	968.6	976.7	984.9	993.1	997.2	1001	1006	1010	1014
2	38.51	39.00	39.17	39.25	39.30	39.33	39.36	39.37	39.39	39.40	39.41	39.43	39.45	39.46	39.46	39.47	39.48	39.48
3	17.44	16.04	15.44	15.10	14.88	14.73	14.62	14.54	14.47	14.42	14.34	14.25	14.17	14.12	14.08	14.04	13.99	13.95
4	12.22	10.65	9.98	9.60	9.36	9.20	9.07	8.98	8.90	8.84	8.75	8.66	8.56	8.51	8.46	8.41	8.36	8.31
5	10.01	8.43	7.76	7.39	7.15	6.98	6.85	6.76	6.68	6.62	6.52	6.43	6.33	6.28	6.23	6.18	6.12	6.07
6	8.81	7.26	6.60	6.23	5.99	5.82	5.70	5.60	5.52	5.46	5.37	5.27	5.17	5.12	5.07	5.01	4.96	4.90
7	8.07	6.54	5.89	5.52	5.29	5.12	4.99	4.90	4.82	4.76	4.67	4.57	4.47	4.41	4.36	4.31	4.25	4.20
8	7.57	6.06	5.42	5.05	4.82	4.65	4.53	4.43	4.36	4.30	4.20	4.10	4.00	3.95	3.89	3.84	3.78	3.73
9	7.21	5.71	5.08	4.72	4.48	4.32	4.20	4.10	4.03	3.96	3.87	3.77	3.67	3.61	3.56	3.51	3.45	3.39
10	6.94	5.46	4.83	4.47	4.24	4.07	3.95	3.85	3.78	3.72	3.62	3.52	3.42	3.37	3.31	3.26	3.20	3.14
11	6.72	5.26	4.63	4.28	4.04	3.88	3.76	3.66	3.59	3.53	3.43	3.33	3.23	3.17	3.12	3.06	3.00	2.94
12	6.55	5.10	4.47	4.12	3.89	3.73	3.61	3.51	3.44	3.37	3.28	3.18	3.07	3.02	2.96	2.91	2.85	2.79
13	6.41	4.97	4.35	4.00	3.77	3.60	3.48	3.39	3.31	3.25	3.15	3.05	2.95	2.89	2.84	2.78	2.72	2.66
14	6.30	4.86	4.24	3.89	3.66	3.50	3.38	3.29	3.21	3.15	3.05	2.95	2.84	2.79	2.73	2.67	2.61	2.55
15	6.20	4.76	4.15	3.80	3.58	3.41	3.29	3.20	3.12	3.06	2.96	2.86	2.76	2.70	2.64	2.59	2.52	2.46
16	6.11	4.69	4.08	3.73	3.50	3.34	3.22	3.12	3.05	2.99	2.89	2.79	2.68	2.63	2.57	2.51	2.45	2.38
17	6.04	4.62	4.01	3.66	3.44	3.28	3.16	3.06	2.98	2.92	2.82	2.72	2.62	2.56	2.50	2.44	2.38	2.32
18	5.98	4.56	3.95	3.61	3.38	3.22	3.10	3.01	2.93	2.87	2.77	2.67	2.56	2.50	2.44	2.38	2.32	2.26
19	5.92	4.51	3.90	3.56	3.33	3.17	3.05	2.96	2.88	2.82	2.72	2.62	2.51	2.45	2.39	2.33	2.27	2.20
20	5.87	4.46	3.86	3.51	3.29	3.13	3.01	2.91	2.84	2.77	2.68	2.57	2.46	2.41	2.35	2.29	2.22	2.16
21	5.83	4.42	3.82	3.48	3.25	3.09	2.97	2.87	2.80	2.73	2.64	2.53	2.42	2.37	2.31	2.25	2.18	2.11
22	5.79	4.38	3.78	3.44	3.22	3.05	2.93	2.84	2.76	2.70	2.60	2.50	2.39	2.33	2.27	2.21	2.14	2.08
23	5.75	4.35	3.75	3.41	3.18	3.02	2.90	2.81	2.73	2.67	2.57	2.47	2.36	2.30	2.24	2.18	2.11	2.04
24	5.72	4.32	3.72	3.38	3.15	2.99	2.87	2.78	2.70	2.64	2.54	2.44	2.33	2.27	2.21	2.15	2.08	2.01
25	5.69	4.29	3.69	3.35	3.13	2.97	2.85	2.75	2.68	2.61	2.51	2.41	2.30	2.24	2.18	2.12	2.05	1.98
26	5.66	4.27	3.67	3.33	3.10	2.94	2.82	2.73	2.65	2.59	2.49	2.39	2.28	2.22	2.16	2.09	2.03	1.95
27	5.63	4.24	3.65	3.31	3.08	2.92	2.80	2.71	2.63	2.57	2.47	2.36	2.25	2.19	2.13	2.07	2.00	1.93
28	5.61	4.22	3.63	3.29	3.06	2.90	2.78	2.69	2.61	2.55	2.45	2.34	2.23	2.17	2.11	2.05	1.98	1.91
29	5.59	4.20	3.61	3.27	3.04	2.88	2.76	2.67	2.59	2.53	2.43	2.32	2.21	2.15	2.09	2.03	1.96	1.89
30	5.57	4.18	3.59	3.25	3.03	2.87	2.75	2.65	2.57	2.51	2.41	2.31	2.20	2.14	2.07	2.01	1.94	1.87
50	5.34	3.97	3.39	3.05	2.83	2.67	2.55	2.46	2.38	2.32	2.22	2.11	1.99	1.93	1.87	1.80	1.72	1.64
60	5.29	3.93	3.34	3.01	2.79	2.63	2.51	2.41	2.33	2.27	2.17	2.06	1.94	1.88	1.82	1.74	1.67	1.58
80	5.22	3.86	3.28	2.95	2.73	2.57	2.45	2.35	2.28	2.21	2.11	2.00	1.88	1.82	1.75	1.68	1.60	1.51
120	5.15	3.80	3.23	2.89	2.67	2.52	2.39	2.30	2.22	2.16	2.05	1.94	1.82	1.76	1.69	1.61	1.53	1.43

$\alpha = 0.01$

$v_2\downarrow$ \ $v_1\rightarrow$	1	2	3	4	5	6	7	8	9	10	12	15	20	24	30	40	60	120
1	4052	5000	5403	5625	5764	5859	5928	5982	6022	6056	6106	6157	6209	6235	6261	6287	6313	6339
2	98.5	99.0	99.2	99.2	99.3	99.3	99.4	99.4	99.4	99.4	99.4	99.4	99.4	99.5	99.5	99.5	99.5	99.5
3	34.1	30.8	29.5	28.7	28.2	27.9	27.7	27.5	27.3	27.2	27.1	26.9	26.7	26.6	26.5	26.4	26.3	26.2
4	21.2	18.0	16.7	16.0	15.5	15.2	15.0	14.8	14.7	14.6	14.4	14.2	14.0	13.9	13.8	13.7	13.7	13.6
5	16.25	13.27	12.06	11.39	10.97	10.67	10.46	10.29	10.16	10.05	9.89	9.72	9.55	9.47	9.38	9.29	9.20	9.11
6	13.74	10.92	9.78	9.15	8.75	8.47	8.26	8.10	7.98	7.87	7.72	7.56	7.40	7.31	7.23	7.14	7.06	6.97
7	12.24	9.55	8.45	7.85	7.46	7.19	6.99	6.84	6.72	6.62	6.47	6.31	6.16	6.07	5.99	5.91	5.82	5.74
8	11.26	8.65	7.59	7.01	6.63	6.37	6.18	6.03	5.91	5.81	5.67	5.52	5.36	5.28	5.20	5.12	5.03	4.95
9	10.56	8.02	6.99	6.42	6.06	5.80	5.61	5.47	5.35	5.26	5.11	4.96	4.81	4.73	4.65	4.57	4.48	4.40
10	10.04	7.56	6.55	5.99	5.64	5.39	5.20	5.06	4.94	4.85	4.71	4.56	4.41	4.33	4.25	4.17	4.08	4.00
11	9.64	7.21	6.22	5.67	5.32	5.07	4.89	4.74	4.63	4.54	4.40	4.25	4.10	4.02	3.94	3.86	3.78	3.69
12	9.33	6.93	5.95	5.41	5.06	4.82	4.64	4.50	4.39	4.30	4.16	4.01	3.86	3.78	3.70	3.62	3.54	3.45
13	9.07	6.70	5.74	5.21	4.86	4.62	4.44	4.30	4.19	4.10	3.96	3.82	3.66	3.59	3.51	3.43	3.34	3.25
14	8.86	6.51	5.56	5.04	4.69	4.46	4.28	4.14	4.03	3.94	3.80	3.66	3.51	3.43	3.35	3.27	3.18	3.09
15	8.68	6.36	5.42	4.89	4.56	4.32	4.14	4.00	3.89	3.80	3.67	3.52	3.37	3.29	3.21	3.13	3.05	2.96
16	8.53	6.23	5.29	4.77	4.44	4.20	4.03	3.89	3.78	3.69	3.55	3.41	3.26	3.18	3.10	3.02	2.93	2.84
17	8.40	6.11	5.18	4.67	4.34	4.10	3.93	3.79	3.68	3.59	3.46	3.31	3.16	3.08	3.00	2.92	2.83	2.75
18	8.28	6.01	5.09	4.58	4.25	4.01	3.84	3.71	3.60	3.51	3.37	3.23	3.08	3.00	2.92	2.84	2.75	2.66
19	8.18	5.93	5.01	4.50	4.17	3.94	3.77	3.63	3.52	3.43	3.30	3.15	3.00	2.92	2.84	2.76	2.67	2.58
20	8.09	5.85	4.94	4.43	4.10	3.87	3.70	3.56	3.46	3.37	3.23	3.09	2.94	2.86	2.78	2.69	2.61	2.52
21	8.01	5.78	4.87	4.37	4.04	3.81	3.64	3.51	3.40	3.31	3.17	3.03	2.88	2.80	2.72	2.64	2.55	2.46
22	7.94	5.72	4.82	4.31	3.99	3.76	3.59	3.45	3.35	3.26	3.12	2.98	2.83	2.75	2.67	2.58	2.50	2.40
23	7.88	5.66	4.76	4.26	3.94	3.71	3.54	3.41	3.30	3.21	3.07	2.93	2.78	2.70	2.62	2.54	2.45	2.35
24	7.82	5.61	4.72	4.22	3.90	3.67	3.50	3.36	3.26	3.17	3.03	2.89	2.74	2.66	2.58	2.49	2.40	2.31
25	7.77	5.57	4.68	4.18	3.85	3.63	3.46	3.32	3.22	3.13	2.99	2.85	2.70	2.62	2.54	2.45	2.36	2.27
26	7.72	5.53	4.64	4.14	3.82	3.59	3.42	3.29	3.18	3.09	2.96	2.81	2.66	2.58	2.50	2.42	2.33	2.23
27	7.67	5.49	4.60	4.11	3.78	3.56	3.39	3.26	3.15	3.06	2.93	2.78	2.63	2.55	2.47	2.38	2.29	2.20
28	7.63	5.45	4.57	4.07	3.75	3.53	3.36	3.23	3.12	3.03	2.90	2.75	2.60	2.52	2.44	2.35	2.26	2.17
29	7.60	5.42	4.54	4.04	3.73	3.50	3.33	3.20	3.09	3.00	2.87	2.73	2.57	2.49	2.41	2.33	2.23	2.14
30	7.56	5.39	4.51	4.02	3.70	3.47	3.30	3.17	3.07	2.98	2.84	2.70	2.55	2.47	2.39	2.30	2.21	2.11
50	7.17	5.06	4.20	3.72	3.41	3.19	3.02	2.89	2.78	2.70	2.56	2.42	2.27	2.18	2.10	2.01	1.91	1.80
60	7.08	4.98	4.13	3.65	3.34	3.12	2.95	2.82	2.72	2.63	2.50	2.35	2.20	2.12	2.03	1.94	1.84	1.73
80	6.96	4.88	4.04	3.56	3.26	3.04	2.87	2.74	2.64	2.55	2.42	2.27	2.12	2.03	1.94	1.85	1.75	1.63
120	6.85	4.79	3.95	3.48	3.17	2.96	2.79	2.66	2.56	2.47	2.34	2.19	2.03	1.95	1.86	1.76	1.66	1.53

(continues)

TABLE D.4 (*Continued*)

$\alpha = 0.005$

$v_2 \downarrow$ $v_1 \rightarrow$	1	2	3	4	5	6	7	8	9	10	12	15	20	24	30	40	60	120
1	16211	20000	21615	22500	23056	23437	23715	23925	24091	24224	24426	24630	24836	24940	25044	25148	25253	253591
2	198.5	199.0	199.2	199.2	199.3	199.4	199.4	199.4	199.4	199.4	199.4	199.4	199.4	199.5	199.5	199.5	199.5	199.5
3	55.55	49.80	47.47	46.19	45.39	44.84	44.43	44.13	43.88	43.69	43.39	43.06	42.78	42.62	42.47	42.31	42.15	41.99
4	31.33	26.28	24.26	23.15	22.46	21.97	21.62	21.35	21.14	20.97	20.70	20.44	20.17	20.03	19.89	19.75	19.61	19.47
5	22.77	18.31	16.53	15.55	14.94	14.51	14.20	13.96	13.77	13.62	13.38	13.15	12.90	12.78	12.66	12.53	12.40	12.27
6	18.62	14.54	12.91	12.03	11.46	11.07	10.79	10.57	10.39	10.25	10.03	9.81	9.59	9.47	9.36	9.24	9.12	9.00
7	16.23	12.40	10.88	10.05	9.52	9.16	8.89	8.68	8.51	8.38	8.18	7.97	7.75	7.64	7.53	7.42	7.31	7.19
8	14.68	11.04	9.60	8.80	8.30	7.95	7.69	7.50	7.34	7.21	7.01	6.81	6.61	6.50	6.40	6.29	6.18	6.06
9	13.61	10.10	8.72	7.96	7.47	7.13	6.88	6.69	6.54	6.42	6.23	6.03	5.83	5.73	5.62	5.52	5.41	5.30
10	12.82	9.43	8.08	7.34	6.87	6.54	6.30	6.12	5.97	5.85	5.66	5.47	5.27	5.17	5.07	4.97	4.86	4.75
11	12.22	8.91	7.60	6.88	6.42	6.10	5.86	5.68	5.54	5.42	5.24	5.05	4.86	4.76	4.65	4.55	4.44	4.34
12	11.75	8.51	7.23	6.52	6.07	5.76	5.52	5.35	5.20	5.09	4.91	4.72	4.53	4.43	4.33	4.23	4.12	4.01
13	11.37	8.19	6.93	6.23	5.79	5.48	5.25	5.08	4.94	4.82	4.64	4.46	4.27	4.17	4.07	3.97	3.87	3.76
14	11.06	7.92	6.68	6.00	5.56	5.26	5.03	4.86	4.72	4.60	4.43	4.25	4.06	3.96	3.86	3.76	3.66	3.55
15	10.79	7.70	6.48	5.80	5.37	5.07	4.85	4.67	4.54	4.42	4.25	4.07	3.88	3.79	3.69	3.58	3.48	3.37
16	10.57	7.51	6.30	5.64	5.21	4.91	4.69	4.52	4.38	4.27	4.10	3.92	3.73	3.64	3.54	3.44	3.33	3.22
17	10.38	7.35	6.16	5.50	5.07	4.78	4.56	4.39	4.25	4.14	3.97	3.79	3.61	3.51	3.41	3.31	3.21	3.10
18	10.21	7.21	6.03	5.37	4.96	4.66	4.44	4.28	4.14	4.03	3.86	3.68	3.50	3.40	3.30	3.20	3.10	2.99
19	10.07	7.09	5.92	5.27	4.85	4.56	4.34	4.18	4.04	3.93	3.76	3.59	3.40	3.31	3.21	3.11	3.00	2.89
20	9.94	6.99	5.82	5.17	4.76	4.47	4.26	4.09	3.96	3.85	3.68	3.50	3.32	3.22	3.12	3.02	2.92	2.81
21	9.83	6.89	5.73	5.09	4.68	4.39	4.18	4.01	3.88	3.77	3.60	3.43	3.24	3.15	3.05	2.95	2.84	2.73
22	9.72	6.81	5.65	5.02	4.61	4.32	4.11	3.94	3.81	3.70	3.54	3.36	3.18	3.08	2.98	2.88	2.77	2.66
23	9.63	6.73	5.58	4.95	4.54	4.26	4.05	3.88	3.75	3.64	3.47	3.30	3.12	3.02	2.92	2.82	2.71	2.60
24	9.55	6.66	5.52	4.89	4.49	4.20	3.99	3.83	3.69	3.59	3.42	3.25	3.06	2.97	2.87	2.77	2.66	2.55
25	9.47	6.60	5.46	4.83	4.43	4.15	3.94	3.78	3.64	3.54	3.37	3.20	3.01	2.92	2.82	2.72	2.61	2.50
26	9.40	6.54	5.41	4.79	4.38	4.10	3.89	3.73	3.60	3.49	3.33	3.15	2.97	2.87	2.77	2.67	2.56	2.45
27	9.34	6.49	5.36	4.74	4.34	4.06	3.85	3.69	3.56	3.45	3.28	3.11	2.93	2.83	2.73	2.63	2.52	2.41
28	9.28	6.44	5.32	4.70	4.30	4.02	3.81	3.65	3.52	3.41	3.25	3.07	2.89	2.79	2.69	2.59	2.48	2.37
29	9.23	6.39	5.28	4.66	4.26	3.98	3.77	3.61	3.48	3.38	3.21	3.04	2.86	2.76	2.66	2.56	2.45	2.33
30	9.18	6.35	5.24	4.62	4.23	3.95	3.74	3.58	3.45	3.34	3.18	3.01	2.82	2.73	2.63	2.52	2.42	2.30
50	8.62	5.90	4.83	4.23	3.85	3.58	3.38	3.22	3.09	2.99	2.82	2.65	2.47	2.37	2.27	2.16	2.05	1.93
60	8.49	5.79	4.73	4.14	3.76	3.49	3.29	3.13	3.01	2.90	2.74	2.57	2.39	2.29	2.19	2.08	1.96	1.83
80	8.33	5.66	4.61	4.03	3.65	3.39	3.19	3.03	2.91	2.80	2.64	2.47	2.29	2.19	2.08	1.97	1.85	1.72
120	8.18	5.54	4.50	3.92	3.55	3.28	3.09	2.93	2.81	2.71	2.54	2.37	2.19	2.09	1.98	1.87	1.75	1.61

$\alpha = 0.001$

$\nu_2 \downarrow$ \ $\nu_1 \rightarrow$	1	2	3	4	5	6	7	8	9	10	12	15	20	24	30	40	60	120
1	405269	500004	540387	562506	576412	585943	592881	598151	602292	605625	610676	615772	620913	623504	626107	628720	631345	633980
2	998.5	999.0	999.1	999.2	999.3	999.3	999.4	999.4	999.4	999.4	999.4	999.4	999.4	999.5	999.5	999.5	999.5	999.5
3	167.0	148.5	141.1	137.1	134.6	132.9	131.6	130.6	129.9	129.3	128.3	127.4	126.4	125.9	125.5	125.0	124.5	124.0
4	74.14	61.25	56.18	53.44	51.71	50.53	49.66	49.00	48.47	48.05	47.41	46.76	46.10	45.77	45.43	45.09	44.75	44.40
5	47.18	37.12	33.20	31.09	29.75	28.83	28.16	27.65	27.24	26.92	26.42	25.91	25.39	25.13	24.87	24.60	24.33	24.06
6	35.51	27.00	23.70	21.92	20.80	20.03	19.46	19.03	18.69	18.41	17.99	17.56	17.12	16.90	16.67	16.44	16.21	15.98
7	29.25	21.69	18.77	17.20	16.21	15.52	15.02	14.63	14.33	14.08	13.71	13.32	12.93	12.73	12.53	12.33	12.12	11.91
8	25.41	18.49	15.83	14.39	13.48	12.86	12.40	12.05	11.77	11.54	11.19	10.84	10.48	10.30	10.11	9.92	9.73	9.53
9	22.86	16.39	13.90	12.56	11.71	11.13	10.70	10.37	10.11	9.89	9.57	9.24	8.90	8.72	8.55	8.37	8.19	8.00
10	21.04	14.91	12.55	11.28	10.48	9.93	9.52	9.20	8.96	8.75	8.45	8.13	7.80	7.64	7.47	7.30	7.12	6.94
11	19.69	13.81	11.56	10.35	9.58	9.05	8.66	8.35	8.12	7.92	7.63	7.32	7.01	6.85	6.68	6.52	6.35	6.18
12	18.64	12.97	10.80	9.63	8.89	8.38	8.00	7.71	7.48	7.29	7.00	6.71	6.40	6.25	6.09	5.93	5.76	5.59
13	17.82	12.31	10.21	9.07	8.35	7.86	7.49	7.21	6.98	6.80	6.52	6.23	5.93	5.78	5.63	5.47	5.30	5.14
14	17.14	11.78	9.73	8.62	7.92	7.44	7.08	6.80	6.58	6.40	6.13	5.85	5.56	5.41	5.25	5.10	4.94	4.77
15	16.59	11.34	9.34	8.25	7.57	7.09	6.74	6.47	6.26	6.08	5.81	5.54	5.25	5.10	4.95	4.80	4.64	4.47
16	16.12	10.97	9.01	7.94	7.27	6.80	6.46	6.19	5.98	5.81	5.55	5.27	4.99	4.85	4.70	4.54	4.39	4.23
17	15.72	10.66	8.73	7.68	7.02	6.56	6.22	5.96	5.75	5.58	5.32	5.05	4.78	4.63	4.48	4.33	4.18	4.02
18	15.38	10.39	8.49	7.46	6.81	6.35	6.02	5.76	5.56	5.39	5.13	4.87	4.59	4.45	4.30	4.15	4.00	3.84
19	15.08	10.16	8.28	7.27	6.62	6.18	5.85	5.59	5.39	5.22	4.97	4.70	4.43	4.29	4.14	3.99	3.84	3.68
20	14.82	9.95	8.10	7.10	6.46	6.02	5.69	5.44	5.24	5.08	4.82	4.56	4.29	4.15	4.00	3.86	3.70	3.54
21	14.59	9.77	7.94	6.95	6.32	5.88	5.56	5.31	5.11	4.95	4.70	4.44	4.17	4.03	3.88	3.74	3.58	3.42
22	14.38	9.61	7.80	6.81	6.19	5.76	5.44	5.19	4.99	4.83	4.58	4.33	4.06	3.92	3.78	3.63	3.48	3.32
23	14.20	9.47	7.67	6.70	6.08	5.65	5.33	5.09	4.89	4.73	4.48	4.23	3.96	3.82	3.68	3.53	3.38	3.22
24	14.03	9.34	7.55	6.59	5.98	5.55	5.23	4.99	4.80	4.64	4.39	4.14	3.87	3.74	3.59	3.45	3.29	3.14
25	13.88	9.22	7.45	6.49	5.89	5.46	5.15	4.91	4.71	4.56	4.31	4.06	3.79	3.66	3.52	3.37	3.22	3.06
26	13.74	9.12	7.36	6.41	5.80	5.38	5.07	4.83	4.64	4.48	4.24	3.99	3.72	3.59	3.44	3.30	3.15	2.99
27	13.61	9.02	7.27	6.33	5.73	5.31	5.00	4.76	4.57	4.41	4.17	3.92	3.66	3.52	3.38	3.23	3.08	2.92
28	13.50	8.93	7.19	6.25	5.66	5.24	4.93	4.69	4.50	4.35	4.11	3.86	3.60	3.46	3.32	3.18	3.02	2.86
29	13.39	8.85	7.12	6.19	5.59	5.18	4.87	4.64	4.45	4.29	4.05	3.80	3.54	3.41	3.27	3.12	2.97	2.81
30	13.29	8.77	7.05	6.12	5.53	5.12	4.82	4.58	4.39	4.24	4.00	3.75	3.49	3.36	3.22	3.07	2.92	2.76
40	12.61	8.25	6.59	5.70	5.13	4.73	4.44	4.21	4.02	3.87	3.64	3.40	3.14	3.01	2.87	2.73	2.57	2.41
50	12.22	7.96	6.34	5.46	4.90	4.51	4.22	4.00	3.82	3.67	3.44	3.20	2.95	2.82	2.68	2.53	2.38	2.21
60	11.97	7.77	6.17	5.31	4.76	4.37	4.09	3.86	3.69	3.54	3.32	3.08	2.83	2.69	2.55	2.41	2.25	2.08
120	11.38	7.32	5.78	4.95	4.42	4.04	3.77	3.55	3.38	3.24	3.02	2.78	2.53	2.40	2.26	2.11	1.95	1.77

APPENDIX E

CONFIDENCE INTERVALS FOR THE MEAN

Table E.1 represents 1000 95% confidence intervals constructed from sample sets selected from a population with a mean (μ) of 25.4 and a variance (σ^2) of 1.69. These data are discussed in Chapter 4. (*Note:* Intervals with asterisks fail to include μ.)

TABLE E.1

(24.21, 25.94)	(24.77, 26.22)	(24.71, 26.01)	(25.43, 27.01)*	(24.22, 26.19)	(25.12, 26.56)	(24.70, 26.54)	(24.50, 25.95)
(23.99, 25.39)*	(25.33, 26.45)	(24.62, 25.89)	(24.79, 26.16)	(24.85, 26.11)	(23.92, 25.67)	(25.30, 26.84)	(24.71, 26.39)
(24.58, 26.10)	(24.47, 25.50)	(25.14, 26.56)	(24.79, 25.72)	(24.22, 25.79)	(25.03, 26.27)	(24.63, 25.83)	(25.09, 26.84)
(24.80, 26.18)	(24.95, 26.07)	(24.61, 26.27)	(25.06, 26.29)	(24.69, 26.25)	(24.64, 25.88)	(24.96, 26.15)	(25.04, 25.92)
(24.59, 26.09)	(24.99, 26.15)	(24.65, 26.29)	(25.08, 26.28)	(24.74, 26.02)	(25.19, 26.43)	(24.54, 25.73)	(24.17, 25.94)
(24.86, 25.81)	(24.40, 25.92)	(24.86, 25.80)	(24.58, 25.91)	(23.69, 25.25)*	(24.64, 25.77)	(24.88, 26.12)	(24.63, 25.92)
(24.84, 25.92)	(24.97, 26.47)	(25.13, 26.38)	(24.39, 26.32)	(24.15, 26.16)	(24.80, 26.01)	(24.87, 26.54)	(24.66, 25.95)
(24.64, 25.93)	(24.79, 26.18)	(24.68, 25.93)	(24.17, 25.83)	(24.29, 26.14)	(24.43, 25.62)	(24.58, 25.99)	(25.11, 26.33)
(24.85, 26.26)	(25.11, 26.44)	(24.58, 25.99)	(24.89, 26.25)	(24.79, 26.36)	(25.23, 26.42)	(25.30, 26.45)	(25.09, 26.53)
(25.23, 26.34)	(24.58, 26.35)	(24.81, 25.88)	(24.37, 25.62)	(24.79, 26.07)	(24.82, 26.19)	(24.66, 26.56)	(24.56, 26.12)
(25.11, 26.18)	(24.73, 25.83)	(24.46, 25.51)	(24.92, 25.87)	(24.46, 26.45)	(24.40, 25.61)	(24.49, 26.45)	(24.61, 26.02)
(25.20, 26.34)	(24.94, 25.99)	(24.53, 25.90)	(24.73, 26.07)	(24.56, 25.75)	(24.49, 25.85)	(24.49, 26.24)	(25.28, 26.26)
(24.26, 25.48)	(24.50, 26.26)	(25.11, 26.56)	(23.96, 25.61)	(24.94, 26.46)	(24.63, 26.31)	(24.37, 25.95)	(24.67, 26.18)
(24.56, 26.17)	(24.48, 25.76)	(24.72, 26.02)	(25.63, 26.40)*	(24.82, 26.09)	(24.97, 26.02)	(24.51, 25.78)	(24.88, 26.18)
(24.72, 25.91)	(24.94, 26.24)	(24.91, 26.39)	(25.46, 26.82)*	(24.84, 26.37)	(25.05, 26.07)	(24.72, 25.78)	(25.20, 26.60)
(24.83, 26.09)	(24.62, 25.96)	(25.10, 26.29)	(24.78, 26.24)	(24.50, 26.30)	(24.60, 26.20)	(25.38, 26.62)	(24.86, 26.20)
(24.31, 25.89)	(24.35, 25.73)	(24.67, 26.07)	(24.86, 26.58)	(24.52, 25.55)	(24.81, 26.47)	(24.45, 26.32)	(23.94, 25.77)
(24.76, 25.56)	(25.04, 26.20)	(24.25, 26.06)	(24.33, 25.69)	(24.99, 26.16)	(25.13, 26.43)	(24.68, 26.26)	(24.69, 25.68)
(25.11, 26.45)	(24.69, 26.00)	(24.45, 25.42)	(24.72, 26.50)	(24.54, 25.80)	(24.57, 25.76)	(25.10, 26.16)	(24.60, 25.90)
(24.21, 25.86)	(25.00, 26.52)	(25.05, 26.55)	(24.47, 25.76)	(24.76, 26.07)	(24.80, 25.95)	(24.45, 25.89)	(24.82, 26.86)
(25.16, 26.21)	(24.62, 26.01)	(24.45, 25.88)	(25.60, 26.70)*	(24.32, 25.83)	(25.11, 26.12)	(24.59, 25.81)	(24.11, 25.86)
(24.53, 25.86)	(24.89, 26.21)	(24.52, 25.83)	(25.08, 26.22)	(25.22, 26.71)	(24.25, 25.39)*	(24.73, 26.34)	(24.67, 26.24)
(24.84, 26.31)	(24.68, 25.55)	(24.69, 26.28)	(24.82, 26.02)	(25.29, 26.44)	(24.84, 26.21)	(24.83, 26.03)	(24.90, 26.26)
(24.84, 26.14)	(25.13, 26.52)	(25.07, 26.44)	(24.52, 25.88)	(24.62, 26.04)	(24.84, 25.88)	(24.34, 25.71)	(24.25, 25.51)
(24.50, 25.88)	(24.49, 25.73)	(25.20, 26.23)	(24.64, 26.30)	(24.73, 26.16)	(25.16, 26.30)	(24.66, 25.87)	(25.20, 26.59)
(24.44, 25.85)	(24.29, 25.29)*	(24.46, 25.94)	(24.46, 25.45)	(24.66, 25.78)	(24.07, 25.62)	(24.48, 25.88)	(24.99, 26.61)
(24.41, 25.97)	(25.04, 26.23)	(24.32, 25.57)	(24.76, 25.85)	(24.60, 26.03)	(25.04, 26.50)	(24.39, 26.27)	(24.97, 26.13)
(25.41, 26.81)*	(24.17, 25.61)	(25.10, 26.48)	(24.78, 26.18)	(24.85, 26.01)	(24.87, 25.88)	(24.62, 26.00)	(25.21, 26.39)

(*continues*)

TABLE E.1 (Continued)

(24.98, 26.33)	(24.11, 26.09)	(24.74, 26.05)	(24.11, 25.65)	(24.47, 25.83)	(24.85, 26.32)	(24.37, 25.63)	(24.66, 25.95)
(24.83, 26.27)	(25.07, 26.29)	(25.24, 26.56)	(24.13, 25.65)	(25.41, 26.47)*	(24.64, 26.04)	(24.42, 25.94)	(24.61, 25.61)
(25.07, 26.38)	(24.90, 26.44)	(24.44, 25.72)	(24.63, 25.95)	(25.28, 26.37)	(24.63, 25.96)	(24.63, 25.71)	(24.98, 26.38)
(24.61, 26.21)	(24.93, 26.12)	(24.41, 25.79)	(24.95, 26.06)	(24.92, 25.88)	(24.69, 26.10)	(24.84, 26.22)	(25.40, 26.67)
(24.41, 25.51)	(24.93, 26.11)	(24.90, 25.91)	(25.18, 26.44)	(24.90, 26.48)	(25.06, 26.12)	(25.23, 26.18)	(24.87, 26.06)
(24.36, 25.85)	(24.45, 25.68)	(24.84, 25.91)	(24.83, 26.40)	(24.69, 26.04)	(24.68, 26.11)	(24.95, 26.17)	(24.94, 26.00)
(24.56, 25.98)	(25.19, 26.24)	(24.27, 25.82)	(24.98, 26.00)	(24.85, 25.98)	(25.00, 25.82)	(24.72, 25.99)	(24.98, 26.48)
(25.19, 26.85)	(24.38, 25.92)	(24.59, 26.19)	(24.76, 26.42)	(24.56, 26.00)	(24.57, 26.29)	(24.60, 26.12)	(24.99, 26.18)
(24.23, 26.01)	(24.28, 25.56)	(24.14, 25.37)*	(24.64, 26.29)	(24.96, 26.40)	(24.89, 26.33)	(24.43, 25.11)*	(24.88, 26.53)
(25.17, 26.78)	(24.52, 26.20)	(24.29, 25.86)	(24.26, 25.97)	(24.74, 26.17)	(24.81, 26.20)	(24.97, 26.15)	(24.85, 26.43)
(24.42, 25.86)	(24.61, 26.04)	(25.13, 26.18)	(24.09, 25.73)	(24.83, 26.19)	(25.18, 26.53)	(25.01, 26.66)	(24.52, 25.97)
(25.43, 26.53)*	(25.00, 26.32)	(24.86, 25.70)	(24.52, 25.65)	(24.67, 26.00)	(25.17, 26.69)	(24.73, 25.98)	(24.58, 25.93)
(24.85, 26.42)	(25.49, 26.78)*	(25.52, 26.38)*	(24.14, 25.61)	(25.03, 26.43)	(24.38, 26.29)	(25.16, 26.73)	(24.92, 26.35)
(24.24, 26.00)	(24.00, 25.11)*	(24.85, 26.03)	(24.73, 26.02)	(25.19, 26.92)	(24.70, 26.18)	(24.85, 26.40)	(24.84, 26.31)
(24.75, 26.11)	(24.82, 26.07)	(24.62, 25.84)	(25.42, 26.48)*	(24.75, 26.03)	(24.32, 25.70)	(25.19, 26.45)	(24.71, 26.31)
(24.42, 25.94)	(24.71, 26.39)	(24.15, 25.84)	(24.72, 25.98)	(25.22, 26.51)	(24.16, 25.76)	(25.18, 25.94)	(25.04, 26.87)
(24.50, 25.73)	(24.96, 26.19)	(25.35, 26.41)	(24.58, 25.92)	(24.23, 25.64)	(25.18, 26.28)	(24.36, 26.29)	(24.18, 25.74)
(25.11, 26.24)	(24.14, 25.70)	(24.94, 25.85)	(24.59, 25.96)	(25.02, 26.60)	(24.67, 26.26)	(24.83, 26.38)	(24.43, 25.38)*
(25.06, 26.28)	(24.26, 25.69)	(24.30, 25.42)	(25.03, 26.56)	(24.47, 26.10)	(24.52, 25.77)	(25.12, 26.44)	(25.55, 26.84)*
(24.53, 26.19)	(24.77, 26.25)	(25.15, 26.49)	(24.96, 26.12)	(25.13, 26.32)	(24.69, 25.99)	(24.63, 26.15)	(25.37, 26.48)
(24.56, 26.21)	(24.44, 25.76)	(24.27, 25.50)	(25.09, 26.03)	(24.76, 26.01)	(24.63, 26.14)	(24.36, 25.59)	(24.36, 26.04)
(24.82, 25.75)	(24.68, 25.98)	(25.01, 26.78)	(24.83, 26.36)	(25.39, 26.46)	(24.39, 25.52)	(25.07, 26.11)	(24.97, 26.26)
(25.05, 26.31)	(24.89, 26.20)	(24.12, 25.66)	(24.79, 26.43)	(24.98, 26.15)	(25.05, 26.57)	(24.81, 26.08)	(24.32, 25.69)
(24.76, 26.37)	(25.30, 26.54)	(25.05, 26.20)	(24.38, 25.72)	(24.89, 26.37)	(24.34, 25.68)	(24.64, 25.75)	(24.54, 26.08)
(24.52, 26.21)	(24.85, 25.86)	(25.16, 25.95)	(24.71, 25.92)	(24.65, 25.36)*	(25.00, 26.48)	(24.58, 25.61)	(24.66, 26.26)
(24.74, 25.92)	(24.45, 25.68)	(24.89, 26.23)	(24.80, 26.04)	(24.84, 26.24)	(24.15, 25.62)	(25.48, 27.00)*	(25.00, 26.34)
(24.71, 26.55)	(25.01, 26.44)	(24.67, 26.20)	(24.94, 26.35)	(24.79, 26.36)	(24.98, 26.31)	(24.61, 26.21)	(23.94, 25.88)
(24.55, 25.94)	(24.79, 26.08)	(24.94, 25.90)	(24.73, 25.87)	(24.40, 25.68)	(25.24, 26.39)	(25.18, 26.61)	(24.42, 25.61)

(24.36, 25.77)	(24.35, 25.70)	(24.85, 26.06)	(24.56, 25.51)	(24.23, 25.69)	(24.41, 25.71)	(24.52, 25.89)	(24.53, 25.55)
(25.57, 26.72)*	(24.85, 26.11)	(24.98, 26.41)	(24.71, 26.00)	(25.26, 26.38)	(24.71, 26.36)	(24.66, 25.99)	(24.89, 25.86)
(25.46, 26.24)*	(25.25, 26.39)	(24.70, 26.22)	(24.59, 25.88)	(24.43, 25.54)	(25.09, 26.50)	(24.83, 26.22)	(24.87, 26.11)
(24.33, 26.14)	(25.00, 26.19)	(23.88, 25.22)*	(24.42, 26.05)	(24.68, 26.02)	(24.93, 26.18)	(24.85, 26.20)	(25.52, 26.77)*
(25.00, 25.95)	(24.51, 26.03)	(24.24, 25.97)	(24.92, 26.35)	(24.45, 26.33)	(24.94, 26.08)	(24.07, 26.38)	(24.84, 26.27)
(24.95, 26.22)	(24.94, 26.28)	(24.75, 26.00)	(24.78, 26.37)	(25.45, 27.04)*	(24.20, 25.91)	(24.71, 26.08)	(24.14, 25.83)
(25.14, 26.48)	(24.63, 26.28)	(24.91, 25.91)	(25.08, 26.23)	(24.47, 25.85)	(24.54, 26.04)	(24.95, 26.74)	(25.00, 26.40)
(24.60, 25.65)	(24.75, 26.31)	(24.23, 26.01)	(24.94, 26.36)	(24.66, 26.01)	(23.84, 25.91)	(24.78, 26.42)	(25.13, 26.34)
(24.97, 26.37)	(24.42, 26.73)	(24.63, 25.53)	(24.48, 25.73)	(24.44, 25.66)	(24.14, 25.11)*	(24.84, 26.33)	(25.05, 26.14)
(24.18, 25.96)	(24.59, 25.61)	(24.73, 25.86)	(24.63, 25.97)	(24.09, 25.90)	(25.47, 26.45)*	(24.69, 25.92)	(24.58, 25.77)
(24.67, 25.80)	(24.59, 26.10)	(23.87, 25.38)*	(24.93, 26.46)	(24.69, 26.16)	(24.66, 25.94)	(24.48, 25.95)	(24.33, 26.04)
(24.30, 25.75)	(25.33, 26.18)	(24.18, 26.00)	(24.58, 26.18)	(25.04, 26.25)	(24.33, 26.15)	(24.58, 26.13)	(24.32, 25.70)
(24.39, 25.48)	(24.89, 26.44)	(24.79, 26.14)	(24.62, 26.04)	(25.29, 26.58)	(24.81, 26.29)	(24.35, 25.28)*	(25.32, 26.60)
(24.04, 25.44)	(24.83, 26.22)	(24.43, 25.72)	(24.38, 26.20)	(24.63, 25.63)	(24.40, 26.17)	(25.09, 26.50)	(24.93, 26.45)
(24.89, 26.28)	(24.05, 25.43)	(24.78, 26.06)	(24.50, 25.94)	(24.54, 26.01)	(24.97, 25.89)	(24.58, 25.87)	(24.86, 26.11)
(24.76, 26.25)	(24.37, 25.53)	(24.85, 25.86)	(24.26, 25.48)	(25.14, 26.41)	(24.17, 25.84)	(24.77, 26.03)	(25.26, 26.53)
(25.08, 26.29)	(24.60, 26.08)	(25.24, 26.50)	(24.79, 26.10)	(25.17, 26.53)	(24.85, 25.88)	(24.79, 26.18)	(24.64, 26.11)
(24.52, 25.99)	(24.89, 26.14)	(24.29, 25.84)	(24.68, 25.99)	(24.41, 26.17)	(25.18, 26.57)	(24.10, 25.54)	(25.23, 26.47)
(24.97, 26.44)	(24.59, 25.63)	(24.80, 26.00)	(24.91, 26.29)	(24.56, 25.92)	(24.65, 26.02)	(24.34, 25.42)	(24.86, 26.32)
(24.54, 25.79)	(25.13, 26.68)	(24.60, 25.78)	(24.49, 25.87)	(24.94, 26.19)	(25.13, 26.39)	(24.47, 25.60)	(24.61, 25.77)
(24.74, 26.67)	(25.03, 26.54)	(24.69, 26.59)	(24.84, 26.37)	(24.81, 26.35)	(24.64, 25.95)	(24.89, 26.07)	(24.50, 25.91)
(24.34, 25.83)	(25.10, 26.12)	(24.87, 26.48)	(25.00, 26.54)	(24.64, 26.00)	(24.71, 26.09)	(24.43, 26.05)	(24.54, 25.69)
(24.37, 25.61)	(24.49, 25.97)	(24.48, 25.60)	(24.65, 26.73)	(25.18, 26.41)	(25.09, 26.44)	(24.95, 26.40)	(24.64, 26.01)
(24.58, 25.81)	(24.47, 25.66)	(24.39, 25.72)	(24.12, 25.71)	(24.58, 25.85)	(25.77, 27.28)*	(24.67, 25.86)	(24.78, 26.43)
(24.44, 26.16)	(24.70, 26.25)	(24.48, 25.68)	(24.32, 25.64)	(24.65, 25.50)	(24.75, 26.21)	(24.25, 25.69)	(24.10, 25.83)
(24.72, 26.08)	(24.73, 26.17)	(25.44, 26.34)*	(24.43, 25.75)	(25.27, 26.74)	(24.13, 25.49)	(25.04, 26.28)	(24.63, 25.80)
(24.27, 25.94)	(24.93, 26.73)	(24.43, 25.84)	(24.29, 25.52)	(24.55, 26.35)	(25.13, 26.54)	(24.27, 25.63)	(25.17, 26.45)
(25.09, 26.67)	(24.44, 25.88)	(24.59, 26.63)	(24.92, 26.37)	(25.10, 26.29)	(24.57, 26.27)	(24.49, 25.90)	(24.76, 26.08)

(*continues*)

579

TABLE E.1 (*Continued*)

(24.64, 26.13)	(24.46, 25.84)	(24.31, 25.89)	(25.34, 26.66)	(24.87, 26.31)	(24.17, 25.47)	(25.28, 26.57)	(24.75, 26.14)
(24.44, 26.00)	(24.46, 25.63)	(24.75, 26.33)	(25.03, 26.03)	(24.87, 26.36)	(24.45, 25.83)	(24.85, 26.15)	(24.12, 25.75)
(25.53, 26.94)*	(23.89, 25.24)*	(24.78, 26.12)	(24.95, 26.19)	(24.81, 26.05)	(24.76, 26.13)	(24.38, 25.77)	(24.45, 26.24)
(25.72, 26.82)*	(24.36, 25.75)	(24.63, 26.25)	(24.65, 26.19)	(25.04, 25.80)	(25.01, 26.66)	(24.48, 25.96)	(24.88, 26.48)
(24.45, 26.06)	(25.19, 26.42)	(24.64, 26.18)	(24.95, 26.25)	(24.43, 26.35)	(24.89, 26.11)	(24.19, 25.61)	(24.52, 26.26)
(24.95, 26.13)	(25.17, 26.46)	(24.70, 26.19)	(24.59, 25.94)	(24.71, 26.06)	(24.84, 26.45)	(24.41, 25.60)	(24.76, 25.96)
(24.83, 26.42)	(24.62, 26.12)	(24.39, 25.41)	(25.18, 26.22)	(24.75, 26.33)	(24.47, 26.37)	(24.45, 25.74)	(25.19, 26.37)
(24.75, 26.15)	(25.33, 26.63)	(24.69, 25.90)	(24.67, 26.29)	(24.89, 26.24)	(24.35, 26.27)	(24.95, 26.01)	(24.77, 25.94)
(24.69, 25.95)	(24.74, 26.22)	(24.61, 26.15)	(25.10, 26.15)	(24.95, 26.40)	(24.46, 26.08)	(25.07, 26.68)	(25.45, 26.28)*
(24.88, 26.52)	(25.04, 26.48)	(24.71, 25.90)	(24.86, 26.18)	(24.45, 25.96)	(24.82, 25.94)	(24.61, 26.09)	(24.87, 26.67)
(24.32, 25.93)	(24.74, 26.00)	(24.42, 25.72)	(25.14, 26.53)	(24.74, 25.97)	(24.35, 26.08)	(24.70, 25.75)	(25.11, 26.29)
(24.55, 25.95)	(24.33, 25.56)	(24.22, 25.61)	(24.59, 26.25)	(25.21, 26.31)	(25.03, 26.40)	(25.09, 26.54)	(24.58, 25.72)
(25.18, 26.27)	(24.10, 25.13)*	(24.64, 25.69)	(25.16, 26.15)	(24.95, 26.03)	(24.91, 26.08)	(24.83, 25.95)	(24.49, 25.69)
(24.60, 26.16)	(24.83, 26.07)	(25.63, 26.64)*	(24.77, 26.01)	(24.41, 25.84)	(25.38, 26.21)	(24.53, 25.86)	(24.17, 25.80)
(24.69, 25.83)	(25.00, 26.41)	(25.06, 26.19)	(24.32, 26.34)	(25.11, 26.42)	(23.99, 25.83)	(24.21, 25.43)	(24.84, 25.99)
(24.89, 26.16)	(24.99, 26.73)	(24.85, 26.24)	(24.91, 26.12)	(24.79, 26.03)	(24.52, 25.83)	(24.74, 26.21)	(24.37, 26.01)
(24.95, 26.36)	(24.37, 25.92)	(24.81, 26.10)	(24.89, 26.40)	(24.80, 25.90)	(24.60, 26.12)	(24.84, 26.32)	(25.02, 26.19)
(24.84, 25.95)	(25.06, 26.51)	(25.09, 25.96)	(25.44, 26.34)*	(25.09, 26.27)	(25.19, 26.86)	(24.97, 26.15)	(25.31, 26.21)
(24.65, 26.09)	(24.96, 26.36)	(24.88, 26.18)	(24.94, 26.05)	(24.70, 25.78)	(25.10, 26.22)	(24.63, 26.06)	(24.77, 26.36)
(24.86, 26.81)	(24.22, 25.71)	(24.98, 26.46)	(24.29, 25.43)	(24.53, 26.00)	(24.58, 26.31)	(24.27, 25.58)	(25.25, 26.68)
(24.65, 26.09)	(25.23, 26.49)	(25.25, 26.83)	(24.97, 25.94)	(24.88, 26.13)	(24.37, 25.41)	(25.37, 26.31)	(24.80, 26.43)
(24.73, 26.32)	(24.42, 25.81)	(24.63, 25.70)	(24.66, 25.43)	(25.58, 26.41)*	(24.22, 26.04)	(24.43, 26.08)	(24.81, 26.14)
(25.14, 26.47)	(24.92, 26.19)	(24.48, 25.91)	(25.13, 26.42)	(24.15, 25.45)	(24.84, 26.14)	(24.77, 26.08)	(25.22, 26.36)
(24.27, 25.36)*	(24.37, 25.56)	(25.35, 26.84)	(24.73, 26.10)	(24.80, 26.20)	(24.35, 25.81)	(24.30, 25.74)	(24.15, 25.64)
(24.70, 26.04)	(24.88, 26.14)	(24.45, 25.86)	(24.30, 26.08)	(24.66, 25.43)	(24.46, 26.12)	(24.38, 25.55)	(25.13, 26.01)
(24.60, 26.25)	(25.52, 26.51)*	(24.68, 26.20)	(24.87, 26.18)	(24.57, 25.65)	(25.60, 26.77)*	(24.91, 26.44)	(24.88, 26.14)
(24.20, 25.79)	(24.52, 25.70)	(25.58, 26.83)*	(25.15, 25.94)	(24.87, 26.34)	(24.70, 26.12)	(24.40, 25.56)	(24.93, 26.21)
(24.62, 25.87)	(24.54, 26.05)	(24.07, 25.80)	(25.01, 26.40)	(25.08, 26.52)	(24.15, 25.80)	(24.59, 26.22)	(24.65, 25.90)

(25.21, 26.21)	(24.85, 26.46)	(25.22, 26.74)	(25.07, 26.45)	(24.05, 25.59)	(24.08, 25.41)	(24.24, 25.36)*	(24.67, 26.08)
(24.40, 25.75)	(25.29, 26.31)	(24.51, 26.07)	(24.80, 26.23)	(24.82, 26.58)	(24.81, 26.06)	(24.74, 26.04)	(24.67, 25.76)
(24.79, 25.71)	(23.90, 25.01)*	(24.61, 26.39)	(24.89, 26.53)	(24.86, 26.49)	(24.54, 25.86)	(24.54, 26.25)	(24.67, 25.82)
(24.22, 25.61)	(24.39, 26.25)	(24.81, 26.08)	(25.41, 26.35)*	(24.62, 25.97)	(25.02, 26.13)	(24.39, 25.76)	(24.53, 25.72)
(24.44, 25.73)	(24.75, 26.44)	(25.03, 25.84)	(24.53, 26.10)	(25.05, 26.12)	(25.21, 26.39)	(24.37, 25.81)	(25.16, 26.25)
(24.46, 25.86)	(24.92, 26.30)	(24.65, 26.25)	(24.85, 26.22)	(24.22, 25.46)	(25.24, 26.42)	(24.63, 25.97)	(24.43, 25.65)
(24.92, 25.76)	(24.65, 26.04)	(24.96, 26.33)	(24.97, 26.55)	(25.30, 26.55)	(25.15, 26.62)	(24.82, 25.93)	(25.05, 26.51)
(25.02, 26.02)	(25.01, 26.38)	(24.71, 26.19)	(24.45, 25.26)*	(24.55, 26.16)	(24.90, 26.09)	(24.71, 26.29)	(25.14, 26.43)
(24.98, 26.21)	(24.78, 26.11)	(24.82, 26.44)	(24.75, 26.34)	(25.21, 26.27)	(24.55, 25.71)	(24.94, 26.42)	(24.34, 25.64)
(25.13, 26.29)	(24.36, 25.74)	(24.38, 25.42)	(24.83, 26.40)	(24.23, 25.42)	(24.34, 25.89)	(24.29, 26.05)	(24.01, 25.61)
(24.80, 26.15)	(24.77, 26.08)	(24.52, 26.03)	(24.42, 25.97)	(25.47, 26.93)*	(24.56, 25.93)	(24.53, 25.74)	(24.72, 26.14)
(25.34, 26.62)	(24.29, 25.38)*	(24.89, 26.23)	(24.89, 26.19)	(24.85, 26.08)	(24.84, 26.24)	(24.89, 26.15)	(24.60, 25.76)
(25.38, 26.61)	(25.28, 26.67)	(24.76, 26.06)	(25.10, 26.19)	(24.65, 25.90)	(24.55, 26.19)	(24.96, 26.05)	(24.48, 25.98)

APPENDIX F

MAP PROJECTION COORDINATE SYSTEMS

F.1 INTRODUCTION

Most local surveyors are well served by using map projections such as the state plane coordinate system. These two-dimensional *grid* systems allow surveyors to perform accurate computations over large regions of land using plane surveying computations. They are the basis for the adjustments discussed in Chapters 14 through 16.

Map projections provide a one-to-one mathematical relationship with points on the ellipsoid and those on the mapping surface. There are an infinite number of map projections. Most map projections are defined by a series of mathematical transformations used to convert a point's geodetic coordinates of latitude, ϕ, and longitude, λ, to xy grid coordinates. Some map projections preserve the shape of objects (conformal); others, areas, directions, or distances of lines. However, since Earth is ellipsoidal in shape and a mapping surface is a plane, all map projections introduce some form of distortion to observations. For example, distances and areas are distorted in a conformal map projection.

To reduce the size of these distortions, the developable surface is often made secant to the ellipsoid and the width of the mapping zone is limited in distance. For instance, when the National Geodetic Survey originally designed the state plane coordinate system during the 1930s, the zone widths were limited to 158 miles so that precision between the ellipsoid distance and the grid distance was no worse than 1:10,000. Since most surveys at that time were only accurate to a precision of 1:5000, this was an acceptable limit. However, with today's modern instruments, observations must be reduced properly if survey accuracy is to be preserved in a map projection system.

All map projections are based on the ellipsoid selected, such as the Geodetic Reference System of 1980 (GRS 80), and defining zone parameters. Typically, the zone parameters define the grid origin (ϕ_0, λ_0); the secant lines of the projection, also known as *standard parallels,* or scale factor, k_0, at the central meridian, λ_0; and the offset distances (E_0, N_b) from the grid origin. Once defined, each map projection has a series of zone constants that are computed using the defining zone parameters. These zone constants are computed only once for each projection. Once the zone constants are computed, the direct and inverse problems can be carried out for any point in the system. The direct problem takes the geodetic coordinates of a point and transforms them into grid coordinates, and the inverse problem takes the grid coordinates of a point and transforms them into geodetic coordinates.

The two primary map projection systems used in the United States are the Lambert Conformal Conic for states that have a long east–west extent and the Transverse Mercator for states that have a long north–south extent. Both map projections are conformal; that is, they preserve angles in infinitesimally small regions about a point. This property is advantageous to surveyors since angles are minimally distorted when using a conformal projection. On the other hand, as shown in Figure F.1, horizontal distances observed must be reduced to the mapping surface to eliminate the distortions of the projection. However, if these reductions are performed properly, the resulting plane computations are as accurate as geodetic computations such as those shown in Chapter 23. In this appendix we look at the mathematics of the Lambert Conformal Conic and Transverse Mercator map projections and demonstrate proper methods in reducing observations before an adjustment.

F.2 MATHEMATICS OF THE LAMBERT CONFORMAL CONIC MAP PROJECTION

The Lambert Conformal Conic map projection was introduced by Johann Lambert in 1772. As its name implies, this map projection uses a cone as its developable surface. The projection is conformal, so angles are preserved but distances are distorted. A Lambert Conformal Conic map projection is defined by two ellipsoidal parameters,[1] grid origin (ϕ_0, λ_0); latitude of the north standard parallel, ϕ_N, and south standard parallel,[2] ϕ_S; false easting, E_0; and false northing, N_b.

---

[1] Typically, an ellipsoid is defined by the length of its semimajor axis, a, and its flattening factor, f. The first eccentricity is computed as $e = \sqrt{2f - f^2}$. The GRS 80 ellipsoid has defining parameters of $a = 6{,}378{,}137.0$ m and $f = 1/298.2572221008$.
[2] The standard parallels are the latitudes of the north and south secant lines for the cone on the ellipsoid.

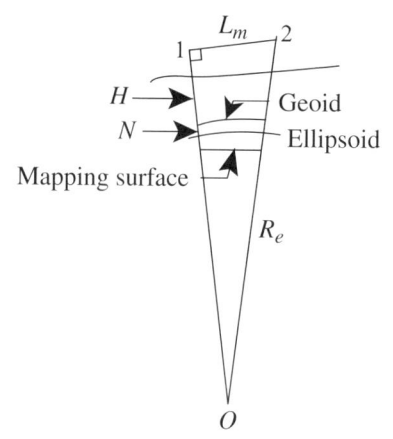

Figure F.1 Reduction of distance to a mapping surface.

F.2.1 Zone Constants

A set of three functions is used repeatedly in computations of the Lambert Conformal Conic map projection:

$$W(\phi) = \sqrt{1 - e^2 \sin^2 \phi} \tag{F.1}$$

$$M(\phi) = \frac{\cos \phi}{W(\phi)} \tag{F.2}$$

$$T(\phi) = \sqrt{\frac{1 - \sin \phi}{1 + \sin \phi} \left(\frac{1 + e \sin \phi}{1 - e \sin \phi} \right)^e} \tag{F.3}$$

Using Equations (F.1) through (F.3), the remaining zone constants are defined as

$$w_1 = W(\phi_S) \tag{F.4}$$

$$w_2 = W(\phi_N) \tag{F.5}$$

$$m_1 = M(\phi_S) \tag{F.6}$$

$$m_2 = M(\phi_N) \tag{F.7}$$

$$t_0 = T(\phi_0) \tag{F.8}$$

$$t_1 = T(\phi_S) \tag{F.9}$$

F.2 MATHEMATICS OF THE LAMBERT CONFORMAL CONIC MAP PROJECTION 585

$$t_2 = T(\phi_N) \tag{F.10}$$

$$n = \sin \phi_0 = \frac{\ln m_1 - \ln m_2}{\ln t_1 - \ln t_2} \tag{F.11}$$

$$F = \frac{m_1}{nt_1^n} \tag{F.12}$$

$$R_b = aFt_0^n = \text{radius of the projection} \tag{F.13}$$

F.2.2 Direct Problem

The direct problem takes the geodetic coordinates of latitude, ϕ, and longitude, λ, of a point and transforms them into xy grid coordinates. Often, the y coordinate is called the point's *northing, N,* and the x coordinate its *easting, E*. Thus, given the geodetic coordinates of a point, the northing, y, easting, x, scale factor, k, and convergence angle, γ, of the point are computed as

$$t = T(\phi) \tag{F.14}$$

$$m = M(\phi) \tag{F.15}$$

$$R = aFt^n \tag{F.16}$$

$$\gamma = (\lambda - \lambda_0)n \quad \text{(where western longitude is considered negative)} \tag{F.17}$$

$$E = R \sin \gamma + E_0 \tag{F.18}$$

$$N = R_b - R \cos \gamma + N_b \tag{F.19}$$

$$k = \frac{Rn}{am} \tag{F.20}$$

F.2.3 Inverse Problem

The inverse problem takes a point's northing and easting coordinates and computes its latitude, longitude, scale factor, and convergence angle. For the Lambert Conformal Conic map projection, the equations for the inverse problem are

$$E' = E - E_0 \qquad (\text{F.21})$$

$$N' = R_b - (N - N_b) \qquad (\text{F.22})$$

$$R = \sqrt{E'^2 + N'^2} \qquad (\text{F.23})$$

$$t = \left(\frac{R}{aF}\right)^{1/n} \qquad (\text{F.24})$$

$$\gamma = \tan^{-1}\frac{E'}{N'} \qquad (\text{F.25})$$

$$\chi = 90° - \tan^{-1}t \qquad (\text{F.26})$$

$$\phi = 90° - 2\tan^{-1}\left[t\left(\frac{1 - e\sin\phi}{1 + \sin\phi}\right)^{e/2}\right] \qquad (\text{F.27})$$

Repeat Equation (F.27) using χ for ϕ in the first iteration. Iterate until the change in ϕ is insignificant; that is, the change should be less than 0.000005″.

$$\lambda = \frac{\gamma}{n + \lambda_0} \qquad (\text{F.28})$$

$$k = \frac{m_1 t^n}{m t_1^n} \qquad (\text{F.29})$$

where m and t are defined in Equations (F.14) and (F.15) using ϕ from Equation (F.27).

F.3 MATHEMATICS OF THE TRANSVERSE MERCATOR

The Transverse Mercator map projection uses a cylinder as its developable surface. It preserves scale in a north–south direction and thus is good for regions with a long north–south extent. This projection was proposed by Johann Lambert, but the mathematics for an ellipsoid were not solved until the early twentieth century. In many countries, this projection is also known as the Gauss–Krüger map projection. The most famous Transverse Mercator map projection is the Universal Transverse Mercator (UTM) developed by the National Geospatial-Information Agency to provide a worldwide mapping system from 80° south latitude to 80° north latitude. This map projection is

F.3 MATHEMATICS OF THE TRANSVERSE MERCATOR

defined by two ellipsoidal parameters,[3] grid origin (ϕ_0,λ_0), scale factor, k_0, at the central meridian, λ_0, false easting, E_0, and false northing, N_b.

There are 60 zones in the Universal Transverse Mercator map projection, each nominally 6° wide. Each zone overlaps its neighboring zones by 30'. The central meridian, λ_0, for each zone is assigned a false easting, E_0, of 500,000 m. The false northing, N_b, is 0.000 m in the northern hemisphere and 10,000,000.000 m in the southern hemisphere. The scale factor at the central meridian, k_0, is 0.9996, which yields a distance precision of 1:2500. The central meridians (λ_0) for each zone start at 177° west longitude and with a few exceptions, proceeds easterly by 6° for each subsequent zone. The grid origins are at 0° and λ_0.

F.3.1 Zone Constants

The Transverse Mercator map projection use the following defining functions:

$$C(\phi) = e'^2 \cos^2\phi \qquad (F.30)$$

$$T(\phi) = \tan\phi \qquad (F.31)$$

$$M(\phi) = a\left[\left(1 - \frac{e^2}{4} - \frac{3e^4}{64} - \frac{5e^6}{256}\right)\phi - \left(\frac{3e^2}{8} + \frac{3e^4}{32} + \frac{45e^6}{1024}\right)\sin 2\phi \right.$$
$$\left. + \left(\frac{15e^4}{256} + \frac{45e^6}{1024}\right)\sin 4\phi - \left(\frac{35e^6}{3072}\right)\sin 6\phi\right] \qquad (F.32)$$

where e is the first eccentricity of the ellipse as defined in Equation (17.5) and e' is defined as

$$b = a(1-f) \qquad (F.33)$$

$$e' = \frac{\sqrt{a^2 - b^2}}{b} = \sqrt{\frac{1-e^2}{e^2}}$$

$$m_0 = M(\phi_0) \qquad (F.34)$$

---
[3] The Universal Transverse Mercator (UTM) uses the WGS 84 ellipsoid, defined in Chapter 17.

F.3.2 Direct Problem

The equations in the Transverse Mercator for the direct problem are

$$m = M(\phi) \tag{F.35}$$

$$t = T(\phi) \tag{F.36}$$

$$c = C(\phi) \tag{F.37}$$

$$A = (\lambda - \lambda_0) \cos \phi \quad \text{where western longitudes are negative} \tag{F.38}$$

$$E = k_0 R_N \left[A + (1 - t + c) \frac{A^3}{6} \right.$$
$$\left. + (5 - 18t + t^2 + 72c - 58e'^2) \frac{A^5}{120} \right] + E_0 \tag{F.39}$$

$$N = k_0 \left\{ m - m_0 + R_N \tan \phi \left[\frac{A^2}{2} + (5 - t + 9c + 4c^2) \frac{A^4}{24} \right.\right.$$
$$\left.\left. + (61 - 58t + t^2 + 600c - 300e'^2) \frac{A^6}{720} \right] \right\} + N_b \tag{F.40}$$

where R_N is the radius in the prime vertical as defined by N in Equation (17.6).

$$c_2 = \frac{1 + 3c + 2c^2}{3} \qquad c_3 = \frac{2 - \tan^2 \phi}{15} \tag{F.41}$$

$$\gamma = A \tan \phi [1 + A^2(c_2 + c_3 A^2)]$$

$$k = k_0 \left[1 + (1 + c) \frac{A^2}{2} + (5 - 4t + 42c + 13c^2 - 23e'^2) \frac{A^4}{24} \right.$$
$$\left. + (61 - 148t + 16t^2) \frac{A^6}{720} \right] \tag{F.42}$$

F.3.3 Inverse Problem

The equations in the Transverse Mercator for the inverse problem are

F.3 MATHEMATICS OF THE TRANSVERSE MERCATOR

$$E' = E - E_0 \tag{F.43}$$

$$N' = N - N_b \tag{F.44}$$

$$e_1 = \frac{1 - \sqrt{1 - e^2}}{1 + \sqrt{1 - e^2}} \tag{F.45}$$

$$m = m_0 + \frac{N'}{k_0} \tag{F.46}$$

$$\chi = \frac{m}{a\left(1 - e^2/4 - 3e^4/64 - 5e^6/256\right)} \tag{F.47}$$

The *foot-point latitude* is

$$\phi_f = \chi + \left(\frac{3e_1}{2} - \frac{27e_1^3}{32}\right)\sin 2\chi + \left(\frac{21e_1^2}{16} - \frac{55e_1^4}{32}\right)\sin 4\chi$$

$$+ \frac{151e_1^3}{96}\sin 6\chi + \frac{1097e_1^4}{512}\sin 8\chi \tag{F.48}$$

Using the foot-point latitude and functions defined in Section F.3.1 and Equation (23.16) yields

$$c_1 = C(\phi_f) \tag{F.49}$$

$$t_1 = T(\phi_f) \tag{F.50}$$

$$N_1 = \frac{a}{\sqrt{1 - e^2 \sin^2 \phi_f}} \tag{F.51}$$

$$M_1 = \frac{a(1 - e^2)}{(1 - e^2 \sin^2 \phi_f)^{3/2}} \tag{F.52}$$

$$D = \frac{E'}{N_1 k_0} \tag{F.53}$$

$$B = \frac{D^2}{2} - (5 + 3t_1 + 10c_1 - 4c_1^2 - 9e')\frac{D^4}{24}$$

$$+ (61 + 90t_1 + 298c_1 + 454t_1^2 - 252e' - 3c_1^2)\frac{D^6}{720} \tag{F.54}$$

$$\phi = \phi_f - \frac{N_1 \tan \phi_f}{M_1} B \qquad \text{(F.55)}$$

$$\lambda = \lambda_0 + \frac{D - (1 - 2t_1 + c_1)(D^3/6) + (5 - 2c_1 + 28t_1 - 3c_1^2 + 8e'^2 + 24t_1^2)(D^5/120)}{\cos \phi_f} \qquad \text{(F.56)}$$

Note that Equations (F.41) and (F.42) can be used to compute the convergence angle γ and scale factor k for the point.

F.4 REDUCTION OF OBSERVATIONS

Most often, the grid coordinates of a point are known prior to the survey and all that is needed is to reduce the observations to the mapping surface. The basic principle to bear in mind is that *grid computations should only be performed with grid observations.* Since the two map projections discussed previously are conformal, observed distances must be reduced to the mapping surface. Similarly, geodetic and astronomical directions must be converted to their grid equivalents.

As discussed in this section, conformality implies that the angles will be only slightly distorted. As will be shown, the *arc-to-chord* correction is applied directions and angles when the sight distances are long. For example, in the state plane coordinate system, this correction should be considered for angles whose sight distances are greater than 8 km. In this section, proper reduction of distance, direction, and angle observations is discussed.

F.4.1 Reduction of Distances

As shown in Figure F.1, an observed horizontal distance must be reduced to the mapping surface. This reduction usually involves using the *grid factor*. The grid factor is the product of the *elevation factor*, which reduces the observed distance to the ellipsoid, and a scale factor (k), which reduces the ellipsoidal distance to the mapping surface.

There are several procedures for reducing an observed distance to the ellipsoid, the most precise being a geodetic reduction. However, surveyed lengths typically contain only five or six significant figures. Thus, less strict methods can be applied to these short lengths. The elevation factor is computed as

$$\text{EF} = \frac{R_e}{R_e + H + N} = \frac{R_e}{R_e + h} \qquad \text{(F.57)}$$

In Equation (F.57), R_e is the radius of the Earth, H the orthometric height, N the geoidal height, and h the geodetic height. All of these parameters are determined at the observation station. The relationship between the geodetic height, h, and orthometric height, H, is

$$h = H + N \qquad (F.58)$$

In Equation (F.57), the radius in the azimuth of the line should be used for R_e. Again since surveyors observe short distances typically, an average radius of the Earth of 6,371,000 m can be used in computing EF. These approximations are demonstrated in Example F.1.

In a map projection system, the scale factor computed using Equation (F.20), (F.29), or (F.42) is for a point. Generally, the scale factor changes continuously along the length of the line. Thus, a weighted mean using two endpoints of the line (k_1 and k_2) and midpoint (k_m) is a logical choice for computing a single scale factor for a line. It can be computed as

$$k_{avg} = \frac{k_1 + 4k_m + k_2}{6} \qquad (F.59)$$

However, as with the elevation factor, this type of precision is seldom needed for the typical survey. Thus, the mean of the two endpoint scale factors is generally of sufficient accuracy for most surveys. In fact, it is not uncommon to use a single mean scale factor for an entire project.

The grid factor, GF, for the line is a product of the elevation factor, EF, and a scale factor, k_{avg}, and is computed as

$$GF = k_{avg} \times EF \qquad (F.60)$$

Thus, a reduced grid distance, L_{grid}, is the product of the horizontal distance, L_m, and the grid factor, GF, and is computed as

$$L_{grid} = L_m \times GF \qquad (F.61)$$

Example F.1 A distance of 536.07 ft is observed from station 1. The scale factors at observing, midpoint, and sighted stations are 0.9999587785, 0.9999587556, and 0.9999587328, respectively. The orthometric height at observing station is 1236.45 ft. Its geoidal height is −30.12 m and the radius in the azimuth is 6,366,977.077 m. Determine the length of the line on the mapping surface.

SOLUTION This solution will compare the grid factor computed using different radii in Equation (F.57) and different scale factors in Equation (F.60). Using the more precise methods, the grid factor is computed as follows. The orthometric height of the observing station in meters is

$$H = 1236.45 \text{ ft} \times \frac{12}{39.37} = 376.871 \text{ m}$$

Using the radius in the azimuth of the line and Equation (F.57), the elevation factor, EF, is

$$EF = \frac{6{,}366{,}977.077}{6{,}366{,}977.077 + 376.871 - 30.12} = 0.999945542$$

From Equation (F.59), the scale factor for the lines is

$$k_{avg} = \frac{0.9999587785 + 4(0.9999587556) + 0.9999587328}{6} = 0.999958756$$

From Equation (F.60), the grid factor for the line is

$$GF = 0.999945542 \times 0.999958756 = 0.99990430$$

Finally, the grid distance for this line is

$$L_{grid} = 0.99990430 \times 536.07 \text{ ft} = 536.02 \text{ ft}$$

Doing the problem again, this time with the mean radius of the Earth and the average of the two endpoint scale factors, yields

$$EF = \frac{6{,}371{,}000}{6{,}371{,}000 + 376.871 - 30.12} = 0.999945577$$

$$k_{avg} = \frac{0.9999587785 + 0.9999587328}{2} = 0.999958756$$

$$GF = 0.999945577 \times 0.999958756 = 0.99990433$$

$$L_{grid} = 0.99990433 \times 536.07 \text{ ft} = 536.02 \text{ ft}$$

Note that using the approximate radius of the Earth and the average scale factor for the endpoints of the line resulted in the same solution as the more precise computations. This is because the length of the distance observed has only five significant figures. The elevation factor computed using the mean radius of the Earth agreed with the radius in the azimuth to seven decimal places. This is also true of the scale factors, which agreed to nine significant figures. Thus, the grid factor was the same to seven decimal places and was well beyond the accuracy needed to convert a length with only five significant figures. This demonstrates why a common grid factor can often be used for an entire project that covers a small region.

F.4.2 Reduction of Geodetic Azimuths

Figure F.2 depicts the differences between geodetic azimuths, T, and grid azimuths, t. Since grid north (GN) at a point is parallel to the central meridian, the convergence angle, γ, is the largest correction between the two geodetic and grid azimuths. Additionally, there is a small correction to convert the arc on an ellipsoid to its equivalent chord on the mapping surface. This is known as the *arc-to-chord correction*, δ. The relationship between the geodetic azimuth and grid azimuth can be derived from Figure F.2 as

$$T = t + \gamma - \delta \tag{F.62}$$

As shown in Figure F.2, this equation works whether the line is east or west of the central meridian. For the Lambert Conformal Conic map projection, the arc-to-chord correction is computed as

$$\delta = 0.5(\sin \phi_3 - \sin \phi_0)(\lambda_2 - \lambda_1) \tag{F.63}$$

An analysis of Equation (F.63) shows that the worst cases for δ are for lines in the northern or southern extent of a map projection. Rearranging Equation (F.63) yields a change in longitude as

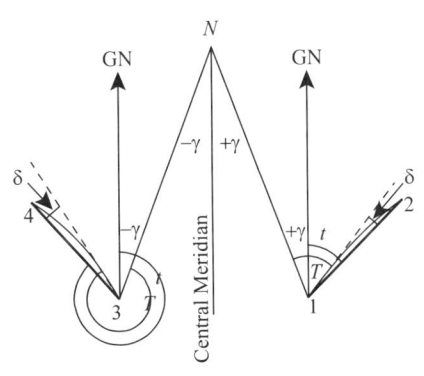

Figure F.2 Relationship of geodetic azimuth (T), grid azimuth (t), convergence angle (γ), and arc-to-chord correction (δ).

$$\lambda_2 - \lambda_1 = \Delta\lambda = \frac{2\delta}{\sin\phi_3 - \sin\phi_0} \tag{F.64}$$

As an example, assume that ϕ_3 is 42°30′. Further assume that the project is in the Pennsylvania North Zone, which has a $\sin\phi_0$ of 0.661539733812. If δ is to be kept below 0.5″, the maximum line in arc-seconds of longitude can be

$$\Delta\lambda = \frac{2(0.5'')}{\sin 42°30' - 0.661539733812} = 71.2''$$

At latitude 42°30′, this corresponds to a line of length of about 5334 ft, or 1.6 km. Few surveyors in northeastern Pennsylvania could find a line of this length to observe. Thus, the arc-to-chord correction is generally ignored in reductions, and Equation (F.62) can be simplified as

$$T = t + \gamma \tag{F.65}$$

APPENDIX G

COMPANION CD

G.1 INTRODUCTION

The companion CD that accompanies this book has several programs and instructional worksheets to aid students in the learning process. The CD contains the software ADJUST, MATRIX, and STATS that were available with the preceding edition of the book. Mathcad worksheets have been added to this edition. For those who do not own Mathcad, html files of these worksheets have been created. These files can be viewed using your computer's Web browser.

All of the software contained on the CD is Windows-based and will run on most Windows-based computers. The CD has an autoinstall program that will allow you to select the packages for installation. Before installing the software on the companion CD, it is important to remove any versions of the software already on your computer. This can be done using the Windows "Add/Remove programs" feature in your Windows control panel. Refer to the Windows help system if you are not familiar with this feature in Windows. If the program fails to autoload on your machine, browse the CD and run the "setup.exe" file from the root directory of the CD. Each program contained on the CD has different system memory requirements; although all will run in less than 2 megabytes of memory.

This software is "freeware" and as such can be freely distributed with this book. However, it is not intended for commercial use, is not guaranteed to be computationally correct, and is not supported in any manner. It is simply provided to aid your understanding of the topics contained in the book. As the software is updated, it will be posted at http://surveying.wb.psu.edu under the FREE GOODIES button. You should visit this site occasionally to download newer versions of the software on this CD.

595

G.2 FILE FORMATS AND MEMORY MATTERS

Most options in the software packages ADJUST, MATRIX, and STATS use formatted ASCII text files for data entry. Each package is equipped with an editor that allows the user to create these files. The help file for each software program describes the format of the data files. It is also possible to create data files using other ASCII editors, such as Notepad. The file-reading routines contained in ADJUST, MATRIX, and STATS can use either commas, spaces, or tabs as delimiters between fields. It is therefore important to avoid commas in large numbers such as coordinates, since the software will read each part of the number individually. For example, a coordinate of 675,301.213 will be read as the numbers 675 and 301.213 since the comma is a delimiter. The use of the TAB delimiter is especially useful in Matrix since this allows you to "cut and paste" values from spreadsheets such as Excel and Quattro Pro.

Starting with Version 4.0, ADJUST has used dynamic memory allocation. That is, data storage structures are not created until their sizes have been defined at runtime. Therefore, users familiar with earlier versions of ADJUST should be aware that some file formats have changed to accommodate this programming change. As a user, this feature means that the size limits of data types in various options in the software are now limited only by your computer's memory resources. Thus, much larger problems can be handled with this software.

G.3 SOFTWARE

G.3.1 ADJUST

ADJUST is the main computational program on the companion CD. It contains programs that either support computation of problems in the book or perform the computations. Several least squares programs and several supporting options are contained in ADJUST. A list of the least squares options and supporting software is provided in Table G.1. For example, the traverse option in ADJUST can compute initial approximations for unknown stations in a horizontal plane survey. The "Horizontal Data" under "Least Squares Adjustments" can perform the least squares adjustment of the data as discussed in Chapters 14 through 16.

File formats required by this software are discussed in the accompanying Help file. At installation, sample data files of some of the numerical examples presented in this book can be loaded onto your computer. By default, these files are loaded in the "Adjustment Computations" subdirectory of the "MY DOCUMENTS" directory. Along with the Help file, these sample data files can be viewed and compared with the accompanying example problem in the book to assist you in creating your own files.

TABLE G.1 Brief Summary of Software Options Contained in ADJUST

Option	Data File Required	Least Squares Adjustment
Astronomical observations		
Reduction for azimuth	Yes	No
Prediction of position	No	No
Coordinate computations		
Forward	No	No
Inverse	No	No
Traverse	Yes	No
Area	Yes	No
State plane coordinates	Dependent on option	No
Universal Transverse Mercator	Dependent on option	No
Geodetic computations	No	No
Coordinate geometry	No	No
Geocentric coordinates	Dependent on option	No
Oblique triangle solutions	No	No
Coordinate transformations		
2D conformal	Yes	Yes
2D affine	Yes	Yes
2D projective	Yes	Yes
3D conformal and affine	Yes	Yes
Estimated errors		
Horizontal/plane data	Yes	No
Differential leveling	Yes	No
Check errors	Yes	No
Fit of points		
Line	Yes	Yes
Circle	Yes	Yes
Parabola	Yes	Yes
GPS data		
Loop closure check	Yes	No
Baseline vector adjustment	Yes	Yes
Simulated adjustment	Yes	Yes
Least squares adjustments		
Differential leveling data	Yes	Yes
Horizontal data	Yes	Yes
3D geodetic network	Yes	Yes

G.3.2 STATS

STATS is a statistical package that computes the basic statistical properties of simple data sets as well as deriving critical values for the normal, t, χ^2, and F distributions.

G.3.3 MATRIX

MATRIX performs simple matrix operations. This software can be used to solve many of the least squares problems in this book. It can read and write files of matrices, perform the operations of scaling, transposition, addition, subtraction, multiplication, and inversing of matrices, and allow the user to view the results of these operations.

When solving least squares problems with this package, a spreadsheet can be used to compute the matrices. These matrices can then be "cut and pasted" into the Matrix editor. Once saved to disk, the data files can be read by the software and manipulated.

G.3.4 Mathcad Worksheets

The companion CD contains an electronic Mathcad book which will be installed in the HANDBOOK directory under Mathcad. The electronic book can be opened from the Mathcad Help menu. For those who do not own Mathcad, html files of the worksheets can be installed on your computer. These files will be located in the "Mathcad HTML" directory in "MY DOCUMENTS." The html files can be viewed using your HTML browser. After installation, a link to these files may be found in the Programs menu.

The worksheets demonstrate most of the numerical examples contained in this book. Most can be modified to compute other problems in the book. Some of the worksheets read ASCII data files. These files can be created by any ASCII editor, such as those contained in ADJUST, MATRIX, and STATS, or by another package, such as Notepad. With many spreadsheets it is possible to save a comma-separated values (.csv) text file. These files can also be read by the worksheets. The format of the file is demonstrated by the numerical example problem being solved in the original worksheet. Thus, when the original worksheet is modified, it should be saved with a different name. You should compare the file formats in these worksheets with the example problem in the book to determine the proper format for your data.

These worksheets also provide a guide as to how various problems are solved using programming. The Mathcad language is very similar to traditional programming languages. However, Mathcad does not provide global variables or functions. Thus, all variables, especially those that will be modified, must be passed from function to function in a worksheet and between worksheets. Mathcad reads the executable commands and variables from the top to the bottom of the worksheet and from left to right on a line. This means that location of a command or variable on the worksheet can be critical to the worksheet performing properly. Another difference between a traditional programming language and Mathcad is that Mathcad has both subscripted variables and array elements. Unfortunately, although the subscript and array element are entered differently, they look visually the same on the worksheet.

Mathcad Azimuth Computation Function:

$$Az(crds,i,1) := \begin{vmatrix} dx \leftarrow crds_{j,1} - crds_{i,1} \\ dy \leftarrow crds_{j,2} - crds_{i,2} \\ \alpha \leftarrow atan2(dy,dx) \\ \alpha \leftarrow \alpha + 2\cdot\pi \text{ if } \alpha < 0 \\ return\ \alpha \end{vmatrix}$$

C Language Azimuth Computation Function:

```
double az(crdstruct crds[10], int i, int j)
{
    double dx, dy, alpha;
    dx = crds[j],[1] - crds[i],[1];
    dy = crds[j],[2] - crds[i],[2];
    alpha = atan2(dy,dx);
    if (alpha<0) alpha = alpha + 2*PI();
    return alpha;
}
```

Figure G.1 Comparison of a Mathcad function and a C function.

Thus, it is easy to confuse a subscripted variable with a matrix element. That is, a_0 may look like a reference to a matrix element when in fact it is a subscripted variable. The user should refer to the Mathcad help system to learn how to distinguish between, and use, subscripted variables and matrix elements. The Mathcad code in Figure G.1 depicts references to matrix elements.

Figure G.1 contains a function to compute the azimuth of a line based on the coordinates of the endpoints. Except for the slight language-specific differences, the code in the Mathcad and C functions is very similar. Thus, with some modifications, the code in the Mathcad worksheets can serve as a model when developing similar code in a traditional programming language.

G.4 USING THE SOFTWARE AS AN INSTRUCTIONAL AID

Many of the problems presented in this book can be solved using the software on the companion CD. However, it would not be wise to solve all problems with this software, since true understanding can only be gained by solving the problem yourself. Still, some problems are so repetitive or long that it is extremely difficult to solve them correctly without the aid of software. In these cases, the reader is often referred to the software on the companion CD. For example, in Appendixes A through C, the matrix operations of addition, subtraction, multiplication, and inversing are presented. Although the student is expected to solve the accompanying problems in these appendices by hand, the MATRIX program should be used to solve the remainder of the problems presented in this book. This frees the reader of the matrix operations so that emphasis can be placed on the topics presented in the chapters. However, this software can be used to check solutions. The Mathcad worksheets are extremely valuable for this since the intermediate steps in the solution can also be viewed and checked against written work.

BIBLIOGRAPHY

Amer, F. 1979. *Theoretical Reliability Studies for Some Elementary Photogrammetric Procedures,* Aerial Triangulation Symposium, Department of Surveying, University of Queensland, St. Lucia, Australia.

Baarda, W. 1967. *Statistical Concepts in Geodesy,* Netherlands Geodetic Commission, Delft.

———. 1968. *A Testing Procedure for Use in Geodetic Networks,* Netherlands Geodetic Commission.

Bjerhammar, A. 1973. *Theory of Errors and Generalized Matrix Inverses,* Elsevier Science, New York.

Box, George E. P., et al. 1978. *Statistics for Experimenters,* Wiley, New York.

Bomford, G. 1980. *Geodesy,* 4th ed., Claredon Press, Oxford.

Bruns, H. 1878. *Die Figur der Erde,* Preuss. Geod. Inst., Berlin.

Buckner, R. B. 1983. *Surveying Measurements and Their Analysis,* Landmark Enterprises, Rancho Cordova, CA.

Burse, Michelle L. 1995. Profile of a Least Squares Convert, *Point of Beginning,* 20(2): 76–82.

Conte, S. D., and Carl de Boor. 1980. *Elementary Numerical Analysis,* 3rd ed., McGraw-Hill, New York.

Dewitt, Bon A. 1994. An Efficient Memory Paging Scheme for Least Squares Adjustment of Horizontal Surveys, *Surveying and Land Information Science,* 54(3): 147–156.

Dracup, Joseph F. 1994. Squares Adjustment by the Method of Observation Equations with Accuracy Estimates, *Surveying and Land Information Science,* 55:2.

El-Hakim, S. F. 1981. A Practical Study of Gross-Error Detection in a Bundle Adjustment, *Canadian Surveyor,* 35(4):373–386.

———. 1984. On the Detection of Gross and Systematic Errors in Combined Adjustment of Terrestrial and Photogrammetric Data, *Commission III, International Archives of Photogrammetry and Remote Sensing,* pp. 151–163.

———. 1986. The Detection of Gross and Systematic Errors in Combined Adjustment of Terrestrial and Photogrametric Data, *Photogrammetric Engineering and Remote Sensing,* 52(1):59–66.

Fubara, D. M. J. 1972. Three-Dimensional Adjustment of Terrestrial Geodetic Networks, *Canadian Surveyor,* 26:4.

George, Alan, and Joseph W.-H. Liu. 1981. *Computer Solution of Large Sparse Positive Definite Systems,* Prentice-Hall, Englewood Cliffs, NJ.

Ghilani, Charles D. 1990. A Surveyor's Guide to Practical Least Squares Adjustments, *Surveying and Land Information Science,* 50(4):287–297.

———. 1994. Some Thoughts on Boundary Survey Measurement Standards, *Surveying and Land Information Science,* 54(3):161–167.

———. 2003a. Statistics and Adjustments Explained, Part 1: Basic Concepts, *Surveying and Land Information Science,* 63(2):73.

———. 2003b. Statistics and Adjustments Explained, Part 2: Sample Sets and Reliability," *Surveying and Land Information Science,* 63(3):141.

———. 2004. Statistics and Adjustments Explained, Part 3: Error Propagation, *Surveying and Land Information Science,* 64(1):29–33.

Hartzell, P., L. Strunk, and C. Ghilani. 2002. Pennsylvania State Plane Coordinate System: Converting to a Single Zone, *Surveying and Land Information Science,* 62(2):95–103.

Harvey, Bruce R. 1994. *Practical Least Squares and Statistics for Surveyors,* School of Surveying, University of New South Wales, Australia.

Heiskanen, W. A., and H. Moritz. 1967. *Physical Geodesy.* W. H. Freeman and Company, San Francisco.

Hirvonen, R. A. 1965. *Adjustment by Least Squares in Geodesy and Photogrammetry,* Frederick Ungar, New York.

Hoffman-Wellenhof, B., et al. 2001. *GPS Theory and Practice,* 5th ed., Springer-Verlag, New York.

Hogg, Robert V., and Johannes Ledolter. 1992. *Applied Statistics for Engineers and Physical Scientists,* Macmillan, New York.

Hotine, M. 1959. A Primer of Non-classical Geodesy, Presented to the *First Symposium on Three-Dimensional Geodesy,* Venice.

Kuang, Shanlong. 1994. A Strategy for GPS Survey Planning: Choice of Optimum Baselines, *Surveying and Land Information Science,* 54(4):187–201.

Leick, Alfred. 2004. *GPS Satellite Surveying,* 3rd edition, Wiley-Interscience, New York.

McMillan, Kent Neal. 1995a. Least Squares: Older and Better Than Barbed Wire, *Point of Beginning,* 20(2):82–84.

———. 1995b. Least Squares Under the Hood, *Point of Beginning,* 20(2):84–88.

Mendenhall, William, and Terry Sincich. 1994. *Statistics for Engineering and the Sciences,* Dellen Publishing, San Francisco.

Mikhail, Edward M. 1976. *Observations and Least Squares,* University Press of America, Washington, DC.

Mikhail, Edward M., and Gordon, Gracie. 1981. *Analysis and Adjustment of Survey Measurements,* Van Nostrand Reinhold, New York.

Millbert, Kathryn O., and Dennis G. Milbert. 1994. State Readjustments at the National Geodetic Survey, *Surveying and Land Information Science,* 54(4):219–230.

Misra, Pratap, and Per Enge. 2001. *Global Positioning System: Signals, Measurements, and Performance,* Ganga-Jamuna Press, MA.

Schwarz, Charles R. 1994. The Trouble with Constrained Adjustments, *Surveying and Land Information Science,* 54(4):202–209.

Schwarz, K. P., E. H. Knickmeyer, and H. Martell. 1990. Assessment of Observations Using Minimum Norm Quadratic Unbiased Estimation, *CISM Journal ACSGC,* 44(1):29–37.

Seber, G. A. F. 1977. *Linear Regression Analysis,* Wiley, New York.

Sideris, Michael G. 1990. The Role of the Geoid in One, Two-, and Three-Dimensional Network Adjustments, *CISM Journal ACSGC,* 44:1:9–18.

Snay, Richard A. 1976. *Reducing the Profile of Large Sparse Matrices,* NOAA Technical Memorandum NOS-NGS 4, republished in *Bulletin Geodésiqué,* 50(4):341.

Strang, G., and Kai Borre. 1997. *Linear Algebra, Geodesy, and GPS,* Wellesley–Cambridge Press, Wellesley, MA.

Vanicek, P., and E. Krakiwsky. 1992. *Geodesy: The Concepts,*. Elsevier, New York.

Veis, G. 1960. Geodetic Use of Artificial Satellites, *Smithsonian Contributions to Astrophysics.* 3(9):95–161.

Vincenty, T. 1979. *The HAVAGO Three-Dimensional Adjustment Program.* NOAA Technical Memorandum NOS NGS 17. National Technical Information Service, NOAA, Silver Spring, MD.

White, L. A. 1987. *Calculus of Observations 380/580,* Western Australian Institute of Technology, Perth, Australia.

Wolf, Paul R., and Charles D. Ghilani. 2006. *Elementary Surveying,* 11th ed., Prentice Hall, Upper Saddle River, NJ.

INDEX

Accuracy, 4, 5
  example of, 5, 6
Adjoints, 537
ADJUST, 596
  options, 597
Adjustment
  analysis of, 492
  conditional, 178, 196
  constrained, 64, 78, 388, 395, 398, 400, 412
  of control stations, 388, 395, 412, 418, 468
  free, 291, 326
  fully constrained, 64, 291, 412, 419
  geodetic network, 454
  leveling, 205
  minimally constrained, 64, 291, 326, 412, 418
  networks, 291–300
  parametric, 178
  traverse, 283
  triangulation, 255
  trilateration, 233
Alternative hypothesis, 68
Analysis of adjustments, 492
Arc-to-chord correction, 593
Associative law, 526

Astronomical observations for azimuth, 116
Atmospheric refraction, 472
Azimuth
  astronomical observations for, 116
  constraint, 295
  equation for, 255, 603
  errors in astronomical observations for, 116

Baarda, 416
Back-substitution, 511
Blunder detection
  a posteriori, 406, 412, 416, 420, 492–502
  a priori, 322–326, 410–412
  data snooping, 416, 494
  example of, 420–428, 493–496
  graphical methods, 411, 499
  handling of control, 326, 388, 418, 468
  network, 291, 327, 409, 492
  numerical example, 420, 493
  procedures, 418, 492
  residuals, 413, 492
  by residual sign, 413, 492, 494
  statistical methods, 44, 416, 501

604 INDEX

Blunder detection (*continued*)
   traverse closure checks, 114, 130, 135, 411
   by traverse closures, 114, 127, 130, 411
Bomford, 474
Bruns, 454

Calibration, EDM, 32, 73, 195
Centering error
   instrument, 100, 106, 121
   target, 100, 104, 121
Central meridian, 583, 587
χ^2 distribution, 52, 564
   goodness of fit test, 300, 412, 496
   test, 74, 300, 412, 496
   use of table, 62, 74
Cholesky decomposition, 508
Class
   example of use, 27
   frequency, 15
   interval, 15, 16
   width, 14
Closure check, angular, 114
Cofactor, definition of, 537
Companion CD, 595
   ADJUST, 596
   file formats, 596
   MATRIX, 598
   STATS, 597
Computer optimization, 504
   processing, 507, 508, 513
   storage, 504, 513
Confidence interval, 50
   interpretation of, 59
   for mean, 56, 59
   for population variance, 61
   ratio of two population variances, 63
   variance, 61
Confidence interval for mean, computation of, 58
Confidence interval for population variance, computation of, 62
Constraint equations, 388
   azimuth example, 395
   control coordinates, 388, 395, 468
   differential leveling example, 398

   by elimination, 394
   geometric, 173
   Helmert's method, 398
Control, adjustment of, 139, 326, 388, 418, 468
Conventional Terrestrial Pole, 316, 480
Coordinate systems
   geocentric, 316, 481
   geodetic, 316, 478
   global, 478, 480
   local, 478, 480
   local geodetic, 455
   NAD 83, 316, 478
   nearly aligned, 479
   rotations between, 480
   satellite, 314
   state plane coordinate system, 233, 582
   WGS 84, 316, 318, 478, 479
Coordinate transformations
   adjustment of, 345
   datums, 480
   eight parameter, 353, 448
   four parameter, 345, 444
   Helmert, 480
   rotations, 357
   scaling, 346, 351, 356
   seven parameter, 356, 449
   six parameter, 350, 447
   statistically valid parameters, 362
   three-dimensional conformal, 356, 449
   two-dimensional conformal, 345, 444
   two-dimensional affine, 350, 447
   two-dimensional projective, 353, 448
Coordinates
   geocentric, 316, 481
   geodetic, 316, 478
   global, 478, 480
   local, 478, 480
   local geodetic, 455
   state plane, 233, 582
Covariance
   definition of, 86
   development of matrix, 221
   matrix, 159

Data
  classes, 14
  frequency of, 14
  graphical representation, 14
  measures of central tendency, 17
  measures of relative standing, 17
  measures of variation, 17, 19
  numerical methods of describing, 17
  range of, 13
  skewed, 17
Datum
  earth-centered, earth-fixed, 480
  global, 478, 480
  International Terrestrial Reference Frame, 478
  local, 478, 480
  North American Datum of 1983, 316, 478
  transformation of, 480
  WGS84, 316, 318, 478, 479
Deflection of the vertical, 472
Degrees of freedom
  in an adjustment, 212
  constrained adjustment, 403
  definition of, 19
  in a sample set, 53
Density function, normal, 36
Determinant, 535, 538
Differential leveling, weights in, 166
DIN 18723, angular error, 103
Discrepancy, 4
Dispersion, 13
Distribution
  bivariate, 109, 369
  χ^2, 52, 564
  F, 55, 568
  t, 54, 566
Distribution function
  normal, 35, 36
  standard normal, 38, 39

Eccentricity, 316, 317, 583, 587
EDM, error sources, 121
Elevation factor, 590
Elimination of constraints, 394
  numerical example, 395

Ellipse
  example computation of, 376, 377
  eccentricity, 317, 583, 587
  flattening factor, 318, 479
  orientation of, 371, 375
  q_{uu}, 375
  q_{vv}, 376
  semimajor axis, 376
  semiminor axis, 376
  t angle, 374
Ellipsoid, 478
  definition of, 479
  eccentricity, 317, 583, 587
  flattening factor, 318, 479
  GRS 80, 318, 478, 479, 583
  radius in the meridian, 461, 468, 484
  radius in normal, 318, 461, 468, 484
  WGS 84, 316, 318, 478, 479
Equations
  conditional, 178, 196
  normal, 181, 184, 185
  observation, 179
Error
  blunder, 4, 409, 417
  definition of, 3, 18
  ellipse, 369, 375–378
  gross error, 4
  instrument, 3
  natural, 3
  percent probable, 43
  personal, 3
  random, 4
  standard, 20, 38
  systematic, 4
  Type I, 69, 416
  Type II, 69, 416
Error ellipse, 369, 427
  advantages, 381
  confidence level, 379
  drawing of, 378
  numerical example of, 376, 377
  percent probability, 379
Error propagation, 2, 84, 99
  in adjusted quantities, 225
  angle, 100–112
  astronomical observation, 116
  azimuths, 169

Error propagation (*continued*)
    in computed quantities, 226
    covariance matrix, 87, 221
    differential leveling, 148
    EDM, 121
    intersection example, 91
    numerical examples, 89
    standard error in a series, 89
    standard error of a sum, 88
    standard error of mean, 89
    traverse, 127
    trigonometric leveling, 152
Error sources
    instrumental, 3
    natural, 3
    personal, 3
    random, 4
    systematic, 4
Estimators, definition of, 52

False easting, 583, 587
False northing, 583, 587
F distribution, 568
    definition of, 55
Fisher distribution, definition of, 52
Fit of points
    circle, 550–554
    line, 192, 437
    parabola, 194
Forward substitution, 508
Free network adjustment, 291, 326
Frequency
    class, 15
    histogram, 15
Fundamental principle of least
    squares, 174

Gauss, 173
General law of propagation of
    variances, definition of, 87
General least squares method, 437
Geocentric coordinates, 316
Geocentric coordinate system, 316, 481
Geodetic network
    adjustment of, 454, 471
    azimuth observations, 457
    differential leveling, 460
    horizontal angle, 459
    horizontal distances, 460
    linearization of equations, 456
    minimum number of constraints, 462
    slant distances, 457
    vertical angle, 459
Geodetic Reference System of 1980, 316, 478
Geographic information systems, 1
Geoid model, 462
Geoidal height, 473
GLOPOV, definition of, 87
Goodness-of-fit test, 300, 412, 496
GPS, 310, 478
    A matrix, 327
    baselines, 312
    carrier phase measurements, 312
    differencing, 312
    errors, 314
    least squares adjustment, 310, 478
    observations, 311
    preadjustment data analysis, 322
    real-time kinematic survey, 501
    reference coordinate systems, 314
    weight matrix, 329
Graphical representation, histogram, 14
Grid origin, 583, 587

Heights
    geodetic, 316, 473
    geoid, 316, 462
    geoidal height, 316, 462, 473
    orthometric, 316, 473
Histogram, 14
    bimodal, 17
    classes, 14
    class width, 14
    common shapes, 17
    definition of, 14
    example construction, 23
    skewed to the right, 17
Horizontal angle
    error sources, 99
    instrumental centering error, 106
    leveling error, 110

linearized observation equation, 259, 459
pointing error, 102
reading error, 100
target centering error, 104
Horizontal time-dependent positioning, 483
Hypothesis testing
  alternative hypothesis, 68
  definition of, 68
  null hypothesis, 68
  population mean, 72
  population variance, 74
  ratio of two population variances, 77
  rejection region, 69
  test statistic, 68
Hypothesis
  alternative, 68
  null, 68
  testing, 68

Inaccessible point, example of, 91
Instructional aids, 595, 599
  ADJUST, 596
  Mathcad worksheets, 598–599
  MATRIX, 598
  STATS, 597
Instrument centering error, example of, 109
Intersection
  angle, 260
  trilateration, 237
Inverse
  by adjoints, 537
  by row transformations, 538
  of 2 by 2 matrix, 535
Iteration termination, 248
  by change in reference variance, 249
  by correction size, 249
  by maximum iteration, 249

Jacobian matrix, 188, 549

Lambert conformal conic, 583
  direct problem, 585
  inverse problem, 585–586
  zone constants, 584
Laplace, 173
Least squares adjustment
  control coordinates, 388
  coordinate transformation, 345, 444
  differential leveling, 205
  general, 437
  geodetic network, 454
  GPS, 310, 478
  line fit of points, 191
  nonlinear system of equations, 188
  parabola, 194
  traverse, 283
  triangulation, 255
  trilateration, 233
Legendre, 173
Leveling
  adjustment of, 205
  in angle measurement, 110
  collimation error in, 144
  Earth curvature and refraction, 146
  random errors in, 148
  in rod plumbing, 148
  systematic errors, 144
Linearized observation equation
  angle, 259, 465
  azimuth, 258, 467
  distance, 237
  slant distance, 464
  vertical angle, 466
Link traverse
  estimated closure of, 136
Local geodetic system, 454
Localization, 483

Map projection, 582–583
  elevation factor, 590
  grid factor, 590
  Lambert conformal conic, 583
  reduction of azimuths, 593
  reduction of distances, 590
  reduction of observations, 590
  scale factor, 590–591
  Universal Transverse Mercator, 586–587
Mapping table, 506
Mathcad worksheets, 598–599

Mathematical model
  conditional, 178
  parametric, 178
Matrices, 520
  addition of, 524
  column, 522
  covariance, 87, 221
  definition of, 520
  diagonal, 523
  dimensions of, 521
  equality of, 523
  idempotent, 415
  identity, 523
  inverse of, 534
  Jacobian, 188, 549
  lower triangular, 508
  multiplication by a scalar, 524
  multiplication of, 525
  partitioning, 229
  rectangular, 522
  rotation, 357
  row, 522
  spareness and optimization, 513
  square, 522
  subtraction of, 524
  symmetric, 523
  systematic formation, 185, 507
  transpose, 523
  unit, 523
  upper-triangular, 508
Mean
  confidence interval of, 56
  definition of, 18
  example computation, 23, 25
  standard deviation in, 21
  weighted, 160, 161
Measurements
  direct, 2
  indirect, 2
Measures of central tendency, 17
  definition of, 17
  mean, 18
  median, 13, 18
Median
  definition of, 18
  example computation, 23
Meridian, radius in the, 461, 468, 484

Minimally constrained adjustment, 64, 291, 326, 412, 418
Mistake, definition of, 3
Mode
  definition of, 18
  example computation, 23
Model
  stochastic, 177
  weighting, 159
Most probable value, 19, 174

National Geodetic Survey, 472, 474, 582
National Geospatial-Information Agency, 586
Network
  adjustment of, 291
  design of, 381, 430
  example of adjustment, 292
Nonlinear equations, solution of, 188, 546, 549
Normal, radius of, 318, 461, 468, 484
Normal distribution
  function, 38
  general principles, 35, 492
  inflection point, 38, 41
Normal equations,
  formation of, 181, 185
  matrix formation, 185
  solution, 187, 189
  tabular formation, 184
Normal error distribution curve, 35
North American Datum
  1927, 345
  1983, 234, 345, 478–480
Null hypothesis, 68

Observation equation
  angle, 258, 459
  azimuth, 255, 457
  control coordinates, 388
  definition of, 179
  differential leveling, 205, 460
  distance, 235, 460
  slant distance, 457
  vertical angles, 459

Observation weight, 160
Optimization
  direct formation of normal
    equations, 507
  storage, 513
Orthometric height, 316, 473
Outliers, detection of, 42, 44, 59, 416

Percent errors, uses of, 43
Percentage points, definition of, 53
Pointing and reading errors
  DIN 18723, 103
  example of, 101–103, 112
Population
  definition of, 13
  sampling, 12
  standard deviation, 20
  variance, 19, 74
Precision
  definition of, 4
  example of, 5
Probability
  definition of, 33
  in compound event, 33
  joint, 34
Probable error
  example of, 44
  50%, 42
  95%, 42
  other percent errors, 43
  standard error, 41
  table of, 43
  use of, 45, 46
Propagation, errors, 84, 221
Pseudorandom noise, 311
Pseudorange, 311, 499

Radius in the meridian, 461, 468, 484
Random error, definition of, 4
Random samples, 50
Range, 13, 16
  example of, 23, 25
Ratio of variances, confidence
    interval of, 63, 74
Reading error
  horizontal angle, 101–103, 112

  leveling, 148
Redundancies, *see* Degrees of
    freedom
Redundancy number, 416, 428
Reference standard deviation
  computation of, 212, 213
  definition of, 211
Reference standard deviation of unit
    weight, 161
Reference variance
  definition of, 160
  goodness-of-fit test, 300, 412, 496
Rejection region, 69
Reliability
  external, 429
  internal, 429
Resection
  initial approximations, 265
  least squares adjustment of, 270
Residual
  analysis of, 442
  comparison of, 493, 500
  definition of, 19
  plots, 500
RTK-GPS, 501

Sample
  definition of, 12
  distribution theory, 52
  mean, 18, 23, 25
  size, 41, 60
  variance, 19, 20, 21, 52
Satellite orbit
  apogee, 314
  argument of perigee, 316
  ascending node, 316
  line of apsides, 314
  perigee, 314
  right ascension, 316
Site calibration, 483
Slant distance, 455, 457
SLOPOV
  definition of, 88
  example of, 88, 127
Software
  ADJUST, 596
  Mathcad worksheets, 598–599

Software (*continued*)
  MATRIX, 598
  STATS, 597
  spreadsheets, 123
Stakeout surveys, RTK-GPS, 501
Standard deviation
  for computed quantities, 226
  definition of, 20
  example computation, 23, 27
  mean, 21, 89
  sample, 20
  series, 89
  sum, 88
Standard error
  definition of, 20
  probability of, 41
Standard parallel, 583
State plane coordinate system, 233, 582
Statistics, 17, 52
STATS, 597
Stochastic model, 177
  importance of, 499
Student distribution, definition of, 54, 562
Survey design, 381, 430
  field specifications, 431
  procedures, 430
  use of maps, 430
  uses of, 94
Systematic errors
  in angles, 472
  collimation, 144
  definition of, 4
  in heights, 473
  in leveling, 144

Tables
  χ^2 distribution, 565
  F distribution, 569–575
  standard normal distribution, 562–563
  t distribution, 567
Target centering error
  example of, 101, 104, 121
Taylor series approximation, 546
t distribution, 54, 566
  definition of, 54
  use of, 56, 72, 114, 131, 134, 363, 416, 493
Terrestrial observation
  with GPS baseline vectors, 478
Test
  χ^2, 74, 300, 412, 496
  F, 55, 419
  goodness of fit, 300, 412, 496
  hypothesis, 68
  population mean, 19, 72
  population variance, 74
  ratio of population variances, 77
  significance of, 70
  statistic, 68
  t, 72
Three-dimensional coordinate transformation
  conformal, 356
Total station, DIN 18723, 103
Transformation
  Helmert, 480
  three-dimensional conformal, 356, 449
  two-dimensional affine, 350, 447
  two-dimensional conformal, 345, 444
  two-dimensional projective, 353, 448
TRANSIT, 310
Transverse Mercator, 586–587
  direct problem, 588
  inverse problem, 588–589
  zone constants, 587
Traverse
  estimated closure, 128, 135
  minimally constrained, 291
  minimum control, 291
  numerical example, 128, 136, 285
  redundant equations, 135, 284
Traverse adjustment
  degrees of freedom, 284
  example of, 285
  observation equations, 284
  redundancies, 284
Traverse closure
  anticipated, 128, 136
  numerical example, 128, 136
Triangular matrices, forward and backward solution, 511

Triangulation adjustment, 255
Trilateration adjustment, 233
True value, 3, 18
Two-dimensional coordinate transformation
  affine, 350, 447
  conformal, 356, 443
  projective, 353, 448
Type I error, 69, 416
Type II error, 69, 416

Universal Transverse Mercator, 586–587

Variance
  confidence interval of, 61
  definition of, 19, 22
  population, 19
  sample, 19
Vincenty, 456

Weighted mean, 161
  angles, 165
  example of, 168
  leveling, 166
Weight, 159
  constrained adjustment, 403
  relationship to corrections, 159
  relationship to variances, 160
  uncorrelated observations, 160
  uses of, 159
World Geodetic System of 1984, 316, 318, 478, 479